Low Carbon Cities

Low Carbon Cities is a book for practitioners, students and scholars in architecture, urban planning and design. It features essays on ecologically sustainable cities by leading exponents of urban sustainability, case studies of the new directions low carbon cities might take and investigations of how we can mitigate urban heat stress in our cities' microclimates. The book explores the underlying dimensions of how existing cities can be transformed into low carbon urban systems and describes the design of low carbon cities in theory and practice. It considers the connections between low carbon cities and sustainable design, social and individual values, public space, housing affordability, public transport and urban microclimates. Given the rapid urbanization underway globally, and the need for all our cities to operate more sustainably, we need to think about how spatial planning and design can help transform urban systems to create low carbon cities, and this book provides key insights. This book is part of the *Earthscan Series on Sustainable Design*.

Steffen Lehmann is Founding Director of the Zero Waste Research Centre for Sustainable Design and Behaviour at the University of South Australia. He is a widely published author and scholar, and is Founding Director of the s_Lab Space Laboratory for Architectural Research and Design (Sydney–Berlin). A German-born architect and urban designer, he was editor-in-chief (2007–14) of the US-based *Journal of Green Building* and an advisor to the Australian and German governments, city councils and industry. Since July 2014 he has been Head of School in the School of Built Environment at Curtin University, Perth.

Earthscan Series on Sustainable Design
Series Editor: Professor Steffen Lehmann (Steffen.Lehmann@curtin.edu.au)

All books in this series are authored and/or edited by leading academics and practitioners in the field of sustainable design.

While there has been an immense amount of theory and technology focused writing published on the topic of sustainable design, many of these books have failed to introduce readers to the wider challenge of what the re-thinking of design, production, operation and recycling of all products, buildings and cities really means.

Sustainability is not a passing fashion and people are constantly searching for more information, ideas and products in this area. This book series will aim to develop a more coherent theoretical framework for how different theories of sustainable design might engage with the practice of architects, designers, urban planners and related professions. The knowledge gained from this book series will equip the readers with the tools for realising the full potential of the good intentions of sustainable design.

The aim is that these books will provide a novel alignment of interdisciplinary perspectives on the problems of global consumerism, sustainable design and strategies to avoid resource waste, on the scales of products, buildings, districts and cities.

The books will become essential reading for architects, industrial designers, urban designers and researchers/students in these disciplines. Potential readers for the books will also include industry and government agencies. Global relevance and the potential for use as textbooks will be essential.

The book series will become a highly useful addition to the literature on sustainable design, urban development and city culture, focusing on the key topics encountered by students and scholars of urban studies, pointing towards related bibliographic material.

If you have an idea for the series then please contact the series editor.

Designing for Zero Waste: Consumption, Technologies and the Built Environment
Edited by Steffen Lehmann and Robert Crocker
December 2011 | Paperback: 978-1-84971-435-8 | Hardback: 978-1-84971-434-1

Motivating Change: Sustainable Design and Behaviour in the Built Environment
Edited by Robert Crocker and Steffen Lehmann
July 2013 | Paperback: 978-0-415-82978-6 | Hardback: 978-0-415-82977-9

Low Carbon Cities: Transforming Urban Systems
Edited by Steffen Lehmann
October 2014 | Paperback: 978-0-415-72983-3 | Hardback: 978-0-415-72982-6

Low Carbon Cities

Transforming urban systems

Edited by Steffen Lehmann

Routledge
Taylor & Francis Group
LONDON AND NEW YORK

First published 2015
by Routledge
2 Park Square, Milton Park, Abingdon, Oxon OX14 4RN

and by Routledge
711 Third Avenue, New York, NY 10017

Routledge is an imprint of the Taylor & Francis Group, an informa business

British Library Cataloguing-in-Publication Data
A catalogue record for this book is available from the British Library

Library of Congress Cataloging-in-Publication Data
Lehmann, Steffen, 1963–
Low carbon cities : transforming urban systems / Steffen Lehmann.
pages cm. — (Earthscan series on sustainable design)
Includes bibliographical references and index.
1. City planning—Environmental aspects. 2. Climatic changes.
3. Sustainable design. 4. Urban ecology (Sociology) 5. Carbon dioxide—
Environmental aspects. I. Title.
HT166.L444 2015
307.1'216—dc23 2014007887

ISBN: 978-0-415-72982-6 (hbk)
ISBN: 978-0-415-72983-3 (pbk)
ISBN: 978-1-315-76600-3 (ebk)

Typeset in Times New Roman
by Keystroke, Station Road, Codsall, Wolverhampton

'For too long, the international debate about green building has been far too object-focused. That is why *Low Carbon Cities* is of utmost importance.'

Ulf Meyer, architecture journalist, Berlin

'From the author who gave us the tour-de-force account of urban sustainability, *The Principles of Green Urbanism*, comes another timely read. Steffen Lehmann has assembled some of the best environmental urban minds in this new anthology that will become essential for all those who are looking for viable and sustainable solutions. *Low Carbon Cities* might just be the inspiration we need in our efforts to rethink and transform our cities towards the new, ecologically sustainable and resilient future.'

Professor Tigran Haas, KTH – Royal Institute of Technology, Stockholm

'The rapid urbanisation underway in Asia is a challenge and an opportunity for urban sustainability. With chapters written by experts from different parts of the world, this book is a timely addition that addresses many aspects of how low carbon cities can contribute to curbing climate change. Its essays and case studies provide useful references for students, practitioners and researchers who are interested in the global issue of sustainable urban development.'

Professor Edwin H. W. Chan, Hong Kong Polytechnic University, Hong Kong

'*Low Carbon Cities* is an inspiring guide to exploring ways to create low carbon cities. The world's cities play a major role in climate change as both sources of problems and pioneering projects for solutions. Low carbon solutions can demonstrate thriving participatory processes of systematic urban transformation. This excellent anthology illustrates how low carbon urban design can help to reduce or mitigate climate change-related emissions.'

Professor Thomas Spiegelhalter, architect, Florida International University, Miami

'This book delivers a holistic view of sustainable cities, bridging a variety of practice, from architecture to environmental studies to urbanism, green building and microclimates. It's a must-read textbook for graduates seeking to understand the complexity of urban environmental issues.'

Professor Wanglin Yan, Keio University, Japan

Other books by Steffen Lehmann

Motivating Change: Sustainable design and behaviour in the built environment

Designing for Zero Waste: Consumption, technologies and the built environment

The Principles of Green Urbanism: Transforming the city for sustainability

Sustainable Architecture and Urban Development: Proceedings, volume I to IV

Back to the City: Strategies for informal urban interventions

Temporary and Permanent: On public space

Flow: Projects review

Absolutely Public: Crossover between art and architecture

Brisbane Towers – Brisbane Bridges: Architecture for the city

Der Weg Brasiliens in die Moderne: Einordnung der modernen Architektur Brasiliens

Rethinking: Space, time, architecture: Ein Dialog zwischen Kunst und Architektur

Tower of Babel: Architecture for the 3rd millennium

Steffen Lehmann: Breite x hoehe x tiefe

Huelle, Schwere, Licht: Lehmann-Riccius-Sterf

Arata Isozaki: The 3rd Generation Art Museum

Contents

List of figures xi
List of tables xvii
List of contributors xix
Acknowledgements xxix

Foreword xxxi
Planning for sustainable and low carbon green cities
DONOVAN STOREY AND K. K. PHILIP KANG

Preface xxxv
Holistic approaches to future cities
JOO-HWA (PHILIP) BAY

Introduction **1**
Low carbon cities: more than just buildings
STEFFEN LEHMANN

PART I
Classic texts on ecologically sustainable cities: conceptual evolutions and different schools of thought **57**

Introduction to Part I **57**

1 **'Ecopolis': the regenerative city** **59**
HERBERT GIRARDET

2 **Much better than climate change adaptation** **75**
RICHARD REGISTER

3 **Imagining biophilic cities** **85**
TIM BEATLEY

4 **Sustainable cities are solar-responsive cities** **97**
NORBERT LECHNER

5 **Trends in low carbon transport and urban development in 33 cities, 1995/1996 to 2005/2006: some prospects for lower carbon transport** 113
JEFF KENWORTHY

6 **How city dwellers affect their resource hinterland: a spatial impact study of Australian households** 131
MANFRED LENZEN AND GREG M. PETERS

PART II
New directions: designing for low carbon cities **149**

Introduction to Part II 149

7 **Future ecological design as urbaneering: new positions on city-making without scale** 151
MITCHELL JOACHIM

8 **A planning approach for the transformation to low carbon cities** 173
NICO TILLIE, ANDY VAN DEN DOBBELSTEEN AND SEBASTIAN CARNEY

9 **Green districts and carbon engineering: increasing greenery, reducing heat island effects and generating energy** 191
STEFFEN LEHMANN

10 **From green buildings to sustainable urban design: two case studies** 210
CHRISTOPH INGENHOVEN, MARTIN REUTER AND BEN DIECKMANN

11 **Urban reclamation and regeneration in Seoul, Republic of Korea** 224
ROBERT CERVERO

12 **From sustainable to low carbon cities: is India's urban transformation triggering a paradigm shift?** 235
SHIPRA NARANG SURI

PART III
Urban microclimates: mitigating urban heat stress **251**

Introduction to Part III 251

13 **The urban heat island and low carbon cities** 253
GERALD MILLS

14 **A holistic view of the effects of urban heat island mitigation** 270
DAVID J. SAILOR

15 Urban microclimates: mitigating urban heat **282**
MATHEOS SANTAMOURIS AND DIONYSIA-DENIA KOLOKOTSA

16 Energy reduction using natural ventilation in city planning **293**
YUKIKO YOSHIDA AND TOSHIAKI ICHINOSE

17 The impact of increasing urban air temperatures on urban planning and building energy consumption in tropical climates **308**
STEVE KARDINAL JUSUF, NYUK HIEN WONG AND ZHUN MIN ADRIAN CHONG

18 Urban heat islands: case studies from Frankfurt am Main, Arnhem and Ho Chi Minh City **331**
LUTZ KATZSCHNER AND SABRINA CAMPE

19 The Green Plot Ratio and the role of greenery in low carbon living **349**
BOON LAY ONG

20 Urban populations' vulnerability to climate extremes: mitigating urban heat through technology and water-sensitive urban design **361**
NIGEL TAPPER, ANDREW COUTTS, MARGARET LOUGHNAN AND DEVNA PANKHANIA

21 Multi-scale analysis of surface-layer urban heat island effect in five higher-density precincts of central Sydney **375**
EHSAN SHARIFI, CONRAD PHILIPP AND STEFFEN LEHMANN

22 Looking ahead: applying low carbon principles to shift urban design paradigms in the Asia-Pacific region towards green urbanism **394**
STEFFEN LEHMANN AND KATHARINE THORNTON

Glossary of urban terms 422
Annotated select bibliography 429
Index 433

Figures

i.1a and i.1b	Aerial images of compact Barcelona and Paris illustrate the optimum density and mixed-use buildings of low carbon cities	3
i.2	The many aspects of the 'sustainable city'	10
i.3a, i.3b and i.3c	Low carbon cities will have new types of infrastructure, decentralized energy systems and concepts of innovative public space	11
i.3d and i.3e	Low carbon cities offer eco-mobility options with easy change of mode	12
i.4	The historical evolution of city theory and different models of thought	16
i.5	Cramped housing conditions in Hong Kong	17
i.6	An innovative urban transport solution in Stuttgart	18
i.7	Humankind is part of the ecosystem, not apart from or above it	21
i.8	A holistic vision of how to strengthen the urban metabolism	23
i.9a, i.9b and i.9c	Urban renewal master plans for Australian cities	25
i.10a and i.10b	The 'Emmen Urban Hybrid' in Switzerland in 2013	26
i.11a	Material flows in a zero-waste city	32
i.11b	A formerly run-down port area in Stockholm is now a model low carbon precinct, Hammarby Sjöstad	32
i.12a	'Stacking Green' house by Vo Trong Nghia in Ho Chi Minh City, Vietnam, 2011	34
i.12b	Seoul's Cheonggycheon Park with a restored river	34
i.13a	A house marked for demolition in Chongqing in 2010	36
i.13b1 and i.13b2	Shanghsa East Village in Shenzhen, China	37
i.13c	Kangbashi New Area, a new district in Ordos, Inner Mongolia, China	37
i.14a	Chengdu eco-city, China	40
i.14b	Wanzhuang eco-city, China	40
i.14c	Qingdao eco-city, China	40
i.14d and i.14e	Masdar eco-city, located outside Abu Dhabi	41
i.15	Population growth in Australia, in millions	44
i.16a, i.16b and i.16c	Good examples of compact urban development in Australia: Barangaroo in Sydney, Docklands in Melbourne and Bowden Urban Village in Adelaide	45

i.17a and i.17b Books previously published in the Earthscan Series on Sustainable Design 47
i.17c The structure of *Low Carbon Cities* 47
1.1 'Agropolis' 61
1.2 'Petropolis' 62
1.3 Solarsiedlung, Vauban, Freiburg 64
1.4 Sanlúcar La Mayor solar power station, Spain 65
1.5 Regenerative cities have a circular metabolism 67
1.6 'Ecopolis' 71
2.1 Flooding of New Orleans by Hurricane Katrina in 2005 84
3.1 Humans are biophilic creatures 86
3.2 Ride a bicycle around your city 88
3.3 A green wall brings nature into the city 90
3.4 New York City's re-vegetated High Line 94
4.1 Easy and inexpensive strategies should be used before the difficult and expensive ones: 'pick the low-hanging fruit first' 98
4.2 East–west streets provide the ideal orientation for buildings' windows 99
4.3 Buildings on north–south avenues can be designed so that their windows will still have the ideal north and south orientation 100
4.4 Especially in cold climates where access to the direct sun is very desirable, the width of streets needs to be larger due to the lower sun angles 100
4.5 To get solar access at the lower floors, streets must be wider as buildings get taller 101
4.6 White roofs have a major impact in reducing the urban heat island effect 102
4.7 Outdoor shading has a major benefit in reducing the urban heat island effect 104
4.8 Because the environment is not static, dynamic shading devices such as awnings perform much better than static ones 105
4.9 The importance of plants in urban areas cannot be overstated 106
4.10 Both the quality and quantity of daylighting is improved by the use of light shelves 107
4.11 China is one of the leading countries in using active solar hot water collectors 109
4.12 Solar zoning ordinances should be used to provide the maximum solar energy to every building in a city 110
4.13 Conceptually clear heliodons are powerful tools for promoting, understanding and designing solar-responsive buildings and cities 111
5.1 Car-use growth trends in prosperous cities from 1960 to 2005 121
6.1 Spatial distribution of greenhouse gases (tonnes CO_2 equivalents) emitted across Australia as a consequence of consumption by a Sydney family 137
6.2 Spatial distribution of greenhouse gases (tonnes CO_2 equivalents) emitted across Australia as a consequence of consumption by a Melbourne family 138

6.3 Spatial distribution of water (ML) consumed across Australia as a consequence of consumption by a Sydney family 140
6.4 Spatial distribution of water (ML) consumed across Australia as a consequence of consumption by a Melbourne family 141
6.5 Spatial distribution of economic turnover (AUD) stimulated across Australia as a consequence of consumption by a Sydney family 142
6.6 Spatial distribution of economic turnover (AUD) stimulated across Australia as a consequence of consumption by a Melbourne family 143
6.7 Spatial distribution of employment (full-time equivalent) generated across Australia as a consequence of consumption by a Sydney family 144
6.8 Spatial distribution of employment (full-time equivalent) generated across Australia as a consequence of consumption by a Melbourne family 145
7.1 The field of urbaneering 151
7.2 to 7.6 Urbaneering Brooklyn 2110, city of the future 154–156
7.2 A spectacle of smart infrastructure highlights the new downtown 154
7.3 All necessities are provided inside the city boundaries 155
7.4 Site plan includes quadrants for energy, food, water and waste 155
7.5 View of Brooklyn Bridge into new downtown 155
7.6 Green brain park space used for algae production 156
7.7 to 7.9 Rapid Re(f)use: waste to resource city 2120 157–158
7.7 History and future of garbage in New York 157
7.8 Every hour, New York City produces enough waste to fill the Statue of Liberty 157
7.9 One-day tower made of 24 hours of compacted waste in New York City 158
7.10 to 7.13 SOFT mobility 161–162
7.10 Blimp bumper bus 161
7.11 The hug-and-kiss sheep car has zero accident victims 161
7.12 Urban stackable car with solar recharge port 162
7.13 Homeway: the great suburban exodus 162
7.14 to 7.20 Super docking, Brooklyn Navy yard 164–167
7.14 View of clean-tech industrial water park 164
7.15 Site with five retrofitted dry docks for green manufacturing 164
7.16 Super-sheds over docks merge landscapes 165
7.17 Clean-tech office towers 165
7.18 Large-scale 3D printer over dry docks 166
7.19 Restorative ecology with phytoremediation systems in surfaces 166
7.20 Overall view with converted warehouses and sewer management 167
8.1 The REAP methodology 176
8.2 The five energy exchange strategies that emerged from REAP2 181
8.3 Heat-cascading grid 181
8.4 Heat cascading via heat and cold storage facility 182
8.5 Technical concepts and spatial implications of heat exchange strategies for Merwe-Vierhavens 183
8.6 Heat maps of The Netherlands and the city of Rotterdam 186

8.7 Results of the inventory for Rotterdam 188
9.1a, 9.1b, 9.1c and 9.1d Recent examples of urban infill using prefabricated engineered timber panels as a low carbon construction system 197
9.2a and 9.2b Thermal photography of urban spaces reveals stored heat 202
9.2c and 9.2d Greenery integrated into façades has been successfully applied in buildings in Singapore and Seoul 202
9.2e Opportunities for mitigating the UHI effect may be found in five areas 203
9.3a and 9.3b Visualization of the new Barangaroo South precinct, East Darling Harbour, Sydney 204
10.1 An aerial view of the redesigned UCD campus 213
10.2 Street view of 1 Bligh, Sydney 216
10.3 Atrium of 1 Bligh, Sydney 217
10.4 Interior of 1 Bligh, Sydney 218
10.5 1 Bligh, Sydney: section 220
10.6 A green high-rise commercial building at 1 Bligh Street, Sydney, in its urban context 222
11.1 Marginal effects of the CGC freeway and urban greenway on non-residential property values by distance intervals, 2001 to 2006 230
11.2 Marginal effects of the CGC freeway and urban greenway on residential property values by distance intervals, 2001 to 2006 231
12.1 Delhi's shift to CNG has played a major role in reducing air pollution and cutting its GHG emissions 242
12.2 The Ahmedabad Janmarg BRTS is a model of low carbon transport for India 243
12.3 Mitigation of emission contributions from various sectors in Business As Usual (BAU) and Low Carbon Society (LCS) scenarios, Bhopal, 2035 246
13.1a and 13.1b Two views of Chicago 254
13.2 The vertical structure of the urban atmosphere 257
13.3 Mean minimum temperature for November 1981 in Mexico City 259
13.4a and 13.4b The properties of urban areas that can enhance or ameliorate the urban effect on temperature 265
14.1 Schematic representation of the key drivers of the urban energy balance 271
14.2a and 14.2b Boundary layer structures in the urban climate system 272
15.1 Urban heat island mitigation techniques in buildings and at city level 283
15.2 How cool roofs work 284
16.1a and 16.1b GDP per capita, population density and heating–cooling degree month 294
16.2 Research flow on environment and energy around building blocks (EEBB) 295
16.3 The Urban Climate Simulation System (UCSS) 296
16.4 Building blocks research process 296
16.5 Research site in Nagoya on Google Earth 297
16.6 Sky view factor of Nagoya 297

16.7 Canopy model of Nagoya 297
16.8 Heat load versus building height (Nagoya) 298
16.9 Vertical average temperature (UCSS in building blocks) 299
16.10 Horizontal average temperature (UCSS in building blocks) 299
16.11 Forecast urban climate (horizontal average temperature) 300
16.12 Forecast urban climate (horizontal maximum temperature) 301
16.13 RayMan1.2 model of parameters 302
16.14 PMV and wind velocity in Nagoya 302
16.15 SET* and wind velocity in Nagoya 302
16.16a SET* and SVF in Nagoya at 10 a.m., 12 September 2008 303
16.16b SET* and SVF in Nagoya at 2 p.m., 12 September 2008 303
16.17 FCU operation time for cooling load in each perimeter zone 305
16.18 Hybrid air-conditioning judgement flow 305
17.1a Building layout model for sensitivity analysis 312
17.1b Building layout model 3 on change of canyon width 312
17.2 Predicted T_{min}, T_{avg} and T_{max} for various values of GnPR 312
17.3 Predicted T_{avg} for various values of number of storeys and GnPR 313
17.4a, 17.4b and 17.4c Predicted T_{min}, T_{avg} and T_{max} for various values of number of storeys and GnPR 314–315
17.5a Maximum temperature map of one-north, Singapore 317
17.5b Section A–A temperature profile 317
17.6a Average temperature map of one-north, Singapore 318
17.6b Section A–A temperature profile 318
17.7a Minimum temperature map of one-north, Singapore 319
17.7b Section A–A temperature profile 319
17.8 Orientation of street grids 322
17.9 A public housing apartment in Singapore with a void deck 322
17.10 Example of varying building heights 322
17.11 Comparison of space load component (total building load) for different soil thickness on roof of a five-storey commercial building 325
17.12 Comparison of annual energy consumption for different types of roofs for a five-storey commercial building 326
17.13 Different building scenarios 1, 2 and 3 326
17.14 Mean Radiant Temperature (Tmrt) for scenarios 1A and 1B of TAS simulation 327
17.15 Cooling load summary for five scenarios 328
18.1 Spatial distribution of urban heat load in Kassel, Germany 332
18.2 Heat load classification in Kassel with corresponding PET values 333
18.3 Increasing number of hot days in Kassel to 2050 334
18.4a Frankfurt's greenbelt concept 337
18.4b Frankfurt's urban climate map 337
18.5a and 18.5b Results from an ENVI-met simulation under existing conditions and with climatic oriented actions 341
18.6 Urban climatic map of the city of Arnhem 342
18.7 Urban climatic map of Ho Chi Minh City with climatic regions 345

19.1 Variation in percentage of radiation transmitted for varying values of k and L 353
19.2 Variation in wind velocity as a function of LAI and k' 354
19.3 Empirical relationships estimated between interception storage and Leaf Area Index for selected trees 355
19.4 The ecological high-rise design 357
19.5 Integration of greenery into water and sewage services 358
20.1 Schematic diagram showing the relationship of exposure, vulnerability and adaptive capacity to heat-related human health outcomes 362
20.2 Mortality anomalies for Melbourne, Australia associated with binned daily mean temperatures, 1979 to 2001 364
20.3 Schematic representation of widespread implementation of stormwater harvesting and water-sensitive urban design elements at the microscale 370
21.1 Urban structure, landscape, land cover and metabolism contribute to the urban heat island effect in cities 377
21.2 Average surface temperature of precincts in Glebe Point, Harris Street, Sydney Harbour, Haymarket and Kings Cross 2010 380
21.3 Street surface temperature in five higher-density precincts in central Sydney, 2010 382
21.4 Roof-top surface temperature in five higher-density precincts in central Sydney, 2010 383
21.5 Streetscape and public space plot ratio in five higher-density precincts of central Sydney 384
21.6 Urban vegetation ratio in five higher-density precincts of Sydney 386
21.7 Annual variation of surface temperature in five higher-density precincts of central Sydney 387
22.1 The eco-city seeks to balance humans' and nature's rural and urban systems 396
22.2 Green urbanism and eco-efficiency are complementary concepts that benefit the economy and the environment by recognizing that nature is at the centre of our existence 402
22.3 The green urbanism wheel shows the 15 principles grouped in four sections 405
22.4 The three pillars of green urbanism and their interactions 406
22.5 Resource-efficient cities and sustainable consumption 409
22.6 Linking the upstream with the downstream in a municipality's decision-making process 413

Tables

i.1 The evolution of green urbanism in association with key texts and recent theories 15
i.2 Leading examples of cities that have implemented low carbon city policies and planning 28
3.1 Some important dimensions and some possible indicators of biophilic cities 92
5.1 The 33 cities surveyed in this update study 114
5.2 Trends in 26 key transport-related variables in US, Canadian and Australian cities, 1995/1996 to 2005/2006 124
5.3 Trends in 26 key transport-related variables in European cities and Singapore/Hong Kong, 1995 to 2005 126
6.1 Characteristics of the household groups selected for Sydney and Melbourne 134
6.2 Total annual impacts per household for Sydney and Melbourne in terms of four indicators 135
8.1 Heat, cold and electricity usage data for Rotterdam's Zuidplein shopping mall in the Hart van Zuid district 179
8.2 Heat, cold and electricity usage data for Rotterdam's Ikazia Hospital 179
8.3 Results of the Greenhouse gas Regional Inventory Protocol (GRIP) for Rotterdam 188
11.1 Five major new towns in Seoul metropolitan area 225
12.1 GHG emissions by sector in 41 Indian cities 244
16.1 Comparison of land use 298
16.2 UCSS case study 300
16.3 Air temperatures inside building blocks predicted using CFD 304
17.1 Impact of climate change on ETTV coefficients 324
17.2 Increase in ETTV for reference building 324
17.3 TAS simulation results 327
18.1 Planning and urban climate scales 335
18.2 Evaluation of legend used in Figure 18.1 340
18.3 Proposed land-use changes and recommendations for different areas of Ho Chi Minh City 346
19.1 The value of *k* for different leaf angle distributions 353
20.1 Maximum temperature $T_{(max)}$ thresholds producing excess mortality for all Australian cities, plus modelled days exceeding that temperature for 1990–2010 (current climate), 2020–2040 and 2060–2080, using a high emissions (A2) scenario 365

20.2 Examples of three major types of human adaptive capacity to urban heat 366
20.3 Strategies to promote adaptive capacity for passive and low energy buildings 368
21.1 Temperature variance of streetscape surfaces and building roof-tops in five higher-density precincts of central Sydney 384
21.2 Street network intensity and average surface temperature in five higher-density precincts of central Sydney 385
21.3 Urban vegetation ratio in five higher-density precincts of central Sydney 386
21.4 Temperature variation of urban land surfaces in five higher-density precincts of central Sydney 388
22.1 The 15 principles of green urbanism 403

Contributors

Editor

Steffen Lehmann, Ph.D. and AADip, is a German-trained architect, urbanist, designer and engineer who has held a chair and been a professor in Australia for the past decade. He is the Founding Director of the Zero Waste Research Centre for Sustainable Design & Behaviour (sd+b) at the University of South Australia, where he is the Chair and Professor of Sustainable Design in the School of Art, Architecture and Design. He is also Director of the China–Australia Centre for Sustainable Urban Development and held the UNESCO Chair in Sustainable Urban Development for Asia and the Pacific from 2008 to 2010. He now continues to advise the United Nations ex-officio. He studied at the Architectural Association School in London and holds a Ph.D. from TU-Berlin. Since July 2014, he has been Head of School of the School of Built Environment at Curtin University in Perth. He is the Founding Director of the s_Lab Space Laboratory for Architectural Research and Design (Sydney–Berlin), which he founded in 1993. With s_Lab he designed and built significant projects in the 1990s and was instrumental in the urban redevelopment of Berlin's city centre. His recent books include: *The Principles of Green Urbanism* (Earthscan, 2010); *Sustainable Architecture and Urban Development, Volumes I to IV* (CSAAR-Press, 2010); and *Designing for Zero Waste* and *Motivating Change*, both co-edited with Robert Crocker (Routledge, 2012; 2013). In recognition of the international significance of his work, he has been an expert advisor on sustainable architecture to policy makers in cities around the world. He has served on numerous government, industry and other boards, and he has held visiting professorships at universities including Tongji Shanghai (2003), TU-Berlin (2005), National University of Singapore (2009), TU-Munich (DAAD-Professor 2009), University of California at Berkeley (2012) and Tianjin University (2014). More information may be found at: www.slab.com.au

Contributors

Joo-Hwa (Philip) Bay has been practising in architecture and urban design since 1986. He has been company director of a large practice and a council member of the Singapore Institute of Architects, and has won several design awards. He received his Ph.D. from Delft University of Technology, in The Netherlands, on design thinking in environmental architecture. He researched and taught at the National University of Singapore from 1997 to 2006. One of his design research consultancies was to advise the Singapore JTC Corporation on new urban housing at the 'new economy' hub One North. He currently runs a practice in Perth and is an Associate Professor at the University of Western Australia. His published works include *Contemporary Singapore Architecture, Cognitive Biases in Design: The case of tropical architecture* and *Tropical Sustainable Architecture: Social and environmental dimensions*. He chaired the International Network for Tropical Architecture (iNTA) from 2004 to 2009, is on the editorial board of two journals, and has been a referee and advisor to major conferences, a jury member for design competitions, a reviewer of journal papers, and an examiner of Ph.D. dissertations.

Tim Beatley is the Teresa Heinz Professor of Sustainable Communities in the Department of Urban and Environmental Planning, School of Architecture at the University of Virginia, where he has taught for the past 25 years. He is the author or co-author of more than 15 books on these subjects, including *Green Urbanism, Native to Nowhere* and *Biophilic Cities*. His book *Ethical Land Use* was declared by the American Planning Association to be one of the '100 Essential Books in Planning'. He directs the Biophilic Cities Project at UVA (http://biophiliccities.org/) and is also co-director, with Reuben Rainey, of UVA's Center for Design and Health, within the School of Architecture. In October 2013, he (with others) launched the global Biophilic Cities Network.

Sabrina Campe is a Research Associate in the Department of Environmental Meteorology at the University of Kassel, Germany, where she is investigating the impact of global climate change on urban and regional atmospheres, which includes the study of Frankfurt's urban climate. She has a Dip.-Ing. from the University of Kassel. In 2009, she was awarded the Faculty of Architecture, Urban Planning and Landscape Planning's best thesis award in the planning and conceptual work category. Her research interests include the climatic impact of urban trees and climate-friendly open space urban planning.

Sebastian Carney has performed research and consultancy on energy policy and carbon reduction in 15 countries. In his doctoral studies in business administration, he developed a carbon and energy accounting process called the Greenhouse gas Regional Inventory Protocol (GRIP), which is a recommended approach of the Covenant of Mayors and is being rolled out globally by ICLEI. He has been appointed as an expert advisor for the Committee of Regions for the EU reporting mechanism on greenhouse gas emissions and on the EU adaptation strategy. He edited the proceedings of the World Bank's fifth Urban Research Symposium, and was an advisor to the UN-Habitat report on cities and climate change. He has been an invited or keynote speaker on carbon accounting and energy policy on six continents. His work has involved extensive stakeholder engagement and the use, and development, of computer tools to perform energy and carbon accounting to assess the impacts of policy and plans.

Robert Cervero is the Friesen Chair of Urban Studies and Professor of City and Regional Planning at the University of California, Berkeley, where he also directs the Institute of Urban and Regional Development (IURD) and the University of California Transportation Center (UCTC). His research centres on the nexus between urban transportation and land-use systems. He has authored or co-authored six books, more than 50 research monographs, and over 200 journal articles in these areas, including the just released book, *Transforming Cities with Transit* (World Bank, 2013). He currently chairs the International Association of Urban Environments and the National Advisory Committee of the Robert Wood Johnson Foundation's Active Living Research Program. He also serves on the International Panel on Climate Change (IPCC), and is a contributing author to the fifth IPCC assessment. He is also a lead author of the 2013 *Global Report on Sustainable Mobility* for UN-Habitat.

Zhun Min Adrian Chong is a faculty member in the Department of Building at the National University of Singapore. He has an M.Sc. in Building Science from the Department of Building at the National University of Singapore and is currently pursuing his Ph.D. degree in Building Performance and Diagnostics at Carnegie Mellon University. His research interests are in the area of building energy simulation and how uncertainties in simulation models can be minimized or mitigated by modelling them into the simulation software. Other interests include the coupling of databases of manufacturers' specifications and design configuration with EnergyPlus so as to generate a model that is both accurate and credible.

Andrew Coutts is a Research Fellow in the School of Geography and Environmental Science at Monash University, Melbourne, Australia. He is an early career researcher currently employed in the Cooperative Research Centre for Water Sensitive Cities. He graduated with a Ph.D. in March 2008 in the area of urban climatology. His current work involves a variety of observational studies aimed at understanding the role of water-sensitive urban design and green infrastructure in improving urban microclimates. This work includes the monitoring of water-sensitive design features, street trees and characteristics of complex urban environments that influence urban microclimates. He has both a policy and scientific research background and his interest is in bringing the two areas together so that research is targeted towards delivering policy-relevant outcomes, which are underpinned by sound science.

Ben Dieckmann has worked for Ingenhoven architects since 1997. He has worked on the design, administration and execution of a variety of projects – research, commercial, industrial, residential and infrastructural. Besides his position as Director of the office, he leads design teams as Design Director through all phases, from competition and master plan to fit-out. His focus is on projects in Asia, the US and Europe, such as the Breezé Tower in Osaka, the Al-Haram Mosque project in Mecca and the recent Google HQ project in Mountain View, California. He was responsible for the European Investment Bank Building, which was the first building in continental Europe to achieve a BREEAM 'excellent' rating. Besides his architectural work, he works on product design, such as the office furniture system Terminal for Vitra.

Andy van den Dobbelsteen is Professor of Climate Design and Sustainability in the Faculty of Architecture at Delft University of Technology in The Netherlands. He chairs the TU Delft Department of Architectural Engineering and Technology. He lectures and leads research projects in various areas of climate design and sustainability in the built environment. He is theme leader for energy in the built environment for the TU Delft Energy Initiative project. He participates in various national committees and juries. He has written numerous publications, both scientific and popular. He has chaired various conferences, among which are the award-winning SASBE2009, Nationale Dubo Dag and Nationaal Congres Energie & Ruimte. In the recent past, he was a visiting fellow at the Melbourne Sustainable Society Institute. He was regional and joint coordinator of the CIB Working Commission W116 between 2006 and 2013. From 2007 until 2011, he was an external examiner at the Manchester School of Architecture.

Herbert Girardet is co-founder and former Director of Programs of the World Future Council. He is the author of ten books, including *Earthrise* (1992), *The Gaia Atlas of Cities* (1992), *Creating Sustainable Cities* (1999), *Cities, People, Planet* (2004), *Shaping our Future* (2005), *Surviving the Century* (2007) and *The Regenerative Cities Report* (2011). In 2003, he was the inaugural Adelaide Thinker in Residence. He has been a consultant to UNEP and UN-Habitat, and has developed sustainability policies for major cities such as London, Vienna and Bristol. He has produced around 50 documentaries on the environment and sustainable development, and is a recipient of a UN Global 500 Award for outstanding environmental achievements. He lives in the English countryside and is a visiting professor at the University of the West of England.

Toshiaki Ichinose is an executive senior research scientist at the National Institute for Environmental Studies in Japan and Professor in the Department of Environmental Engineering and Architecture at Nagoya University. He completed his postgraduate degrees in urban engineering at the University of Tokyo. With a background in geography and civil engineering, he is interested in various aspects of urban environmental systems, ranging from urban climatology, energy systems and GIS to their application in urban planning. He has undertaken many field surveys in China and other Asian countries and published the results widely. His publications include 'Counteracting urban heat islands in Japan' (*Urban Energy Transition*, 2008). He has served as a board member of the International Association for Urban Climate (IAUC) and a member of the WMO-Expert Team on Training on Urban Climatology. His main research areas are strategic urban planning and the assessment of low carbon cities.

Christoph Ingenhoven founded Ingenhoven architects in 1985, and today it is one of the world's leading practices in sustainable architectural design. Among his best-known projects are the RWE headquarters in Essen, Germany (one of the world's first ecological high-rises); the Lufthansa Aviation Centre at Frankfurt Airport; the European Investment Bank building, Luxembourg; the headquarters of Daniel Swarovski Corporation in Zurich; and the high-rise tower 1 Bligh in Sydney, which has received several awards since its opening in 2011. The main train station in Stuttgart, Germany and the office and apartment high-rise tower at Marina One, Singapore are current projects that are both under construction.

Mitchell Joachim is Co-President of Terreform ONE and an Associate Professor at New York University. He was formerly an architect at Gehry Partners and at Pei Cobb Freed. He is a 2011 TED Senior Fellow and has been awarded fellowships with Moshe Safdie and Martin Society for Sustainability, MIT. He won the Victor Papanek Social Design Award, Zumtobel Group Award for Sustainability and Humanity, History Channel and Infiniti Award for City of the Future and *Time Magazine* Best Invention of 2007 with MIT Smart Cities Car. His project Fab Tree Hab has been exhibited at MoMA and widely published. He was chosen by *Wired* magazine for its '2008 Smart List: 15 People the Next President Should Listen To'. *Rolling Stone* magazine honoured Mitchell by including him in its list: 'The 100 People Who Are Changing America'. *Popular Science* magazine featured his work as a visionary for 'The Future of the Environment' in 2010. *Dwell* magazine featured Mitchell as one of 'The NOW 99' in 2012. He earned a Ph.D. at Massachusetts Institute of Technology, an MAUD at Harvard University and an M.Arch. from Columbia University.

K. K. Philip Kang is an Economic Affairs Officer, affiliated with the Sustainable Urban Development Section, Environment and Development Division of the United Nations

Economic and Social Commission for Asia and the Pacific (UN ESCAP), Bangkok, where he has served for more than nine years. His portfolio includes sustainable urban development, low carbon green cities, urban infrastructure, energy security, energy efficiency and renewable energy as they relate to sustainable development issues. He has been working on energy issues with a particular interest in institutional capacity building and sustainable energy systems in urban planning and design. He has served with the United Nations for 13 years, including assignments to the United Nations Development Programme and the United Nations Transitional Administration in East Timor (UN TAET), dealing with numerous topics including energy, environment, natural resources, rural development and programme management.

Steve Kardinal Jusuf has a Ph.D. degree in Building Science from the Department of Building at the National University of Singapore. He received the World Future Foundation Ph.D. Prize in Environmental and Sustainability Research for his Ph.D. work on an air temperature prediction model, known as Screening Tool for Estate Environment Evaluation (STEVE), within urban climatic mapping methods. His research interests are in the area of urban microclimates and urban climatic mapping with geographical information systems (GIS). He has worked on a number of research projects with various Singaporean government agencies, mainly on urban climatic mapping for sustainable urban development. He serves as an editorial board member for the ASCE *Journal of Urban Planning and Development* and has received several recognitions as an outstanding reviewer. He has also published a number of articles in international journals and chapters of books on the topic of the urban heat island.

Lutz Katzschner, Ph.D., is a meteorologist and Professor of Environmental Meteorology in the Faculty of Architecture and Urban Planning at the University of Kassel, Germany. His main research interest is urban climatic mapping from meso- to microscales and their implementation from an urban planning perspective and the impact of global climate change in cities. He is Chairman of the Urban Climate and Planning Guidelines Committee for the Association of German Engineers. He is currently carrying out projects on aspects of global warming and its effect on urban climatology in different countries (Hong Kong, Singapore, Germany and Brazil). He presents Masters-level courses on urban climate studies at universities in Hong Kong, Vietnam and Italy.

Jeff Kenworthy is Professor in Sustainable Cities at the Sustainability Policy Institute, Curtin University, in Perth, Australia. He has spent 35 years in the transport and urban planning field and currently teaches courses and supervises postgraduate students in urban sustainability fields, focusing on transport and particularly car dependence. His research focuses on large international comparisons of cities from a transport and land-use perspective. He is author or co-author of a number of books as well as over 200 other book chapters and journal publications in the area of urban planning, transport and sustainability. He has received two awards from the German government for his research and teaching, including a Mercator Guest Professorship (Goethe University), and is currently a Guest Professor at the University of Applied Sciences in Frankfurt.

Dionysia Denia Kolokotsa is an Assistant Professor in the Department of Environmental Engineering at the Technical University of Crete. She has a degree in Physics from the National Kapodestrian University of Athens (NKUA), an M.Sc. degree in Environmental Physics from NKUA, an M.Sc. degree in Architecture Environmental Design and

Engineering from University College, London and a Ph.D. from the Technical University of Crete's Department of Electronics and Computer Engineering. Her research interests include monitoring and measurement for energy management, indoor environment, energy efficiency in buildings, solar energy and cool materials, distributed energy management systems and building automation. She has participated in more than 25 European projects and has coordinated two EU projects. She is the author of more than 100 papers published in scientific journals and conference proceedings. She is a member of the *Advances in Building Energy Research* editorial board and guest editor of the *International Journal of Low Carbon Technologies* and of *Sustainable Cities and Society*. She is the co-founder and President of the European Cool Roofs Council.

Norbert Lechner is an architect and Emeritus Professor at the University of Auburn, Alabama. He is an expert in energy-responsive architectural design with an emphasis on solar responsive design, including daylighting. He is the author of *Heating, Cooling, Lighting: Design methods for architects* (1991), which is in its fourth edition and used by half of all architecture schools in the US, and is a very popular handbook among professionals. His book has been translated into Chinese, Indonesian, Spanish and Farsi. He has been a Visiting Professor at Chiang Mai University in Thailand. He is the inventor of new types of heliodons (sun simulators used by architects to design solar responsive architecture) that are much clearer and easier to use than any existing models.

Manfred Lenzen is Professor of Sustainability Research with the ISA group in the School of Physics at the University of Sydney (www.isa.org.usyd.edu.au). He has a Ph.D. in Nuclear Physics and 15 years' experience in renewable energy technologies, life-cycle assessment and carbon footprinting. His current research interests revolve around high-resolution impact mapping, novel computational approaches to industrial ecology and its application to global environmental problems (www.nature.com/nature/journal/v486/n7401/full/nature 11145. html). He is Associate Editor for the *Journal of Industrial Ecology*, and is Editor-in-Chief of *Economic Systems Research*.

Margaret Loughnan is a Research Fellow in the School of Geography and Environmental Science at Monash University in Melbourne, Australia. She is a health geographer with expertise in climate–health relationships specializing in heat stress and spatial determinants of population vulnerability. Her recent work includes developing maps of population vulnerability for Melbourne for the Victorian Department of Health, identifying heat–health thresholds for all Australian capital cities and describing the spatial vulnerability of urban populations and related emergency service demands during extreme heat events. She has recently extended this research to examine heat-adaptive behaviours and community resilience in regional towns. She worked as a visiting scientist with the SIMMER project at the US National Centre for Atmospheric Research – Research Applications Laboratory in 2010.

Gerald Mills is a Senior Lecturer at University College Dublin in the School of Geography, Planning and Environmental Policy. He is a physical geographer with a particular interest in climates near the ground and especially those of cities. He has published on aspects of urban energy budget modelling, the global effect of cities and the history of the study of urban climates. He has been centrally involved in the development of the International Association for Urban Climates (IAUC) since its formation in 2000 (www.urban-climate.org). IAUC is the main international organization for those interested in the climates of cities. He has acted as Editor of its newsletter, *Urban Climate News*, and most recently as its President. He organised the eighth International Conference on Urban Climates in 2012 under the auspices of the IAUC.

Shipra Narang Suri is an urban planner with a Ph.D. in Post-War Recovery Studies from the University of York, UK, and has a keen interest in the issues of liveability and sustainability of cities, good urban governance, urban planning and management, as well as post-conflict and post-disaster recovery. She has worked with the United Nations, specifically, UN-Habitat, UNDP and UNESCO, for over 15 years. She has worked in Asia, Africa, Southeastern Europe and the Middle East, and has lectured at several universities in India, Europe and the US. She is Editor of *CITY*, and a member of a global think-tank on 'Livable Cities' set up by Philips. She has been involved in the development and execution of a variety of development, research and training projects and programmes, is a regular public speaker at national and international fora, and has several publications to her credit.

Boon Lay Ong's work spans across academia, research and practice, revolving around the notion of architecture as environment modifier. His latest work is in thermal aesthetics, promoting thermal experience itself as an aesthetic experience. After completing his Bachelor of Architecture, he proceeded to a Master of Architecture in Acoustics. His Ph.D. covered the role of greenery in architecture, incorporating a thermal model for plants. He also developed the Green Plot Ratio concept and the notion of 'nurtured landscapes'. At the National University of Singapore, he developed and was Director of the Master of Landscape Architecture programme and, later, Director of the Master of Science (Environmental Management) programme. He co-edited *Tropical Sustainable Architecture: Social and environmental dimensions*, with Joo-Hwa (Philip) Bay. For the past three years, he has been with the University of Melbourne, focusing on environmentally sustainable design. His latest book, *Beyond Environmental Comfort*, was published by Routledge in 2013.

Devna Pankhania is a Research Assistant in the School of Geography and Environmental Science at Monash University, Melbourne, Australia. She is associated with the Cooperative Research Centre for Water Sensitive Cities. Her Honours research involved remote sensing to assess the role of landscape irrigation in cooling the Australian city of Dubbo under extreme temperatures. She graduated in 2013 with first-class Honours in a B.Sc.

Greg M. Peters is an Associate Professor in Chemical Environmental Science but a chemical engineer by training. He spent seven years employed in the water industry where he was principally concerned with planning engineering innovations in recycled water, desalination and bio-solids management. He has developed research teams at the University of New South Wales and more recently at Chalmers University of Technology. His current research interests include life-cycle assessment, sustainable decision making and the transport and fate of contaminants. His team is currently working on projects for the textiles, agriculture, chemicals and water industries.

Conrad Philipp graduated with a diploma in geography and urban planning from the Friedrich Schiller University Jena, Germany in 2009 and has worked for the Jena-Geos-Ingenieurbuero GmbH as a deputy team leader. In 2013, he received his Ph.D. from the

Institute of Urban Planning and Urban Design (ARUS) of the University of Duisburg-Essen, Germany with a dissertation entitled 'The impact of a green belt on the urban development of a megacity: Interdisciplinary research on the example of Seoul'. During his Ph.D., he held a scholarship from the Federal Republic of Germany and from the University of Duisburg-Essen. As a postdoctoral researcher at sd+b centre at the University of South Australia, he is a team member of the project 'Urban Micro Climates: Comparative study of major contributors to the Urban Heat Island effect in three Australian cities (Sydney, Melbourne, Adelaide)'.

Richard Register is a pioneering theorist and practitioner of ecologically healthy city design and planning who has led local waterway restoration, urban orchard and 'Slow Streets' design/build projects, and championed local general plan and zoning ordinances. He is based in California and has spoken internationally for over 40 years in 31 countries and counting. He has designed and built passive solar greenhouse, and advised the City Council of Oakland California on energy policy. He is probably best known for authoring and illustrating books on ecological city design, planning and history, notably his *Ecocities: Rebuilding cities in balance with nature* (New Society, 2006). He also co-founded the non-profit organization Arcology Circle in 1975, which became Urban Ecology, and founded another NGO, Ecocity Builders, in 1991, of which he is still President. In 1990 he launched the first of the now ten international Eco-City conferences, which average between 300 and 2,000 participants and have been held, in order, in Berkeley, California; Adelaide, Australia; Yoff and Dakar, Senegal; Curitiba, Brazil; Shenzhen, China; Bangalore, India; San Francisco, California; Istanbul, Turkey; Montreal, Canada; and Nantes, France.

Martin Reuter has worked for Ingenhoven architects since 1998 and has been responsible for the design, administration and execution of a variety of projects, including high-rise buildings, commercial, retail, residential and university projects. Besides his position as Managing Director, he leads teams as a design director from competition to completion. His focus is on a wide range of international projects in the Far East, Australia and Europe such as Marina One in Singapore and the Breezé Tower in Osaka. The most recent project he has been responsible for is 1 Bligh in Sydney, which achieved the Australian Six-Star Green Star and Five-Star NABERS Energy rating. Currently he is in charge of the mixed-use high-rise towers Marina One in Singapore, which are under construction, and the fit-out for the Australian Prime Minister's office in 1 Bligh, Sydney.

David J. Sailor received his Ph.D. in 1993 from the University of California at Berkeley, where he conducted research with the Energy and Environment Division at Lawrence Berkeley National Laboratory. His early research focused on atmospheric modelling of urban areas with an emphasis on heat island mitigation strategies. After ten years in the engineering faculty at Tulane University, he moved to Portland State University where he became Founding Director of the Green Building Research Laboratory (GBRL). His research encompasses scales ranging from energy analysis of individual buildings to measurements and modelling of the urban climate system. He was a contributing author to the fourth IPCC Assessment Report and has authored more than 60 peer-reviewed articles. He has served as Chair of the AMS Board on the Urban Environment (2008–2011) and is currently (2011–2015) serving as Secretary of the Board of the International Association for Urban Climate.

Matheos Santamouris is Professor of Energy Physics at the University of Athens and serves as Visiting Professor at the Brunel University in London, Metropolitan University of London, Tokyo Polytechnic University, National University of Singapore, Bolzano University and the Cyprus Institute. He is Editor of the Earthscan book series on buildings, energy and solar technologies. He is Editor-in-Chief of the *Journal of Building Environmental Research*, Editor of the *Energy and Buildings* journal, Associate Editor of the *Solar Energy* journal and member of the editorial boards of seven additional journals. He has coordinated many international research programmes and authored over 200 papers published in international peer-reviewed scientific journals. He is past president of the National Energy Centre, past chairman of the Cool Roof Council of Europe, past vice-chairman of the Air Infiltration and Ventilation Centre, and vice-chairman of the World Society for Sustainable Energy Technologies.

Ehsan Sharifi has a Master of Architecture (2002) and completed a Master of Sustainable Design at the University of South Australia in 2011, focusing on local influences of pedestrian-friendly urban design. His research interests include the quality, liveability and sustainability of public space. Previously he worked as a Lecturer at Shiraz University in Iran, focusing on the history and theory of urban design and rehabilitation of existing urban precincts. He was the recipient of the University of South Australia's Presidential Scholarship in 2012, and a National AAD Travel Grant in 2013 for his Ph.D. at the sd+b Centre at the University of South Australia. He is currently researching the urban heat island effect for his Ph.D. at the sd+b Centre. He is a workshop supervisor of sustainable urban design in the School of Natural and Built Environment at UniSA.

Donovan Storey is currently Chief of the Sustainable Urban Development Section, Environment and Development Division of the United Nations Economic and Social Commission for Asia and the Pacific (UN ESCAP). Prior to joining the United Nations, he was a researcher and academic at several universities, specializing in development management, social development, and urban planning and governance. His most recent position was at the University of Queensland, Australia. Over his professional career, he has coordinated projects related to urban planning and governance, environmental management, social inclusion and sustainable development, principally in relation to the urban experience in Asia and the Pacific. He has authored or co-authored over 30 journal articles, book chapters and books on his areas of specialization. Much of this work has argued for a balance between sustainable development and inclusive and pro-poor outcomes.

Nigel Tapper is Professor of Environmental Science in the School of Geography and Environmental Science at Monash University, Melbourne, Australia. He leads the Climate Research Group within the Monash Weather and Climate Program. He is also Programme Leader in the Cooperative Research Centre for Water Sensitive Cities and an Associate Director of Monash's Water for Liveability Centre. Within the CRC, his group undertakes fundamental work on heat, health and thermal comfort and on the performance of different elements of water-sensitive urban design in improving urban climate to create more liveable cities, in the face of climate change.

Katharine Thornton is a research assistant in the Zero Waste Research Centre for Sustainable Design and Behaviour (sd+b) at the University of South Australia, Adelaide. She is interested in both the rhetoric and practice of sustainability, so she has solar panels on her roof and a compost bin in her garden, tries to grow food and uses public transport.

Nico Tillie is a researcher and teacher in landscape architecture and sustainable development at Delft University of Technology and has lectured in various places around the world. In addition to this, he is currently completing a doctorate (Ph.D.) on liveable low carbon cities, focused on integrating sustainable development and energy planning into the overall planning process. In the city of Rotterdam, he works on city-wide projects as a leading advisor on energy, green, water and resilient city strategies. He worked on the 'Museumpark' and the Architecture Biennale 2012 project 'People make the inner city' and 'Urban by nature' (Urban Metabolism, 2014). He is a Senior Fellow of the Global Cities Institute of the University of Toronto and representative for the Rotterdam region in the German Marshall Fund of the United States in Washington, DC. He is the Dutch representative in the ISO sustainable and resilient cities indicators project. Currently he is coordinating several European Union projects on sustainable development, energy transitions and ecosystem services. He is initiator of the GIS resilience profile which helps to accelerate transition into more resilient cities. Within these fields, he is Programme Leader for the sustainability strategy for the southern part of Randstad Holland Metropolis. He previously worked as a landscape architect in Amsterdam and holds two M.Sc. degrees from Wageningen University in Plant Breeding and Botany, and in Landscape Architecture.

Nyuk Hien Wong is a Professor in the Department of Building at the National University of Singapore. His areas of expertise and research interests include the urban heat island effect, urban greenery, thermal comfort in the tropics and building energy simulation. He is the principal investigator of a number of collaborative research projects with various government agencies in Singapore to study the urban heat island effect in Singapore and to explore the various mitigation measures such as the effective utilization of urban greenery and cool roof materials. Over the years, he has researched how greenery can be effectively integrated with building façades through the incorporation of roof-top gardens and vertical greenery systems to reduce the heat gain into the buildings and also to reduce the urban heat. He has also been involved in an international research project commissioned by the Planning Department of Hong Kong to develop an air ventilation assessment system which is now in place for Hong Kong.

Yukiko Yoshida is an Assistant Professor in the Department of Environmental Engineering and Architecture at Nagoya University. She completed her Ph.D. in Engineering at Nagoya University in 2013. Her educational background is in architecture. She is interested in how to reduce energy consumption through systems such as the energy system, air-conditioning controls and urban planning. She completed her architectural studies at Tokyo University of Science (Kazuhiro Kojima Lab) in 1999. She worked as an assistant at Japan's National Institute for Environmental Studies from 2000 to 2008. She has been a member of ASHRAE since 2004.

Acknowledgements

This book has benefited from the advice and support of many people who have contributed to its development, either through their direct assistance or indirect support. It is the result of a collaborative effort. I am very grateful to all the authors who contributed to this collection of essays. I am particularly grateful to Donovan Storey and K. K. Philip Kang of UN ESCAP, Bangkok for the wise Foreword; and to Joo Hwa (Philip) Bay of Curtin University, Perth for contributing a thought-provoking Preface. These texts help to frame an innovative and interdisciplinary discussion, with a diversity of approaches that makes this a special book in the field; one that brings architecture, urban design and the disciplines of planning and engineering into conversation about the low carbon transformation of cities and new approaches to urbanization.

While I am now Head of School at Curtin University, this book was produced during my time as Research Professor at the University of South Australia; it was supported by the University of South Australia, where it received significant resources.

There are many people who deserve to be personally acknowledged. First, I wish to thank all my exceptional colleagues at the University of South Australia in Adelaide, at Curtin University in Perth and at Tianjin University in the People's Republic of China for their valuable contributions and insights during the development of this publication. I would also like to thank Zero Waste SA and its chief executive, Vaughan Levitzke, who again enthusiastically supported the idea of this book. I hope the contributing authors will find the book a worthy reflection of their contributions. I am grateful for the input of Kate Leeson in reviewing draft chapters. My grateful thanks to Katharine Thornton for her dedication; her editing has made this a better and more readable book. Without Katharine's project management and writing skills, this book would not have been possible. She has also made a contribution to the text of the part introductions.

The research in this book could have never been undertaken without the tireless support of many people – too many to name here – in universities, government departments and industry partner organizations, who have generously shared information and supplied their insight. My thanks for wise advice and big thinking to (in alphabetical order): Pal Ahluwalia, Cida de Aragon, Simon Beecham, Cees de Bont, Peter Brandon, Robert Cervero, Edwin H.-W. Chan, Guanyi Chen, Rae Kwon Chung, Tim Dolin, Peter Droege, Hisham Elkadi, Majella Franzmann, Herbert Girardet, Keith Hampson, Manfred Hegger, Peter Herrle, Richard Horden, Christoph Ingenhoven, Arata Isozaki, Françoise-Hélène Jourda, Hermann Kaufmann, Norbert Lechner, Annie Leonard, David Lloyd, Bart Lootsma, Ken McCluskey, Peter Newman, Gustav Peichl, Igor Peraza, Deo Prasad, K.T. Ravindran, Nigel Relph, Saffa Riffat, J. Alexander Schmidt, Arjan van Timmeren, Kelvin Trimper, David Turnbull, Alec

Tzannes, John Wardle, Nyuk Hien Wong, Siegfried Zhiqiang Wu, Ken Yeang and George Zillante.

My most sincere thanks go to my publishing team at Routledge (an imprint of Taylor & Francis, the Informa Group) in the UK, who were very helpful in establishing this (then) new series on sustainable design and in this book's production. I especially wish to thank Nicki Dennis and Alice Aldous for their patience and support. I also wish to thank all my dedicated colleagues at the School of Built Environment, at the Zero Waste Centre for Sustainable Design and Behaviour, at the China-Australia Centre for Sustainable Urban Development, and the CRC for Low Carbon Living, for their ongoing and valuable support.

Steffen Lehmann
Perth/Berlin, June 2014

Foreword

Planning for sustainable and low carbon green cities

Donovan Storey and K. K. Philip Kang[1]

The urban challenge

Over half of the world's population lives in urban areas and this proportion is expected to grow to 67 per cent in 2050, or 6.2 billion urban citizens. This transformation is being driven in large part by urban growth in the developing world, and particularly by the unprecedented demographic transition taking place in the Asia-Pacific region. Indeed, in 2050 the developing world will account for 5.1 of the 6.2 billion people estimated to be living in urban areas, with Asia-Pacific alone accounting for 3.3 billion – or approximately the equivalent total global urban population of today (United Nations 2012).

This will constitute a dramatic and historic shift in human settlement, and one that will place cities at the very centre of our well-being. Our response in preparing for and shaping this urban future is therefore urgently required and of critical importance. This book makes a valuable addition to the knowledge we need to design, plan and build cities so that they are sustainable and livable. It is obvious that sustainable cities need to be low carbon cities, and that their economic structure and growth must be driven by a sustainability agenda, but how do we transform our urban systems in the context of rapid population growth and given their increasing role as regional and global economic centres based upon competition and high levels of resource use?

Several local but significant events have highlighted the fragility of cities that have not decoupled resource exploitation from economic growth. In January 2013, Beijing recorded its highest level of air pollution – well over 30 times the WHO's recommended limit. For several days in July 2012, Indian cities suffered complete blackouts which caused chaos for an estimated 600 million people. There have also been severe flood crises in Bangkok, Brisbane, Jakarta, Manila and Karachi, among other major cities in the Asia-Pacific region. Other serious and well-documented challenges include traffic congestion, the urban heat island effect, waste management, and water and food security, as well as a number of health-related hazards that pose threats to both current residents and the urban generations to come.

There has been increasing attention in recent years paid to the role of cities as engines of economic growth (producing an estimated 80 per cent of global GDP). But cities as drivers of prosperity is only one side of the coin. Cities have a vast and growing ecological and resource footprint; they account for 67 per cent of all energy use and for 71 per cent of all greenhouse gas emissions (UNEP 2012). Buildings alone account for 40 per cent of a city's carbon emissions, and recent research on the carbon footprints of infrastructure materials suggests that the impact of the built environment is much greater. Cities also consume 75 per cent of energy supply used (IEA 2008) – and this impact is being magnified by the emergence of new middle classes, especially in the Asia-Pacific region, which is driving unprecedented shifts in consumption and production patterns.

Driving an agenda of change

While it is often recognized that today's level of urbanization is unprecedented and the pressures on cities, especially those in the Asia-Pacific region, will continue to grow, we are yet to see the necessary paradigm shift required to decouple urban-driven growth from resource exploitation and exhaustion. The patterns of resource exploitation currently fuelling urban development, and the inefficiencies that underpin such systems, pose considerable challenges to the delivery of cost-effective and affordable services and the realization of environmental sustainability. We need to identify practical pathways to low carbon and sustainable cities. This should not be seen as limited to a 'green agenda'. Traffic congestion costs and poor-quality infrastructure have been found to have significant negative influences on the competitiveness of urban areas. The profitability of investments, possibilities for economic growth and quality of life (including attractiveness to professional migrants) are compromised without a greater commitment to effective urban planning. By contrast, opportunities multiply through the creation of green industries. Clearly, the business-as-usual scenario and 'conventional' development and planning models of planning are insufficient for (if not incapable of) leading away from unsustainable production/consumption patterns, and will create inefficient economic systems in the long run (UNESCAP 2012).

In order to achieve sustainable development, building sustainable cities is essential; the United Nations Secretary General Ban Ki-Moon has stated that 'our struggle for global sustainability will be won or lost in cities'. The chapters in this book outline what is required to bring about the transformations that are essential for a more sustainable and prosperous urban-based future. These all indicate the need to close the widening gaps between current growth paradigms, and what is needed to realize sustainable cities in the context of a green economy. These gaps not only exist within but also between sectors. Sectoral approaches to sustainability need to be replaced by more holistic and integrated paradigms, if we are to make the changes that are necessary in the key areas identified through this book and to create low carbon cities that introduce advanced environmental technologies; have a sound material cycle, including the 3R (reduce, reuse and recycle) approach to sustainable consumption and production; are resilient against natural disasters; and are human-centred with a focus on quality of life and inclusive planning and development processes. Such lessons and strategies must also be seen as global ones. Eco-efficient urban planning and smart green cities can pave the way for developing cities to plan for more sustainable and inclusive urban futures before it is too late to change course and irreversible investment decisions have permanently 'locked in' urban infrastructure (buildings, bridges, factories, homes, railways, roads and public facilities) for decades. This is especially relevant for small and medium-sized cities, which will be the fastest growing urban areas and account for the majority of urban residents in the twenty-first century.

Towards 'the urban future we want'

The United Nations Rio +20 conference on sustainable development reaffirmed a global commitment to promote sustainable cities and human settlements through integration of the three dimensions of sustainability underpinned by effective and participatory (local) governance. In particular, the outcome document, *The Future We Want*, states that the global community commits itself to promoting 'an integrated approach to planning and building sustainable cities and urban settlements, including through supporting local authorities' (p. 1). Member states, in also recognizing 'the important role of municipal governments in setting a vision for sustainable cities, from the initiation of city planning through to revitalization of older cities and neighbourhoods', highlighted the essential place of planning in the realization of sustainable cities. As is evident throughout the chapters that follow, there is not only a compelling case for change but also a rich tapestry of experiences and ideas for citizens, policy makers and urban leaders to draw upon in making these changes.

This book is a valuable and timely contribution to the urgent task which confronts us – the realization of an urban future that is inclusive and sustainable. Through a discussion of conceptual models and frameworks, and in presenting a number of case studies, it provides a greater understanding of the fundamental elements of green urbanism and the associated policy opportunities that could be implemented for a more sustainable and low carbon urban future.

Note

1 Sustainable Urban Development Section, Environment and Development Division, United Nations Economic and Social Commission of Asia and the Pacific. The views expressed in this article are the responsibility of the authors alone, and should not necessarily be considered as reflecting the views or carrying the endorsement of the United Nations.

References

IEA (International Energy Agency) 2008, *World Energy Outlook 2008*, International Energy Agency, Paris.

United Nations, Department of Economic and Social Affairs, Population Division 2012, *World Urbanisation Prospects: The 2011 revision*, United Nations, New York.

United Nations 2012, *The Future We Want – Outcome Document*, United Nations, New York.

United Nations Economic and Social Commission of Asia and the Pacific (UNESCAP) 2012, *Low Carbon Green Growth Roadmap for Asia and the Pacific: Turning resource constraints and the climate crisis into economic growth opportunities*, UNESCAP, Bangkok.

United Nations Environment Programme (UNEP) 2012, *Sustainable, Resource Efficient Cities: Making it happen!*, United Nations Environment Programme, Nairobi, Kenya.

Preface

Holistic approaches to future cities

Joo-Hwa (Philip) Bay

In the old days, cities were so small and compact that they caused insignificant damage to the vast mother Earth. Cities today not only consume many more resources, they also create much heat and pollution within the urban area, with damaging repercussions on the hinterlands, other regions and the Earth's atmosphere. Not many cities can be like Costa Rica, which has a small ecological footprint within a larger region of natural environment.

In 2000, two businessmen and an architect visited the then Governor of Phnom Penh. The Governor said, 'Gentlemen, before you introduce any new trade and industry, please remember that Cambodia is agriculturally based, and anything you introduce should preferably enhance this culture.' He also did not want his city to model itself after Singapore or Shanghai and then have to compete on their terms.[1]

For many developing regions, the patterns of consumption seem to be modelled after the developed world's, or the industries seem designed to feed the higher demands of the developed world. If consumers who can afford high prices need alternative fuels (made from sugar cane, for instance), farmers may forgo the production of food on available land, even though food shortage is one of the biggest problems for humankind. Given that the production of red meat already contributes to one-third of the ecological burden, if the people in developing cities grow wealthier and demand a lot more red meat, this will only aggravate the problem. This also applies to the consumption of air-conditioners, cars, bigger houses, larger facilities, longer hours of operation and longer distances travelled.

Living simply, treading on the Earth lightly, and giving back something in return for what we take to keep nature and society in balance are ideal ways to live. From devastated countries like Cambodia and Afghanistan, to rapidly developing, emerging cities, to developed cities; from the tropical regions where most of the rapidly developing cities are, to the temperate zones where old facilities and infrastructures form most of the redevelopment challenges; what are the series of questions we need to ask regarding the social and environmental dimensions of urban development to arrive at more holistic and sustainable solutions now and into the future? How do we go backwards far enough, zoom up high enough, to reframe the problem? The cognitive framework of specialization has given way to more inclusive and integrated thinking in dealing with the complex problem of low carbon design and sustainability; instead of the single building, we now consider the urban scale; instead of an urban system, we look at the regional network, and physical environmental attributes are no longer considered apart from social dimensions, which include cultural habits, community and social capital, and economics. These new perspectives raise several questions:

- Does the rate of healing match the rate of decay, or are we already at the point of no return?

- How have cities been planned, designed and regulated towards being more low carbon and sustainable places in relation to their surrounding regions? What are the various schools of thought and new approaches that have not yet been tested? What are the success stories and the constraints? What are the crucial areas needing more research, policy development and better governance? What are the models of managing with regulations and incentives that work well?
- Can there be information systems that track consumption and development patterns reflecting the related social and environmental damage, to make it easier to identify problem areas needing more attention? Perhaps the concise and clear presentation of the negative and positive impacts that result from consumption patterns in relation to various policies, planning and design strategies can assist in more reflective and critical decision making.
- Can the principle of 'planting one or two trees for every tree felled' be an approach we may also apply to the consumption of other resources? Imagine farmers having to plant 15 times more food crops for every kilogram of red meat sold. Imagine miners having to reforest the equivalent areas of land they destroyed. Imagine developers having to grow back the amount of leaf area they destroyed for each plot of land cleared. Is there a mechanism beyond the carbon tax system for something more direct, immediate and tangible that compensates for the damage caused to the environment while also creating real jobs? What are the possible areas where we could apply similar principles of replacement, and what scientific methodologies may be used to inform practical regulations and encourage such efforts?
- Can cities that are 'guzzlers' be adapted to lower energy and lower pollutant systems? What and how can alternative energy systems, with fewer side effects, be introduced into existing cities? What are the current methods proposed for mitigating the heat island effect, and are they effective enough? What critical areas, that are likely to have a greater positive impact, need more research attention? Which urban models make the best use of passive design principles in relation to solar heat control, natural ventilation, evaporative cooling and the integration of vegetation to reduce energy consumption by one-third or more? What are the differences needed in such models for tropical or temperate regions? What are the different comfort indices, allowing for greater latitudes of tolerance to temperature and humidity, to save energy by using appropriate design criteria and avoiding unnecessary cooling, heating and insulation?
- Can cities be positive contributors to the health of their ecological and social systems instead of damaging them? How can developments be more regenerative in nature? Can non-profit organizations go beyond merely providing emergency shelters and small houses to assisting with the creation of building systems and related decentralized industries (such as biomass energy systems and suitable permaculture) accompanied by matching education, for longer-term sustainability? Can contemporary cities be used to produce food and biodiversity? Can big cities be planned and designed to collect and recycle water for both recreation and consumption? Can all the waste products be channelled back into various cycles of use? Can the differential heat levels of facilities, machinery and natural open spaces be used and reused between buildings, groups of buildings and districts? Can old facilities that amount to about three-quarters of the volume of the built environment in many cities be seen as reclamation and regenerative resources, both socially and environmentally?
- When higher-density and compact city options are preferred over suburban sprawl, what are the new design alternatives? What is the motivation for wanting to live further out

of the city? Can there be new housing types that offer a private 'piece of green' or land in the sky, that give residents a strong sense of belonging and security, and that are comfortable without having to turn on air-conditioners? What other models can provide the balance of green and hard building materials, and the social community dimensions? What architectural and urban design alternatives can provide for flexibility of usage and adapt to changes of lifestyles and the need for smaller spaces as one grows older, or when a family's children want their own spaces? Can dwellings be live, work and play spaces? Can strict zoning make way for more diversity, and with a more intimate social scale? What are the elements of the compact city models and transportation systems that attract more occupants to live in the city? What are the implications of this for tropical cities? Are there urban ways for pedestrian and cyclist movement that are simultaneously connectors, recreational spaces and green lungs? Can public spaces and urban canyons provide 'tree canopy' comfort and be a catalyst for more intimate social events and a better sense of security, thus contributing in a more holistic and integrated way to a sustainable urban environment?

In a previous book, Steffen Lehmann introduced a set of principles for green urbanism that went beyond mere green buildings with a more holistic approach to transforming urban systems. In this book, he has brought together experts from around the globe, who have contributed 22 powerful chapters that offer valuable answers to many of the questions above, and yet more critical questions to inspire future research and modes of practice towards lower carbon and more sustainable urban development. The issue of sustainability is multifactorial and this collection offers many multidisciplinary approaches, via theoretical frameworks, research and practice, for healing and proactive regeneration.

Given the pace of environmental change in the world around us, we need well-thought-out theories and tools, in time to adapt our cities, societies and economies. I believe this important publication will succeed in engaging practitioners, students and scholars of low carbon cities in discussions of the fundamental urban challenges and solutions ahead.

Note

1 Joo-Hwa (Philip) Bay, the author of this Preface, was the architect who met Chea Sophara, the then Governor of Phnom Penh, Cambodia, on 21 February 2000.

Introduction

Low carbon cities: more than just buildings

Steffen Lehmann

Transforming cities: shifting the paradigm to prevent environmental degradation and reduce greenhouse gas emissions

At the start of the twenty-first century, humanity is a predominantly urban species. *Low Carbon Cities* is about the challenges of tomorrow's cities and what we need to do to make urban settlements healthier places while radically reducing their greenhouse gas emissions. It is about how to inspire and implement long-term systemic change for urban sustainability.

Cities are the powerhouses of our economy and they can be generators of wealth, innovation and social inclusion; they provide economic opportunities and a good quality of life, and workers with specialized skills flock to cities to be near to the sorts of firms that hire them (New York City, San Francisco and London are good examples of this). More people now live in cities than in rural areas, and urbanization is expected to continue, most notably in cities in the emerging and developing world (led by booming economies in China, India, the Middle East, South America and Africa). To manage this process of transformation and urban growth, cities will need to be designed, retrofitted and managed to decarbonize their energy supply and minimize waste in all forms, encourage urban biodiversity, allow ecosystems to flourish and provide inhabitants with the basic elements of well-being in a resource- and energy-efficient manner. So it comes as no surprise that the World Business Council for Sustainable Development notes that 're-envisioning the design and management of cities, green buildings and infrastructure systems will be central to the urban evolution' (World Bank 2010; WBCSD 2010, p. 39).

Each city has a basic underlying (partially intangible) system that gives it its identity and sense of place. It is impossible to come to an agreed definition of what constitutes good urban design for vibrant cities, and what constitutes the ability to express, in a single building, bigger urban ideas. Great cities are not always practical or efficient; they can be romantic, multilayered and poetic places, but they must also, increasingly, be resilient. The essentially visible structure of any city is laid out by its pattern of streets, boulevards, parklands and building blocks, and its network of public spaces, waterfronts, gardens and squares. Walkability and connectivity (horizontally and vertically) have again become very important urban design principles; streets are for people and trees, not just for cars (Gehl 2010). In the twentieth century, streets, squares and public spaces have lost their prominence and relevance in shaping the culture and use of cities. If a city is walkable or can easily be experienced by bicycle – as in Copenhagen, Stockholm, Barcelona, Paris or Munich – it adds greatly to its liveability and quality. Thousands of tiny details make up the character of a place, overlaid by cultural, spiritual and social qualities that form its identity. The cities we admire have evolved over a long time and, in line with economic, technical, cultural and social changes, have human-scaled urban solutions (often designed in an era when one could still cross town on foot and enjoy neighbourly relationships) that have been modified gradually as human society progressed. The idea of the city as a mixing pot for people of different cultural, ethnic and class backgrounds has never lost its significance. Mixed-use buildings with appropriate density create a sense of 'urbanness' and vibrancy, which are essential criteria for any liveable city. Mixed-use buildings make the most of an urban area and can better complement what is already there. Much of this urban feel is dictated by the 'urban grain', the fine scale and the mix of all kinds of activities, both private and public. Public space is shaped by the width of the streets, boulevards and laneways, and the dimension of blocks, building heights and gaps between buildings, creating a pleasant scale and rhythm with a rich layer of surfaces and textures, details, and materiality to be experienced and enjoyed as one walks or cycles through the city. The ultimate goal of urban design should be rich experiences like these and opportunities for human relationships, which activate the potential for intervention at the neighbourhood level, enriching our lives as citizens. We all know such places – where urban design has improved quality of life and brought health, joy, social relationships and equity to urban living; as others have said before me, a liveable city is a happy one. This liveability is also enhanced when buildings are made from durable, locally sourced construction materials and when the reuse of old, meaningful structures is incorporated into their design.

Sociologist Richard Sennett (1974) has extensively discussed the relationship between private and public life, asking what has happened to public life and public space in cities today. Why can't these qualities become the basis of our urban thinking and future developments once again? Since it is very difficult to change urban form retrospectively, we need to shape the next generation of cities on the basis of such substance of urban form and a high-quality public space network that allows for informal and random social interaction between pedestrians with easy access to a mix of uses; with appropriate density, mixed-use programmes and walkability, accepting these as drivers of our urban strategies. In this book, the contributors will argue that energy efficiency, low carbon emissions, reduced embodied energy and sustainable life cycles should also be added as drivers of our future urban planning.

Urban design is intimately dependent on the economic cycle. Usually, much of the worst development emerges from a building boom (think of Dubai), when there is too much development and not enough reflection. For centuries or even millennia, good and thoughtful urban design has combined and balanced artistic with scientific and technological knowledge to shape civic places that are rich in cultural diversity and history, and has sought to continue the richness and complexity of the older city districts. The aim was always the creation of 'place', a concept defined as a meaningful specific location with emotional, spiritual and symbolic dimensions (Sitte [1889]1945; Jacobs 1961; Rossi 1966; Rowe and Koetter 1978; Alexander 1979; Norberg-Schultz 1980; Madanipour 1996; Kostof 1999; Krier 2003; Bosselmann 2008).

The attributes of compactness, mixed use and walkability are a city's elegant and enduring qualities, where monumental civic buildings touch us and where quality density manifests itself through diversity in variations of three- to ten-storey urban blocks, supporting the public realm and streetscape (see Figures i.1a and i.1b). They are also well-known principles of timeless urban design that should be applied to all new developments, and an experienced urban designer should always be aware of how to apply them generously to existing urban situations to ensure pleasant, human-scale, and compact yet comfortable, mixed-use precincts

Figures i.1a and i.1b The optimum density and mixed-use buildings of low carbon cities do not need to be invented; they are already manifested in the traditional European city model, which has no high-rise towers. Aerial images of compact Barcelona and Paris illustrate this point well: traditional urbanism is ecological urbanism.

Source: Supplied by the author.

and neighbourhoods (avoiding monotonous, repetitive buildings, which are so easy to create). Today, many developments lack the monumental grandeur, aesthetic unity and continuity that we associate with the great urban achievements of earlier eras. We can build on these distinctive characteristics of our cities, while maintaining their sense of place, cultural diversity and walkability, to produce meaningful public-spirited works. In arguing for a new ethics of the urban condition, we can point out that the traditional urbanism of European cities (such as Barcelona, Paris, Berlin or Athens) is also ecological urbanism (Lehmann 2005; Brugmann 2009).

Cities and their public spaces: always evolving

The commercialization of urban public space and the increasing involvement of the private sector in the design of public space have frequently been criticized. In the age of communication technologies, old public space typologies are being retrofitted to contemporary needs while new types of public spaces are emerging. In this regard, Mehta notes, 'while modern societies no longer depend on the town square or the piazza for basic needs, good public space is required for the social and psychological health of modern communities' (Mehta 2014, p. 53).

In all this, we should remember that cities were never intended to be completed. Any city is inherently evolutionary, in constant transformation, and much in its character lies in the complexity and diversity of its urban spaces. However, with the impact of globalization, population growth, demographic change, climate change and the urgency of global warming, achieving sustainable urban development has become significantly more relevant as well as urgent and complex (see the Preface by Bay; and the Foreword by Storey and Kang). The questions that characterize the contemporary city have now shifted; the spaces between buildings keep changing (see Chapter 4 by Lechner, which deals with these spaces and their optimum shape and orientation). Since the Industrial Revolution, the process of urbanization has become ever more resource-intensive. The urbanization process, constantly regenerated, dramatically affects all energy, water and material consumption (see Chapter 1 by Girardet, which also explains cities as ecological and economic systems). For these reasons, the urban apartment block as housing type has many advantages: apartments share a circulation system, construction and wall spaces, and one can easily obtain good thermal performance and save construction materials by building blocks three to ten storeys high. My research over the past decade has shown that energy use can drop by 30 per cent if people move from free-standing suburban houses to inner-city apartments (Lehmann 2010; see also Chapter 9 by Lehmann).

Cities are the single largest contributor to climate change. In the past century, the availability of cheap fossil fuels has been a driving force for rapid urbanization and the shaping of urban form. It has also enabled the increase of urban footprints and the development of car-dependent suburbs. Today we recognize that, in order to deal with the new set of challenges, a shift in scale and ambition is necessary, from individual building scale to entire urban precincts and neighbourhoods, with a renewed focus on public space. Thus, the challenges vary from region to region; in Europe, the development task is less about building new, but rather about improving and upgrading the existing urban fabric. European cities have been built and their form is deeply rooted in a past that often freezes it in time and place. On the other hand, in the US, Canada and Australia, the issue is to overcome car dependency and the legacy of sprawl (Hall and Tewdwr-Jones 2009). The situation is quite different again in the Asia-Pacific region, where building entire new cities from scratch is an option. With

China's rapid urbanization process, we find new concepts of city emerging, allowing for radical new thinking about urban precincts and architectural group forms, and advocating for an end to context as the all-determining factor. Here, the future is not about what buildings look like, but how groups of buildings support the public realm, perform and interact, and how they connect with each other to balance their operational needs (on this interrelationship between clusters of buildings, see Chapter 8 by Tillie, van den Dobbelsteen and Carney).

Mega-trends shaping our urban futures: new city typologies emerging

According to many experts, the impact of humanity on the Earth is already overshooting the Earth's capacity to supply humanity's needs – exceeding the carrying capacity – which is an unsustainable position (Rees 2006; see also Chapters 1 and 2 by Girardet and Register). As the Global Footprint Network (2013) points out, humanity is now using ecological resources and services at a rate it would take over 1.5 Earths to renew. We are on track to require the resources of two Earths well before 2050. Today, over 80 per cent of the world's population lives in countries that use more than the ecosystems within their borders can provide and renew. These 'ecological debtor' countries either deplete their own ecological resources or obtain them from elsewhere. For instance, Japan's residents consume the ecological resources of 7.1 Japans; it takes 4 Italys to support Italy; and Egypt uses the ecological resources of 2.4 Egypts. Not all countries demand more than their ecosystems can provide, but even the reserves of 'ecological creditors' like Brazil, Indonesia and Sweden are shrinking over time. We can no longer sustain a widening gap between what the environment is able to provide and how much our infrastructure, economies and lifestyles require. The need to move to low carbon city planning is obvious.

Urban areas currently account for 60 to 80 per cent of global energy consumption, 75 per cent of carbon emissions and more than 75 per cent of the world's natural resources. A book on the appropriate transformation of urban systems is therefore important and timely, as it is essential to deal with the future increase in urban populations, current over-consumption and cities' growing footprints despite finite resources and limited availability of land. Therefore, this book presents the urgent issue that we are facing by highlighting the need to take steps to address greenhouse gas emission reductions and the global nature of the challenge (a challenge that may even include abandoning new fossil fuel-based projects given the opportunity).

In 2007 and 2013, the pivotal IPCC reports pointed out that human-induced greenhouse gas emissions are mainly a result of burning fossil fuels and land-use changes, but that there are also a range of other indicators impacting upon emission levels (for instance, the link with population and economic growth, consumption, energy and water usage, industrial production, waste management, food supply, land use planning, transport patterns and policy decisions, which are all critical). In the past 20 years, progress in the area of international climate policy has been modest at best. Annual greenhouse gas emissions have increased by over one-third since 1992 and keep rising. Acute conflicts of interest among industrialized, emerging and developing countries remain persistent obstacles to a comprehensive global climate treaty.

Experts have identified the following mega-trends affecting cities and shaping our urban future:

- Globalization: global economic power is shifting from Europe and North America to Asia and Latin America.

- Demographic change: urban populations are increasing, with an ageing population in many countries.
- We are moving towards a global knowledge economy, and the rise of this knowledge-intensive economy is reshaping our cities and workplaces.
- The global middle class is rapidly expanding, increasing consumption levels, especially in developing countries; at the same time, urban sprawl is still happening in many places, without regard for its consequences on urban infrastructure and loss of agricultural land.
- Climate change: greenhouse gas emissions keep rising; reductions are not happening as planned, nor as fast as necessary.
- The world could be running out of certain resources, while we still waste so much energy, water, raw materials, food and space.
- Environmental degradation and fossil fuel-dependent systems jeopardize people's quality of life. For instance, the supply of drinking water has become critical and sea levels may rise very quickly.
- Urbanization continues, causing challenges of social inclusion and equity, while there is more and more competition between cities to attract investment and a skilled workforce (which is the most important resource of all). With rising energy and fuel costs, car-dependent sprawl becomes unaffordable – will this lead to a suburban exodus?
- Social media can reveal issues and quickly circulate stories; for instance, large corporations polluting the environment.
- Mobility has emerged as a new paradigm. Mobility now characterizes and dominates our lives, as we travel faster and farther, but still spend more time in transit (between home and work), which will lead to new roles for public spaces. Why does today's mobility still mean high carbon travel?
- Smart environments: surveillance cameras, smart meters and computers are being installed everywhere, inside buildings and outside in public space – but unseen: ubiquitous computing is already with us, with computers in our cars, phones, fridges and toys. But we still have to use keyboards to input information. Information-processing devices are steadily getting smaller and will soon be virtually everywhere, completely invisible and without the need for keyboards. By 2030, we will be surrounded by computers that perform autonomously, but we will be utterly unaware of their presence. 3D-printers will allow for personal fabrication in every home. The ability to produce our own products in our very own homes will upset and completely change traditional models of manufacturing.
- Experts are moving away from simplistic linear approaches to urban development, to complex dynamic network systems and systems thinking, understanding the importance of interlinked and inter-reliant systems.
- More and more municipalities require an evidence base for their urban policies; they have a strong focus on low carbon mobility, renewable energy, waste management, resource recovery and the reintegration of biodiversity in the urban environment; some cities even have ambitious targets such as being 'climate neutral' or becoming the 'greenest city' in their region.

These mega-trends are changing the way we perceive cities and how we understand the relationship between our own lifestyle, the city and the environmental crisis. Clearly, now that we are heading towards a global population of nine or ten billion people by the end of this century, we will not be able to keep our high-energy-consumption lifestyles.

A new set of questions and challenges – and a definition of a low carbon city

While the knowledge of good urban design allowed us for centuries to design cities that functioned well and had beautiful proportions, now an entirely new set of questions about optimal city form and urban management have emerged that have not previously been asked, such as:

- How can we use new development approaches to transform our existing cities to emit much less greenhouse gas?
- What are the behavioural, technological and urban design options for better managing this change?
- Which city form and size is best for enabling affordable mobility using public transport? For instance, can we establish an optimal range of density for sustainable city form and decentralized systems generating on-site energy?

This thinking about urban futures goes far beyond conventional or traditional ideas of aesthetics (of the 'City Beautiful' type) or functional and organizational city form (of the 'Ville Radieuse' type); this is about the long-term sustainability of urban settlements and communities. It forces us to look at cities in a completely new way, and new types of city are likely to emerge. The urgency seems appropriate; as the atmosphere has moved beyond 400 ppm of carbon in 2013, the time has come for a new boldness in tackling environmental problems through revised urban planning. However, there will be no quick urban fix to the problem of global warming. This means that we need to apply incremental approaches as well as seeking major economic and social change with strategies that will work, to envisage and apply a new, daring and ambitious environmentalism to radically re-engineer our urban settlements.

There are now over 50,000 published peer-reviewed articles on climate change predicting global warming (data 2013). Extreme heatwaves, urban flooding, coastal erosion, poisoned water and dwindling water supplies, polluted air, melting permafrost releasing greenhouse gases, loss of biodiversity, prolonged droughts and crop failures could all become more frequent if global warming and rising temperatures are not brought under control. In addition, we already face increasing energy costs, transport congestion and housing shortages in most cities (Haas 2008; Berners-Lee and Clark 2013; Urry 2013). Despite all this, energy demands and the use of primary energy from fossil fuels – and consequently carbon emissions – are rapidly rising. Unfortunately, the green dream of weaning the world off fossil fuels remains far off.

The list of new challenges opens up many possibilities for innovative design thinking at different scales; for instance, we are likely to develop new kinds of infrastructural systems that better engage with the social and environmental conditions that continuously reconfigure the city today. In all of this, urban design still warrants a very high priority as it sets the framework for any future urban development at an early stage and remains central to successful low carbon outcomes in which urban form, public space, density and infrastructure all have a strong interrelationship. However, a large number of cities (especially in Asia) are suffering from severe environmental degradation caused by too rapid, out-of-control urbanization, or outdated patterns of growth and decline that cannot deal with the new challenges. Their environmental problems include severe air and water pollution, excessive waste generation (and lack of modern treatment facilities), severe traffic congestion – due to lack of efficient public transport – and changes in the urban microclimates; these cities usually also struggle to provide affordable housing and a secure supply of energy, water and easy access to sanitation for all residents. Clearly, cities need a robust infrastructure to generate economic growth and improve quality of life; an inadequate, outdated, fossil fuel-dependent and underperforming infrastructure can reduce economic output and be a hindrance to better living conditions.

Connected with this, further challenges are posed by demographic shifts of an ageing population (which first started as a particularly Western and Japanese problem). As people age, they want to remain as engaged, active and independent as possible for as long as possible. 'Ageing in place' means being able to stay at home, not having to age in an institution or nursing home. Therefore, what kind of housing and inter-generational precincts will we need to provide in the city of tomorrow to allow for affordability, social inclusion and ageing in place?

In this context, I am often asked: What exactly are eco-cities? The basic principal behind eco-cities is living within the means of the environment and resources. For the benefit of the people living in the city, eco-cities reduce their greenhouse gas emissions by producing energy through renewable sources such as solar, wind, hydropower and biomass, use low carbon public transport, have a green infrastructure and reduce the embodied energy of buildings (Roseland 1997). Eco-cities produce energy on site, so they require little or no energy from the outside and favour local, renewable and easily recyclable materials. Resources are conserved and recovered through waste management, recycling and the natural bio-filtration of stormwater. Eco-city residents even grow food and plant new green areas within the boundaries of the city as urban farming. Eco-cities can be new cities, built from the ground up, or redesigns of existing cities, which are transformed and retrofitted district by district. While global city hubs usually evolve naturally over time, the success of Curitiba, Freiburg, Copenhagen and Singapore illustrates that sustainable city models can be deliberately created to establish a positive reputation. The ambitious ultimate goal of these cities is self-sufficiency. Rather than being places just to work, eat and socialize, we want future urban districts of eco-cities to be more autonomous and perform 'systemic' roles: as part of the energy system (generating more power than needed to operate), waste system (collecting and recycling waste locally), water harvesting (collecting and purifying stormwater), inter-generational exchange (delivering social benefits, with socially restorative qualities) and high-performance environmental control systems. Above all, eco-cities are not built on ecologically important or useful arable land. They emit few or no greenhouse gases; they are also 'low carbon cities'.

Defining the term 'low carbon city'

The general concept of sustainable urban development appeared around 1990, but without any agreement regarding a clear definition, an assessment framework or standards (Hopkins 2008). More recent theories of compact cities, solar cities and green urbanism (when less ideologically driven) encapsulate the belief that urban revitalization and the future of the city can only be achieved by 're-compacting' the city (Jenks and Burgess 2000). Similarly, the term 'low carbon city' is not a uniform notion, but an umbrella term that covers various approaches to sustainable urbanism and brings together multiple forms of sustainable urban development theories applied at different scales and locally contextualized.

In order to clarify the term 'low carbon city' (see also the Glossary at the end of this book), the contributors have provided chapters that, taken together, offer a systematic framework for a theory of low carbon city creation. In a nutshell, low carbon cities are compact and mixed-use cities that have significantly reduced their carbon-intensive energy consumption; they are based on sustainable urban development principles and supply chains, emphasize the avoidance and reuse of resources and aim to exist in balance with the ecosystem. A low carbon city has minimal output of greenhouse gas emissions into the environment, especially the greenhouse gas carbon dioxide (CO_2). While there is no generally accepted and applicable definition of a low carbon city, cities that embark on a low carbon transformation will ultimately also become more efficient, competitive, liveable and sustainable. In this book, we will argue that the 'low carbon city' is an urban model incorporating previous concepts of ecological and sustainable urbanism and resilience theory, combined with practical solutions. The Compact City, the Green City, the Regenerative City, the Eco-City, the Smart City, and so on – they can all be the same place (see Figure i.2).

The concept of green urbanism arose in the late 1990s, promoting compact, energy-efficient urban development with green spaces, green roofs, gardens and urban farming, seeking to transform and re-engineer existing city districts and regenerate the post-industrial city centre (Lehmann 2005). This principle-based holistic concept of sustainable urban systems enables existence and change (growth or shrinkage) without negatively impacting upon planetary life support systems and ecosystems. It is a particular form of urbanism that is based on a metabolic view of cities as living organisms, concerned with a healthy balance between the city and its hinterland (see Chapter 22 by Lehmann and Thornton). Green urbanism also underpins practical action in order to reshape the urban environment in a sustainable way.

Green urbanism is one of a series of conceptual frameworks for urban planning. Depending on the complexity of context and methodology, these frameworks prioritize different aspects of social, economic and environmental sustainability. However, compared to other frameworks, green urbanism principles offer a more open-ended approach to the problem from a higher level. Some of the other frameworks, to name just a few, are as follows:

- UN-Habitat's Eco-City Planning;
- the World Bank's Eco2 Cities;
- Bio Regional's One Planet Community;
- the Chinese government's Low Carbon City in China Initiative;
- Siemens' Green City Index;
- ILBI's Living Building Challenge;
- the Green Building Council of Australia's Green Star Sustainable Communities.

Other frameworks have been developed by C40, the Clinton Climate Initiative and ICLEI.

The principles of green urbanism (outlined in detail in Chapter 22 by Lehmann and Thornton) need to be adjusted and tailored to the different dimensions and scale, and to local policy and economic conditions (depending, for instance, if the project is at precinct-wide level or at site-specific level; see Figures i.3a to i.3e for examples of decentralized precinct systems and innovative infrastructure). As a result, the relevance and weighting of principles varies, depending on whether they are intended to serve short-term or long-term goals, and on the specific context in which they are to be applied.

Most of our buildings are out of tune with the environment, relying too much on air-conditioning and other systems to keep us comfortable (see Chapter 10 by Ingenhoven, Reuter and Dieckmann). While cities are, of course, much more than just buildings, there is a strong link between individual low carbon buildings as urban components and low carbon city master planning. According to the International Council for Local Environmental Initiatives (ICLEI 2012), the fastest and most cost-effective gains can be made by focusing on the following 'low-hanging fruits', and by better anticipating future needs prior to building: urban public transport, better water/wastewater management and waste management, energy-efficient buildings, LED lighting and solar hot water systems. This is also supported by the McKinsey Abatement Curve, the curve for cost-effective greenhouse gas reductions (McKinsey & Company 2009). Other areas where specific buildings can deliver significant gains include looking at cooling system capacity, where plant capacity can be reduced (smaller pumps, fans, chillers, etc., requiring engineers to stop installing the same large, oversized capacity in low carbon buildings); and a better understanding of daylight quality (not just quantity), driven through intelligent façade design to avoid glare and overheating.

Rethinking infrastructure is essential. Sustainable infrastructure for the twenty-first century operates at the precinct scale; smart grids, solar roof-tops, cycling path networks, light

Figure i.2 The many aspects of the 'sustainable city' for a more integrated approach to urbanization. Urban sustainability can mean very different things conceptually and practically, as illustrated in this diagram.

Source: Diagram by Lehmann (2013a).

Figures i.3a, i.3b and i.3c Low carbon cities will have new types of infrastructure, decentralized energy systems and concepts of innovative public space. Above left: energy tower with urban farming in Madrid. Above right: Plus-energy district Solarsiedlung in Vauban, Freiburg, generating its own energy. Below: Foster + Partners' 2014 plan for an elevated car-free 'SkyCycle' bike route in London: a 220-kilometre-long network of bike paths suspended above railway lines to allow for safe cycling, liberated from the congested city; each of ten routes could accommodate 12,000 cyclists per hour and cut down travel times.

Source: Images by Ecosistema Urbano, R Disch and Foster + Partners, supplied by the author.

Figures i.3d and i.3e Low carbon cities offer eco-mobility options with easy change of mode, such as light railway networks with high frequency of trams, car-sharing initiatives and an infrastructure for electro-vehicles. Left: The tram is well integrated in Nantes' urban life. Right: Oslo has implemented many incentives for electro-vehicles, such as free parking, free charging and tax-free registration.

Source: Photos by Lehmann (2013), supplied by the author.

rail, wastewater recycling plants, resource recovery stations and community gardens are all decentralized parts of the urban system that operates at the scale of the community or neighbourhood. Districts could be turned into power stations generating their own energy as 'plus-energy precincts', where citizens become their own energy producer (see Chapter 6 by Lenzen and Peters). Upscaling sustainability solutions from buildings to green districts and low carbon precincts allows for the introduction of decentralized infrastructure, making the district the most suitable scale for intervention.

In this book, we will present some innovative ideas and solutions that deal with the above-mentioned new challenges. The contributors present strategies that will create highly liveable, economically vibrant and resilient cities of the future. A global group of urban experts aim to deliver answers to the new set of urgent questions confronting us, discussing how we can increase the resilience and durability of cities against heatwaves and extreme weather events. The chapters provide plenty of data on these as an evidence base for design recommendations and policy formulation. For instance, urban infill resulting in higher density can actually increase the walkability of a city and create the possibility of a better quality of life at a lower carbon footprint through more efficient infrastructure and land use; however, higher density is often unpopular as it can be a threat to privacy (see Chapters 5 and 11 by Kenworthy and Cervero). In addition, the interplay between density and urban heat islands must be well understood (see Chapters 14 and 16 by Sailor and by Yoshida and Ichinose). I believe that some of the thinking around high-density precincts in China, South Korea or Singapore could translate to other countries and inform new, more acceptable approaches to increasing inner-city density and reducing urban sprawl.

Green networks are part of a growing trend to create comprehensive cycle networks that encompass not only city centres and ring roads, but also connect the city with the suburbs (see Figure i.3c). For instance, the City of Hamburg has decided on an ambitious plan to create car-free 'park connectors' between all major parks, community gardens and playgrounds by 2030. The resulting green network will cover 40 per cent of the city and will enable commuters to cross the once car-dominated city entirely by bicycle or on foot. The network of green spaces will contribute to the overall health of inhabitants, lower CO_2 emissions and keep the city cooler; in addition, it will help to alleviate flooding in the event of heavy rainfall.

Different schools of thought: the evolution of low carbon city theory

Reviewing the literature reveals the breadth of thinking that surrounded ecologically sustainable urban development in the twentieth century and identifies the origins of the sustainable design movement. Publications on sustainable cities and green urbanism have been available since Ebenezer Howard's 1902 book, *Garden Cities of Tomorrow*. Howard introduced the concept of the human–environment link, and the political and social agendas of his theories have recently made a comeback. Frank Lloyd Wright's Broadacre City (1932–1958) was a plan for an urban utopia in the United States of America; it emphasized a self-sufficient lifestyle in harmony with nature and profoundly influenced twentieth-century American urban planning theory. New housing estates by Bruno Taut, Mart Stam and Ernst May in Europe were realized as social visions of resourceful, liveable cities in the 1920s.

Kevin Lynch made a plea for recognition of the human scale of urban form and for respect for local particularities in his book *Image of the City* (1960); while Reyner Banham (1969) pioneered the idea that technology, human needs and environmental concerns should be considered as integral parts of architecture and urban design. Arguably no historian before him had so systematically explored the impact of environmental engineering and services on the design of buildings and cities. Rachel Carson's *Silent Spring* (1962) was the first publication to document the detrimental effects of pesticides on the environment.

From *Silent Spring*, to Victor Olgyay's *Design with Climate* (1963), to Reyner Banham's *Architecture of the Well-Tempered Environment* (1969), to Ian McHarg's *Design with Nature* (1969), authors of pivotal publications have sought to reconnect urbanism with local environments and ecology, describing a blueprint of a better future where urban life is sensitively and sensibly re-embedded in biogeochemical systems (this has also been supported by the writings of other leading architects, from Alvar Aalto, to Glenn Murcutt, to Norman Foster; and by Koenigsberger *et al.* 1980; Drew and Fry 1976; Szokolay 1973).

McHarg recognized early on that increased urbanization leads to fewer trees, less green space and less natural air movement in cities. *Design with Nature* brought environmental and urban planning concerns together, influencing the modern environmental movement and making the environment one of the central concerns of architecture and urban design today. In 1965, Alexander Mitscherlich wrote the important book *Die Unwirtlichkeit unserer Städte: Anstiftung zum Unfrieden* (*The inhospitability of our cities: A deliberate provocation*) that dealt with the societal and psychological consequences of post-war urban planning and critiqued functionalist architecture and the inhospitality of public space. The oil crisis in the early 1970s spurred further interest in renewable energy sources and energy-efficient cities (see Table i.1).

From the remarkable *The Limits to Growth* report by Meadows and colleagues (1972) and *The Brundtland Report* (Brundtland 1987); to the novel *Ecotopia* by Ernest Callenbach (1975); to the important contributions from Robert and Brenda Vale (1991), and the *Charter for Solar Energy in Architecture and Urban Planning* (Herzog 1995), to *Urban Future 21: A Global Agenda for 21st-Century Cities* by Hall and Pfeiffer (2000), the fields of sustainable city theory and climate-responsive urbanism have recognized the profound interconnections between urbanization and climate change (see also, to name a few, Register 1987; Moewes 1995; von Weizaecker *et al.* 1997; Satterthwaite 1999; Girardet 1999; Wheeler and Beatley 2004; Cuthbert 2006; Lovelock 2009).

While *The Limits to Growth* (Meadows *et al.* 1972) was a wake-up call for the over-consuming, sleepwalking societies of the 1970s, our current time is still characterized by wastefulness and uncertainty. As Ulrich Beck has noted, we have arrived in 'a new era of uncertainty', where energy, water and food supply, transport and information flow are critical, but their secure availability is questionable. 'We live in a world of increasingly non-calculable uncertainty that we create with the same speed as its technological developments' (Beck 2000, p. 76).

In 1992 William McDonough and Michael Braungart developed the Hannover Principles, which outlined a framework for sustainable architectural design. The concepts were extended from cities and buildings to materials and products in their pivotal book *Cradle to Cradle: Remaking the Way We Make Things* (2002), which became an influential environmental manifesto. Architect Ken Yeang explored the possibility that high-rise buildings could be sustainable in *The Skyscraper Bio-Climatically Considered* (1996) and in *The Green Skyscraper: The Basis for Designing Sustainable Intensive Buildings* (1999).

More recent theories of 'compact cities' and 'solar cities' encapsulate the belief that urban revitalization and the future of the city can only be achieved through 're-compacting' the city. The authors of the chapters in Part I of this book are all urban theorists who have contributed milestone texts to the conceptual evolution of compact and low carbon city theory. Richard Register's book *Eco-City Berkeley: Building Cities for a Healthy Future* (1987) and Herbert Girardet's many early documentaries and *The Gaia Atlas of Cities* (1992) were pioneering works, laying the foundation for a better understanding of sustainable urbanism. Since the late 1990s, Tim Beatley and I have written extensively on the concept of 'green urbanism' (however, in doing so, Beatley's view was always more from the American context, whereas my focus was on Europe and the Asia-Pacific region). But since then, the focus has shifted away from technological concerns. While most of the technology is available and becoming more affordable and efficient, in much of the recent literature it appears that the biggest challenges are now on the consumer side, in the need to recognize the behavioural dimension; to change behaviour and persuade people to adopt and properly use these new technologies. We have gained the understanding that simply adopting new technology will not be enough; it needs to be fully integrated into the social context in order to deal with the range of complex societal factors (see Chapters 7 and 12 by Joachim and Suri). There is also a large body of work investigating the metabolic aspects of low carbon cities; for instance, Gissen and others have recognized the need to analyse cities as socio-natural spaces of flows in order to understand the complex nature–society relations mediated through the urban environment (Gissen 2009).

The need for behaviour change is widely recognized. In the light of runaway consumerism, changing individuals' behaviour is at least as relevant as changing public policy or technology (Lehmann and Crocker 2013). For instance, increasing the knowledge and practice of recycling is integral to changing human behaviour for sustainability. Pro-environmental behaviour change can make a big difference; it has been recognized as a key factor in the transition towards low carbon cities and the acceptance of new low carbon technologies. But

Table i.1 The evolution of green urbanism in association with key texts and recent theories

Publications	*Authors (date of publication)*
Garden Cities of Tommorrow	Ebenezer Howard (1902)
Image of the City	Kevin Lynch (1960)
Technology, human needs and environmental concerns	Reyner Banham (1969)
Silent Spring	Rachel Carson (1962)
Design with Climate	Victor Olgyay (1963)
Architecture of the Well-Tempered Environment	Reyner Banham (1969)
Biogeochemical systems	Koenigsberger *et al.* (1980); Drew and Fry (1976); Szokolay (1973; 1981)
The Limits to Growth	Meadows *et al.* (1972)
The Brundtland Report	Brundtland (1987)
Ecotopia	Ernest Callenbach (1975)
Green Architecture	Robert and Brenda Vale (1991)
Solar City Charter	Herzog *et al.* (1995)
Urbanisation and climate change	Register (1987); Moewes (1995); Von Weizaecker, Lovins and Lovins (1997); Satterthwaite (1999); Lovelock (2009)
Compact Cities, Solar Cities	Burton (1997); Jenks and Burgess (2000); Lehmann (2005)

achieving behaviour change can be very difficult and some policies to encourage 'green behaviour' may even have negative side effects. For instance, researchers have found that after performing a positive action people may feel 'entitled' to more consumptive or self-indulgent behaviour, which is known as the 'moral licensing' effect (this is different from the 'rebound effect', which relies on a financial incentive). The effect shows that resource conservation in one area could lead consumers to be more wasteful in another. All green awareness campaigns should carefully consider the risk of unintended effects beforehand.

We are hopelessly locked into these unsustainable socio-technological (over-)consumption systems fuelled by capitalism, whether we like it or not; over-consumption is made to look 'normal' and 'compulsory'. German philosopher Peter Sloterdijk defines capitalism as 'a big acceleration project, to accelerate change through city-based cultural projects' (Sloterdijk 2013). In *Motivating Change*, I wrote that 'the once popular idea that the individual consumer would be persuaded to change and turn from consumer into citizen by information campaigns alone has been shown not to work, and needs to be replaced by design-led social innovations' (Lehmann and Crocker 2013, p. 419). Compellingly, Ann Thorpe has noted that 'our consumer-driven economic growth mentality is one of the biggest obstacles to real sustainability that prioritizes wellbeing and confronts consumerism' (Thorpe 2012, p. iix). Challenging our unsustainable consumption patterns and proposing more sustainable behaviour alternatives are some of the most difficult tasks policy makers, planners and designers are facing today.

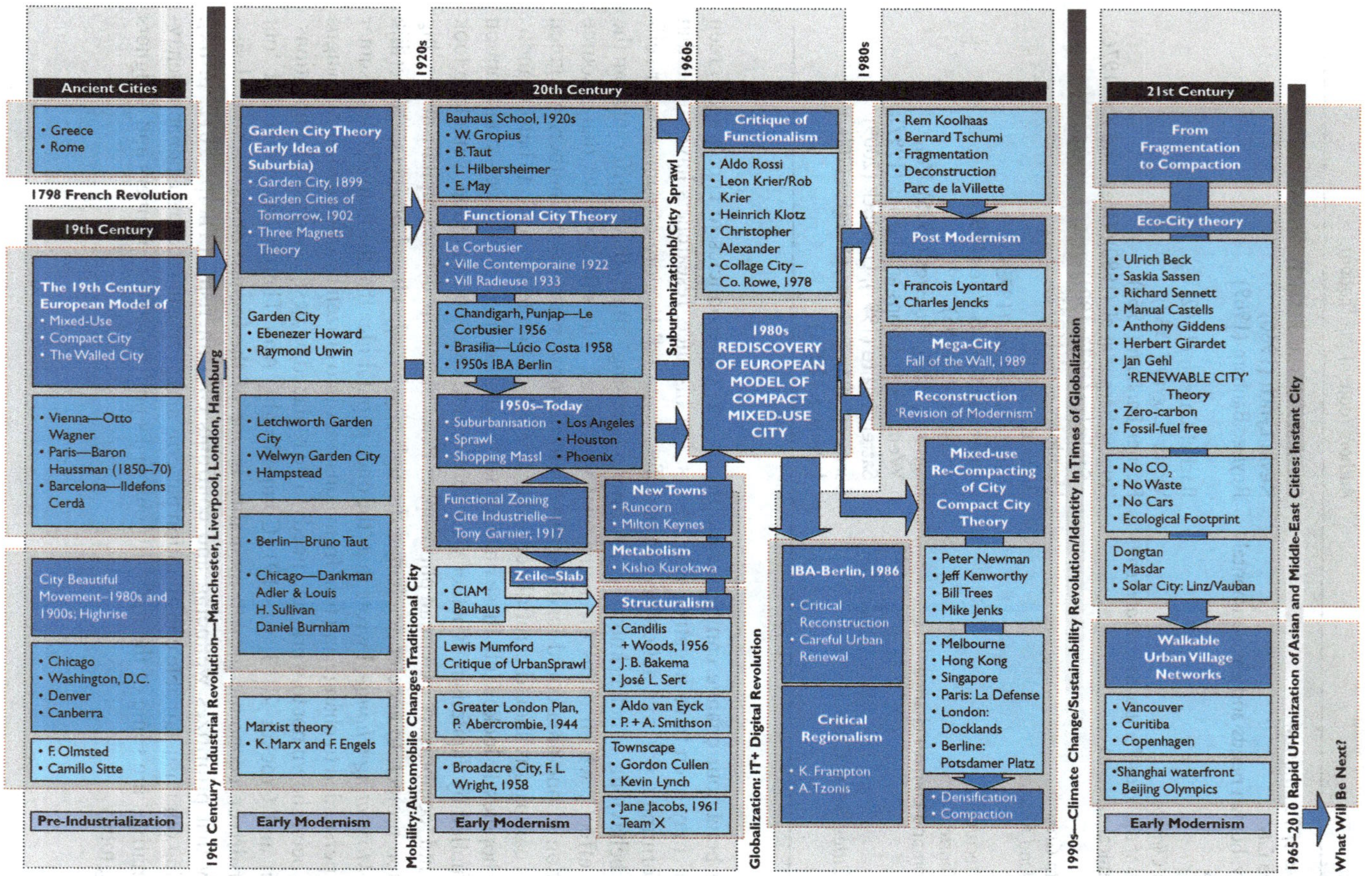

Figure i.4 The historical evolution of city theory and different models of thought. At the beginning of the twenty-first century, we are again on the brink of new city models emerging, including the low carbon city.

Source: Diagram by Lehmann and Dong (2010).

Compact and green: containing urban sprawl, reducing carbon intensity, enabling walkable neighbourhoods

Ending climate change has to begin in our cities. All growing cities face critical challenges connected to population growth, land-use changes caused by urbanization, rising consumption despite finite resources and increasing temperatures. Outdated urban development patterns must be adjusted towards sustainable growth and low carbon development, including the transformation of our complex urban systems and infrastructure for water, waste, energy systems, mobility and food supply (see Chapter 22 by Lehmann and Thornton on green urbanism).

Urban development is forecast to increase in the foreseeable future, raising questions about the kind of urban change that is desirable and necessary, and the preferred patterns of these changes. Which development patterns will allow our cities to become more sustainable, resilient and adaptive? An enormous effort is needed to reduce the developed world's high rates of consumption in cities if we are to make them more resilient and sustainable over long time frames. To meet the challenges ahead – climate change, population growth, social equity, food supply, affordable housing, a healthy built environment, providing job opportunities and efficient green infrastructure, among others – we will need to re-imagine how we plan, design, construct, manage, operate and redevelop built environments (Appleby 2011; see Figures i.5 and i.6).

Figure i.5 Cramped housing conditions in Hong Kong are a sad reality for many.

Source: Photo by Benny Lam (2013), supplied by the author.

Figure i.6 Changing transport modes from cycling to light railway: an innovative urban transport solution that allows for easy interconnection in Stuttgart.

The provision of appropriate housing plays an important role in this. Worldwide, more than one billion urban citizens still live in inappropriate housing (characterized by poor indoor environmental conditions such as extremely low or high temperatures, poor ventilation and sanitation, and lack of space); even in the developed world, a high percentage of people live in low-income households with poor living conditions (Kolokotsa *et al.* 2013).

Sustainable development is inextricably tied to the development of cities and their infrastructural systems. According to the United Nations, there are four interconnected dimensions of sustainable development: economic development that is within the limits of ecosystems; social inclusion that ensures health and well-being; environmental sustainability for more resilient cities; and good urban governance (UNEP 2011). However, not all cities will be able to evolve on these principles and succeed in the twenty-first century – this is a simple fact of global competition. Cities that do not have a diverse ecosystem, or socio-cultural diversity, or that are dependent on a single industry, run the risk of decline and could become victim of what is called the 'Detroit syndrome' (much has been said about Detroit's steady urban decline and loss of population since the 1950s, and its shrinking city status, which does not need to be repeated here). For instance, a number of North American and East European cities, former manufacturing centres, have suffered such dramatic losses in population and employment that urban experts have put them in a class by themselves, calling them 'rustbelt cities' or 'shrinking cities'; the decline has led to property disinvestment, extensive demolition and abandonment (Dewar and Thomas 2012). However, when cities start to decline, investment in public space and economic diversity are two of the strategies that can save them.

Most cities around the world are trying to achieve sustainable conditions. North American and Australian cities are among the least sustainable globally, given their high per capita fossil fuel and car dependency, excessive waste creation and greenhouse gas emission levels. In terms of development pattern, the Australian and North American situation is quite different from the situation in Europe or Asia. While many residents in mono-functional suburbs simply live too far away from their jobs (possibly leading to a future suburban exodus in US cities), in the majority of Australian cities we can now observe a population shift towards inner-city living. Demographic trends show that younger people, downsizing retirees, students and company executives who are keen to walk to their offices are increasingly opting to settle back in the city centre where they can be closer to cultural activities and public transit, for more contemporary lifestyles. There are now more developments on redeveloped brownfield land and urban renewal sites – as people want to live closer to the city centre or in the inner suburbs, to reduce their need to commute or look after a large garden (see Chapter 2 by Register). On the other hand, suburbs have huge retrofitting potential and the advantage of enough space to introduce household-based solar power technology to commence the 'decarbonization transition'; while some other suburbs could even emerge as polycentric clusters or satellite towns with a denser city form. Increasing urban density means building on infill lots. Higher-density living can be a good solution as long as these developments also provide for new treed parks; but a well-used park a drive away cannot be a substitute for new green space. Change is already happening. Through gentle densification, medium-density neighbourhoods are becoming increasingly vibrant, mixed use and fun to live in, without the need to resort to high-rise towers.

The increasing supply of city apartments and townhouses will allow more people to live closer to existing infrastructure, transportation hubs and inner-city services, reducing the pressure for outward sprawl in the peri-urban ring (what Thomas Sieverts termed the *Zwischenstadt*, or 'in-between city' (1997), the ubiquitous terrain that many architects refuse to see or deal with). The peri-urban is an urbanized landscape that exists between inner suburbs and open countryside, the result of disintegration of the once compact historical European city, which has led to new forms of contested urbanization in Europe: large dispersed urban regions, such as the Randstad in the Netherlands or the Ruhr area in Germany; but also Chinese urban mega-clusters such as the Beijing–Tianjin–Hebei multi-core city cluster with over 100 million people.

For too long, urban expansion at the outer edge of cities has been driving people further out, creating unsustainable sprawl and further loss of precious farmland. Instead, urban growth should focus on already built-up areas, offering more diverse housing choices through intelligent infill and careful densification strategies for the existing fabric, where abundant services and infrastructure already exist. This is why many urban theorists suggest that city centres and inner suburbs will have to accommodate future population growth, and any further expansion of outer suburbs needs to be actively discouraged through growth boundaries and a zeroing out of all new greenfield development. And the evidence speaks for itself: building in existing urban areas with existing infrastructure, rather than developing new areas, can easily save up to 30 per cent of the operational energy of a city (Lehmann 2010; see also Chapter 10 by Ingenhoven, Reuter and Dieckmann).

Urban sprawl makes cities less efficient, less resilient and therefore less competitive. Allowing for greater density within key strategic locations will bring more people back to the city as residents and curb urban sprawl.

The answer to rising population lies in better use of the existing infrastructure, by retrofitting our existing districts and increasing their density, making the most of what we have already got. Cities have already been retrofitted and slowly infilled worldwide, but this has often been quietly done and relatively invisible; complementing what is already there has been a more sustainable – but less spectacular – choice, compared to developing new greenfield developments on the fringe of cities. In this process, existing urban areas have been upgraded, with the transformation of post-industrial river banks and harbour fronts or former industrial sites, but also more discreetly as an ongoing renewal process. From the IBA Emscher Park, to the High Line in New York, to Seoul's removal of a freeway, to Melbourne's '1200 Buildings Retrofit' initiative, there are plenty of positive examples to learn from (see Chapter 11 by Cervero).

For some time now, planners and policy makers have used smart densification strategies with mixed-use infill projects to create more sustainable cities. The intensification of compact, medium-density, mixed-use developments along transit corridors, and the urban renewal and revitalization of city centres or post-industrial brownfield sites, including adaptive reuse strategies and an appropriate public space network for walking and cycling, are among the key ingredients of any sustainable city. Green infrastructure, community garden spaces and urban regeneration will help growing cities to become more human and improve their entire urban metabolism (see Figures i.7 and i.8). Green spaces, urban forests, meadows and community gardens play an instrumental role in reintroducing biodiversity and maintaining human well-being. Systems thinking for low carbon cities includes a focus on closer links with nature and synergies across multiple scales and between various systems, to establish self-reliant networks of green urbanism. Areas earmarked for the production of food in the city take short supply chains into consideration: the quality of food, returning nature to the urban environment and the needs of an ageing population – a series of benefits that emerge from landscape strategies of productive open spaces.

It has now become obvious that limiting global warming to just 2° Celsius (+3.6° Fahrenheit), as previously called for by the UN Framework Convention on Climate Change in 1992, has become patently unrealistic today. For a long time, that upper limit of two degrees was considered to be the threshold to 'dangerous climate change'. But political will is weak, and emissions continue to increase, so that target has become unrealistic and needs to be revised. Our task now is to adapt cities to the more likely and projected 4 to 6° Celsius increase in temperatures by 2100 to try to soften the impact of global warming.

This leads us to another major focus of this book: the changing urban microclimate. The interplay between density, public space and the increased risk of the urban heat island (UHI) effect is still poorly understood. In urban development we must always ensure that there is enough room for nature in our cities, for instance, through roof gardens and 'green walls', so that we can mitigate trapped heat, but also keep people in touch with the natural world, which is especially important for healthy ageing (of course, keeping humans in touch with the natural world is important for all ages). The integration of trees, public green spaces and gardens in particular helps to keep the built environment cool, and reduce trapped heat (see Chapters 13, 15 and 19 by Mills, Santamouris and Kolokotsa, and Ong). Otherwise, urbanization leads to the UHI effect, the subject of Part III of this book. It is essential that urban dwellers stay connected to nature by reintroducing high-quality gardens and green spaces back into the urban, restoring beneficial connections between people and nature in cities (e.g. with tree size, trees grow in environmental value over time; see Chapter 3 by Beatley on biophilic cities). Urban gardens act as unique ecosystems that reintroduce wildflowers and native trees and plants into urban green spaces. There is much scientific evidence to support the proposition that contact with nature fosters human health and a sense of well-being, by reducing obesity and stress, improving healing, and increasing creativity and productivity (Kellert 2011).

Of course, it would be wrong to blame urban density per se for the UHI effect, as it is ironically the most effective way to achieve an efficient city with a reduced need for transport (as Kenworthy argues in Chapter 5; see also Chapter 20 by Tapper, Coutts, Loughnan and Pankhania). For a long time, planners and architects have not been doing enough to address urban heat and explain the benefits of green roofs and medium density with three- to ten-storey blocks. I have frequently encountered an Australian 'anti-urban vernacular distaste' for denser environments in past debates, when I had advocated strong urban streetscapes and compact, mixed-use districts; but things are now changing.

Figure i.7 Humankind is part of the ecosystem, not apart from or above it. This diagram depicts this simple fact beautifully.

Source: Supplied by the author.

Figure i.8 The holistic vision of creating 'buildings like trees and cities like forests' would ensure a healthy microclimate and zero-waste ecosystems, strengthening the urban metabolism.

Source: Image by the author.

What does the low carbon paradigm mean for urban planning?

Cities account for 80 per cent of the global economy, and this is likely to rise even further as Latin America, Asia and Africa continue to urbanize. Sanjeev Sanyal, Global Strategist at Deutsche Bank, noted that he expects 'the urbanisation of India to be the single largest human event of this century' (Sanyal 2009).

However, cities are also responsible for over 65 per cent of all greenhouse gas emissions globally (some researchers suggest that it is even over 75 per cent). Consequently, the main driver of urban change has to be climate change. The former mayor of Barcelona, Joan Clos, points out that 'urbanisation is characterised by the demographic and economic dominance and transformation of cities' (2011, p. 4). The interdisciplinary and complex process of urban transformation instigates the reshaping of city precincts, districts and their wider context. As previously described, a paradigm shift is guiding this urban transformation to make cities more productive, liveable and sustainable; worldwide, urban design's new focus is on mixed-use infill development as it moves away from outdated models of greenfield development, which have caused inefficient sprawl and car dependency. The next generation of urban developments will not lock cities into a particular technology or inappropriate, low-density built form; rather they will create a flexible framework for constant low carbon innovation to reduce greenhouse gas emissions.

Environmental improvement should also mean an improved quality of life, well-being, productivity and increases in a city's competitiveness, creating economic and social benefits. 'Achieving more with less' is therefore a key paradigm of low carbon cities, and there are plenty of positive developments. Good (and much publicized) examples of this are initiatives by the successful cities of Copenhagen, Stockholm, Malmo, Vaxjo, Munich, Freiburg, Hannover, Lyon, Curitiba, Portland and Singapore that have managed to reduce their overall greenhouse gas emissions despite economic pressures and population growth. Their governments have engaged leading experts and practitioners in urban sustainability to help solve challenges such as how municipalities can better incorporate sustainability principles into strategic plans.

New decentralized infrastructure for the smart city

Cities thrive and, contrary to expectations that communications technology would make location less relevant, we are finding that twenty-first-century life is increasingly dependent on the clustering of diverse information technologies, a dense network of urban amenities (from playgrounds to swimming pools and libraries), institutions, innovation and human capital. Maintaining innovation all the way through to project delivery is important. We can see this with Masdar (UAE) and Tianjin Eco-City (P.R. of China), two outstanding demonstration projects that have set themselves up as hubs for green technology enterprises and research, so-called 'living laboratories'. Social inclusion and the use of ICT are perhaps the most important aspects of Tianjin Eco-City, where 25 per cent of all housing is subsidized for low-wage workers and their families. Once completed, it is hoped that in Tianjin Eco-City 90 per cent of all travel will be made on foot, by bicycle or via public transport. We can also find interesting ideas about 'smart city' development in Songdo; this newly built city on reclaimed land, 50 kilometres southwest of Seoul, is the test site for a new type of ICT infrastructure, offering effortless access to high-quality services, knowledge transfer, mobility, communication and connectivity to social infrastructure. The Amsterdam Smart City project is another good example of how information technologies may be applied to optimize infrastructure and provide substantial services to the population. Ideally, a smart city integrates the flows and supply chains of all key resources – energy, waste, water, food and transport – using innovative technologies to achieve a more resource-efficient city (Lehmann 2012b). In many ways, infrastructure again plays a key role. The recent United Nations Environment Programme report *City-level Decoupling: Urban Resource Flows and Governance of Infrastructure Transitions* (UNEP 2013) explores the key role of cities as societal 'nodes' where much of the current unsustainable use of natural resources takes place (and where the greatest potential exists for sustainability-oriented innovations), focusing on the key role of infrastructure in directing material flows and therefore resource use in the urban context. However, there is also critique of the smart city concept and the accumulation of 'big data'. Experts ask: Doesn't the increased dependency on technology actually make the city less resilient? While new technology can be advantageous, it does not replace compact, walkable, mixed-use communities and a good relationship between housing and employment.

Today, we are able to capture big data about cities and their urban systems through the use of ICT that allows us to gain an analytical insight into different urban scenarios, either existing or likely to evolve. This enables us to better monitor and lead the operation of our urban systems and to use these predictive models to identify potential risks or future demands. It also allows for better, evidence-based decision making through anticipation of likely demands and events, rather than merely reacting to demands or pressures. IT experts point out that in our post-Fordist networked society, social exchange and productivity are increasing with density of communication and information flow. Obviously, services cannot be designed in isolation if infrastructural co-benefits are to be harnessed. Forward-looking urban infrastructure decisions are essential, as these have long-lasting ramifications.

Greenhouse gas emission reductions and good health must be the core aspirations of any city's development plan, and decisions made in infrastructure development today will determine the effectiveness of cities in delivering services for decades to come. Smart grid and distributed energy generation technologies have become cost-effective alternatives. This is why the district and precinct scale is the appropriate scale for decisions relating to the relationship between public space (supporting walkability) and infrastructure (supporting the provision of services).

Reinventing the urban block as a low carbon precinct and part of the diverse, healthy city

The automobile's dominance causes major social problems and I have frequently proposed designs where cars were to be left in parking lots on the outskirts of the town or at the fringe of a precinct, and getting around in the centre would be limited to walking, cycling, electric scooters or light railway. While this may be uncomfortable for a few people (e.g. those used to parking close to shops), everyone would get a much better feeling for the town centre and be able to enjoy public space. Making buses, trams and bike rental free of charge (subsidized by the car registration tax) would increase acceptance and passenger numbers, opening up space on the streets for essential traffic and freeing up time by reducing road congestion. Some of these ideas were used in my designs for precincts in Australian cities (see Figures i.9a to i.9c).

With the rising popularity of car sharing in many cities, the devaluation of the car has commenced, and the automobile is losing its role as status symbol. As collaborative consumption becomes increasingly mainstream, many young people are starting to see vehicle ownership as inconvenient and old-fashioned. The trend is towards 'membership' instead of 'ownership'; young people today join car-sharing clubs as they want to be mobile and a cellular phone gives them completely different possibilities when it comes to a vehicle.

In all of this, innovative infrastructure will emerge, like elevated bike lanes, smart energy grids that integrate renewable sources and a network of charging stations for electric cars. Future power supply will be decentralized and embedded within the urban fabric and buildings (see Figures i.3a to i.3e). A single-source energy supply will soon be a thing of the past; an innovative energy mix will integrate various renewable energy sources through smart grid technology, moving away from fossil fuel sources. Urban infrastructure and public space will take on new roles. Public space in low carbon cities will have to contribute much more than just being a pleasant place to meet others. Public space will become a productive urban landscape, contributing to the mitigation of urban heat stress, making walking safe and pleasant, generating renewable energy, growing food locally, collecting and cleaning stormwater, and improving natural ventilation (see Chapter 8 by Tillie and van den Dobbelsteen and Carney, and Chapter 11 by Cervero). Good examples of the potential of new public space are the author's urban renewal plans for the cities of Newcastle and Taree in New South Wales, and the city of Mildura in Victoria, Australia (see Figures i.9a to i.9c). In Newcastle, for instance, the barrier of a heavy railway line will be removed to create a 'green connector', reconnecting the city centre with the public waterfront of the harbour, fostering adjacent real-estate development and increasing the competitiveness of the city. In Taree and Mildura, the master plans activate the riverfront for active living, recreation and cycling, making green urbanism principles a reality.

In *The Principles of Green Urbanism: Transforming the City for Sustainability* (2010), I argued that for this urban transformation to be effective, it will need to be principle-based with clear indicators and benchmarks; and with demonstration projects that are replicable and adaptable to other conditions. Mindful of the advantages accruing from economies of scale, I have also argued for the need to focus on districts and precincts, as this scale allows for more effective options for sustainable urban transformation, including infrastructural systems, the step-by-step improvement of public space networks and the engagement of communities (Lehmann and John 2014). Observing that 'the most sustainable building is the one not built; and the second most sustainable building is the one which already exists and is adaptively reused' (Lehmann 2010, p. 120), I described new urban development strategies where all existing structures are retained, highly insulated, adaptively reused and integrated into new precinct master plans.

Another interesting infill development currently under construction in Switzerland could become a model for the future of inner-city living. Instead of the large monolithic housing block asked for by the brief, MVRDV Architects designed a mixed, irregular urban hybrid with small apartment buildings at the corners, townhouses along the streets, and garden and patio houses inside the block. The diversity of 16 different housing types, which vary in size from 30 to 130 sqm and from two to four floors, will offer good housing choice and naturally attract a mixed group of inhabitants, an important factor in creating a diverse and vivid precinct. The project consists of 150 units (9,000 sqm of housing) and 2,000 sqm of services and shops (see Figures i.10a and i.10b). Each house or apartment will have its own façade colour, emphasizing its individual ownership. A pastel range of colours was chosen based on the colours traditionally found in historic Swiss town centres in the area. The exterior of the irregular block is a varied urban street front while the interior offers the quality of a green

Figures i.9a, i.9b and i.9c Urban renewal master plans in which the author was recently involved; i.9a for the city of Newcastle, 2008; i.9b for the city of Taree, 2009; and i.9c for the city of Mildura, 2009.

Source: i.9a. Smith, Hulme and Lehmann; i.9b and i.9c. Lehmann with McGregor, Suters and Arup. Supplied by the author.

Figures i.10a and i.10b Can we build an urban block of 150 units that would usually develop slowly over years and simulate urban infill evolution from scratch, in a way that avoids the sterile look of new monolithic developments? To this end, Dutch architects designed the 'Emmen Urban Hybrid' in Switzerland in 2013, with 16 different housing types, which vary in size and in height.

Source: Photos courtesy of the architects, supplied by the author.

and intimate village. An important aspect of the project is the idea that residents will be able to buy a more or less finished, basic house – with options leading up to almost full fit-out. Home owners with less money can therefore do the finishing work themselves, and still live in a high-quality, newly built home. The interior of the block is divided into private and public spaces, with dividing walls used to hang tables or benches. A cohesive landscaping plan foresees a wide variety of fruit trees in the courtyard, in both the private and public areas. The garden and patio houses in the centre of the courtyard have their own entrance doors at the outer perimeter of the block. This convincing infill model could easily be adopted by other cities.

Topics for further research: reshaping existing cities, rethinking urban planning

Reshaping existing cities involves more than aesthetics and function; it is about the longer-term sustainability of all urban settlements and pathways to decarbonization, where we rethink the future of our cities and their supply systems. However, there is still a lack of knowledge about tools and stages in the decision-making process that would help to create districts and precincts that are both compact and mixed use. The process of knowledge-based decision making and data on density in urban development projects is still unclear and under-researched. For instance, little scholarship exists about which density levels are generally considered appropriate or acceptable by the community. Many cities in the developed world are now at an important crossroads and need to rethink the way they will continue to develop. Systems thinking for sustainable urban transformation includes upscaling from green buildings and the adaptive reuse of existing buildings to the retrofitting of entire precincts or districts within cities. What is needed is multidimensional city-wide change in planning and governance (WWF 2010). Experts have long recognized that low carbon development strategies must be innovative enough to transform the urban sector towards low emissions, but must also be practical enough to be usable, affordable and achievable, and robust enough to be adaptable to different conditions.

Hence, *Low Carbon Cities* examines how we can best and most effectively reshape our existing cities so that they become more sustainable, productive and liveable, and emit less carbon. Obviously, progress towards mitigation of and adaptation to climate change will be heavily dependent on the approaches we select to design and re-engineer our existing cities.

Approaches to decarbonizing the energy supply are discussed in several chapters of this book (see Chapter 18 by Katzschner and Campe, and Chapter 6 by Lenzen and Peters). Much has been achieved already. The *Energiewende* ('energy turning point') policy in Germany, for instance, will reshape its entire energy system and increase the amount of renewable energy in the energy mix from 23 per cent in 2013 to 35 per cent in 2020. The prioritization of biomass, wind and solar energy has lessened demand for polluting coal and gas. Denmark is making similar changes; it is now 25 per cent powered by wind (in 2013) and will double capacity by 2020. Large solar farms often have the disadvantage that they take up a lot of land, competing with agriculture and cities for space; the best places to build solar farms are therefore closed landfills near the city to reduce transmission losses.

Cities can lead the way in cutting greenhouse gas emissions. Some recognized examples of cities that have implemented low carbon city planning best practice are listed in Table i.2.

Table i.2 Leading examples of cities that have implemented low carbon city policies and planning. Comparative city research shows that successful cities need good leaders who have the ability to take strategic decisions for the long term, and not be swayed by electoral cycles

Cities	*Strategies implemented*
Adelaide and San Francisco	Zero-waste concepts
Melbourne and Berlin	Urban renewal, energy-efficient buildings
Sydney and Malmo	Water management, energy co-generation
Stockholm Hammarby- Sjöstad	Waterfront development, renewable energy
Freiburg Vauban	A holistic policy promoting low energy consumption
Hannover Kronsberg	A district with very high ecological standards
Nanjing and Hangzhou	Public transport and free bike schemes
Songdo	Smart city concept, intelligent buildings with sensors and meters
Singapore	Integration of urban biodiversity, low carbon mobility, water management, good public housing
Helsinki, Leipzig and Curitiba	Affordable mobility, waste recycling
London	Holistic strategies, providing community spaces
Munich	Public transport integration, low energy consumption
Copenhagen	Safe cycling and air quality
New York City	Urban green spaces and public space, urban blocks
Vancouver and Portland	Quality urban density and public transport, green construction

Co-benefits from low carbon energy systems, waste management and compact city form

Most of the world's cities have a serious waste problem, which demands that their authorities find economically viable and sustainable ways to dispose of gigantic amounts of solid residential and commercial waste every single day. For a long time, important aspects of cities such as changing citizens' behaviour and waste management have been overlooked in urban planning; and most urban planners still fail to engage with long-term waste management strategies, or the challenge that increasing consumption causes growing amounts of waste. Waste is essentially a grossly 'misallocated resource' (Lehmann 2010), closely entangled with our expansionary consumer society and inefficient processes.

Currently, the world's cities generate about 1.3 billion tonnes of solid waste per year and the volume is expected to increase to 2.2 billion tonnes by 2025 (Hoornweg and Bhada-Tata 2012). Waste generation rates will more than double over the next 20 years in lower-income countries.

The construction industry is part of this waste problem. Is construction without waste a utopian vision? We need to change the construction industry's inefficiencies and voracious appetite for virgin materials and building components that are sourced and manufactured far away and at great cost to the environment. It is now common sense that it would be much better to make use of locally obtained materials that can easily be recycled and reused at the building's end-of-life. Timber has recently been rediscovered as a sustainable construction material, because it can easily be recycled and reused, stores carbon and is a fast-regrowing resource (Lehmann 2012d). Concrete, on the other hand, is an extremely energy-intensive construction material, difficult to recycle and currently responsible for an estimated 6 to 8 per cent of human-produced greenhouse gas emissions.

The prevailing waste management paradigm of high-consumption societies is based upon linear systems using inefficient manufacturing processes, which convert precious resources into mostly non-recoverable, poorly degradable wastes that ultimately get dumped in landfills. Thinking beyond simplistic linear resource extraction means focusing on the knowledge economy; the adaptation of cities and society is going to depend on design innovation and better ways of production and resource recovery. Technologies and policies for waste management need to be adapted to the local context (Haggar 2007; UN-Habitat 2010). For instance, in developing countries 50 to 70 per cent of the waste is organic matter, which could easily be used for composting, closing the loop as fertilizer or as biomass for energy generation. Waste management experts agree that transporting waste in trucks to faraway landfills has little future; instead, it is clear that we must dramatically increase recycling and composting efforts, and encourage alternatives to throwing things away. In the future, high-density precincts will allow us to prioritize precinct-wide waste management systems due to relatively low transportation costs for waste collection; whereas low-density precincts mean longer distances to travel and therefore a focus on minimizing waste generation from source and treating the collected waste locally in smaller, decentralized facilities. This includes offering public recycling facilities in every district and neighbourhood, with smaller, decentralized processing facilities as part of precinct planning, and composting facilities for organic waste and food scraps, in connection with community gardens, to resolve the problem close to the source. What is needed to achieve this is new public policy that genuinely attempts to minimize the amount of waste created by addressing the throwaway mentality and wasteful lifestyle that cause so much waste in the first place, by changing our behaviour and values (Lehmann and Crocker 2013). Waste reduction and significantly increased recycling diversion rates have to become the cornerstones of any urban development strategy.

Innovation in the construction process through prefabrication and the use of recycled materials creates new opportunities for sustainable urban development. Prefabrication has many environmental advantages compared to conventional construction; it can radically reduce construction time, minimize the waste of materials and reduce the embodied energy in building components. Prefabrication allows for the use of recycled materials, the reuse of elements and flexibility or adaptability of buildings, thereby improving their life cycle. Using recycled materials reduces the amount of waste going to landfill. 'Waste' is actually a precious resource, and resource recovery and the reuse of building materials is part of a circular metabolism, which minimizes and avoids the creation of waste, optimizes resource efficiency and significantly reduces greenhouse gas emissions in the process. Using local materials means there are fewer greenhouse gas emissions from transport and less embodied energy.

Two Swedish cities are particularly interesting examples of waste and energy solutions: the Western Harbour district in the city of Malmo and Hammarby Sjöstad in Stockholm. Both show what can be achieved in a developed city. Here, decaying industrial port areas (brownfields) have been transformed into autonomous eco-districts that have their own clean energy supply and waste treatment systems, with very few cars – and plenty of satisfied residents (Hammarby Sjöstad 2010; see Figures i.11a and i.11b). These districts use biogas from food waste and highly efficient wastewater treatment technology and are self-supporting, powered by renewable energy; they are even exporting surplus energy to adjacent districts by generating more than their own energy needs. Planned as stand-alone communities with close access to services, the districts have virtually no cars, and pedestrians and bicycles have priority. Most residents park their vehicles outside the core area and then walk to their homes or use the light railway and biogas buses. Scandinavian cities have been using district heating systems and combined-heat-and-power generation for residential districts for many years, so they have gained a good understanding of the interdependencies within city infrastructure, the operation of their energy systems, supply and demand sides, and opportunities to use waste heat for district heating.

From low carbon city design towards zero-waste cities: longevity, not landfill

How much longer can we sustain the high-consumption lifestyle we are used to? To get the full picture on consumerism, we need to understand interconnected global systems of production and consumption, and how urban development locks us into 'systems of consumption', whether we like it or not.

Waste is the symbol of inefficiency of any modern society. Over the past five years, my research team has developed the 'zero-waste city' concept, which is based on closed-loop circular urban systems thinking, where all waste materials are fully recovered through the use of a series of processes and criteria, including highly efficient production processes, selection of easily recyclable low carbon materials, design for disassembly and durability, and products that are based on modular systems thinking and that therefore can be repaired (Zaman and Lehmann 2011, 2013).

Cities take resources from nature and from places far beyond their boundaries. Research in resource-efficient urbanism includes examining the sustainable flow of resources, materials, energy, transport, water, biodiversity, nutrients and food – cycling energy and material (waste) flows (this is, of course, not new in nature: see Figure i.8). Resource-efficient urbanism will best be achieved through closed-cycle urban systems on the resource and material level, while harnessing renewable energy sources to minimize environmental impact. Girardet suggests that we need to go even further when he demands 'regenerative cities', which he defines as 'cities that do not just become resource-efficient, but become cities that restore damaged ecosystems, increase their bio-capacity and positively enhance the ecosystem services they receive from beyond their boundaries' (2010, pp. 13–14). The new challenge for cities is to find ways to continuously help regenerate natural systems from which they draw resources.

From a global perspective, urban consumption patterns are clearly unequal and wasteful. For instance, take US cities: the US has just 5 per cent of the world's population, but consumes 30 per cent of the world's resources and creates 30 per cent of the world's waste. The US is not alone in its wastefulness; a small slice of the world's population consumes most of the resources and produces most of the greenhouse gas emissions. Material resources are depleting at such a rate that soon we are likely to face shortages in rare earths and other materials we currently dump in landfill: lead, copper, cadmium, wolfram (tungsten) and zinc, to name a few. In this context, plastic waste and toxic e-waste are ticking time bombs; they constitute a new kind of waste (Strasser 1992). Why do electronic devices break so soon after purchase and why are they always cheaper to replace than repair?

In 1960, cultural critic and consumerism theorist Vance Packard published his pioneering book *The Waste Makers*, a critique of planned obsolescence. He pointed out that consumers who learn that the manufacturer invested money to make the product obsolete faster might turn to a producer (if any exists) who offers a more durable alternative. We need to support a move towards more single-material, recyclable components in all industry sectors. Urban centres rely heavily on inward flows of food, water, consumer goods and services that may all result in greenhouse gas emissions from areas outside the city. Collectively, we ignore the

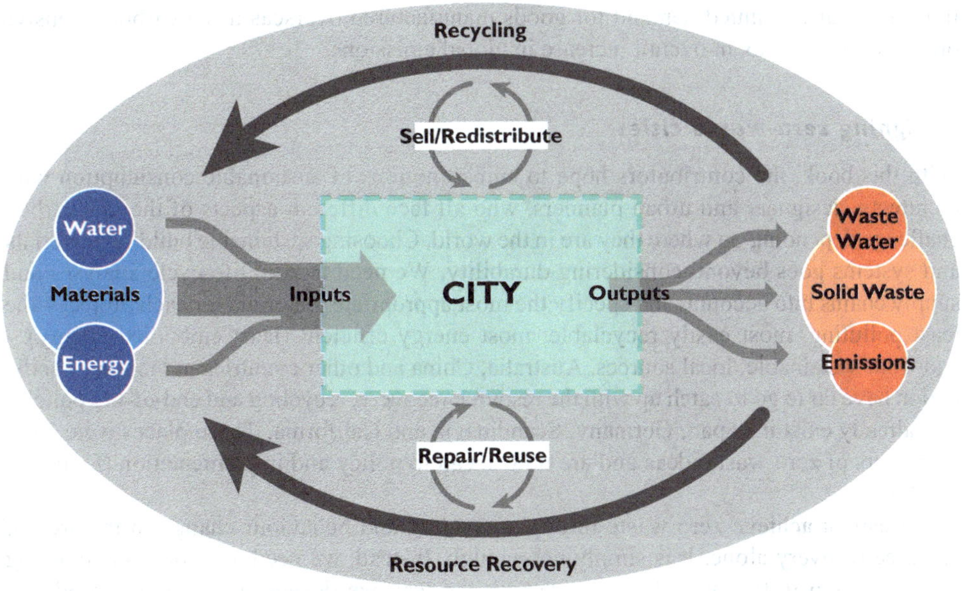

Figures i.11a and i.11b Above: Material flows in a zero-waste city. Below: The Swedish green district Hammarby Sjöstad has transformed a formerly run-down port area in Stockholm into a model low carbon precinct.

Source: i.11a: Adapted by Zaman and Lehmann (2013) from Girardet (1992); i.11b: Photo by Lehmann (2012).

impact of our continued demand for goods manufactured overseas using carbon-intensive processes, leading to an overall increase in global emissions.

Designing zero-waste cities

With this book, the contributors hope to embed notions of sustainable consumption with architects, designers and urban planners, who all face different aspects of the low carbon challenge depending on where they are in the world. Choosing sustainable building materials and systems goes beyond considering durability. We need to take life-cycle analyses and supply chains into account, and specify the most appropriate materials for each project – the least polluting, most easily recyclable, most energy efficient (least embodied energy) – and from sustainable, local sources. Australia, China and other countries in the Asia-Pacific region have far to go to catch up with the resource recovery, recycling and end-of-life policies that already exist in Japan, Germany, Scandinavia and California. Those places were early embracers of zero-waste ideas and are now ahead in policy and implementation (Lehmann 2012c).

We cannot achieve zero waste through recycling and behaviour change in the area of resource recovery alone. It is simply not enough. Instead, we need to refocus on avoiding waste creation in the first place and rethink the way we design and construct products, buildings, precincts and cities to facilitate reuse and disassembly at their end-of-life. This change of focus makes the concept of zero waste both powerful and controversial (see Figure i.11a). From a purely economic point of view, producing waste and wasting resources is unproductive. If we can begin at the beginning and design waste out of the picture, we can recover not only the final product, but also the energy, materials and time embodied in the product or building. However, reversing the existing, wasteful business system and changing manufacturing practices is not a fast, easy or cheap process. The zero-waste ethos is a big call, radical in its ramifications, and it requires more than a top-down, government-imposed approach. To be successful, zero waste needs to be embraced and implemented by citizens and community groups, business and industry. It is already technologically possible to build a zero-waste and low carbon city. The question now is: Are we willing to do so and transform from consumers to citizens?

The Asia-Pacific region and China's rapid urbanization

More than in any other region, cities in the Asia-Pacific are experiencing rapid urbanization and face immense pressure to meet projected needs, for instance, for housing and transportation. In 2013, the Asia-Pacific region was home to 4.2 billion people, representing 60 per cent of the world's population. While the region is comparatively under-urbanized, with approximately 46 per cent of people living in urban areas (at a current urban growth rate of 2 per cent p.a. predicted for the next two decades), the Asia-Pacific will account for 60 per cent of the demographic expansion of all urban areas across the world by 2030 (McKinsey and Company 2009; UN ESCAP 2011, p. 6; UN-Habitat 2012, p. 24). By 2030, over 50 per cent of the population in the region will be living in cities and urban settlements – twice as many as in the year 2000. The urbanization of this dynamic and populous region over the past two decades has been at a most impressive scale and speed. The final chapter in the book discusses cities in the Asia-Pacific region in detail (see Chapter 22 by Lehmann and Thornton), taking a wider view than just the usual approach of looking at Singapore's success in sustainable city planning. Singapore's context as a city-state is unique and very different from that of a city in a big country, or a country with many cities. Because Singapore's case is unique, many of its ideas cannot be easily replicated elsewhere. Therefore, a more differentiated view is required. As expected, the immense urbanization activity has also brought some milestone green buildings and innovative urban design to the Asia-Pacific region, as illustrated in Figures i.12a and i.12b.

Electricity demand in Asia-Pacific countries is growing at 4 per cent annually, but most of this electricity (over 90 per cent) is still generated from fossil fuels (wind power is likely to power a large amount of China's electricity needs from 2030). The most intensive urbanization process in the region is currently happening in China, which makes this country an interesting case study. It now has 120 cities with over one million people, and 36 cities of over two million, and it is predicted that China's urban population will be one billion by 2030 (with 70 per cent of its population urbanized). By then, China is projected to have a staggering 200 cities with populations of over one million; and one-eighth of the world's population

Figure i.12a The naturally ventilated 'Stacking Green' house by Vo Trong Nghia in Ho Chi Minh City, Vietnam, 2011.

Source: Photo by Oki Hiroyuki, reproduced by permission.

Figure i.12b In 2006, Seoul (Korea) removed a 6-kilometre-long section of a motorway and created a new linear public park with a restored river. During the summer, the Cheonggycheon Park keeps the surrounding area approximately 3 degrees Celsius cooler than before.

Source: Photo by Lehmann (2013).

will be residing in a Chinese city. In comparison, the United States has just nine cities with over one million people, while Australia has five. Today, 23 cities in China have populations of over five million while there are only three such cities in the whole of Europe. Chinese urbanization is also five times faster; what took 100 years to happen in Europe has taken place in China in just 20 years. This scale and pace of urban growth is a defining feature in the twenty-first century, with profound implications for people everywhere. If we can identify practical and successful solutions for China, these will be beneficial to others and may be replicable elsewhere.

China's traditional culture considers the city and the countryside as two different ways of living, each with unique advantages and disadvantages. Urbanization has been one of the major factors in China's phenomenal economic growth over the past three decades. Cities are an important part of the country's goal to develop into a modern state with a strong middle class, while lowering its greenhouse gas emissions. However, rapid urbanization can easily overwhelm municipalities, landscapes and communities; and China is urbanizing faster than any country in history. Urbanization in China is predicted to rise from 51 per cent in 2013 to 70 per cent in 2030. Chinese cities are expected to swell by another 350 million residents in the next 25 years, according to World Bank (2012) estimates. The social challenges in such times of rapid change are immense, including the affordability of housing and the availability of age-friendly neighbourhoods. In the past two decades, China's urban boom has also become synonymous with traumatic individual experiences, with poor farmers leaving the countryside for cities in the hope of better job opportunities, an air-conditioned flat, private car and supermarkets full of goods; but ending up as a new urban underclass. There is now an increasing awareness of socioeconomic inequality and the need to base the economy on innovation and knowledge rather than simply on cheap labour and environmental exploitation. Now that the need to protect agricultural land is being acknowledged, urbanization models have started to move away from Western sprawl concepts. Hangzhou, for instance, a city close to Shanghai, now has a Bus Rapid Transit (BRT) system modelled on Curitiba's, a new subway system with eight efficient lines and the world's largest free bike scheme, with over 70,000 bikes.

While China's twelfth Five-Year Plan, issued in 2010, includes important CO_2 emission reduction targets, rapidly rising consumption levels and greenhouse gas emissions remain serious challenges. According to the report *Trends in Global CO_2 Emissions* (Olivier *et al.* 2012), China's average per capita carbon dioxide emissions increased by 9 per cent in 2011 alone, to 7.2 tonnes per person. This figure is similar to the per capita emissions in the European Union (of 7.5 tonnes in the same year). China, the world's most populous country, has already caught up with the developed world and is now well within the 6 to 19 tonnes/person range spanned by the major industrialized countries. In comparison, in 2011, emissions in the United States and Australia (still some of the world's largest emitters of CO_2) were at 17.3 and 19 tonnes per capita respectively. Consequently, China now accounts for over a quarter of all global greenhouse gas emissions (at the time of writing, China is planning to introduce a national emissions trading scheme that is likely to become the largest in the world).

Currently, China uses 33 per cent of the world's overall consumption of reinforced concrete. However, the construction quality is often low; standards of materials and technologies are frequently sacrificed to lower production costs and to speed up construction times.

Growing disenchantment among Chinese citizens

In its twelfth Five-Year Plan (2011–2015), China announced a 17 per cent reduction target for carbon intensity, compared to 2005 emission levels. Chinese authorities are increasingly interested in altering the course of unsustainable development and setting new benchmarks by introducing better standards and environmental values in their urbanization programmes. Today, not every project has to be rushed (as was the case a decade ago). A younger generation of urban dwellers is asking for higher living standards, a better quality of life and the kinds of environmental protection they have seen when visiting Seoul, Singapore, Sydney or Tokyo; they are demanding social well-being, not just growth. However, one has to be quite realistic and aware of competing interests and targets, as a recent report by the World Bank notes: 'The imperative to reduce carbon intensity of Chinese cities is only one of many competing priorities for Chinese government officials amidst unprecedented urbanisation, modernisation and economic development' (2012, p. xx).

China is also seeking advice from around the world on more environmentally sustainable ways to support the transformation of its cities, to resolve the serious challenges caused by too rapid urbanization. These challenges include conflicts with citizens due to central and local governments' top-down 'command and control' approach to urbanization, which has led to inequality, overcrowding and environmental pollution. Land disputes, as a result of growing urban populations, have the potential to become a cause of social unrest. For instance, in Beijing alone, 60 urban areas were demolished in 2010, mainly to erect high-rise complexes; more than 180,000 residents were affected by forced demolitions and there was a series of ugly clashes between government officials and residents. The *China Daily* reported on 4 August 2010 that in Beijing alone, 4.43 million square metres of old courtyards have been demolished since 1990 – equivalent to around 40 per cent of the downtown area. The main beneficiaries of this rapid urban development have been developers, state-owned enterprises and local governments; by some estimates, property has accounted for half the revenues of local governments. In the past three decades, China's farmers have 'contributed' 200,000 square kilometres (about 77,000 square miles) of former agricultural land to development, with little compensation. In 2009, cultural heritage officials warned that urban development had destroyed tens of thousands of historic sites and villages in the past three decades. Thousands of urban residents have been intimidated and forced to move, or have been evicted to make way for the destruction of old neighbourhoods.

Understandably, there is growing disenchantment with and opposition to China's pollution and lack of democratic participation models (see Figures i.13a to i.13c). In February 2013, for instance, Deng Fei, an investigative journalist and influential activist, initiated a campaign; he invited Chinese 'netizens' to take photos of polluted rivers in their home towns and upload them to Weibo (a popular Chinese internet platform), launching the 'China Water Crisis Independent Investigation'. His activities received thousands of responses and for the first time provoked a national debate on groundwater safety. For instance, cities in the Beijing–Tianjin–Hebei region have severe air pollution from coal burning in plants and power stations and from road transport; approximately 9,900 premature deaths in China in 2011 are estimated to have been due to air pollution (Chinese Ministry of Environmental Protection 2012). High air pollution levels have sparked widespread public anger and officials concerned about social unrest have responded by implementing tougher policies. For instance, Hebei, the country's biggest steel producer, is home to as many as seven of China's ten most polluted cities, according to statistics published monthly by the Ministry of Environmental Protection. Pollution in Hebei often spreads to neighbouring Beijing and Tianjin. While China has drawn up dozens of laws and guidelines to improve the environment, it struggles to enforce them in the face of powerful enterprises.

The challenge for Chinese planners is to create healthier and more liveable cities and places that do not make people sick. There is good evidence that ambient air pollution can create the acute symptoms of, and exacerbate, both cardiac and respiratory disease; for instance, there is a causal link between the development of asthma in children and exposure to traffic air pollution. Children and elderly people, and those with chronic cardiac and respiratory diseases such as asthma, are generally more likely to be affected by air pollution and heat stress. A recent study of temperature, air pollution and total mortality in summers found that maximum temperature and sulphur dioxide air pollution had significant effects on mortality rates (Dennekamp and Carey 2010).

While some new legislation and initiatives are promising, many new projects are still not particular sustainable; instead, projects that are spectacularly out of scale are favoured, as highlighted by the award-winning Chinese architect Wang Shu (see his critique when receiving the Pritzker Architecture Prize in 2012). An architect who profoundly disagrees with China's rush to urbanization and gigantic developments, Wang has found a way to criticize it through his own more careful work and his diverse and sophisticated urban renewal projects. He encourages social innovation rather than technology-driven solutions; the adaptive reuse of existing buildings (rather than their demolition) and the regeneration of city

Figure i.13a The residents of this 'nail house' (a Chinese expression for homes belonging to people who refuse to move out) marked for demolition in Chongqing refused to make space for a new development in 2010; urban residents who refuse to accept compensation are often evicted or bullied into leaving.

Source: Supplied by the author.

Figures i.13b1 and i.13b2 China's new slums: illegally constructed ten-storey apartment buildings at Shanghsa East Village in Shenzhen; the density is so high that there are only narrow gaps left between the buildings, not enough for sufficient air and daylight.

Source: Supplied by the author.

Figure i.13c As a result of an out-of-control building boom, China has some brand-new and abandoned cities. Kangbashi New Area, a new district in Ordos (Inner Mongolia), was commenced in 2003 for one million residents, but only 25,000 people were living there in 2013.

Source: Supplied by the author.

centres would not only save energy and materials but also help maintain each city's identity, he argues. 'Only 10 per cent of historical buildings in China have survived to the present day' (Wang Shu, Pritzker Architecture Prize 2012). He notes that what is missing is critical architecture that goes against the present-day consumerist situation and superficial or nostalgic replicas of Western vernacular architecture.

Recycling and up-cycling of old buildings through adaptive reuse means creating cultural continuity and new value through the reuse of already existing, derelict structures; this stands in sharp contrast to the Modernist paradigm of always constructing new buildings and cities. But then, the city has always recycled its own urban fabric.

Led by activists such as Deng Fei, Wang Shu and Ai Wei Wei, outrage over environmental pollution and the frenetic destruction of cultural heritage has spread widely among both prominent and ordinary Chinese people, and attracted attention on social media channels throughout the world. Numerous real estate development problems are bubbling under the surface and criticism is mounting in blogs and online forums. For instance, so many urban complexes with major shopping precincts have sprung up across the country that the property markets in second- and third-tier Chinese cities are facing saturation and even collapse (see Figure i.13c). An urban complex can usually have a direct influence on about 300,000 residents, which means three or four complexes are enough for a city with a population of one million people. But in Hohhot, in the Inner Mongolia Autonomous Region, more than 30 urban complexes are being currently built for an area whose population is only three million.

Chinese people are becoming fed up with the growing incidence of primary disease including cancer, as well as daily food scandals, disgraceful corporate governance (or the lack of it) and a truly frightening scale of air, water and soil pollution. China's state-run media is now full of stories about the country's massive pollution problem and the central question that runs through many articles is whether, given its environmental impact, China's growth can be maintained. For a long time it has seemed that, in China, more growth simply means more pollution. There are appeals to speed up legislation to improve environmental protection, and the Chinese public is increasingly vocal and determined in response to the country's environmental crises.

China is now in the process of building a middle-income society by 2020, and plans to basically complete its modernization process by 2050. In this process, the development of mega-city regions (urban clusters) in coastal areas, such as the Beijing–Tianjin–Hebei region, plays an important role. It is estimated that China's population will peak around 2035, with 1.6 billion people. After 2035, population growth is predicted to flatten out, while urbanization will continue to rise until 2050, to 70 per cent. It is expected that the newly added urban population will aggregate in mega-city regions, thus forming eight to ten poly-centric metropolitan regions near coastal areas, each with a population of more than 40 million (with residential densities of up to 8,000 people per sq km).

It is now clear that the migration from rural to urban areas will continue for some time to come. Urbanization and migration are an integral part of economic development and societal transformation, and historical experience shows that they cannot be stopped. Poor migrants continue to arrive in cities where they contribute cheap labour for manufacturing and services; typically, they do the dirty, dangerous and degrading jobs that locals do not want to do. In addition, the challenge of an ageing population has reached China; the country is ageing before it becomes wealthy – today (2014), more than 185 million Chinese are over 60 years old and a quarter of the elderly live in poverty, few of them cared for by their children, a result of the one-child policy.

China has started to ban the construction of new coal-fired power plants in several key industrial regions to combat air pollution, and is committed to an action plan; it aims to raise the share of non-fossil fuel energy, such as wind, biomass and solar power to 13 per cent by 2017 (in 2012, the share of renewables was 9 per cent). However, whether the region is innovative enough to meet the challenges remains a major question. Governments and municipalities in the Asia-Pacific region will need to develop better participatory models to involve their communities, and to implement incentives for people to take better protection of the ecosystem, which we risk destroying if we follow the consumption patterns of the past three decades.

Eco-cities in China: the ability to adapt is critical

The transformation of its cities is central to China's effort to revitalize its economic model for sustainable growth. By adopting new approaches to urbanization, its leaders can ensure more balanced investment, address a major source of debt and clean up the country's environment. Today, the idea of a 'master plan' is increasingly seen as too rigid and inadequate, because the concept is too inflexible to deal with the speed of change and rapid urbanization of developing cities; instead, flexible frameworks that can accommodate change are needed. There are promising major developments that are about rethinking cities and planning instruments, and exploring urban solutions based on walkability and public transport. Self-contained, compact eco-city projects are underway in China (for instance, in Chengdu, Wanzhuang and Qingdao; see Figures i.14a to i.14c), which could become promising prototypes for other countries.

There are generally two types of eco-city initiatives in China: newly built eco-city projects (e.g. Sino-Singapore Tianjin Eco-City and Qingdao Eco-Park) and eco-remodelling and retrofitting of existing cities (e.g. Wanzhuang and Huainan). Although laudable, eco-city efforts are at risk of being just a drop in the ocean of China's rapidly looming environmental crisis, and a lot more needs to be done than to have a few show-pony eco-cities. Taking into account anticipated changes, the master planning done today is in danger of being outdated tomorrow; what counts is the adaptability of a master plan to change.

While China is working hard to create its sustainable future and to enhance the well-being of its urban citizens, actually implementing the various eco-city projects has proven more difficult, putting several projects in serious question. For instance, Tianjin Eco-City, currently under construction and believed to be the most advanced of China's eco-cities, is widely seen as too conventional in its plan and energy supply and, lacking genuine green ambition, not living up to its early expectations.

Tianjin Eco-City is an initiative by the Chinese and Singaporean governments for a new-build flagship development that has identified 26 key performance indicators, mainly technical indicators around resource efficiency; it is a demonstration project where the city of Tianjin is pro-actively taking steps ahead of other municipalities. The Sino-German Qingdao Eco-Park initiative in Qingdao has recently adopted Tianjin's framework (Xie 2012). Both developments are expected to accommodate a large number of residents upon final completion in 20 years (350,000 in Tianjin and 60,000 in Qingdao); and both share a common sustainability framework based on four dimensions: economy, society, environment and resources. Other visionary eco-city projects across China have seen little progress when facing the grey, harsh reality of investment decisions and lack of policy continuity; for instance, the pioneering green city project Dongtan (close to Shanghai) was stalled after

Figures i.14a, i.14b and i.14c Three Chinese eco-city projects: artists' impressions of (from top to bottom) Chengdu, Wanzhuang and Qingdao eco-cities.

Source: Images courtesy of the architects, supplied by the author.

Figures i.14d and i.14e Masdar eco-city, located outside Abu Dhabi, had to be scaled down and postponed due to a range of problems in realizing the vision of a sustainable city in the desert. Only a small part has been built to date.

Source: Images courtesy of the architect, supplied by the author.

political leadership changes; and other projects have only been realized as ordinary real estate developments and greenfield developments, or even as gated communities for high-income households, in conflict with true eco-city principles. Chinese 'green-washing' advertisements for conventional developments have cast the eco-city concept in the wrong light, which is a pity given the importance of China's role in leading the way for other developing countries to pursue sustainable development (Lehmann 2013a).

Chinese cities will remain the engine of the country's growth story, but they must continue their transformation towards real long-term sustainability and better social integration (Lehmann 2013b). The upside is that China now has more researchers than any other country. With this shift to a knowledge-based economy and the global centre of gravity shifting to the Asia-Pacific, it is a timely moment to learn from sustainability efforts and experiments in China.

Another unconvincing eco-city project is Masdar, 20 kilometres south-east of Abu Dhabi in the United Arab Emirates, currently under construction (see Figures i.14d and i.14e). Masdar was supposed to be the future of sustainable urban living built in the present. Six square kilometres in size, financial and political realities ruined the vision of the eco-city in the desert when reality hit the overly ambitious project; after the global financial crisis, promises of the first zero-carbon city were just not economically feasible anymore, forcing the investor to scale back his ambitions. Now it is not zero carbon, but low carbon. Other promises included that Masdar was going to be zero waste, car-free, and built on a nine-metre-high platform. However, the investor had to backpedal on all of these ideas, postponing the city's completion date to 2025. Masdar has no low-income, subsidized or affordable housing, meaning that most of its workforce must drive to their jobs. What good is a car-free city if everyone is commuting from somewhere else?

The situation in Australia: making cities low carbon, less car dependent

Australians are fortunate to enjoy some of the most liveable cities in the world. There is obviously more to life than the cold numbers of GDP growth and economic statistics, and several organizations publish an annual Liveability Index; Australian cities usually rate very high in these. However, Australian cities are vulnerable as they are feeling the impact of global warming earlier and stronger than others; in the past decade, rising sea levels, coastal erosion, heatwaves, urban flooding, severe storms and other extreme weather events have significantly increased and their impact has been strongly felt.

Cities in Australia have a series of particular struggles: cities and regional areas have vast distances between them; Australian cities are characterized by low-density sprawled urban form and their centralized infrastructure for energy, water and waste; there is a high car dependency, with over 85 per cent of all trips made by private car, often leading to traffic congestion, productivity damage and other disadvantages; lack of cycle paths and walkability, leading to the rise of obesity; high dependency on air-conditioning and fossil fuels, combined with increased energy costs; more severe bush fires; water scarcity; high proportion of inefficient urban sprawl and low density (often as low as only 10 or 15 dwellings per hectare), particularly severe in cities with expansive suburbs and a lack of effective growth boundaries, such as Melbourne, Sydney, Brisbane and Perth; housing shortages, lack of genuine housing choice and diversity, as well as a loss of affordability; a lack of political commitment to implement long-term strategies, strategic policy directions and visions that have been published but are not being realized; and a lack of respect for built heritage, just to name a few.

However, the fourth *State of Australian Cities* report published by the Australian government indicates that heatwaves and increases in extreme temperatures (caused by climate change) are the major threat to the future of Australian cities (2013, p. 6), which could face up to a fourfold increase in heat-related deaths by 2050. Australia has been experiencing record-breaking heat in recent years; and between 1890 and 2013, heatwaves have been more deadly than any other natural disaster. From 2012 to 2013, the *State of Australian Cities* report notes that Australia 'registered the hottest September to March period on record, the hottest summer on record, the hottest month on record and the hottest day on record and the longest national-scale heatwave. It was also the hottest summer on record for Australian sea-surface temperatures' (2013, pp. 189ff.). Some areas already suffer temperatures reaching 50º Celsius and face severe water shortages; they risk becoming uninhabitable in the future.

The other issues identified relate to population growth and an ageing population. Like those around the world, Australian cities' populations are also ageing; by 2056, around 20 per cent of most of Australia's capital cities' populations will be over 65 – a demographic shift that could further increase numbers of heat-related deaths. It is predicted that the number of annual heat-related fatalities will rise to 2000 by mid-century, compared to the approximately 500 recorded in 2011. The urban heat island effect puts cities particularly at risk of heat-related deaths, since the prevalence of dark, heat-absorbing surfaces and lack of trees often cause cities to have significantly higher average temperatures than surrounding suburban and rural areas. The *State of Australian Cities* report found that the cities of Brisbane and Perth will see the most heat-related deaths, with predictions of nearly 800 in each city by 2050, compared with fewer than 200 in 2011. Sydney could experience close to 350 heat-related deaths by 2050, compared to about 70 in 2011. The victims of intense urban heat are likely to be the most vulnerable residents: the elderly, sick, poor and homeless.

Australian cities are in transition. For instance, Perth is Australia's fastest growing city; with over 1,000 people arriving per week, the city is predicted to reach three million people by 2025. Ongoing population growth has led to ever-expanding cities (Forster 2004). Most cities are coastal cities and the development of their urban waterfront is therefore a particularly important topic (John *et al.* 2013). According to projected population growth figures published by the Australian Bureau of Statistics (ABS 2012), the eight largest Australian cities will almost double in size by 2050. Most of Australia's population growth will occur in these eight cities (see Figure i.15). We must ensure that this growth of cities is in population, but not in ecological footprint, and that low carbon infrastructure is in place. Adopting compactness as a strategy for sustainable urban growth and easier access to jobs will ensure that further population increases occur within the existing fabric, particularly by using all opportunities for urban infill on brownfield sites.

Currently, Australia has the largest house sizes in the world, with the largest average floor area of new dwellings (in 2010, new houses were on average 248 sqm); but smaller houses in urban infill and more diversity in sizes and typology will characterize the future. Urban infill means valuing more what already exists and taking advantage of the existing fabric. Promoting development in already established core areas means housing more people within the existing urban fabric and infrastructure. The shift from low-density housing in the outer suburbs to more concentrated urban areas is a cultural change that will continue, as people want to live closer to amenities like transport, theatres and shops with easy access to jobs and parks; preferring places where they can share amenities such as gardens, shops and mobility choices. People are increasingly time-poor, looking for inner-city solutions that allow short travel times to the workplace and need less upkeep than detached homes.

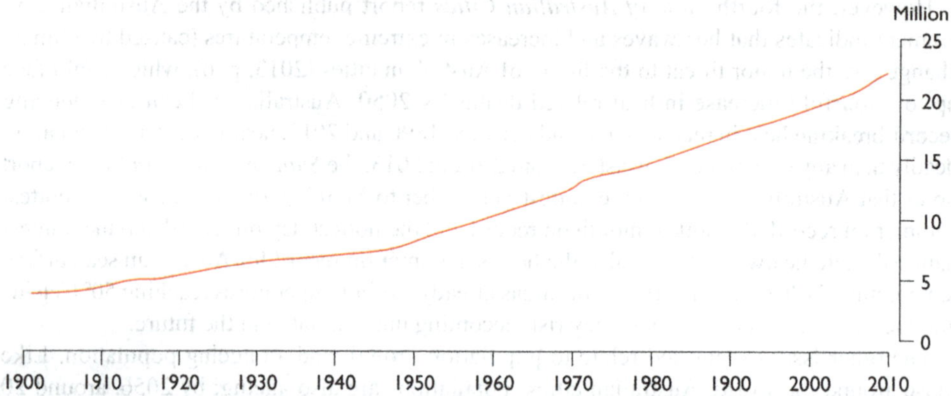

Figure i.15 Population growth in Australia, in millions; Australia's population is estimated to be almost 27 million by 2025.

Source: ABS (2012), supplied by the author.

Many architects and developers are now looking at housing diversity and propose compact housing, but not necessarily apartments. The reliance on large apartment buildings as the main house type to increase densities is often not viable, as the costs of development are higher than with other typologies. Townhouses, studios and flats built above the existing garages of houses (with their own strata titles) have increased in popularity; and municipalities allow more frequently for 'zero lot line' dwellings, which are houses built right on one site boundary but not attached to the neighbouring house, in the hope of finding more affordable and acceptable solutions for careful, gentle densification.

Australian cities need more and better infill developments. Public transport options and affordable inner-city housing are closed, intertwined issues. Australian cities need to be transformed to be more compact and less carbon-intensive cities by implementing strong growth boundaries to prevent urban sprawl. Low-density dispersed suburbs are extremely resource intensive and consume large portions of precious farmlands and forests.

Australia's large urban renewal projects, such as Barangaroo and Green Square in Sydney, Southbank and Docklands in Melbourne, or Bowden Urban Village and Tonsley Park in Adelaide, are now utilizing urban infill models that avoid too much negative impact on privacy or heritage character (see Figures i.16a to i.16c). It will be interesting to see to what extent these medium- to high-density mixed-use projects are able to create a sense of 'urbanness', ensure well-serviced communities and use far less energy, while allowing for highly efficient public transport (such as light rail) and the integration of new types of public green spaces; and especially to see the impact of these blocks at street level. For instance, a comprehensive and holistic action plan to transform the existing city district has been realized in the revitalization of Victoria Harbour at Docklands in Melbourne, which is now the urban district with the highest concentration of green buildings in Australia (see Figure i.16b).

In the Australian commercial building sector, the use of electricity is responsible for 89 per cent of the buildings' greenhouse gas emissions; these include cooling (28 per cent), air handling (22 per cent), lighting (21 per cent) and heating (13 per cent). Taken together, cooling, heating, ventilation, air-conditioning and lighting account for 84 per cent of commercial building sector greenhouse gas emissions, which is similar to emissions in the residential building sector (DIICCEE 2010). Therefore, some of Australia's biggest opportunities

Figures i.16a, i.16b and i.16c Growing compact: good examples of this trend in Australian urban renewal projects are the development of (from top to bottom) Barangaroo in Sydney, Docklands in Melbourne and Bowden Urban Village in Adelaide.

Source: Images courtesy of the architects, supplied by the author.

to reduce emissions (with lowest associated costs) relate to the availability of building technology and materials to support more sustainable lifestyles and to infrastructure for low carbon transport. Australia has an abundance of sun and wind power and wave power which could be used. There is plenty of good news too. Australia has undergone a 'quiet' revolution in the use of solar power; in 2013, over one million households – approximately 2.5 million people – were obtaining their energy from solar panels, up from only 8,000 people in 2007 (mainly due to reduced costs for PV panels and the energy feed-in tariffs).

Governments, planners and developers are now giving more priority to urban infill (with apartment buildings, terraces and townhouses), increasing the proportion of medium-density construction and moving away from greenfield 'house-and-yard' developments and stand-alone houses. This relates to the importance of urban infill as a catalyst not only for architecture, but also for the city at large; infill projects are becoming an accepted way to activate the city. In addition, the opportunity to experiment with low carbon materials and systems of assembly, using timber as a fast-regrowing resource, and the idea of demountable and reusable structures, are seen as having huge potential for future greenhouse gas reductions. Reducing costs for low and zero carbon developments is key for getting the concept into the mainstream. So it is time to rethink how Australian cities should evolve and what exactly Australian urbanism should stand for in the future.

Low Carbon Cities: structure, scope and purpose

By recognizing that cities are more than just buildings, and combining different approaches, we can bundle our multidisciplinary efforts to create sustainable low carbon cities in the future. This is an ongoing project, neither easy nor quickly done. I am editing this series of books on sustainable design to keep up with the latest developments in these interrelated fields. All books in this series bring diverse viewpoints together and are useful reading for professionals and students working on the future of cities.

In 2012, Routledge published the first book of this series, *Designing for Zero Waste*, which uses case studies to describe resource recovery and the optimization of material flow (see Figure i.17a). Zero-waste cities can only be achieved alongside and through behaviour change to reduce both the creation of material waste and wasteful consumption.

Motivating Change, which was published in 2013 and co-edited with Robert Crocker, takes up the problem of the need for behaviour change at many levels, from the city to the individual (see Figure i.17b). It explores the underlying dimensions of behaviour change and the diversity of current approaches to creating change on scales from individuals' daily lives to the urban environment. Encouraging people to adopt sustainable behaviour just because it will save them money is a flawed idea and a potentially counter-productive strategy. The book explains why we must develop better ways of promoting pro-environmental behaviour that go beyond the economic argument. It is mainly morals and values, not money, that ultimately underlie our behaviours; people are motivated by a positive self-concept and values of responsible citizenship.

With *Low Carbon Cities: Transforming Urban Systems*, the third volume in the series, I am presenting the different approaches to ecologically sustainable cities, low carbon city design and how we can best manage urban microclimates to mitigate urban heat stress, with a specific focus on Asian case studies (see Figure i.17c).

The books in the series are all authored and edited by leading academics and practitioners in the field of sustainable design and urban development. Although there has been an

Figures i.17a and i.17b The volumes previously edited by Steffen Lehmann and published by Routledge in the Earthscan Series on Sustainable Design: *Designing for Zero Waste* (2012) and *Motivating Change* (2013).

Figure i.17c The structure of *Low Carbon Cities*.

immense amount of theory and technology-focused writing published on the topic of sustainable urban design, many of these books have failed to introduce readers to the wider challenge of what the rethinking of design, production and operation, and the reuse/recycling of products, buildings and cities really mean. Sustainability is not a passing fashion, and people are constantly searching for more reliable information and relevant ideas about emerging systems in this area.

These books demonstrate that any urban precinct has the potential to greatly reduce its greenhouse gas emissions. Together, these publications offer a comprehensive range of teaching materials for universities and explain clearly why the battle against climate change is likely to be decided in cities. They describe the connections between low carbon, zero-waste cities and decreasing consumption, and explore the need to motivate behaviour change by shifting values.

Low Carbon Cities aims to inspire change towards more integrated approaches to transforming urban systems among all stakeholders in the complex field of urban development. The book aims to provide city planners, architects and decision makers with practical tools, knowledge and methodologies to address the challenges of developing sustainable cities. Key concepts are introduced that underpin theories of green urbanism and methods of low carbon urban development, including international case studies and practical planning approaches. It brings together a selection of classic and new readings by different writers who share a concern with low carbon sustainable cities. Going forward, we must take advantage of the knowledge gathered here. *Low Carbon Cities* is essential reading, structured in three major sections, each with six or more chapters:

- Part I: Classic texts on ecologically sustainable cities: conceptual evolutions and different schools of thought.
- Part II: New directions: designing for low carbon cities.
- Part III: Urban microclimates: mitigating urban heat stress.

Part I features a selection of classic texts on ecologically sustainable cities by authors renowned for their pioneering work on the topic of low carbon cities. Pioneers are always a little ahead of the others. While some of these texts were written some time ago and helped establish the legitimacy of sustainable urbanism, they are still as valid as when they were first published; and there are also new, previously unpublished pieces written especially for this book (by Beatley, Lechner and Kenworthy) that reflect back on earlier texts.

Infrastructure is going to be a critical issue in any urban transformation. Part II presents new directions in urban transformation, including case studies of low carbon city design. This part aims to help the reader understand the low carbon city primarily as an urban design and architectural task; it features a series of cutting-edge case studies and best practices, and also looks at Asian urbanization. Several of the case studies discuss the role of ICT and the 'smart city' concept, where, in the future, woven seamlessly into the fabric of our lives, the built environment will harvest energy and small distributed technologies will generate energy in the city from biomass, solar and wind power and from waste, store it effectively and distribute it wisely. It is likely that small, decentralized technologies are inherently more amenable to innovation than large, centralized and capital-intensive technologies.

Part III is dedicated to the problem of urban microclimates and the increasing need to mitigate urban heatwaves. The chapters in Part III build on the classic texts in Part I and on the best practice examples and case studies in Part II. Cities are warmer than the surrounding countryside due to human activities, and the mitigation of increasing city temperatures has emerged as a major topic for our cities, including the latest advances in the development of cool materials and techniques for a hotter built environment.

The urban heat island effect means that rising temperatures in built-up areas, and also in the suburban fringes, amplify climate change. The built environment's surfaces store heat, which is released later during the night. Urban areas usually dry up very quickly, with drainage rapidly taking water away; this, and having few green areas, means that less heat is lost by evaporation or transpiration by plants than in the countryside. And air pollution makes the health impacts of heatwaves even worse, with the biggest impacts on the elderly and those with existing heart, lung and other health problems. Part III features chapters by world-renowned experts on urban microclimates.

Heat stress-related deaths: why is the urban microclimate such an important topic?

The higher density of cities can bring efficiency gains, but there is interplay between the increased risk of an urban heat island effect and higher densities that needs to be well understood. Pavement surfaces and vegetation play a primary role in creating urban heat island effects (Oke 1982; see also Chapter 17 by Jusuf, Wong and Chong). The urban microclimate of any public space is determined by the density, geometry, materiality and other factors of a precinct, such as the amount of green space. Why do we accept that the street trees, which once lined our streets and boulevards, have gradually disappeared?

We need more tree-planting and green roof programmes to cool the city. Urban heat islands, combined with an ageing population and increased urbanization, will increase the vulnerability of urban populations to heat-related health impacts (such as dehydration and heatstroke). Daily death counts in cities increase markedly when temperatures reach levels that are very high relative to what is normal in a given location. Heat stress-related deaths in metropolitan areas, created by heatwaves and extreme temperatures, are estimated to rise by over 20 per cent in the coming decade (Li *et al.* 2013; World Bank 2013).

It is well established that water bodies and vegetation, such as street trees, gardens and green roofs, are natural air-conditioners providing shade and cooling for the urban environment, while reducing surface temperatures in our cities. However, it is less clear what to do with the one- and two-storey suburban buildings, where houses cannot carry the additional load of green roofs and green, planted walls. Here, using white or light-coloured roofs, coated surfaces and low-impact pavements, increasing the solar reflectance (called 'albedo'), are valid strategies to mitigate the urban heat island effect. It now seems time to change the building code in countries with hot climates and to legislate against black roof tiles. Leafy, green streets, irrigated open space areas and light-coloured paving and roofs are usually much cooler than built-up urban areas without green infrastructure. Ideally, cities should provide over 15 per cent of their land area with tree cover; over 30 per cent of their footprint should be green spaces (more is better, up to 50 per cent) and a minimum of 50 square metres of green space per resident has worked as a general guide. Urban farming inside the city helps too, growing food on left-over public spaces, in community gardens or by vertical farming.

The successful mitigation of the urban heat island effect leads to a significant reduction in greenhouse gas emissions city-wide. In heatwaves, residents power up their air-conditioners, putting pressure on the electrical grid. With the cost of electricity projected to rise over the next few years, the impact of heat islands is of increasing interest to all. Every degree Celsius reduction in urban air temperature is equivalent to a 5 per cent saving in energy used for cooling (Wong 2009). A study of Tokyo (Ichinose *et al.* 2008) shows that if the temperature goes down by 0.8°C, electricity bill savings are achieved that are equivalent to over US$2 million per day, with an important reduction in peak load. Total calculated savings in Tokyo over a year were up to US$10 billion. Another study of Singapore by Wong (2009) estimates that a 6 per cent reduction in greenhouse gas emissions in Singapore's centre is achievable only through the comprehensive implementation of recommended measures in mitigating the urban heat island effect. We may assume a similar figure for Sydney and other global cities. However, more research is needed to deliver evidence-based guidance for policy formulation in this important field, for mitigation of the urban heat island effect and reduction of greenhouse gas emissions; and to advocate for the importance of urban microclimates as a quantifiable guiding principle for urban development policy and reform, promoting sustainable and healthy built environments in cities.

Conclusion: moving into an unknown urban world?

Today we stand at a crossroads in urban development. As outlined above, the traditional tools of urban development are unsuited to deal with the new challenges posed by climate change. The most important questions planners need to ask are about creating practical alternatives to business-as-usual models. Given the rapid urbanization already underway in the Asia-Pacific and Latin American regions (as well as in Africa), and the need for all our cities to operate more sustainably, it is time to think about how spatial planning and design can help transform urban systems to create low carbon cities (Rydin 2010). The mega-trend of urbanization will undoubtedly continue and a blueprint is required for future urbanization efforts. We now find that rapidly urbanizing countries, such as China, are leaving the path of stubborn insistence on the primacy of economic growth, which has been shown to be unsustainable, and instead starting to prioritize citizenship and social well-being (e.g. policy makers need to know how actions taken today might affect *future* well-being). Measuring whether well-being is likely to be sustainable over time is especially challenging because it requires an in-depth understanding of what will matter for our well-being in the future and how it can be measured. Prioritizing well-being requires governments to develop new, thoughtful metrics to measure progress in social health.

This book makes a uniquely useful contribution to the discussion about the future of our cities, because:

- It takes a different perspective, combining texts by the pioneers ('classics') in Part I with real-world case studies in Part II, and cutting-edge new research in the area of urban microclimates in Part III.
- It takes an institutional perspective by seeing urban sustainability and design as embedded in broader social, cultural, technical and political structures.

While there is a growing body of research related to low carbon cities, this book identifies strategies to more rapidly synthesize and translate evidence-based knowledge to inform policies, technologies and the development of sustainable design. Consolidating research into practical models that may be utilized in our cities requires integrated solutions and comprehensive policy thinking; but to better understand low carbon cities, there is an ongoing need for research into the comparability of sustainable city indicators and frameworks. There is now little doubt that climate change mitigation in cities must proceed at a much faster pace than it currently is – a reality also acknowledged in the latest IPCC report (2013). In this, it is essential for cities not simply to put right what is currently wrong, but also to introduce innovative urban solutions in an overall sustainability agenda. Instead of putting our energy into making brand-new cities, let us focus first on retrofitting and making our existing cities more sustainable, identifying pathways to sustainability that can be replicated in other cities.

Low Carbon Cities presents a collection of over 20 important chapters as a primary resource that will be useful for students, practitioners and scholars of architecture, urban planning, engineering and design. The book explores some of the underlying dimensions of the transformation of existing cities and the design of low carbon cities in theory and practice, with reference to sustainable design, productivity and well-being, social and individual values, behaviour change integrated with technology, public space, housing affordability, public transport, urban microclimates and consumption (especially in relation to the required transformation of urban systems and planning practices), the policies and processes that shape urban development, and the abilities of individuals or groups to actively influence change, at a collective and systemic level. That is what makes this book so useful to the reader.

References

Alexander, C. 1979, *The Timeless Way of Building*, Oxford University Press, New York.

Appleby, P. 2011, *Integrated Sustainable Design of Buildings*, Earthscan, London.

Australian Bureau of Statistics (ABS) 2012, *Population Size and Growth, Australia*, Australian Bureau of Statistics, Canberra, ACT.

Australian Government, Major Cities Unit, Department of Infrastructure and Transport 2013, *State of Australian Cities 2013*, 4th report, www.minister.infrastructure.gov.au/aa/releases/2013, viewed 1 September 2013.

Banham, R. 1969, *Architecture of the Well-Tempered Environment*, University of Chicago Press, Chicago, IL.

Beck, U. 2000, *Risk Society: Towards a new modernity*, Sage, London.

Berners-Lee, M. and Clark, D. 2013, *The Burning Question*, Profile Books, London.

Bosselmann, P. 2008, *Urban Transformation: Understanding city design and form*, Island Press, Washington, DC.

Brugmann, J. 2009, *Welcome to the Urban Revolution: How cities are changing the world*, Bloomsbury Press, New York.

Brundtland, GH/UN 1987, *The Brundtland Report: Our common future*, Oxford University Press, New York.

Burton, E. 1997, *The Compact City: Just or Just Compact?*, unpublished PhD thesis, Oxford Brookes University, Oxford.

Callenbach, E. 1975, *Ecotopia*, Bantam Books, New York.

Carson, R. 1962, *Silent Spring*, Houghton Mifflin, Boston, MA.

Chinese Ministry of Environmental Protection 2012, *2012 State of Environment Report*, MEP, Beijing, http://english.mep.gov.cn/.

Clos, J. 2011, 'Introduction', in UN-Habitat and UNESCAP, *Are We Building Competitive and Liveable Cities?*, UNESCAP, Bangkok, www.unescap.org/esd/environment/infra/, viewed 15 March 2013.

Cuthbert, A.R. 2006, *The Form of Cities: Political economy and urban design*, Blackwell, Oxford/Carlton.

Dennekamp, M. and Carey, M. 2010, 'Air quality and chronic disease: why action on climate change is also good for health', *NSW Public Health Bulletin*, vol. 2, nos 5–6, pp. 115–121.

Department of Industry, Innovation, Climate Change, Science, Research and Tertiary Education (DIICCEE) 2010, *Barriers to Effective Climate Change Adaptation*, www.climatechange.gov.au/climate-change/publications, viewed 15 March 2013.

Dewar, M. and Thomas, J.M. (eds) 2012, *The City after Abandonment*, University of Pennsylvania Press, Philadelphia, PA.

Drew, J. and Fry, M. 1976, *Architecture and the Environment*, George Allen and Unwin, London.

Forster, C. 2004, *Australian Cities: Continuity and change*, Oxford University Press, Melbourne, Vic.

Gehl, J. 2010, *Life between Buildings: Using public space*, 6th edn, Danish Architectural Press, Copenhagen.

Girardet, H. 1992, *The Gaia Atlas of Cities: New directions for sustainable urban living*, Anchor, London.

—— 1999, *Creating Sustainable Cities*, Green Books, Totnes, UK.

—— 2010, *Regenerative Cities*, World Future Council and HCU, Hamburg, Germany.

Gissen, D. 2009, *Subnature: Architecture's other environments*, Princeton Architectural Press, New York.

Global Footprint Network 2013, *Earth Overshoot Day 2013*, www.footprintnetwork.org/en/index.php/gfn/page/earth_overshoot_day/ , viewed 25 August 2013.

Haas, T. (ed.) 2008, *New Urbanism and Beyond: Designing cities for the future*, Rizzoli, New York.

Haggar, S.E. 2007, *Sustainable Industrial Design and Waste Management*, Elsevier Academic Press, Amsterdam.

Hall, P. and Pfeiffer, U. 2000, *Urban Future 21: A global agenda for 21st-century cities*, E & FN Spon, New York.

Hall, P. and Tewdwr-Jones, M. 2009, *Urban and Regional Planning*, Routledge, London.

Hammarby Sjöstad, City of Stockholm 2010, *Hammarbysjostad.se*, www.hammarbysjostad.se, viewed 15 March 2013.

Herzog, T. (ed.) 1995, *Charter for Solar Energy in Architecture and Urban Planning*, Prestel, Munich.

Hoornweg, D. and Bhada-Tata, P. 2012, *What a Waste: A global review of solid waste management*, http://go.worldbank.org/BCQEP0TMO0, viewed 15 March 2013.

Hopkins, R. 2008, 'What are resilience indicators?', in *Transition Handbook: From oil dependency to local resilience*, Green Books, Totnes, UK.

Howard, E. 1902, *Garden Cities of Tomorrow*, ATC Books/Faber and Faber, London.

Ichinose, T., Matsumoto, F. and Kataoka, K. 2008, 'Counteracting urban heat islands in Japan', in P. Droege (ed.), *Urban Energy Transition: From fossil fuels to renewable power*, Elsevier, Amsterdam, the Netherlands.

Intergovernmental Panel on Climate Change (IPCC) 2013, *The Fifth Assessment Report (AR5)*, IPCC, Bilthoven, Netherlands, 25 October, www.ipcc.ch.

International Council for Local Environmental Initiatives (ICLEI) 2012, *Local Governments for Sustainability Report: Building an eco-city, building a sustainable city*, ICLEI, Bonn, www.iclei.org/resources/publications.html, viewed 15 March 2013.

Jacobs, J. 1961, *The Death and Life of Great American Cities*, Random House, New York.

Jenks, M. and Burgess, R. (eds) 2000, *Compact Cities: Sustainable urban forms for developing countries*, Spon Press, London.

John, M., Lehmann, S. and Sivam, A. 2013, 'The sustainable design and renewal of water's edge public spaces in the Asia-Pacific region: Sydney, Hong Kong and Singapore', *Journal of Sustainable Development*, vol. 6, no. 8, pp. 26–52.

Kellert, S. (prod.) 2011, *Biophilic Design: The architecture of life*, Tamarack Media, USA.

Koenigsberger, O. 1980, *Manual of Tropical Housing and Building: Climatic design*, Longman, London.

Kolokotsa, D., Santamouris, M. and Akbari, H. 2013, *Advances in the Development of Cool Materials for the Built Environment*, Bentham Science Publishers, Oak Park, IL.

Kostof, S. 1999, *The City Shaped: Urban patterns and meaning through history*, 2nd edn, Thames & Hudson, New York.

Krier, R. 2003, *Town Spaces*, Birkhäuser, Basel/Berlin.

Lehmann, S. 2005, 'Towards a sustainable city centre: integrating ecologically sustainable development principles into urban renewal', *Journal of Green Building*, vol. 1, no. 3, pp. 83–104.

—— 2010, *The Principles of Green Urbanism: Transforming the city for sustainability*, Earthscan, London.

—— 2012a, 'Green urbanism: formulating a series of holistic principles', in T. Haas (ed.), *Sustainable Urbanism and Beyond: Rethinking cities for the future*, Rizzoli/Random House, New York.

—— 2012b, 'Urban metabolism and the zero-waste city: transforming cities through sustainable design and behavior change', in M. Lindfield and F. Steinberg (eds), *Green Cities*, Asian Development Bank, Mandaluyong City, Philippines.

—— 2012c, 'Zero waste city', *Kirk-Othmer Encyclopedia of Chemical Technology*, John Wiley & Sons, New York.

—— 2012d, 'Low carbon construction systems using prefabricated engineered solid wood panels for urban infill to significantly reduce greenhouse gas emissions', *Sustainable Cities and Society*, no. 4, pp. 57–67.

—— 2013a, 'Working with China on urban challenges', *Eco-Business*, 15 March 2013, www.eco-business.com/opinion/working-with-china-on-urban-challenges/.

—— 2013b, 'Low-to-no carbon city: lessons from western urban projects for the rapid transformation of Shanghai', *Habitat International*, vol. 37, pp. 61–69.

Lehmann, S. and Crocker, R. (eds) 2012, *Designing for Zero Waste: Consumption, technologies and the built environmental*, Routledge, London.

—— 2013, 'Epilogue: the consumption dilemma – from consumer to citizen', in R. Crocker and S. Lehmann (eds), *Motivating Change: Sustainable design and behaviour in the built environment*, Earthscan from Routledge, Oxfordshire, UK.

Lehmann, S. and John, M. 2014, 'New directions: green urbanism and ecological restoration', in J. Byrne, J. Dodson and N. Sipe (eds), *Australian Environmental Planning*, Routledge, London.

Li, T., Horton, R.M. and Kinney, P.L. 2013, 'Projections of seasonal patterns in temperature-related deaths for Manhattan, New York', *Journal of Nature Climate Change*, no. 3, pp. 717–721.

Lovelock, J. 2009, *The Vanishing Face of Gaia: A final warning*, Basic Books, New York.

Lynch, K. 1960, *Image of the City*, MIT Press, Boston, MA.

McDonough, W. and Braungart, M. 1992, *The Hannover Principles: Design for sustainability*, William McDonough Architects, Charlottesville, VA.

—— 2002, *Cradle to Cradle: Remaking the way we make things*, North Point Press, New York.

McHarg, I. 1969, *Design with Nature*, Doubleday, New York.

McKinsey & Company 2009, *China's Green Revolution*, www.mckinsey.com, viewed 15 March 2013.

Madanipour, A. 1996, *Design of Urban Space*, John Wiley, New York.

Meadows, D.H., Meadows, D.L., Randers, J. and Behrens, W.W. 1972, *The Limits to Growth*, Club of Rome, New York.

Mehta, V. 2014, 'Evaluating public space', *Journal of Urban Design*, vol. 19, no. 1, pp. 53–88.

Mitscherlich, A. 1965, *Die Unwirtlichkeit unserer Städte: Anstiftung zum Unfrieden* (*The Inhospitality of Our Cities: A Deliberate Provocation*), Suhrkamp Verlag, Frankfurt a.M.

Moewes, G. 1995, *Weder Huetten noch Palaeste: Architektur und Oekologie in der Arbeitsgesellschaft*, Birkhäuser, Basel.

Norberg-Schultz, C. 1980, *Genius Loci: Towards a phenomenology of architecture*, Rizzoli, New York.

Oke, T.R. 1982, 'The energetic basis of the urban heat island', *Journal of the Meteorological Society*, vol. 108, pp. 1–24.

Olgyay, V. 1963, *Design with Climate: Bioclimatic approach to architectural regionalism*, Taylor & Francis Group, London.

Olivier, J., Janssens-Maenhout, G. and Peters, J. 2012, *Trends in Global CO_2 Emissions: 2012 report*, PBL Netherlands Environment Assessment Agency, The Hague, www.pbl.nl/sites/default/files/cms/publicaties/PBL_2012_Trends_in_global_CO2_emissions_500114022.pdf, viewed 5 April 2013.

Packard, V. 1960, *The Waste Makers*, D. McKay & Company, New York.

Pritzker Architecture Prize 2012, *Wang Shu, 2012 Laureate*, www.pritzkerprize.com/2012/biography, viewed 5 April 2013.

Rees, W.E. 2006, 'Ecological footprints and bio-capacity: essential elements in sustainability assessment', in J. Dewulf and H. Van Langenhove (eds), *Renewables-Based Technology: Sustainability assessment*, John Wiley & Sons, Chichester, UK.

Register, R. 1987, *Eco-City Berkeley: Building cities for a healthy future*, North Atlantic Books, Boston, MA.

—— 2006, *Ecocities: Rebuilding cities in balance with nature*, New Society Publishers, Gabriola Island, BC.

Roseland, M. (ed.) 1997. *Eco-city Dimensions: Healthy communities, healthy planet*, New Society Publishers, Gabriola Island, BC.

Rossi, A. 1966 [English translation 1982], *The Architecture of the City*, MIT Press, Cambridge, MA.

Rowe, C. and Koetter, F. 1978, *Collage City*, MIT Press, Cambridge, MA.

Rydin, Y. 2010, *Governing for Sustainable Urban Development*, Earthscan, London.

Sanyal, S. 2009, 'Building cities for 21st-century India', *Business Standard*, 9 July.

Satterthwaite, D. (ed.) 1999, *Reader in Sustainable Cities*, Earthscan, London.

Sennett, R. 1974, *The Fall of Public Man*, Faber, New York.

Sieverts, T. 1997 [2004], *Zwischenstadt: Cities Without Cities: An interpretation of the Zwischenstadt*, Routledge, London.

Sitte, C. 1889 [1945], *The Art of Building Cities: City building according to its artistic fundamentals*, Random House/Reinhold, New York.

Sloterdijk, P. 2013, *In the World Interior of Capital: Towards a philosophical theory of globalization*, Polity Press, Cambridge.

Strasser, S. 1992, 'Waste and want: the other side of consumption', in *German Historical Institute Annual Lecture Series No. 5*, Berg Publishers, Washington, DC, www.ghi-dc.org/publications/ghipubs/annual/al05.pdf, viewed 15 March 2013.

Szokolay, S.V. 1973, 'Using solar energy in housing', *Royal Institute of British Architects Journal*, vol. 80, pp. 177–179.

Thorpe, A. 2012, *Architecture and Design versus Consumerism*, Routledge, London.

United Nations Economic and Social Commission for Asia and the Pacific (UN ESCAP) 2011, *Are We Building Competitive and Liveable Cities? Guidelines for developing eco-efficient and socially inclusive infrastructure*, UN ESCAP, Bangkok.

United Nations Environment Programme (UNEP) 2011, *Towards a Green Economy: Pathways to sustainable development and poverty eradication*, UNEP, Geneva, viewed 15 January 2013, www.unep.org/greeneconomy/.

—— 2013, *City-Level Decoupling: Urban resource flows and governance of infrastructure transitions*, UNEP, Paris.

UN-Habitat, 2010, Tibaijuka, A., 'Foreword', in United Nations Human Settlements Programme, *Solid Waste Management in the World's Cities*, Earthscan, London.

—— 2012, *Time to Think Urban*, Nairobi, United Nations Human Settlements Programme, www.unhabitat.org/pmss/getElectronicVersion.aspx?nr=3456&alt=1, viewed 15 August 2013.

Urry, J. 2013, *Societies beyond Oil*, Zed Books, London.

Vale, R. and Vale, B. 1991, *Green Architecture: Design for an energy-conscious future*, Little, Brown & Co, Boston, MA.

von Weizaecker, E.U., Lovins, A.B. and Lovins, L.H. 1997, *Factor Four: Doubling wealth, halving resource use*, Earthscan, London.

Wheeler, S.M. and Beatley, T. (eds) 2004, *The Sustainable Urban Development Reader*, Routledge, London.

Wong, N.H. 2009, *Tropical Urban Heat Islands: Climate, buildings and greenery*, Taylor & Francis, New York.

World Bank 2010, *Cities and Climate Change: An urgent agenda*, World Bank, Washington, DC.

—— 2012, *Sustainable Low Carbon City Development in China*, World Bank, Washington, DC.

—— 2013, *Turn Down the Heat: Climate extremes, regional impact and the case for resilience*, World Bank, Washington, DC.

World Business Council for Sustainable Development (WBCSD) 2010, *Vision 2050: The new agenda for business*, www.wbcsd.org/vision2050.aspx, viewed 15 March 2013.

World Wildlife Fund (WWF) 2010, *Reinventing the City: Three prerequisites for green urban infrastructures*, WWF, Gland, Switzerland, http://awsassets.panda.org/downloads/wwf_low_carbon_cities_final_2012.pdf, viewed 15 January 2013.

Xie, H.T. 2012, 'Tianjin Eco-City in China', in S. Joss (ed.), *International Eco-Cities Initiative: Tomorrow's city today: eco-city indicators, standards & frameworks*, Bellagio Conference Report, University of Westminster, London.

Yeang, K. 1996, *The Skyscraper Bio-Climatically Considered*, Academy Press, London.

—— 1999, *The Green Skyscraper: The basis for designing sustainable intensive buildings*, Prestel, Munich.

Zaman, A. and Lehmann, S. 2011, 'Challenges and opportunities in transforming a city into a "zero waste city"', *Challenges*, vol. 2, no. 4, pp. 73–93.

—— 2013, 'The zero waste index: a performance measurement tool for waste management systems in a "zero waste city"', *Journal of Cleaner Production*, vol. 50, pp. 123–132.

Part I

Classic texts on ecologically sustainable cities

Conceptual evolutions and different schools of thought

Introduction to Part I

Low carbon cities are sustainably built, within the means of finite environmental resources. The ultimate goal of eco-cities is to eliminate all greenhouse gas emissions and waste; to produce energy on site through renewable sources such as solar, hydro and wind power and biomass; and to ensure residents' optimal health and well-being. Low carbon cities have an efficient public transport system and appropriate densities, so that they are compact, mixed use and walkable; they operate as a largely self-contained economy, with resources found and products supplied locally. Resource conservation and maximizing efficiency of energy, water and material resources is essential, as is avoiding waste creation and supporting a waste management system of reuse and resource recovery that recycles materials in a zero-waste system.

Low carbon cities do not damage the natural environment or consume a disproportionate amount of resources; they do not waste energy, water or materials due to short life, poor design or inefficient construction and manufacturing procedures. Consistent with eco-efficiency principles, these cities are affordable, manageable and sustainable in the long term.

By comparing everything people consume in a single year, including energy, food, travel, goods and services, researchers have proved that suburbs account for more greenhouse gas emissions per household compared to more compact urban areas. Inner-city residents tend to have much lower carbon footprints, because they live in smaller homes and use more public transport.

The six chapters in Part I, written by some of the pioneers of urban thinking, examine ways to create such ecologically sustainable cities. Pioneers are always a little ahead of others in the way they think about a problem. This range of perspectives on the practice of urban design, planning and infrastructure challenges us to rethink what a 'good city' is (or might be) to encourage us to transform our urban systems and begin the transition to low carbon cities.

Herbert Girardet outlines his vision of 'ecopolis', a city that is able to regenerate itself. Richard Register argues that, if cities are to sustain themselves in the face of environmental change, they need to go beyond merely adapting and instead should look at how such challenges have been dealt with in the past. Tim Beatley suggests ways to achieve and measure biophilic cities. Norbert Lechner advocates architectural and urban design that is cognizant of the sun's power. Jeff Kenworthy presents the latest information on a longitudinal global study of the connections between urban planning and public transport. Manfred Lenzen and Greg M. Peters show the links between cities and surrounding areas by studying urban residents' use of resources in a case study of Australian households.

Chapter 1

'Ecopolis'

The regenerative city

Herbert Girardet

Summary

Cities tend to absorb resources from nature and discharge their wastes without taking responsibility for the consequences. In the last 100 years, the process of urbanization has become ever more resource-intensive, and it now significantly contributes to climate change, loss of soil carbon, decreased natural fertility of farmland and the loss of biodiversity across the world. A comprehensive approach – beyond established concepts of sustainable development – is needed. Regenerative development solutions to many challenges have already been implemented. These include land-use planning in favour of compact, transport-efficient communities; energy efficiency in buildings; smart grids and renewable energy production; waste recycling and reuse. However, we need to think beyond these challenges, ensuring a comprehensive understanding of the global impacts of urbanization on the planet as a living system. This chapter argues that the established concept of *urban ecology* should be expanded to include all the territories involved in sustaining urban systems. The *regenerative development* of cities thus takes on the meaning of *eco-regeneration*.

Introduction

In early December 2012, some 24,000 government representatives, academics, NGO campaigners and journalists got together in the latest round of negotiations to try to tackle global climate change (UN 2012). COP18 was the eighteenth annual 'conference of parties' trying to get to grips with an existential challenge facing humanity: how to avoid a global climate catastrophe. The year 2012 also saw the 20-year follow-up conference to the 1992 Rio Earth Summit aiming to address the wider issue of the ever-increasing human impacts on the ecosystems of our home planet (Rio 12 2012). At both events it became clear that none of these issues can be tackled without addressing the consumption patterns of cities. This has become a central responsibility of urban policy makers, developers and city dwellers.

Cities take vast quantities of resources from nature. The new challenge is for cities to help to continuously regenerate the natural systems from which they draw resources.

Humanity has become a predominantly urban species and this historic development represents a systemic change in the relationship between humans and nature. Cities are constructed on only 3 to 4 per cent of the world's land surface, but the ecological footprints of cities cover much of the planet (Girardet 1999).

Modern cities are an astonishing human achievement. Their buildings and infrastructure systems are unprecedented in their technical complexity. Their transport systems have developed a global reach. Cities are centres of economic activities and intense human interaction. They can offer a great variety of services at comparatively low per capita cost. However, the environmental impacts of an urbanizing humanity have become a great cause for concern. Apart from a near monopoly on the use of fossil fuels, metals and concrete, urbanizing humans now consume nearly 40 per cent of nature's annual primary production as well (Haberl *et al.* 2010).

These ecological, economic and social externalities of our urban systems need to be addressed in new ways. The ravenous appetite of our fossil fuel-powered urban lifestyles for natural resources has enormous consequences for all life on Earth, including human life. Cities tend to absorb resources from nature and discharge their wastes without taking responsibility for the consequences. In the past 100 years, the process of urbanization has become ever more resource-intensive, and it now significantly contributes to climate change, loss of soil carbon, decreased natural fertility of farmland and the loss of biodiversity across the world.

Of major new concern are the rapidly growing resource demands of cities in developing countries. As cities become larger and richer, they draw increasingly upon nature's global bounty rather than on resources from their own local hinterland. The footprints of individual cities can be hundreds of times larger than their own surface area. In a rapidly urbanizing world, we need to wean cities off their systemic fossil fuel dependence, and get them to actively contribute to the restoration of damaged ecosystems upon whose efficient 'services' they ultimately depend. However, this is an as yet unfamiliar challenge for urban politicians, planners, managers, architects, developers and city dwellers.

A comprehensive approach beyond established concepts of sustainable development is needed. Regenerative development solutions for many challenges have already been implemented. These include: land-use planning in favour of compact, transport-efficient communities; energy efficiency in buildings; smart grids and renewable energy production; waste recycling and reuse. However, we need to think beyond these, ensuring a comprehensive understanding of the global impacts of urbanization on the planet as a living system.

The transformative changes that are required are challenging because they call for long-term perspectives as compared to the short-term compromises and patchwork solutions that tend to characterize commercial and political decision making.

From urban regeneration to regenerative cities

In recent years there has been a proliferation of *urban regeneration* or renewal initiatives in run-down cities of industrialized countries. These have aimed to restore the urban fabric, economy and well-being of urban citizens – the 'inner-urban environment'. Such initiatives have received much funding and media attention, and they have improved the lives of millions of people ('Urban renewal' 2013).

However, the concept of *regenerative cities* goes further – seeking to address the relationship between cities and their local and global hinterland. We need to re-enrich the landscapes upon which cities depend, and this includes the need to increase their capacity to absorb carbon emissions. We need to find ways of creating a *restorative relationship* between cities and the world beyond, making use of new opportunities in technology, policy and business practice.

This chapter argues that the established horizon of *urban ecology* should be expanded to include all the territories involved in sustaining urban systems. The *regenerative development* of cities thus takes on the meaning of *eco-regeneration*.

Creating *regenerative cities* thus means *developing comprehensive financial and technological strategies for an environmentally enhancing, restorative relationship between cities and the ecosystems from which they draw resources for their sustenance.*

Agropolis: the city and its local landscape

Traditional cities had an intimate connection to their local landscapes. In his book *The Isolated State*, the prominent nineteenth-century economist Johann Heinrich von Thünen described the way in which human settlements, in the absence of major transport systems, are systemically tied into these landscapes through various logically arranged modes of cultivation (Von Thünen 1966). In fact, they ensure the landscapes are continuing productivity and fertility by returning appropriate amounts of organic waste to them. I call this traditional form of settlement 'Agropolis' (see Figure 1.1).

In many parts of the world, towns and cities – lacking efficient transport systems connecting them to the outside world – had these kinds of systemic relationships to the landscapes from which they emerged. They depended for their sustenance on nearby market gardens, orchards, forests, arable and grazing land, and local water supplies. Until very recently, many Asian cities were still largely locally self-sufficient in food as well as fertilizer, using human and animal wastes to sustain the fertility of local farms (King 1911). Can we learn from these traditional arrangements in the future while using more up-to-date methodologies and technologies?

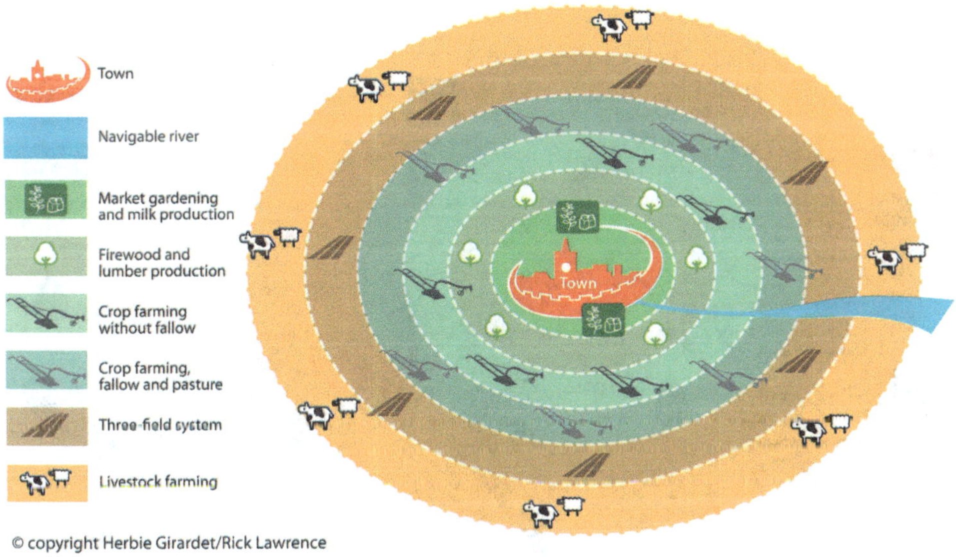

Figure 1.1 'Agropolis'.

Source: Supplied by the author.

The rise of 'Petropolis'

The Industrial Revolution caused an explosion of urban growth that continues to this day (Edmondson 2007). Steam-engine technology enabled the unprecedented concentration of industrial activities in urban centres. Cities increasingly declared independence from their local hinterlands and became centres of consumerism and global economic and transport hubs. This process undermined local economies, as new modes of transportation made it easier to supply food, raw materials and manufactured products from ever greater distances.

The phenomenal changes in human lifestyles that emerged are reflected in new concepts of land-use planning, particularly for accommodating the road space needed for motor cars. The modern city could be described as 'Petropolis'; all its key functions – production, consumption and transport – are powered by massive injections of fossil fuels (see Figure 1.2). However, there is growing evidence that the resulting dependencies are ecologically, economically and geopolitically unsustainable, particularly because the fossil fuel supplies upon which modern cities depend are, most definitely, finite.

It took around 300 million years for oil, gas and coal to accumulate in the Earth's crust and we are on track to burn much of it in just 300 years – now at a rate of well over a million years per year. Cities are particularly responsible for this. Despite taking up only 3 per cent of the world's surface area (Earth Institute 2005), they use approximately 75 per cent of its resources and also discharge similar proportions of waste (United Nations Environment Programme [UNEP] 2012).

These highly problematic patterns of fossil fuel-dependent urbanization are still expanding across the world – the rapid growth of cities such as the Gulf cities of Dubai, Abu Dhabi,

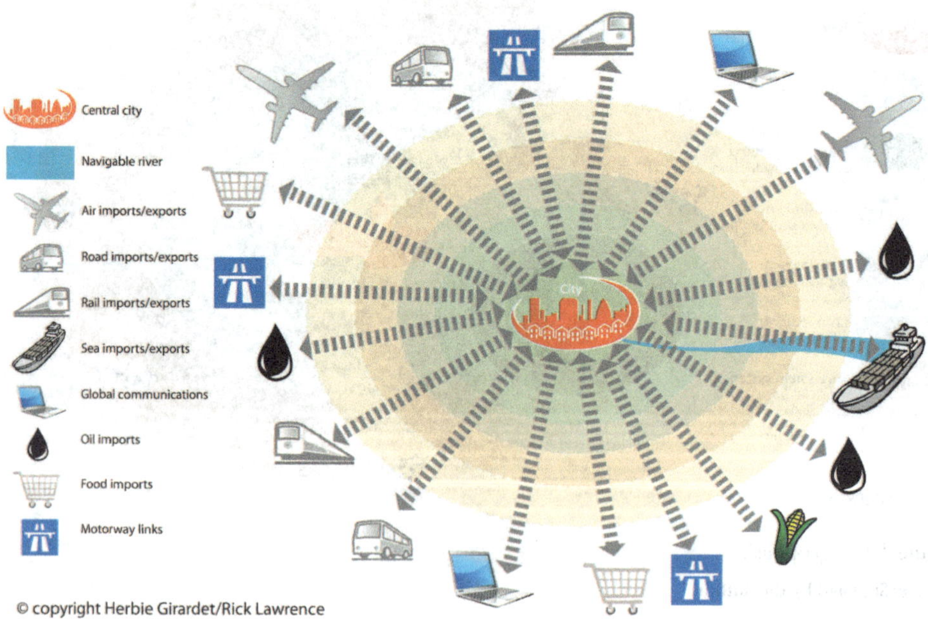

Figure 1.2 'Petropolis'.

Source: Supplied by the author.

Doha and Jeddah with their amazing collection of skyscrapers, their artificial islands and their low-density desert suburbs is the latest and most astonishing example of this. Here the reliance on desalinated water is an issue of particular concern.

Today, urbanization and economic and financial globalization are closely connected. Cities have become globalized centres of production as well as consumption, with throughputs of unprecedented quantities of resources and industrial products becoming the norm. In developing countries, urbanization is closely associated with ever-increasing per capita use of fossil fuels and with impacts on ever more distant eco-systems (WWF 2012).

What goes in must come out again. Cities are *dependent systems* that rely on external inputs for their sustenance and they discharge vast quantities of solid, liquid and gaseous wastes without much concern about the consequences. The challenge now is to reduce this dependence before the risks of food insecurity, storms and sea level rises start to undermine the very existence of our modern cities.

A large part of the increase of carbon dioxide in the atmosphere is attributable to combustion in and on behalf of the world's cities (Pratt 2012). Two hundred years ago, atmospheric CO_2 concentrations were around 280 parts per million, but since then they have risen to over 390 ppm (CO_2Now 2013).Until recently, it was widely assumed that we could get away with such increases of pre-industrial concentrations, but gradually it has become clear that this could cause the planet to overheat, with dire consequences for all life. A recent report, published not by a campaigning environmental NGO but by the World Bank, warns that we are on track for a 4°C warmer world marked by extreme heatwaves, declining global food stocks, loss of ecosystems and biodiversity, and life-threatening sea level rise (World Bank 2012).

In recent years, the most dramatic population growth has occurred in giant coastal cities, particularly those in Asia and Africa. In fact, with the expansion of global trade, coastal populations and economies have exploded on every continent. Of the 17 megacities of over ten million people around the globe, 14 are located in coastal areas. Forty per cent of the world's cities of between one and ten million people are also located near coastlines. Careless

The value of ecosystems services

> We cannot manage what we do not measure and we are not measuring either the value of nature's benefits or the costs of their loss. We seem to be navigating the new and unfamiliar waters of ecological scarcities and climate risks with faulty instruments. Replacing our obsolete economic compass could help economics become part of the solution to reverse our declining ecosystems and biodiversity loss.
>
> We need a new compass to set different policy directions, change incentive structures, reduce or phase out perverse subsidies, and engage business leaders in a vision for a new economy. Holistic economics – or economics that recognises the value of nature's services and the costs of their loss – is needed to set the stage for a new 'green economy'.
>
> (Pavan Sukhdev, drawing on his 2010 report *The Economics of Ecosystems and Biodiversity*)

development practices have degraded or destroyed important habitats such as wetlands, coral reefs, sea grasses and estuaries. And with substantial sea level rises expected by the end of the twenty-first century, major northern coastal conurbations such as London, New York, Shanghai, Calcutta, Dhaka and Lagos could well become the primary victims of global urban greenhouse gas emissions, with huge impacts on property values.

All of this indicates the systemic flaws of Petropolis, the fossil fuel-dependent city that is the current global 'urban archetype', which need to be fundamentally challenged.

Creating 'Ecopolis'

Energy

What measures are needed for modern cities to minimize their systemic dependence on fossil fuels? To start with, it is crucial to improve the energy efficiency of existing buildings and to ensure high efficiency standards for new buildings. In many countries, appropriate policy measures are now being taken (GreenandSave 2011; California Energy Commission 2013). Such measures are a strong incentive to all property developers – particularly to those building properties for rent – to ensure superior energy efficiency standards which might otherwise not be a high priority for them. Their tenants will save money while CO_2 emissions are significantly reduced at the same time.

Can cities that are the product of combustion technologies be powered by renewable energy (RE) instead? The good news is that in the past few years rapid strides have been made in the development and use of RE technologies, particularly wind and solar. While a large proportion of renewable energy for cities will need to come from peri-urban areas, much can be done within cities themselves.

Figure 1.3 Solarsiedlung, Vauban, Freiburg.

Source: Supplied by the author.

Figure 1.4 Sanlúcar La Mayor solar power station, Spain.

Source: Supplied by the author.

For example, there is the Solarsiedlung in Freiburg, Germany, consisting of 60 dwellings, completed in 2007, which shows the feasibility of building houses that can produce surplus energy beyond their inhabitants' own requirements (see Figure 1.3). The highly energy-efficient 'plus-energy' buildings with south-facing solar roofs are widely regarded as a model for intra-urban RE production (Solarsiedlung 2013). The Solarsiedlung was made possible by the German renewable energy feed-in tariff legislation, which was introduced as national policy in 2000 (Hagemann 2007). The policy also enabled the retrofitting of millions of houses across Germany, which is now making a significant contribution to the country's electricity supply.

However, across Europe regional peri-urban RE installations are also growing in number and size. So far, the most notable examples are the 'concentrated solar power' stations that have been built near Seville in Spain (see Figure 1.4). They feature arrays of mirrors that focus beams of sunlight onto the top of a tower through which liquid is circulated that is turned into steam, which then powers turbines and generators. The latest versions of these installations can store solar heat in a salt solution, which enables them to produce electricity for up to eight hours after dark. Seville is the world's first large city to obtain most of its electricity supplies from solar power stations in its hinterland, as well as from installations within the city (Power-Technology n.d.).

Solar energy may also be used to produce hot water from roof-top installations, and China is the world leader in this technology. Some 25 per cent of its households now have solar hot water systems, often in households that had not had the benefit of hot water before. Canadian environmentalist Brian Gordon (2008) reported predictions that

> by 2010 the number of solar water heaters installed in China will equal the thermal equivalent of the electrical capacity of 40 large nuclear power plants. Globally, solar water heaters have the capacity to produce as much energy as more than 140 nuclear power stations.

The 75,000-square-metre 'sundial' building that hosted the fourth World Solar Cities Congress in September 2010 is a Chinese showcase of solar technology that has become highly influential in a country more usually known for its rapid expansion of coal-fired power station capacity (Koerner 2009). 'Solar thermal' technology has also been used for many years in the Mediterranean. It is becoming commonplace even in less sunny countries such as Austria and Germany.

Wind power is also a solar technology because the Earth's air currents are driven by the heat of the sun. Many technological breakthroughs in this field have been facilitated by government policies. Denmark was the first country to introduce feed-in tariffs for wind energy in the 1980s. At that time 50 KW wind turbines were the norm, but by 2010 their output had risen to as much as 5 megawatts – 100 times greater. In countries with long coastlines such as Britain, large-scale wind farm development is now well under way.

While it is desirable for cities to produce much of their energy from within their own territory or from their immediate hinterland, very large cities may require additional renewable energy supplies from further afield. Networks of interconnected solar, wind, hydropower and geothermal systems are now under development. The Desertec project, supported by major European companies, aims to link up RE installations in North Africa, the Middle East and Europe (Desertec Foundation 2013). Other projects propose supplying electricity across continents like North America and Asia via new direct-current 'smart supergrids'.

Meanwhile, new transport technologies are also advancing rapidly. Just ten years ago, car manufacturers could barely imagine making cars that did not run on petrol or diesel. Today, all mainstream manufacturers are developing and building hybrid, electric or fuel cell-powered vehicles, which are promising to become the norm in a matter of years.

All of these measures, taken together, can dramatically change urban energy production and consumption patterns while also creating major new economic sectors in our cities.

The urban metabolism

Similar to nature's organisms, cities, often described as 'super-organisms', have a definable metabolism – the transformation of resources into vital functions (Girardet 2004a, 2008). Nature essentially has a circular, zero-waste metabolism; every output by an organism is also an input which replenishes and sustains the whole living environment.

In contrast, the metabolism of many modern cities is essentially linear, with resources flowing through the urban system without much concern about their origin, or about the destination of wastes (see Figure 1.5). Inputs and outputs are considered to be largely unrelated. Fossil fuels are extracted from rock strata, refined and burned, and the waste gases discharged into the atmosphere. Raw materials are extracted, combined and processed into consumer goods that ultimately end up as rubbish that cannot be beneficially reabsorbed into living nature. In distant forests, trees are felled for their timber or pulp, but all too often forests are not replenished.

Similar processes apply to food; nutrients and carbon are taken from farmland as food is harvested, processed and eaten. The resulting sewage, with or without treatment, is then discharged into rivers and coastal waters downstream from population centres, and usually not returned to farmland. Rivers and coastal waters all over the world are 'enriched' with sewage and toxic effluents, as well as the run-off of mineral fertilizer applied to farmland used for feeding cities.

In an urbanizing world, aiming for long-term viability, this linear, open-loop approach cannot continue. The environmental externalities of urban resources use can no longer be ignored. Unless we learn from nature how to create *circular systems,* an urbanizing world will continue to be an agent of global environmental decline.

In an urbanizing world, it is vital for cities to adopt a circular urban metabolism, to ensure the long-term viability of both the cities and the ecosystems upon whose health they ulti-

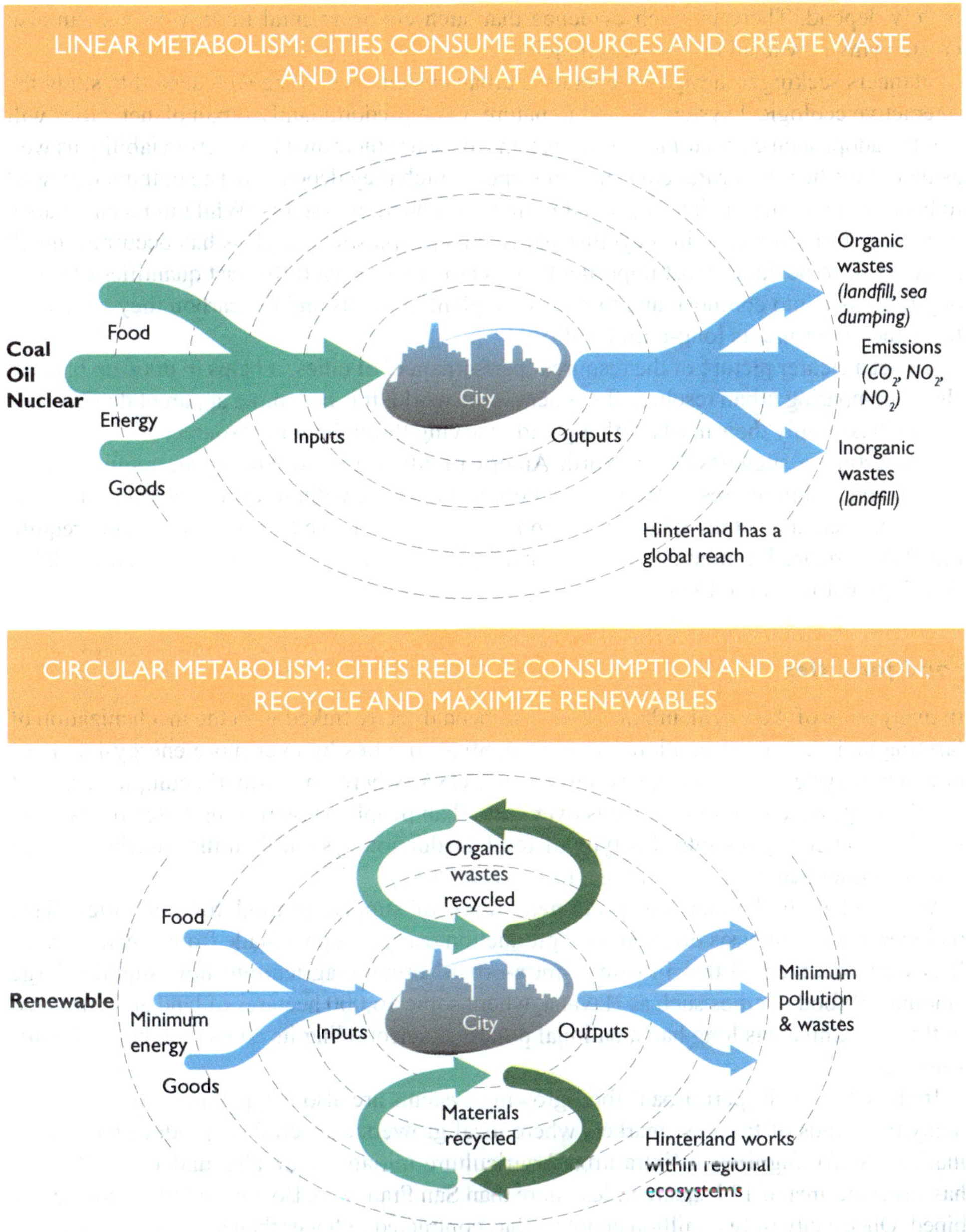

Figure 1.5 Regenerative cities have a circular metabolism.

Source: Supplied by the author.

mately depend. There is much evidence that such environmental improvements can also create many new local businesses and jobs.

Planners seeking to design regenerative urban systems would be well served to study the interactive ecological systems found in nature. On a predominantly urban planet, cities will need to adopt such circular metabolic systems to assure their own long-term viability as well as that of the health of rural environments upon which they depend. Waste outputs will need to become useful inputs into local and regional production systems. While in recent years a very substantial increase in recycling paper, metals, plastic and glass has occurred, much more needs to be done. Most importantly, it is crucial to convert the vast quantities of urban organic waste into compost, and to return the plant nutrients and the carbon they contain to farmland, to ensure its long-term fertility.

To get a clearer picture of the resource 'performance' of cities, it helps to draw up balance sheets comparing urban resource flows across the world. It is becoming apparent that similar-sized cities supply their needs with a greatly varying throughput of resources.

One estimate suggests that a North American city with 650,000 people requires some 30,000 square kilometres of land to meet domestic needs, without even including the environmental demands of its industries. In comparison, an Indian city of this size would require just 2,800 square kilometres, or less than 10 per cent of an American city's land (Global Development Research Center 2013).

Food for cities

In many parts of the world, urban growth has been directly linked with the mechanization of farming and rural depopulation. Food is supplied to cities by ever more energy-intensive production systems. The prime example is the USA, where one farmer's complex array of fossil fuel-powered equipment feeds over 100 urban people. However, up to ten times more fossil fuel energy goes into this type of food production system than the calories that are actually contained in the food we get to eat (Lott 2011).

We need to find much more efficient ways of supplying food for our cities. This includes a new emphasis on local food production using compost made from organic waste. It is well documented that in Cuba, 'intra-urban' organic agriculture now supplies large amounts of food to cities such as Havana, where some 16,000 hectares of land are cultivated in the city. China has long had a national policy of surrounding its cities with belts of cultivated land.

In the USA, such 'peri-urban' food-growing systems are also reappearing. There are now many thousands of farmers' markets where local growers can sell their produce ('Farmers' market' 2013). Significant 'intra-urban' agriculture initiatives are also under way. Detroit has a surface area of 139 square miles, more than San Francisco, Boston and Manhattan combined. Once a city of two million people, it has contracted to fewer than 900,000 people, with vast derelict areas. In a recent report, the American Institute of Architects came to this conclusion: 'Detroit is particularly well suited to become a pioneer in urban agriculture at a commercial scale' (Whitford 2009). Already hundreds of urban farmers are growing crops on previously derelict land in Detroit, and in low-income areas of Milwaukee, Oakland and East New York (McMillan 2008). Similar projects exist in New Orleans, St Louis, Cleveland and Newark.

In Denver, the Living City Block project goes beyond urban agriculture. It aims to create an example of a replicable, scalable and economically viable framework for the resource-efficient redevelopment of existing cities. The organization's website explains: 'Starting with a block and a half of Denver's historic Lower Downtown district, Living City Block will create a demonstration of a regenerative urban centre . . . a paradigm for the new urban landscape' (Living City Block 2013).

These examples show clearly how derelict urban land can be put to good use. Even large cities can source substantial amounts of the vegetables and fruit they require from their urban territory and the surrounding countryside. However, grain and animal fodder supplies require much larger areas of land and the majority will always have to be brought in from farmland further afield.

The ecosystems beyond

Renewable energy, urban agriculture and resource efficiency are only part of the story of creating truly regenerative cities. We also need to address the relationship between cities and the distant ecosystems on which they currently depend.

As already stated, the ecological footprints of cities cover much of the world's productive land. How can cities reduce these footprints and try to develop beneficial relationships with the world's ecosystems?

One interesting example is the measures taken by one mega-city, Seoul, to try to protect itself from sandstorms. The capital of South Korea imports meat from the herdsmen of the Gobi Desert 800 miles away, but it has also been suffering from increasingly severe yellow dust storms from the Gobi. A Seoul-based initiative, Future Forest, has set out to improve matters; it recruited Korean and Chinese students to help plant some four million trees in the Gobi to stabilize eroding land and to try to prevent the eastward spread of the desert, the source of the dust storms. The originator of this regeneration initiative, former South Korean Ambassador to China Kwon Byong Hyon, has received a major award for this work from the UN Convention to Combat Desertification (Hattam 2012).

A few other cities are now involved in similar initiatives. Internationally, the Billion Trees Campaign initiated by UNEP in 2007 recorded that over ten billion trees had been planted by 2010, a quarter of these by urban community groups, NGOs and local governments (UNEP 2008). Meanwhile, the International Union for the Conservation of Nature has started the Plant a Pledge initiative to regenerate no fewer than 150 million hectares of degraded land across the planet by 2020. Many national and local governments are part of this historic initiative (International Union for the Conservation of Nature 2013).

Other interesting examples of large cities that have adopted significant ecosystem regeneration initiatives include the watershed protection and restoration measures initiated by New York City and New York. The Catskills watershed is the largest unfiltered water supply in the USA and is one of New York City's most important natural resources, providing about 1.3 billion gallons of clean drinking water to approximately nine million people every day. Filtration for that much water would be prohibitively expensive, so New York City decided to invest millions of dollars to protect and regenerate the watershed and maintain its high water quality (New York State Department of Environmental Conservation 2013). The city of Manila has also realized the crucial importance of watershed forests. To ensure its water security, it is undertaking the reforestation of the St Tomas watershed area, which covers several thousands of hectares of forest land, deforested by decades of logging from the 1950s to the 1990s, as well as slash-and-burn farming methods.

The regenerative development of ecosystems serving such large cities is a tall order, but must be pursued with vigour. If cities do not proactively regenerate run-down ecosystems – including watershed forests, farm soils and marine ecosystems – from which they draw essential resources, their long-term prospects look bleak. While cities continue to burn fossil fuels, they will also need to find ways to sequester their carbon dioxide emissions through 'bio-sequestration' in soils and forests (Girardet and Mendonca 2009).

The CO_2 output of cities is far too large for trees within their territories to be able to absorb. We are now discharging over ten billion tonnes of carbon per year (Science Daily 2008) of which four to five billion tonnes are not being reabsorbed into the world's ecosystems but are accumulating in the atmosphere. This is the primary cause of climate change.

The world's 19 mega-cities of over ten million people only accommodate just under 10 per cent of the world's urban population (United Nations Department of Economic and Social Affairs 2012). The ecological footprint of these vast mega-cities will invariably extend far beyond their immediate hinterland. They have little intrinsic bio-capacity, and must rely upon territories elsewhere ('Ecological footprint' 2013). A key question to which we still need to find answers is whether these footprints can be reduced to reasonable proportions.

From Petropolis to Ecopolis

Shifting from urban systems that damage and degenerate ecosystems to ones that renew and sustain the health of ecosystems upon which they ultimately depend requires a fundamental rethinking of urban systems design.

One of the primary tasks we face at the start of the twenty-first century is to try to map out what is *necessary* in order to try to expand the boundaries of what becomes politically *possible*. It is in cities where creativity is most vibrant and where people can interact most vigorously in the search for solutions. The primary challenge is to find cost-effective ways to make our cities function in an environmentally responsible manner, and for long-term perspectives to prevail.

Urban analysts often say that cities are the places where solutions to the world's environmental and climate problems can most easily be implemented; cities, as places where people live closely together, have the potential to make efficient use of resources. Can we create spatial structures that satisfy the needs of urban dwellers while also ensuring their ecological and economic viability? Can we create secure, clean and pleasant habitats that also ensure positive human interaction with natural systems?

The idea of 'Ecopolis' – the ecologically and economically restorative city – aims to draw together the various themes discussed above into one comprehensive concept (see Figure 1.6). The proposal is to reintegrate contemporary cities into their surrounding environment – not only drawing upon regional biologically productive land but also developing the potential for renewable energy supplies.

Since modern cities tend to be much larger than traditional human settlements, their reintegration into their local hinterland can be rather difficult. In addition, the reality that far more people have to be accommodated in cities today than in the days of 'Agropolis' needs to be taken into account in developing relevant concepts.

As suggested above, creating environmentally regenerative cities is an as yet unfamiliar challenge for urban administrators and educators. They need to develop existing policy measures, using both 'sticks', such as waste disposal taxation and carbon taxes, and 'carrots',

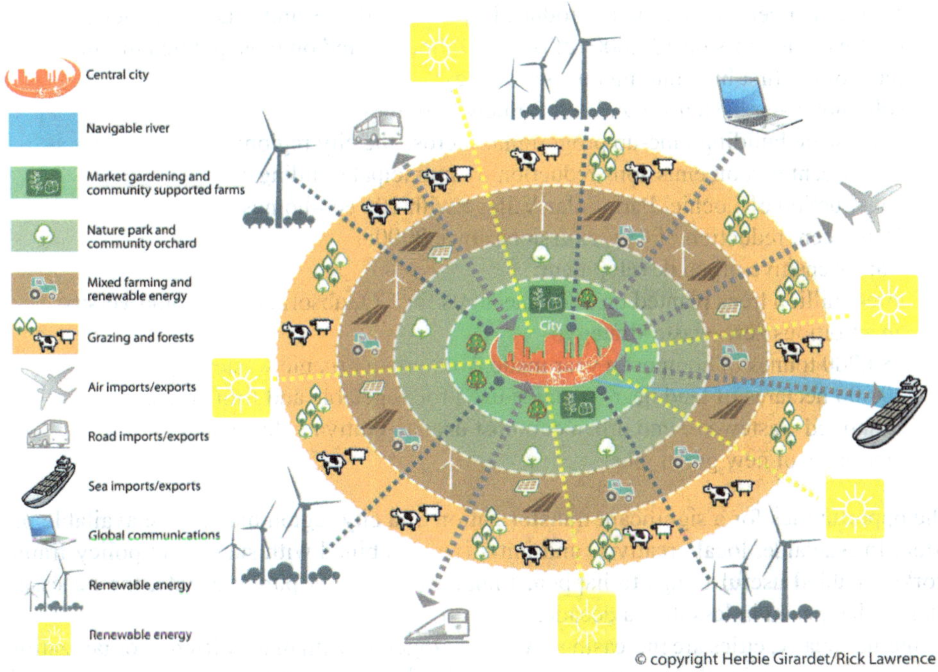

Figure 1.6 'Ecopolis'.

Source: Supplied by the author.

such as feed-in tariffs for renewable energy. The scenario also includes support schemes for local food production already in place in some cities.

This challenge for city managers has been made more difficult, since the privatization of many urban services in recent years has reduced their capacity to create integrated urban systems. However, awareness is growing that integrated, restorative planning and management of cities presents major new opportunities for reviving urban economies and creating new businesses and jobs. Positive outcomes promise to be beneficial for both global ecology as well as for the local urban economy.

Creating parameters for appropriate action will involve both political and business decisions – from transnational, to national and urban levels of decision making. It will involve drawing up novel legal frameworks as well as addressing the profit logic of developers and other commercial enterprises.

The concept of regenerative urban development is relevant both for retrofitting existing cities as well as for cities that are being newly built. By and large, the concepts defining Masdar Ecocity in Abu Dhabi (Mubadala 2012) would comply with regenerative city principles. Unfortunately, only a small proportion of this ambitious project has been realized.

The best examples of regenerative urban development, thus far, have come from smaller cities. Perhaps the most successful example is Adelaide, South Australia. In 2003 I had an opportunity to develop plans for the transformation of this city region of 1.2 million people for South Australia Premier Mike Rann (Girardet 2004b). The following bullet points (Girardet 2012) list the changes that have been initiated in Adelaide to date:

- over 26 per cent of electricity produced by wind turbines and solar PV panels;
- photovoltaic roofs on 120,000 (of 600,000) houses, and on most public buildings;
- the world's first bus running on solar energy;
- solar hot water systems mandated for new buildings;
- large-scale building tune-up programmes across the city region;
- 60 per cent carbon emissions reduction by municipal buildings;
- construction of Lochiel Park Solar Village with 106 eco-homes;
- 15 per cent reduction of CO_2 emissions since 2000;
- water-sensitive urban development;
- three million trees planted on 2,000 hectares for CO_2 absorption and biodiversity;
- an ambitious zero-waste strategy;
- 180,000 tonnes of compost a year made from urban organic waste;
- 20,000 hectares of land near Adelaide used for vegetable and fruit crops;
- reclaimed wastewater and urban compost used to cultivate this land;
- thousands of new green jobs.

The opportunities for a significant transformation of a city region may not be available that often. In Adelaide, local creativity and initiative, combined with state-level policy frameworks, enabled useful things to happen. Enhanced by lively public debate, the necessary changes happened in less than a decade.

Internationally, cities are increasingly working together with many different organizations and associations to develop and implement policies for regenerating regions across the world that have been damaged and depleted by urban consumption patterns. The concept of regenerative cities is receiving a lot of traction. The challenge now is to get it widely adopted as a plausible concept that can help to future-proof our urbanizing world.

References

California Energy Commission 2013, 'Building energy efficiency program', www.energy.ca.gov/title24/, viewed 5 April 2013.

Co_2Now 2013, *Atmospheric CO_2 data for February 2013*, www.co2now.org/, viewed 5 April 2013.

Desertec Foundation 2013, *Desertec Foundation: start*, www.desertec.org, viewed 5 April 2013.

Earth Institute 2005, *The growing urbanisation of the world*, www.earth.columbia.edu/news/2005/story03-07-05.html, viewed 4 April 2013.

'Ecological footprint' 2013, *Wikipedia*, wiki article, http://en.wikipedia.org/wiki/Ecological_footprint, viewed 4 April 2013.

Edmondson, R. 2007, 'The Industrial Revolution and public health', www.myuniversalfacts.com/2007/06/industrial-revolution-and-public-health.html, viewed 4 April 2013.

'Farmers' market' 2013, *Wikipedia*, wiki article, http://en.wikipedia.org/wiki/Farmers%27_market, viewed 4 April 2013.

Girardet, H. 1999, *Creating Sustainable Cities*, Green Books, Dartington, UK.

—— 2004a, *Cities, People, Planet: Urban development and climate change*, Wiley, London.

—— 2004b, *Creating a sustainable Adelaide*, www.thinkers.sa.gov.au/lib/pdf/Girardet_Report.pdf, viewed 4 April 2013.

—— 2008, *Cities, People, Planet: Urban development and climate change,* 2nd edn, Wiley, London.

—— 2012, 'Regenerative Adelaide', *Solutions*, vol. 5, no. 3, pp. 46–54.

Girardet, H. and Mendonca, M. 2009, *A Renewable World*, Green Books, Dartington, UK.

Global Development Research Center 2013, *Urban and Ecological Footprints*, www.gdrc.org/uem/footprints/, viewed 5 April 2013.

Gordon, B. 2008, 'China's solar hot water capacity will soon be equivalent to 40 nuclear power plants', *Environmental Graffiti*, www.environmentalgraffiti.com/sciencetech/china-solar-hot-water-capacity-soon-to-be-equivalent-to-40-nuclear-plants/822, viewed 5 April 2013.

GreenandSave 2011, 'European Union to set high efficiency standards for buildings', www.ledsavingsolutions.com/international-trends/european-union-set-high-efficiency-standards-buildings, viewed 5 April 2013.

Haberl, H., Erb, K-H. and Krausmann, F. 2010, 'Global human appropriation of net primary production (HANPP)', *Encyclopedia of Earth*, www.eoearth.org/article/Global_human_appropriation_of_net_primary_production_(HANPP), viewed 4 April 2013.

Hagemann, I. 2007, 'Solarsiedlung am Schlierberg, Freiburg (Breisgau), Germany', Fraunhofer-Institut für Solare Energiesysteme, www.pvupscale.org/IMG/pdf/Schlierberg.pdf, viewed 5 April 2013.

Hattam, J. 2012, 'Planting trees in the Mongolian desert to fight dangerous dust storms in Seoul', *Treehugger*, 9 March, www.treehugger.com/natural-sciences/planting-trees-mongolian-desert-fight-dangerous-dust-storms-seoul.html, viewed 5 April 2013.

International Union for the Conservation of Nature 2013, www.plantapledge.com/, viewed 5 April 2013.

King, F. 1911, *Farmers of Forty Centuries: Or, permanent agriculture in China, Korea, and Japan*, Courier Dover Publications, http://soilandhealth.org/01aglibrary/010122king/ffc.html, viewed 4 April 2013.

Koerner, P. 2009, 'Massive solar powered sundial building now complete in Dezhou China', *Jetson Green*, www.jetsongreen.com/2009/12/dezhou-solar-cities-sundial-building-china.html, viewed 5 April 2013.

Living City Block 2013, *Living City Block*, www.livingcityblock.org/, viewed 5 April 2013.

Lott, M. 2011, '10 calories in, 1 calorie out: the energy we spend on food', *Scientific American*, web log post, 11 August, http://blogs.scientificamerican.com/plugged-in/2011/08/11/10-calories-in-1-calorie-out-the-energy-we-spend-on-food/, viewed 5 April 2013.

McMillan, T. 2008, 'City farmers' crops go from vacant lot to market', *New York Times*, 7 May, www.nytimes.com/2008/05/07/dining/07urban.html?ex=1210824000&en=9d6a23b0418d45a4&ei=5070&emc=eta1&_r=0, viewed 5 April 2013.

Mubadala 2012, 'Sustainability in Masdar City', www.masdar.ae/en/#city/detail/sustainability-in-masdar-city, viewed 5 April 2013.

New York State Department of Environmental Conservation 2013, 'Facts about the NYC Watershed', www.dec.ny.gov/lands/58524.html, viewed 5 April 2013.

Power-Technology n.d., *Power Technology Projects*, Kable, www.power-technology.com/projects/Seville-Solar-Tower, viewed 1 April 2013.

Pratt, S. 2012, 'Carbon and the city: tracking emissions from megacities', *Earth Magazine*, 22 May, www.earthmagazine.org/article/carbon-and-city-tracking-emissions-megacities, viewed 4 April 2013.

Rio 12 2012, 'World Climate and Energy Event', *Rio 12*, www.rio12.com/, viewed 5 April 2013.

Science Daily 2008, *Global Carbon Emissions Speed Up, Beyond IPCC Projections*, www.sciencedaily.com/releases/2008/09/080925072440.htm, viewed 5 April 2013.

Solarsiedlung 2013, 'Solarsiedlung', *Solarsiedlung*, www.solarsiedlung.de/, viewed 5 April 2013.

Sukhdev, P. 2010, 'Putting a value on nature could set scene for true green economy', *Guardian*, 10 February, www.guardian.co.uk/commentisfree/cif-green/2010/feb/10/pavan-sukhdev-natures-economic-model, viewed 5 April 2013.

United Nations (UN) 2012, *UN Climate Change Conference COP18*, www.cop18.qa/, viewed 4 April 2013.

United Nations Department of Economic and Social Affairs 2012, *World Urbanisation Prospects: The 2011 Revision Highlights*, http://esa.un.org/unup/pdf/WUP2011_Highlights.pdf, viewed 4 April 2013.

United Nations Environment Programme (UNEP) 2008, *Plant for the Planet: The Billion Tree Campaign*, www.unep.org/pdf/BTC_Booklet.pdf, viewed 4 April 2013.

—— (UNEP) 2012, 'UNEP Launches Global Initiative for Resource Efficient Cities', 18 June, www.unep.org/newscentre/default.aspx?DocumentID=2688&ArticleID=9179, viewed 4 April 2013.

'Urban renewal' 2013, *Wikipedia*, wiki article, http://en.wikipedia.org/wiki/Urban_renewal, viewed 4 April 2013.

Von Thünen, J.H. 1966, *The Isolated State; An English edition of Der isolierte Staat*, trans. C Watenberg, edited with an introduction by Peter Hall, Pergamon Press, Oxford.

Whitford, D. 2009, 'Can farming save Detriot?', *CNN Money*, 29 December, http://money.cnn.com/2009/12/29/news/economy/farming_detroit.fortune/, viewed 4 April 2013.

World Bank 2012, 'Climate Change Report warns of dramatically warmer world this century', http://climatechange.worldbank.org/content/climate-change-report-warns-dramatically-warmer-world-century, viewed 4 April 2013.

World Wildlife Fund (WWF) 2012, 'Rising consumption, increased resource use by a growing population puts unbearable pressure', wwf.panda.org/wwf_news/?204732, viewed 4 April 2013.

Chapter 2

Much better than climate change adaptation

Richard Register

Summary

Is it a naive idea to think that cities could actually contribute to nature's biodiversity and help build soils, tap into the renewable energy flows of sun, wind and water, and recycle other resources essentially forever? Or maybe the notion is the foundation for the only approach that makes any sense. Cities are the largest things human beings build, and cities may have the power to prevent further crises. But to get to such a strategy, of thriving, not just maintaining, in a deteriorating situation, we have to start our thinking there, not in reaction to anything. Better to create something actively good in the first place rather than respond to problems that urban design has created. Looking at cities in this frame of mind is crucial because ecologically informed urban design coincides with, and is in no way contrary to, the best solutions brought forward by an attempt at adaptation. Actions for building cities that solve the climate change problem at the root cause together with actions at the level of adaptation to what residuals of the problem remain open us up to the full spectrum of tactics for a strategy of thriving into the future.

For many years now it's been somewhat depressing to see all the focus and words of intense concern over climate change and hear almost nothing about urban form and design in the debate. As I often say, cities are the largest things human beings create – how did they miss that? A few of the climate-concerned mention something cursory about cities but not much, and it certainly doesn't roll off the tongue as part of the solution like, 'Hey, let's amp up solar and wind!' At least on the supply side we have that. But on the demand side, designing cities, towns and villages that require very little energy to operate – and eco-cities are the poster child for that – is the largest part of the solution there is. And so when the *New Yorker* published the usual article in early January 2013 (Klinenberg 2013) on adapting to climate change, I went ballistic all over again. Can't anyone get to grips with trying to prevent the problem?

My contention is that there are a few important essays that change people's perspectives by encapsulating a crucial issue. One was William James' 'The Moral Equivalent of War' (1906). Another was Garrett Hardin's 'The Tragedy of the Commons' (1968). There needs to be one on getting climate change strategy right and the importance of ecological city design. What follows is my best attempt at that topic.

On 27 October 2009 I got off Vancouver's SkyTrain at the Lake City Way Station. Before me was a sign I'd seen several times flashing by through the window of my moving train. It turned out to be part of a campaign by the Vancouver Aquarium. Featuring an image of a polar bear floating on a slab of ice in an almost open ocean, the copy read: 'Adaptation is not an option. Canada's Arctic. In the grip of change. Visit the new exhibit to understand the impact.' The polar bear, on all fours, gazing at the horizon, looked like any you'd expect on a poster, calendar or in a coffee table book . . . except that it had the black stripes on white of an African zebra. Obviously telling it to adapt was writing it off to extinction. Some changes are exterminating and that's all there is to it. Working to promote 'adaptation' may not be such a good idea. It's something we should contemplate.

Klinenberg's (2013) article for *The New Yorker*, 'Adaptation: How Can Cities Be "Climate Proofed"?', took the usual approach to solving climate problems and rising sea levels. But why start our thinking with adapting to something that disastrous when we could focus our thinking and strategies on preventing it? Our efforts at adaptation may be just distraction enough to lose us the opportunity to prevent far worse disasters, and maybe even prevent us from figuring out how to reverse the heating trend.

Could we build our cities and towns not only to prevent the problem, but to build the absolute best for human purposes as well? Is it a naive idea to think that cities could actually contribute to nature's biodiversity and help build soils, tap into the renewable energy flows of sun, wind and water, and recycle other resources essentially forever? Or maybe the notion is the foundation for the only approach that makes any sense.

But to get to a strategy of thriving (not just maintaining in a deteriorating situation) we have to start our thinking there, not in reaction to anything, including rising seas. Call it the creative instead of 'resilient' frame of mind. 'Resilient' is OK, but it is still reactive, if in a good way. Better to create something actively good in the first place rather than build dykes around what's left over. We need cities good enough that they don't bring on a suite of problems that require resilience simply to cope. Looking at cities in this frame of mind is crucial because ecologically informed city design coincides with, and is in no way contrary to, the best solutions brought forward by an attempt at adaptation. Together, actions for building cities that solve the climate change problem at the level of the causes of the problem and actions at the level of adaptation to what residuals of the problem remain opens us up to the full spectrum of tactics for a strategy of thriving into the future.

To defend cities from storms and rising seas, there are quite obviously dykes, dams and barrages (moveable walls like gates in the locks of canals but very much larger). They are designed to prevent high tides and storm surges from inundating estuaries and bays. As with the Thames Barrage that opened in 1984, these structures link land masses across reasonably narrow river banks or straits to the sea and are offered as a definitive solution in the adaptation strategy —for billions of dollars each. They have nothing to do with prevention of the problem of sea level rise and human-amplified storms in the first place. Building on high ground, it becomes evident, is about the only cheap, one might say *unsubsidized* solution, but it is too late for many of the existing low-lying coastal cities.

Or is it? Why isn't the subject of prevention, why isn't the approach of 'whole-systems' analysis, why isn't looking into the deep future and deep past, and why isn't the guidance of ecological principles the first priority when trying to solve the climate change and sea level rise problems? We were warned about the impending situation more than 30 years ago by the likes of New Yorker James Hansen, Chief of NASA's Goddard Institute for Space Studies. New York itself runs on about 60 per cent of the energy per person of the average US city – we can learn something from that compact transit-oriented city for starters. Why aren't these things talked about as a first priority instead of being barely heard of? Why is adaptation with dykes the main, the much talked-about strategy – while hoping for the best of what's coming over the horizon to assault us?

Let's start as close to the beginning of this story of rising waters as makes sense. After all, some basic problems were solved a long time ago, long before the information glut became overwhelming and distracting, and expertise was compromised by the quest to get the next job, entrenched connections and stiff habits of thinking. When it comes to flooding, why not ask the Sumerians, who built the world's first settlements that were large and complex enough to be called cities 4,500 years ago, in the valley of the Tigris and Euphrates rivers?

They built on elevated, artificial mounds of earth. Their cities were simply built to rise up over the floods. They adapted gradually – they lived there for many centuries – augmenting the mounds on which they worked, walked, presumably had some good times, reproduced, brought up many generations . . . as the cities rose from the flatlands, largely on worn-out building materials but, in many cases, on purposefully elevated earth. They didn't just raise their houses higher. They raised the whole city higher, streets and all. When the floods came, it must have been an exciting experience to see what looked like a whole ocean sliding by, your city like a large ship moving upstream in a vast current of water.

But what is most important is that the compact city – the pedestrian city of streets for people, of almost everything within walking distance, the city of high population per acre, of higher economic flow-through per square foot – fits beautifully with the elevated artificial mounds solution to floods. That's the solution that simply lets the populace and their property float along high above the deluge: nobody hurt, very inexpensive, lower insurance rates, no crying over drowned children and old folks. The car-dependent suburbs? No way to elevate that kind of thinly developed enormous acreage. Moving earth, over a fairly long period of time, basket by basket, even before the wheel, carts and draft animals for traction and transport existed, the people in many parts of the world were clustered in enough density that they could elevate their towns. But to raise suburbia? Far too massive a project and with counter-productive results: the car/sprawl/paving/cheap energy infrastructure is the main engine of climate change and needs to be left behind. The city designed for cars, not people, is the major cause of our current problem in the first place.

Mesopotamia isn't a rare case lost in ancient history either. It responded to an eternal truth, if a seemingly simplistic one: in a flood, get to high ground. *If there is no high ground, build it.* Now why in these days don't more people think of that? With bulldozers on our side, it's several hundred times easier than it was when civilizations were busily cranking up.

Rising above the flood is not 'simplistic', it's basic. 'Basic' is a notion that comes after one does something called sensible proportionalizing: considering the big and important factors, then prioritizing to deal with them first. This is not the 'low-hanging fruit' approach that procrastinates on the difficult work until too late. The important factors are usually the fundamental physics, chemistry and biology of our world. The case in point: getting to higher ground, natural or artificial, in flood country is a good, basic idea. Get that far in your thinking and a whole complex of good ideas flows forth.

I know something about all this from personal experience. Ecocity Builders, the organization I work with, prides itself on 'basic thinking' and promotes what we call ecological city, or 'ecocity' design and planning. The term is broadly used in China and Korea, if not too common in the United States as yet. In any case, interest is steadily increasing in the ecocity approach, netting me many invitations for talks around the world. After Hurricane Katrina, one such invitation took me to the city of New Orleans, four months after the storms. When I faced the true shock of what lay before me, the Sumerians came to mind. Of course: rebuild as the ecocity on elevated fill. That lesson from the ancient cities makes sense into the deep future. Why not a compact new development covering very little land, with apartments and commerce similar to the French Quarter, which is built on a natural levee well above sea level, but in the replacement development areas built on elevated mounds of earth? They already love their streetcars in New Orleans and it is an exceptionally flat city perfect for bicycles, so the sustainable transport system to fit compact ecocity design was almost ready to go. Density, the lesson that gives Manhattan considerable liberation from the automobile, would be the key to making a new ecological development pattern practical in New Orleans. Nor would you need Manhattan density to gain most of the advantages of living, working, exchanging and re-creating in an area where things are close together, as many European small cities and city cores demonstrate. Twenty-five or 30 feet above sea level should last a mid-range estimated sea level rise for 100 or 200 years, if we can make serious progress in avoiding further global heating as opposed to trying to adapt to it.

But this is not the way reconstruction went in New Orleans. So many people sympathized with the desires of displaced residents to try to get back to normal that governments simply paid for and built stronger, taller dykes. Now their car-dependent, low-density way of living below sea level will be attempted all over again. It was a sad mistake that ignores the fact that the elevated mounds approach works worldwide. If rebuilt with ecocity design teamed up with elevated fill – in compact centres with the functional diversity of living, working and enjoying life all close together on artificially raised earth – New Orleans would have taken the lead in solutions not only for cities prone to flooding but for all cities by demonstrating the power of ecological principles in urban design. In the aftermath of Superstorm Sandy, could New York and New Jersey take that lead?

Building on mounds worked for the American Indians along New Orleans' very own Mississippi River, and upstream as far as 2,000 miles by water, as well as along the Ohio River. There were over 10,000 mounds scattered over the central and southeastern area of what is now the United States. They were constructed as early as 5,400 years ago – before the Sumerians – as reported in Charles Mann's book *1491: New Revelations of the Americas before Columbus* (2005). The earliest is now called the Ouachita Ring, a circle of mounds up to two storeys high in Northeastern Louisiana. Around 1000 CE, Cahokia (close to today's St Louis) had about 15,000 inhabitants and was the largest city north of the Rio Grande until Boston and New York reached that population. It was centred on and partially built as a town and partially as a ceremonial centre. Although the people there suffered through many floods, Mann (2005 p. 351) reported, 'There is little indication that Cahokia floods killed anyone.' Of course not. The people would simply hunker down up on their mounds. If they were farming the bottomlands when the floods came, they would simply take a short walk up the mounds. The mounds were their vertical bomb shelters from the disasters of flood, not war, though the elevated prospect would undoubtedly help in defence against unkindly people as well as hostile waters.

Zoom forward in time from ancient native North and South American examples to the worst natural disaster in the United States' history. After the hurricane of 1900 that killed an estimated 6,000 to 9,000 people in Galveston, Texas – nobody knows the true figure, since the city was growing so fast and had so many transients – most of the city was rebuilt on a raised platform averaging about 20 feet higher and made of pumped sand and mud from the bottom of the Gulf of Mexico and Galveston Bay. The constructed mound was edged on the Gulf side by a concrete sea-wall to break the huge waves. Some farm villages in Bangladesh have survived the floods on the delta of the Ganges and Brahmaputra on elevated fill. Then there is Venice, if built on wooden pilings rather than earthen fill. The idea is not historically unusual, but is probably forgotten now because it requires longer-range planning than we are used to in our distracted, hurry-up times. But it works. It's basic.

A key concept here is proportionality. For example, cars are on average about 25 or 30 times as heavy as a human being, travel about ten times as fast in normal operation and consume about 60 times the volume of space when standing still, much more when moving. Designing cities to accommodate both sets of criteria – the 'needs' of cars and of people – at the same time and in the same space is a hopeless task if one is hoping for climate change solutions, or even just hoping to create a good environment for people. The attempt, dominant in 'developed and developing' societies since the Second World War (that is, in the era of extremely cheap fuel energy, produced sprawl development, paved farms and natural landscapes), drove species to extinction – probably the polar bear eventually – consumed massive amounts of energy for transportation . . . and still does. That way of building cities also wastes heat and cooling energy because single-family houses, unlike apartment units with common walls and shared floors and ceilings, lose climate-moderating energy with just one use.

Then there are the full costs of car ownership, not just the money but the time it takes to earn the money plus the time in and repairing and pampering the car itself. If you divide the time into the distance travelled, said Ivan Illich in *Energy and Equity* (1974), you are hurtling along at less than five miles per hour. With bigger traffic jams since then, it has only got worse. Then there is the associated death rate on the transportation system, private cars having several times the fatalities and injuries of transit modes and aeroplanes per vehicle mile travelled. Bicycles are much safer – unless they are hit by cars. Cities for pedestrians are much safer than cities designed around the automobile.

The low-density city, the one that cannot be elevated to avoid floods, is a gigantic infrastructure compared to even, say, Manhattan, when considered in relation to population served. I first heard this notion from architect/philosopher Paolo Soleri when he was visiting Los Angeles in 1965. He was the earliest and then essentially the only strong advocate for ecological city design. At one of his public talks, he said that if you look around the thinly scattered city, it appears to be made mainly of small buildings – single-family homes, one-storey shopping areas, small office buildings, franchise restaurants, gas stations, parking lots (there weren't many multi-storey parking structures then). But in reality that infrastructure was far larger in every way than the compact city mainly accessible by foot, such as the old city cores of many cities in Europe built before the influence of cars, or even streetcars. What we were seeing in Los Angeles was a case of gigantism, a term Soleri used: an enormous expanse of asphalt and concrete; redundant, unshared walls serving only one family or one individual; thousands of miles of wire and pipes for electricity, telephone connections, gas, water, sewerage; enormous acreage in sentimental lawns producing no food and eliminating natural habitat. The compact city of larger buildings in mixed-use centres with towering skylines was paradoxically much smaller in materials, energy, money and time to build and keep going. Why? Shared infrastructure and shortened linkages of all sorts. Soleri had a very important point.

But the Angelenos paid no heed. When they looked up and noticed murky grey skies and beheld nights without stars from smog and glare, when they gasped for air with burning throats and cheeks streaked with tears, they decided to solve their air pollution problem by cleaning up the car. They fixed the car by affixing a smog device, stripping out about 90 per cent of the air pollution and proving the car could be your friend. LA became the city of the future, leading the car-dependent cities that followed and increasingly dominate the world. They fixed their local air pollution problem with the wrong kind of leadership – responding in a piecemeal and adaptive way instead of using whole-systems thinking.

The lesson here – closely related to getting proportionality right, then prioritizing for the necessary work – is that we need to recognize that we are dealing with whole systems. We

Angelenos fixed the car, but not the city. The city is a whole system, made up of its buildings, open spaces, networks of transport with their vehicles and supply of material, energy and information, recycling and waste disposal; all systems integrally relate to one another, like organs in a complex living organism. Treating the car as if it were an independent item, without understanding its connections to the urban system, was a disastrous mistake. The system of transport is intrinsically linked to the problems or solutions implied by the design of the whole.

Comparing cities, towns and villages to living organisms, which I call the anatomy analogy, is very helpful. Living organisms of any complexity are three-dimensional. That's largely because, in living organisms, the compact form makes possible short links between physical and electro-chemical connections, for the movement of nutrients and energy, fluids and information, within the body. The sprawling, 'flat' city battles against that healthy logic of short distances by design. The distances to be covered in the flat city and the expense of infrastructure and vehicles and fuel to cover them can be truly mind-boggling.

A key principle is 'access by proximity'. The shortest distance between two points is a straight line, we hear. Not quite. It's designing those points close together. This applies to biological organisms' internal systems and to cities. Compact structure allows efficient, low-energy connections throughout the organism, whether it is a product of biological evolution or the result of human construction.

Why are we not proportionalizing in the way I'm suggesting here? My guess is that it takes extra commitment and energy, and some risks, and it raises the spectre that we might need to change some things substantially. We'd rather pick the aforementioned low-hanging fruit. But it is time to leave the low-hanging fruit for the children.

With the growing universal agreement that climate change may be the largest problem world society needs to deal with, with implications on the scale of evolution's trajectory, how *could* the solutions be easy? No one won the day with the battle cry, 'Relax! Do something easy first . . .'. I think one of the reasons we have settled for perseverating on the adaptation issues is that those most concerned have not insisted, as I am doing here, on always looking for the effective long-range solutions and not just the easy or short-term changes. Maybe we have to start thinking about designing cities without *my* car. Maybe we should entertain the idea of changing *my* neighbourhood. Maybe even to rethink *my* career choices as to whether they fit with the much more pedestrian city, daring to imagine a very different kind of city into the future. Then there is the way *I* vote, make purchases and even talk in everyday conversations with my friends.

That then becomes an intriguing challenge: if we start thinking in really basic terms, we may realize that the city that is best for adaptation adapts least – because in the not too distant future, it doesn't have to. Mild adaptations while giving first priority to reshaping the city in major ways for humans and other animals alike is radically preferable to stupendous works of defence against 'nature' and ignoring the deeper causes, principles and likely futures. Jaime Lerner, the mayor who launched Curitiba, Brazil into world leadership in ecologically healthy city design, reminds us that great changes in any city can happen in as few as two years.

We can begin looking at this contrast between fundamental change and trying to defend what we have now – adaptation – by thinking through some of the recommendations Klinenberg (2013) canvassed. Klinenberg interviewed geophysicist Klaus Jacob, author of a 2009 study of storm surge risks for New York City whose predictions mostly came true when Superstorm Sandy hit in December 2012. Jacob produced a map to inform the *New Yorker*'s feature article, with blue zones of increasing development in New York City and pink zones of decreasing population and development activity. The blue zones of increase happened to be in the areas on the sea level fringe with water where techniques of new building are more easily exercised than the techniques of convincing people in their already built-up neighbourhoods to accept new development. The neighbourhoods generally safe from flooding are, simply, on higher ground.

What will actually work? As usual, everyone turns eyes to the Netherlands. The city of Rotterdam has a 'Climate-Proof Programme' that Klinenberg (2013) thought was pretty good. Rotterdam has built a floating pavilion in the city centre supported by three enormous floating silver half-spheres complete with tennis courts, exhibition space and water playground. These look to me like specialized expensive public works not in the nature of 'floating the city'. Are they good solutions for housing or offices, manufacturing facilities, warehousing and shopping? I would guess in a very wealthy society that the people might build like that, floating on the rising seas, but building on water, especially corrosive salt water, is intrinsically more expensive than building on land. Where poorer people live over water, it is generally because of a lack of other options on crowded land, or land being too far from their employment or that they make their living right there: fishing, trading, raising their children *on the water*. The situation – floating on a moving medium – requires near constant maintenance. Stressing wind, corrosion, biological agents in the water, waves and heaving swells and moist conditions twist, turn and gnaw away at the structure with tiny barnacle teeth, which is why boats and ships require more maintenance than buildings that provide housing and work spaces on land.

Smart design (information technology to help coordinate physical actions) helps, but that too implies greater expense in linking and manipulating the infrastructure. An electronically coordinated Dutch electric power grid is 'circular' instead of the radial sort that branches out from one central power source. The idea is that power can come in from several directions depending on availability in case some power sources go down or lines are cut, such as when wires are blown down with their poles by hurricanes. This 'circular' solution requires redundant wiring, more lines coming in from more directions and more power company information infrastructure, though in some cases it can save on some physical infrastructure. Certainly this is not that difficult, but it is more expensive in both construction and maintenance.

Another solution: place the electrical equipment above the highest potential flood level. Only reasonable. Or bury wires in waterproof pipes. A good practice but more expensive than wiring on poles. Another, this one from Singapore: raise the entrances of your subway system up out of the ground like projecting tubes – walk up a few extra steps to their thresholds, then back down to the level of the street and on down into the system. When the waters wash down the street, they can't make it into the underground but you can. Again doable, an adaptation, more expensive, but beginning to sound a little bizarre.

What is interesting here is that all of these adaptations, like the improvement of the car in Los Angeles, do nothing substantial to reduce the cause of the problem in the first place. But if we were to build the compact ecocity for people, including on artificially elevated land, we would be changing the major cause of the climate change problem and vastly improving our prospects while also adapting to what's left of the problem after good moves for prevention.

The best of all possible cities? Is it too idealistic to strive for that? Is the attempt impractical? Or is it just that we haven't really tried yet and don't know how possible it really is? Certainly humanity has more knowledge and power than ever before. It is not too difficult to see that by recycling urban organic waste into the soil, from kitchen and garden waste to human waste and clippings from public parks and dung from the zoo, by way of best composting and including such systems in the agricultural landscape that feed the people, there is no reason why we can't build soils. We can sustain biodiversity and even enhance it in and near urban areas – and do so in many places already. When I was in Central Park a few years ago, I was walking over one of those high, arching foot-bridges and a hawk flew by no more than 15 feet from me. In a city! We can intentionally create environments to bring back vital biodiversity.

The massive general impact of cities is presently very destructive to nature but there is a thin scattering of ecocity features that everybody senses vaguely but meaningfully is the 'green' and 'sustainable' city of the future already everywhere around us. There are roof-top gardens, solar passive designed buildings and multi-storey solar greenhouses that aren't really recognized as such as yet but exist as atriums, foyers and gallerias. There are bridges between buildings in higher-density areas, including spectacular features like the High Line, bicycle-only streets and car-free areas. There are bountiful community gardens and thriving restoration projects for bringing local species back from the brink in adjacent mangroves and reedy wetlands. There are glass exterior elevators from San Francisco, California to Ningbo, China that make vertical transport something of an adventure. There are moveable awnings and solid sheltered walkways for the pedestrian in rain, snow and blasting sun. There are excellent public transport systems that could get even better, and even the stop-gap sharing of our pedestrian city's old arch enemy, the car itself. Various rental companies provide 'wheels when you want them', as the catch phrase for Zip Car goes. These can help commuters eventually break free from the car habit altogether as their city and town centres become more dense and mixed in uses.

So the issue is not that the ecocity cannot exist; it already does but only in small bits and pieces presently overwhelmed by the infrastructure of the sprawling city for cars. Those bits and pieces need to proliferate vastly in number and come together in good designs: basic principles everywhere the same, local conditions everywhere different, the universal and the unique united.

But the seas are in fact rising. Unlike circumstances when entire neighbourhoods are destroyed by water as in much of low-lying New Orleans during Hurricane Katrina and by fire in the Breezy Point neighbourhood of Queens during Superstorm Sandy, parts of cities that are going to be inundated if dykes are not built cannot be easily removed. We cannot easily and economically remove big, useful buildings that remain anchored firmly to the ground, certainly not in the short term. Should we fill their basements with concrete and move the first floor up to the third floor? How to elevate the street too and do the same thing with other buildings in the area? Perhaps we could create such an area a block wide that becomes a ring around the rest of the city that remains at the old street level, in the future well below sea level. This is in fact the solution implemented in Old Sacramento, California where the downtown area was frequently flooded by the Sacramento River during springtime mountain snow melt and late spring rain run-off.

And, if we do seriously think through our options of 'working with nature', New York comes to the rescue. In 1961, Gerald Lloyd wrote a paper for the Urban Land Institute calling for a means to transfer development rights from one place to another so that much more flexibility in city design and development could be planned, approved by the authorities and built. Cities could then be changed more rapidly for any number of good reasons: preservation of a historic landmark, pollution avoidance, building a better cluster of buildings, opening up a buried waterway, repairing past mistakes or building a piece of the best we can imagine. Lloyd (1961) called the notion 'transferable density'. 'In the midst of this scholarly activity, New York City adopted a new Landmark Preservation Law in 1968', wrote Rick Pruetz (1997), past Director of Planning for Burbank, California in his book on transfer of development rights

(TDR). The railroad company wanted to build a 55-storey office tower on top of Grand Central Terminal. Instead, TDR was created as a legal tool to allow the shifting of rights to develop from one location to another. The innovation was upheld by the US Supreme Court in 1978, the terminal building was preserved and TDR became a zoning tool not only in cities to save landmarks but also for saving natural areas and agricultural land all over the United States.

The kind of flexibility provided by tools like TDR, along with supportive zoning laws and other policies, community fundraising for major projects, appropriate taxes and the like, could make possible the gradual withdrawal from areas known to be soon flooded and a transfer of development facilitated to higher ground, some of which might be created as modern-day versions of the Indian mounds that once dappled much of the US central and southeastern territories and raised cities in ancient Mesopotamia.

But here is the most important point: the fact is that the much-talked-about adaptive strategy does very little to reduce impacts leading to climate change relative to the scale of the problem. Planting trees by the millions actually does – by absorbing carbon out of the atmosphere. Grasslands regenerated with the best grazing techniques can turn millions of acres of degraded soils into deep reservoirs of dead organic carbon matter by the gigatons stripped from the atmosphere. Rancher Alan Savory has demonstrated what can be done in the grazing lands of Zimbabwe. To accomplish this, he has reproduced the clustering effects of large numbers of cattle modelled on the defensive bunching of zebras and wildebeest into tight herds that act like giant self-replicating, solar-powered, agricultural fertilizing and seed-planting machines: mixing their fertilizer, earth and seeds with their hooves. Then there are the peat lands (if we could save most of them from cutting, or artificially create more of them) that build up carbon almost magically without the need of soil, sphagnum mosses just growing ever thicker on top of dying and dead moss underneath, laying down ever thicker and thicker layers of stored carbon below.

Solar and wind energy to replace fossil fuels helps greatly on the supply side. Building cities that consume radically less energy in the first place pairs beautifully with such renewable energy sources on the demand side, dramatically reducing demand for land and materials as well as energy. Do all of the above and we are well on our way to reversing global heating and obviating the more extreme mitigations and massive walls to keep out the seas that everyone is beginning to talk about as the big guns of adaptation. But even that is not enough if population and consumption, one or both, continue to grow and put ever more pressure on natural systems, especially the climate system and the biosphere. And even that will probably not be enough unless we move to a form of agriculture that builds soils organically, recruiting cities for much of the fertilizer – in a similar way to what Alan Savory does when he recruits cattle to build the soil in Zimbabwe. The evolving ecocity, in shrinking back from the land that car cities have paved over with concrete, asphalt and inedible lawns, helps in this way too; millions of new acres of land close to the city can be liberated for both farming and the return and flourishing of native species. And even that may not be enough if we don't phase out war and expensive weapons systems and phase in much more serious peace efforts.

Thus the lesson of whole-systems thinking, using the anatomy analogy and ecological thinking, is not just to see cities as analogous to living organisms, or to understand their role in the whole system of their ecology and cycles of resource use and recycling, but also to begin imagining how all those large components come together in a plan resembling Lester Brown's 'Plan B' ideas (Brown 2009). We need to understand what's needed, learn about limits and literally build a better future. We need the new ecologically healthy city built for people, a far better agricultural and dietary system and a strategy to gain control of our vast

Figure 2.1 Flooding of New Orleans by Hurricane Katrina in 2005.
Source: Image supplied by the author.

numbers. We will need to invest generously in three major areas: ecocities, organic agriculture and population moderation. Another 'whole system' we need to think about, and develop a strategy for, is a commitment to generously sharing the planet with all the people and species that have made life on Earth such a rich and varied phenomenon. Could this be the one thing powerful enough to constitute William James' (1906) notion of the much-needed 'moral equivalent of war' which he wrote about in one of the most influential essays of the early twentieth century?

I believe so – and now we may understand better why adaptation to climate change in our cities is a very limited idea. The strategic system needs to be prioritized: first, work for the best for all; second, for preventing the disaster – then third, mop up as we must with adaptations to our failures of the first two.

References

Brown, L. 2009, *Plan B 4.0: Mobilizing to save civilization*, W.W. Norton, New York.

Hardin, G. 1968, 'The tragedy of the commons', *Science*, vol. 162, no. 3859, pp.1243–1248, www.cs.wright.edu/~swang/cs409/Hardin.pdf, viewed 22 March 2013.

Illich, I. 1974, *Energy and Equity*, Calder & Boyars, London.

James, W. 1906, 'The moral equivalent of war', www.constitution.org/wj/meow.h, viewed 22 March 2013.

Klinenberg, E. 2013, 'Adaptation: how can cities be "climate-proofed"?', *The New Yorker*, 7 January, www.newyorker.com/reporting/2013/01/07/130107fa_fact_klinenberg, viewed 22 March 2013.

Lloyd, G. 1961, 'Transferable density in connection with density zoning', *Urban Land Institute Technical Bulletin*, no. 40, pp. 136–137.

Mann, C. 2005, *1491: New revelations of the Americas before Columbus*, Alfred A. Knopf, New York.

Pruetz, R. 1997, *Saved by Development: Preserving environmental areas, farmland and historic landmarks with transfer of development rights*, Arje Press, Hermosa Beach, CA.

Chapter 3

Imagining biophilic cities

Tim Beatley

Summary

Nature provides many emotional, spiritual and health benefits to residents of cities. *Biophilia* argues that we have co-evolved with nature, and that we have a deep need to affiliate with the natural world. The human species has 'grown up with nature', as E.O. Wilson says, and little wonder that we are happier, more productive, more creative and even more generous in the presence of nature. Nature in cities offers the promise of lives that are wondrous and connected, lives attentive to the natural magic around us. Ensuring daily and hourly contact with nature in dense, growing cities becomes a challenge, but there is often remarkable urban nature and biodiversity nearby, and there are many creative ways in which to integrate nature into the urban fabric. Here, Beatley describes the new planning and design agenda of *biophilic urbanism* and the need to shift towards a vision of *biophilic cities*. He identifies some of the key dimensions of biophilic cities, as well as some of the main strategies and innovative practices of emerging biophilic cities around the world.

Introduction

That we need daily contact with nature to be healthy, productive individuals, and indeed have co-evolved with nature, is a critical insight of Harvard myrmecologist and conservationist E.O. Wilson. Wilson popularized the term *biophilia* two decades ago to describe the extent to which humans need connection with nature and other forms of life (see Figure 3.1). More specifically, Wilson (1993, p. 32) describes it this way: 'Biophilia . . . is the innately emotional affiliation of human beings to other living organisms. Innate means hereditary and hence part of ultimate human nature.'

To Wilson, biophilia is really a 'complex of learning rules' developed over thousands of years of evolution and human–environment interaction:

> For more than 99 per cent of human history people have lived in hunter-gatherer bands totally and intimately involved with other organisms. During this period of deep history, and still further back, they depended on an exact learned knowledge of crucial aspects of natural history. . . . In short, the brain evolved in a biocentric world, not a machine-regulated world. It would be therefore quite extraordinary to find that all learning rules related to that world have been erased in a few thousand years, even in the tiny minority of peoples who have existed for more than one or two generations in wholly urban environments.
>
> (Wilson 1993, p. 32)

Figure 3.1 According to E.O. Wilson, humans are biophilic creatures; we have an innate emotional affiliation with other organisms.

Source: Image supplied by the author.

Stephen Kellert of Yale University reminds us that this natural inclination to affiliate with nature and the biological world constitutes a 'weak genetic tendency whose full and functional development depends on sufficient experience, learning, and cultural support' (Kellert 2006, p. 4). Biophilic sensibilities can atrophy, and society plays an important role in recognizing and nurturing them.

The *nature* of cities

While we are already designing biophilic *buildings* and the immediate spaces around them, we must increasingly imagine biophilic *cities*, and should support a new kind of biophilic *urbanism*. As the planet barrels rapidly down the path of urbanization, the need for green and *nature-ful* cities is an ever more urgent need.

There is already much nature in cities, of course, more than we realize. It is both large and small, visible and hidden. It is intricate, yet sweeping. It is amazing in its biological functioning, ever present yet highly dynamic, and vastly underappreciated for its urban ubiquity. In understanding the nature of cities it is necessary to think beyond our usual approach to visualizing or imagining space and place, and to understand that nature is everywhere in cities if we look; it is above us, flying or floating by, it is below our feet in cracks in footpaths, or in the diverse micro-organic life of soil and leaf litter. Nature reaches our senses, well beyond sight, in the sounds, smells, textures and feelings of wind and sun. Understanding the natural history of a city helps us to see cities as ever-changing, ever-evolving palettes of life.

In the higher reaches of our cities, the roof-tops and façades also harbour nature, sometimes by design, and sometimes by accident and natural volunteerism. New forms of nature are being created in cities all over the world in the form of ecological roof-tops and roof-top gardens, hosting grasses and sedum, and increasingly found (over time and with the right design elements) to harbour great diversity of invertebrates, bird and plant life. We know, for instance, that butterfly species will visit roof-tops on high-rise structures, and food, for humans and nature alike, can be grown there as well.

This nature in cities is the raw ingredient for a new global urban society organized around wonder. Few have made a more compelling and eloquent plea for the importance of wonder in the natural world than Rachel Carson more than half a century ago. In a 1956 essay entitled 'Help Your Child to Wonder', she describes the value and pleasures of exposing her young nephew to the nature found along the Maine coast:

> If I had influence with the good fairy who is supposed to preside over the christening of all children I should ask that her gift to each child in the world be a sense of wonder so indestructible that it would last throughout life, as an unfailing antidote against the boredom and disenchantments of later years, the sterile preoccupation with things that are artificial, the alienation from the sources of our strength.
>
> (Carson 1956, p. 46)

Carson counsels looking at the sky, taking walks and uncovering and experiencing nature, even if (as parents) we are not able ourselves to identify a species or a constellation. It is about cultivating an awareness of the sights, sounds and natural rhythms surrounding us, paying attention to and learning to see the mystery and beauty in everything around us.

We need wonder and awe in our lives, and nature has the potential to amaze us, stimulate us, propel us forward to want to learn more about our world. The qualities of wonder and fascination, the ability to nurture deep personal connection and involvement, visceral engagement in something larger than and outside ourselves, offer a potential for meaning in life few other things can provide.

My landscape architecture colleague Beth Meyer argues that with matters of environment and sustainability, we need also to emphasize the beauty and pleasure and enjoyment we derive from them. We often forget about the aesthetics, or try to reduce them to monetary values. At the end of the day, watching that circling hawk or turkey vulture, walking or bicycling through an urban wood, harvesting and eating produce from one's garden, and listening to the sounds of Kadydids and tree frogs on a humid August evening are deeply pleasurable experiences; they are the building blocks of a life enjoyed (see Figure 3.2). We climb trees as children because this is a fun and enjoyable thing to do, and as adults unfortunately we often forget these pleasures (and of course rarely climb trees!).

In our recent documentary film *The Nature of Cities* (dir. Davis and Beatley 2009), we spent several stimulating days in Austin, Texas filming the 1.5 million Mexican free-tailed bats that have inhabited the underside of the city's Congress Avenue Bridge during the summer months. People line up hours before nightfall to get a good look at the wondrous columns of bats emerging from the bridge. Merlin Tuttle, founder of Bat Conservation International (BCI), dutifully recites the many environmental (and economic) benefits provided to the city by these bats. And they are considerable, including the millions of mosquitoes they eat each day. But ultimately the sight of thousands of bats flying off, in distinct columns that may be seen for several miles, is an immense and beautiful thing. It is the raw beauty of the natural

Figure 3.2 Being able to ride a bicycle around one's city should be a pleasure.

Source: Image supplied by the author.

world, a primordial spectacle unfolding against a backdrop of high-rise buildings and a human-dominated (at least so we think) urban environment.

In many American cities, the biodiversity is aquatic and sometimes offshore, as in Seattle, which has abundant and wondrous life in the not-far depths of the bay and sound. Much of the biodiversity of King County, in which the City of Seattle lies, is found in the 'deep subtidal habitat' of Puget Sound, in some places almost 900 feet below the surface, and including 'over 500 benthic and 50 pelagic invertebrates' (King County 2008, p. 57). And while some are known and recognizable to residents, such as the king crab, many are not. That the Seattle metropolitan region is also home to such unique marine critters as the giant Pacific octopus and giant acorn barnacle suggests a wildness and mystery very close at hand.

New forms of nature can be fostered in the many leftover spaces of the city. A visit to the Green Roofs Research Centre, in Malmo, Sweden, shows the extent of possibilities; here they have planted and monitor hundreds of green roof test plots, testing different plant and soil combinations. Some of these plots are for so-called brown roof-tops – places in urban environments (there are many) where plants may be used to restore and even take up pollutants in highly contaminated and degraded settings (phytoremediation). And the Centre's immense research roof-top also shows the potential of different, sometimes surprising delivery methods; their standard green roof, as Trevor Graham who runs many of the Centre's green city efforts explains, is made from recycled polyurethane car seats, and in several places there are small mounted frames, with sedum growing vertically, showing the potential for a kind of natural artwork suitable for hanging in one's living room!

These new forms of nature are catching on, and are now encouraged and in some places mandated by codes, and we will see more of this happening in every city around the world. And new creative developments in cities – such as Via Verde (the Green Way), a 200-unit complex of affordable housing planned for a 1.5 acre site in the South Bronx of New York – will find many ways to insert and grow nature. In this case, nature takes the form of a connected multi-functional garden 'that begins at street-level as a courtyard and plaza, and spirals upward through a series of programmed, south facing roof gardens that end in a sky terrace' (New York City Department of Housing Preservation and Development 2007). Increasingly, biophilic cities will understand roof-tops, courtyards and façades as places to cultivate nature.

What is a biophilic city?

Exactly what is a biophilic city, and what are its key features and qualities? Perhaps the simplest answer is that it is a city that puts nature first in its design, planning and management; it recognizes the essential need for daily human contact with nature as well as the many environmental and economic values provided by nature and natural systems.

A biophilic city is at its heart a *biodiverse* city, a city full of nature; a place where in the normal course of work and play and life, residents feel, see and experience rich nature – plants, trees, animals. The nature is both large and small – from tree-top lichens, invertebrates, even micro-organisms, to larger, natural features and ecosystems that define a city and give it its character and feel. Biophilic cities cherish what already exists in and near cities (and there is much, as we have already seen), but they also work hard to restore and repair what has been lost or degraded, and to integrate new forms of nature into the design of every new structure or built project. We need contact with nature, and that nature can also take the form of shapes and images, integrated into building designs (see Figure 3.3).

A biophilic city should be judged by the existence of nature and natural features, but also in some way by its biophilic sensibilities or *spirit*; how important is nature and how central is it to the lives and *modus operandi* of the city, its leaders and its populace? Rather harder to quantify, this *biophilic spirit*, or sensibility, suggests a value dimension, the sense that residents and public officials alike recognize the importance and centrality of nature to a rich and sustainable urban life. This quality could easily fit as both an activity and an approach to governance.

Every city will have its natural spectacles – some large, others more nuanced – but a biophilic city is one that pays attention, a city that sees and conveys this sense of beauty and wonder and caring. It may be the running of the steelhead trout in Niagara River, or the appearance of orcas in Prince William Sound, or the migratory return of robins along the east coast of the US. A biophilic city celebrates this wonder and sees in these events the opportunity to connect, to strengthen bonds, to mark the cycles of life and seasonality. This celebration often involves the direct experience of that biodiversity and nature, such as watching migratory birds, or visiting a park or green area, or it may be a more referential form of biophilic expression.

Figure 3.3 A green wall brings nature into the city.

Source: Image supplied by the author.

As Table 3.1 suggests, how actively citizens enjoy the nature around them and actively participate in this nature is also an important measure of a biophilic city. 'Participation' is an interesting word to use here because it implies a level of active engagement beyond just passively observing something; it suggests a keen and active interest in the subject. Citizens of a biophilic city, and their leaders, are not removed from the nature around them, but are highly aware of it and present in its midst. A biophilic city is a city in which a large percentage of its population is actively enjoying nature. This enjoyment and engagement can take many different forms of course, from walking and hiking in natural areas, to bird-watching and plant and tree identification, to organized nature events and activities, from fungi forays to nature festivals.

Biophilic cities help to make it easier to enjoy nature and reflect an understanding that exposure to and enjoyment of nature are key aspects of a pleasurable and meaningful life. There are many potential outlets and venues for our need to connect with nature, and most are also intensely social. Facilitating contact with nature has the potential to help create new friendships and build social networks, in turn helping to make urbanites healthier and happier. In San Diego, the activities of a number of 'friends' of the canyons groups help to conserve and protect the canyons as a neighbourhood and community resource, but also provide opportunities for neighbours to interact and socialize, in a way and to an extent that would otherwise not occur. In Rose Canyon, for instance, residents from different sides of the canyon have places and opportunities to converse and come together, something that would have been difficult without the pull of nearby nature.

Cities must also begin to see the value and importance of facilitating such connections with nature, and perhaps offering help and support similar to in the Australian 'Bushcare' model. Here local groups of citizens and community volunteers organize around a specific urban ecosystem – a patch of green space, a stream, a park – and, with the help of a member of municipal staff (bushcare officer), spend weekends and spare hours cleaning up, repairing and tending these spaces. The result is not only ecological repair, but also making friends and the rebuilding of community, as well as becoming more embedded in place and environment.

Creatively involving citizens in the conducting of science is another way to intimately engage people with the nature around them. In San Diego, citizens have been trained to become 'parabotanists' (like paralegals), helping to collect plant specimens in this highly biodiverse county. There are now 200 citizens serving as parabotanists, working to collect plant data for the San Diego County Plant Atlas Project (begun in 2002). The project records plants on a 3-square-mile grid. Parabotanists are now steered to collecting on grid squares where fewer plant data exist. Once they sign up for a square, they are mailed maps and permits from the San Diego Natural History Museum. A biodiversity 'hotspot' and the most floristically biodiverse county in the US, recording and protecting this biodiversity takes on special importance. The Plant Atlas will eventually result in an 'internet-accessible, data-based plant atlas based upon vouchered specimens'. There are more than 1,500 native species of plants in San Diego County and so there is much to document and record, and citizens here play an important role. Volunteers go through training by the museum, and once trained, collect and press the plants and record data about the plant's location. A museum botanist verifies the plant's identification.

A biophilic city, then, is a city with an extensive and robust social capital, to extend Robert Putnam's (2001) concept. Evidence is compelling that we need extensive friendships and social contact to be healthy and happy, as well as our contact with nature, so finding creative ways to combine these needs becomes an important goal in the biophilic city. I have been

Table 3.1 Some important dimensions and some possible indicators of biophilic cities

Some important dimensions of biophilic cities	*Some possible indicators*
Biophilic conditions and infrastructure	Percentage of population within a few hundred feet or metres of a park or green space
	Percentage of city land area covered by trees or other vegetation
	Number of green design features (e.g. green roof-tops, green walls, rain gardens)
	Extent of natural images, shapes, forms employed in architecture, and seen in the city
	Extent of flora and fauna (e.g. species) found within the city
Biophilic behaviours, patterns, practices, and lifestyles	Average portion of the day spent outside
	Visitation rates for city parks
	Percentage of trips made by walking
	Extent of membership and participation in local nature clubs and organizations
Biophilic attitudes and knowledge	Percentage of residents who express care and concern for nature
	Percentage of residents who can identify common species of flora and fauna
Biophilic institutions and governance	Priority given to nature conservation by local government
	Percentage of municipal budget dedicated to biophilic programmes
	Existence of design and planning regulations that promote biophilic conditions (e.g. mandatory green roof-top requirement, bird-friendly building design guidelines)
	Presence and importance of institutions, from aquaria to natural history museums, that promote education and awareness of nature
	Number/extent of educational programmes in local schools aimed at teaching about nature
	Number of nature organizations and clubs of various sorts in the city, from advocacy to social groups

Source: Beatley (2010), supplied by the author.

calling this *natural* social capital, acknowledging that there are many ways that learning about and experiencing nature can also help to nurture friendships and overcome the increasing levels of social isolation felt (at least by Americans). How many social organizations or clubs, or community events or activities, focus explicitly on the unique nature of cities? The extent of creative social possibilities is almost limitless: weekend fungi forays, wildlife tracking clubs, 'bioblitzes' and nature festivals, and wildflower and bird-watching clubs, among many others.

Nudging that happens is often a function of the range of organizations, some public and some private, that exist in a city and that can help in supporting the educating and engagement of citizens. One measure of a biophilic city is the extent of the organizational support, the quality and the reach of the biophilic organizations that exist in a city that can actively work to nudge us towards nature. Bird-watching and nature hikes through the city may be one option, but there should be many: swimming, canoeing and kayaking in urban waters; visiting parks near and far; experiencing nature on a sidewalk or roof-top or building façade as one walks to work or to the subway, among many others.

Many cities around the world are located on or near water bodies and a measure of their biophilic tendencies is how easy it is for residents to enjoy these aquatic environments. In some cities, such as Boston, non-profit organizations have worked to make it cheap(er) and easy to learn how to sail. In that city, a junior sailing programme run by the non-profit organization Community Boating Inc. offers children the chance to learn how to sail for only US$1, for the entire summer season. Many American cities, moreover, have worked hard to re-establish direct physical contact and connection with rivers, creeks and harbours, through waterfront parks and trails, and opportunities to get out on a kayak or canoe.

Biophilic cities are cities that work to expand the opportunities to spend time outside and in close proximity to nature. Partly this means rethinking the ways in which parks and green spaces are used. New York City has been a leader in creating opportunities for urbanites to camp on weekends in city parks. The programme runs during the summer months and is quite popular. In 2009, family camping took place in every borough of the city. These camping evenings are – especially from the perspective of children – quite enjoyable and exciting. The City's Parks and Recreation Department provides the tents and sleeping bags, and there are typically barbeques, night hikes and sky-watching.

Biophilic cities are to be identified not just by the presence or absence of nature, of green spaces and green infrastructure, but also by other forms of investment that facilitate a biophilic life. A biophilic city invests in a robust network of public (and private) institutions that will educate about, restore and protect, and nudge residents towards enjoying nature. These include traditional environmental education and natural science institutions such as local botanical gardens, zoological parks and natural history museums, among others. Environmental education centres have been very effective in some cities, in certain cases being based in urban neighbourhoods.

Biophilic cities are also concerned about and work to protect nature beyond their borders. Each city has opportunities to express care about the environment and other life in the world. Large cities exert a tremendous pressure on global biodiversity through their material flows and consumption patterns, and one measure of a biophilic city is the extent to which it seeks to moderate or reduce those impacts.

New York City, for instance, has recently acknowledged that it purchases a large amount of tropical hardwoods, an estimated US$1 million worth each year. The city uses this wood – South American species such as *Ipe* and *Garapa* – for such things as benches, boardwalks and ferry landings. The ten-mile-long Brooklyn Bridge Promenade is constructed of *Greenheart*, another South American hardwood. In recognition of the destructive impact of such purchases, NYC Mayor Michael Bloomberg announced a plan in 2008 to significantly reduce the city's purchasing of such wood – a 20 per cent reduction immediately and larger reductions later as the city researches and pilots alternative wood sources and alternative materials that could be used. Describing tropical deforestation as an 'ecological calamity', and noting that it may be responsible for as much as 20 per cent of greenhouse gas emissions, Mayor Bloomberg has made an eloquent plea for cities to become better stewards of the global environment: 'New Yorkers don't live in the rain forest. But we do live in a world that we all share. And we're committed to doing everything we can to protect it for all of our children' (New York City Office of the Mayor 2008). City purchasing policies and decisions are an important opportunity for the expression of biophilic values (Davis and Beatley 2009).

Biophilic cities in our future?

What constitutes a biophilic city is still very much a matter of discussion and debate. Less a definitive list or set of principles, the categories described here are meant to identify at least some of the potential building blocks of a biophilic city. It is unlikely that a singular coherent vision of a biophilic city will emerge. Rather, perhaps there are many different *kinds* of biophilic cities, many different expressions of urban biophilia. And they may be expressed through different combinations of and emphases on the qualities and conditions described here. At the simplest level, though, a biophilic city is a city that seeks to foster a closeness to nature – it protects and nurtures what it has, and understands that its abundant wild nature is usually considerable. It also actively restores and repairs the nature that exists, while at the same time finding new and creative ways to insert and inject nature into the streets, buildings and urban living environments. A biophilic city is an outdoor city, a city that makes walking and strolling and daily exposure to the outside elements and weather possible and a priority (see Figure 3.4).

But as the above discussion also indicates, a biophilic city is not just about its physical conditions or natural setting, and it is not just about green design and ecological interventions – it is just as much about a city's underlying biophilic spirit and sensibilities, about its funding priorities, and about the importance it places on support for programmes that entice urbanites to learn more about the nature around them, for instance. A biophilic city may be measured and assessed more by how curious its citizens are about the nature around them, and the extent to which they are engaged in daily activities to enjoy and care for nature, than by

Figure 3.4 New York City's re-vegetated High Line has increased people's contact with nature.

Source: Image supplied by the author.

the more physical qualities or conditions, or, for instance, the number or acres of parks and green spaces per capita that exist in a city.

There are a variety of important research questions about designing and planning biophilic cities in the future. We still have, for instance, relatively little knowledge of the *cumulative* recuperative and healing powers of urban nature. How do the many smaller green features in a city or urban neighbourhood contribute to our closeness with nature and what are their interactive effects? Is having access to a large forest more effective than living in a neighbourhood full of smaller green features, such as street trees and green roof-tops? What is the actual daily minimum level of nature needed by urbanites, and in what form, to live a healthy life?

There are also a host of research questions that relate to how effective our biophilic strategies in fact are. What are the most effective planning and policies for getting people outside? What will it take to nudge urban populations to adopt a more outdoor nature-oriented lifestyle? In addition, our very understanding of the science and ecology of cities remains quite limited and there is much work to be done here as well. Research is needed to better understand the biology and life cycle of fauna found in cities and how it changes or is modified in an urban setting (e.g. think of coyotes in American cities!), as well as the management implications therein. There are many, almost countless, research questions and opportunities that arise from the agenda of biophilic cities.

Much of the task in the future, certainly for those in city planning and urban design, will be in reoffering an alternative vision of the cities and urban neighbourhoods of tomorrow. As Stephen Kellert has said:

> We need to do more than just avoid all the bad things that we have done in terms of our adverse effects on natural systems. We also have to create the context for thriving, for development, for meaningful exchange with the world around us, and the people around us. And for that we need to restore that sense of relationship with the natural world which has always been the cradle of our creativity.

That vision will be of dense, sustainable, walkable cities, places that are full of nature, and that are profoundly restorative, magical and wondrous: *biophilic cities.*

References

Carson, R. 1956, 'Help your child to wonder', *Woman's Home Companion*, July, pp. 24–27, 46–48, http://library.fws.gov/Carson/WHC-july56.pdf, viewed 14 November 2013.

Davis, C. and Beatley, T. (dirs) 2009, *The Nature of Cities*, Throughline Productions, Boulder, CO.

Kellert, S. 2006, *Building for Life: Designing and understanding the human–nature connection*, Island Press, Washington, DC.

King County 2008, *King County biodiversity report*, King County, Washington, www.kingcounty.gov/environment/animalsAndPlants/biodiversity/king-county-biodiversity-report.aspx, viewed 21 June 2013.

New York City Department of Housing Preservation and Development 2007, *City of New York and American Institute of Architects Announce Winner of First Juried Competition for Sustainable and Affordable Housing*, media release, www.nyc.gov/html/hpd/html/pr2007/pr-01-17-07.shtml, viewed 21 June 2013.

New York City Office of the Mayor 2008, *Mayor Announces Plan to Reduce the Use of Tropical Hardwoods*, media release, www.nyc.gov/portal/site/nycgov/menuitem.c0935b9a57bb4ef3daf2f1

c701c789a0/index.jsp?pageID=mayor_press_release&catID=1194&doc_name=http%3A%2F%2Fwww.nyc.gov%2Fhtml%2Fom%2Fhtml%2F2008a%2Fpr045-08.html&cc=unused1978&rc=1194&ndi=1, viewed 17 February 2009.

Putnam, R. 2000, *Bowling Alone: The collapse and revival of American community*, Simon & Schuster, New York.

Wilson, E.O. 1993, 'Biophilia and the conservation ethic', in S.R. Kellert and E.O. Wilson (eds), *The Biophilia Hypothesis*, Island Press, Washington, DC.

Chapter 4

Sustainable cities are solar-responsive cities

Norbert Lechner

Summary

To be sustainable, a city must be energy efficient, and for a city to be energy efficient its buildings must be energy efficient. Because most building energy consumption is for heating, cooling and lighting, all of which are greatly impacted by the sun, buildings and therefore cities must be solar responsive. The orientation of buildings, which is largely a function of the street layout, is the single most important factor in designing with the sun, whereby the energy consumption for heating, cooling and lighting can be reduced by as much as 40 per cent. The best street layout for new cities or developments is oblong city blocks with the long axis running east–west so that most buildings will have only the ideal north and south exposures. The second most important strategy is bulk zoning so that access to the winter sun and year-round daylighting is available for as many buildings as possible. All roofs, walls, streets, parking areas, sidewalks and plazas should be white or of a light colour to significantly reduce the heat island effect and to improve the amount and quality of daylighting. It is vital to design cities and buildings to be solar responsive, but that requires a good understanding of solar geometry. Because of its complexity, solar geometry is best learned with the help of a heliodon, which is also a great tool for learning and testing solar strategies.

Introduction

A sustainable future is one where new cities and developments are solar responsive and existing cities are made as solar friendly as possible. It is therefore critical that urban designers and planners make responding to the sun a priority in their work.

Energy is the number one sustainability issue because the enormous consumption of fossil fuels is the main cause of global warming (climate change). Since buildings consume about 50 per cent of all energy, and since most of that energy is used to heat, cool and light buildings, it is vital to design buildings and even more importantly entire cities that use less energy, especially for their thermal and lighting requirements. Furthermore, since heating, cooling and lighting are all greatly impacted by the sun, very low-energy buildings and cities can be created by making them solar responsive. Fortunately, we have the knowledge and tools necessary to design such low-energy and solar-responsive buildings and cities.

Contrary to popular belief, the greatest benefit of solar-responsive design is not from the use of photovoltaics (solar cells) but from free and low-cost solar strategies such as orientation, surface colour, window design, passive solar, shading and daylighting. Only when these 'low-hanging fruits' have been picked does it make sense to harvest the 'fruit' that is

Figure 4.1 It is common sense that the easy and inexpensive strategies should be used before the difficult and expensive ones, especially if their impact is also significant. Thus, 'pick the low-hanging fruit first'.

Source: Supplied by the author.

on higher branches, i.e. active solar hot water heating and the production of electricity with photovoltaics (see Figure 4.1).

As the planet urbanizes, the focus of solar-responsive design must expand from individual buildings to the urban level. Because the ability of any building to benefit from solar responsive design is greatly affected by its surroundings, cities must be designed so that each building has as much access as possible to the sun.

Orientation of buildings and streets

The single most cost-effective solar strategy for buildings is orientation, which is largely a function of street orientations. Consequently, the urban street layout is of critical importance. The goal of orientation is to maximize north and south windows while minimizing those facing east and west. In the temperate climates of the northern hemisphere, the best orientation is usually south because south-facing windows collect much more winter than summer sun, thereby helping in both heating and cooling. North-facing windows, however, are best for buildings that need little if any heat in the winter as is typical of large office buildings, especially in mild climates. As one moves south in the northern hemisphere into the tropics, the solar impact on north windows approaches that of south windows until at the equator north and south windows become identical. For simplicity, unless otherwise noted, all discussion in this chapter is in regard to the northern hemisphere. Of course, the situation is the same in the southern hemisphere, with certain reversals of time and orientation. Since east and west windows receive mostly summer sun everywhere on the planet, they are a major source of overheating, and are therefore best avoided. Because the east and west sun is quite low for part of each day, it is almost impossible to fully shade east- and west-facing windows.

Without doubt, the best street layout for creating low-energy sustainable communities is an orthographic grid with avenues running north–south and streets running east–west. The resulting rectangular blocks should be oblong with the short dimensions facing the north–south avenues and the long dimensions facing the east–west streets, because that results in most buildings having only south and north windows (see Figure 4.2). The benefit is most striking in dense urban areas where buildings are contiguous along the streets as, for example, row houses on an east–west street. Such houses will have only the ideal northern and southern windows. Although there will be a few buildings facing the north–south avenues, their windows do not have to face east and west. Figure 4.3 shows how the east and west façades can undulate so that windows can still face north and/or south.

The width of east–west streets, in regard to solar-responsive design, is determined by the planned height of the buildings, their functions, climate and latitude. For example, in cold, northern climates, streets must be wider to provide access to the low winter sun for both heat and daylight (see Figure 4.4). Because lighting consumes about 40 per cent of the energy in office buildings, it is important to provide sufficient daylight to the lower floors of buildings, even if they do not need solar heating. Consequently, the higher the office buildings, the greater the required street width (see Figure 4.5). Since higher latitudes also have a lower daylight resource, the streets must be wider again to compensate. The access to the sun and sky may be further increased by appropriate solar bulk zoning ordinances, where the height and form of buildings is controlled to give maximum access to the winter sun and to daylighting. Solar bulk zoning will be described in more detail later.

Utilizing orientation to reduce energy consumption is not only a 'free lunch', but a 'free lunch that one is paid to eat'. In regard to buildings, a free lunch is a building that costs the same as a standard building but uses less energy and therefore costs less to operate. A free

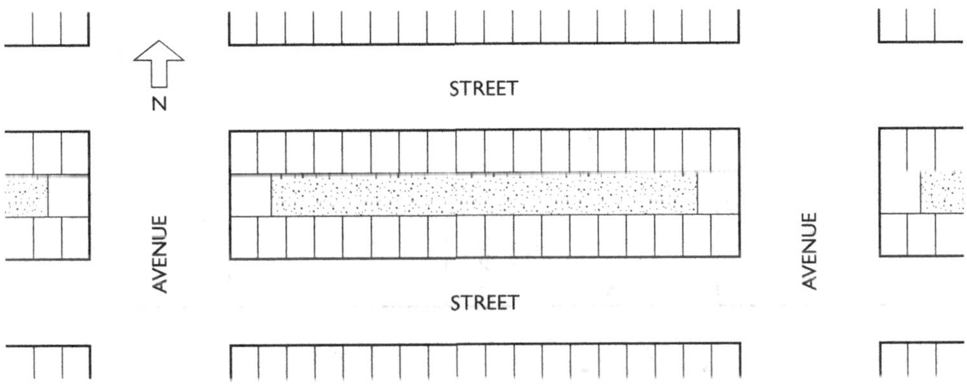

Figure 4.2 East–west streets provide the ideal orientation for buildings so that most windows will face north and south.

Source: Supplied by the author.

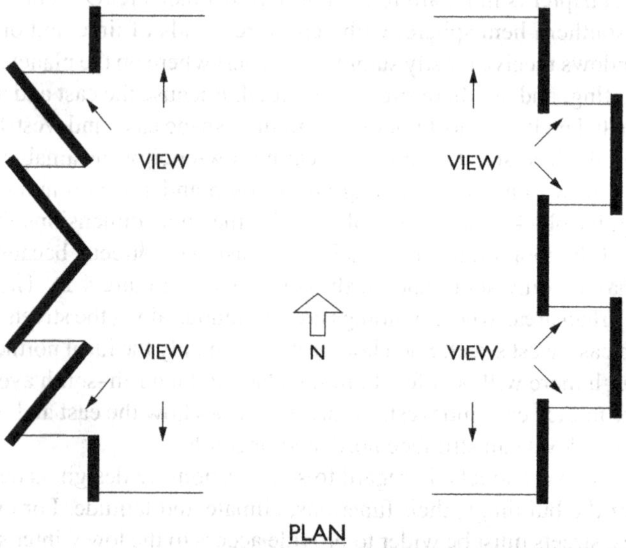

Figure 4.3 Even though buildings on north–south avenues will have their façades facing east and west, they can be designed so that their windows will still have the ideal north and south orientation.

Source: Supplied by the author.

lunch that you are paid to eat would be a building that uses less energy (money) to operate and costs less to build than a standard building. This amazing opportunity is a result of the fact that buildings with no or very few east and west windows will have a smaller cooling load, and therefore need a smaller and less expensive air-conditioning system than a standard building. Since the maximum heating load occurs on winter nights, the size of the heating system is not affected by solar collection during the day.

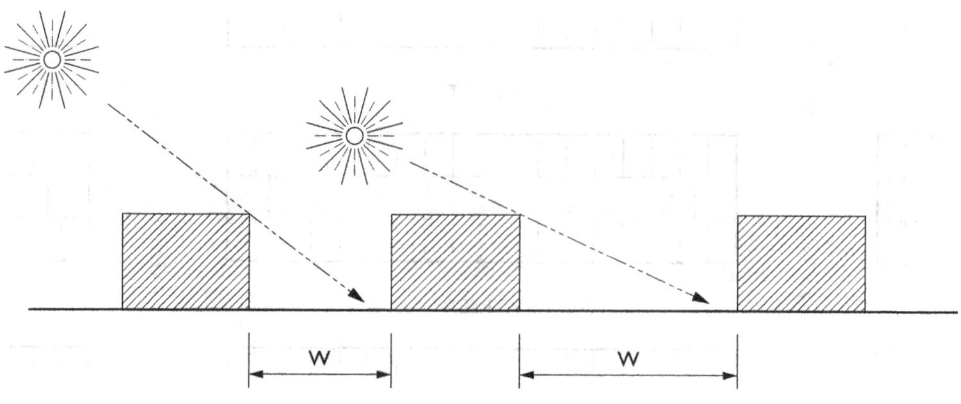

Figure 4.4 Especially in cold climates where access to the direct sun is very desirable, the width of streets needs to be larger due to the lower sun angles.

Source: Supplied by the author.

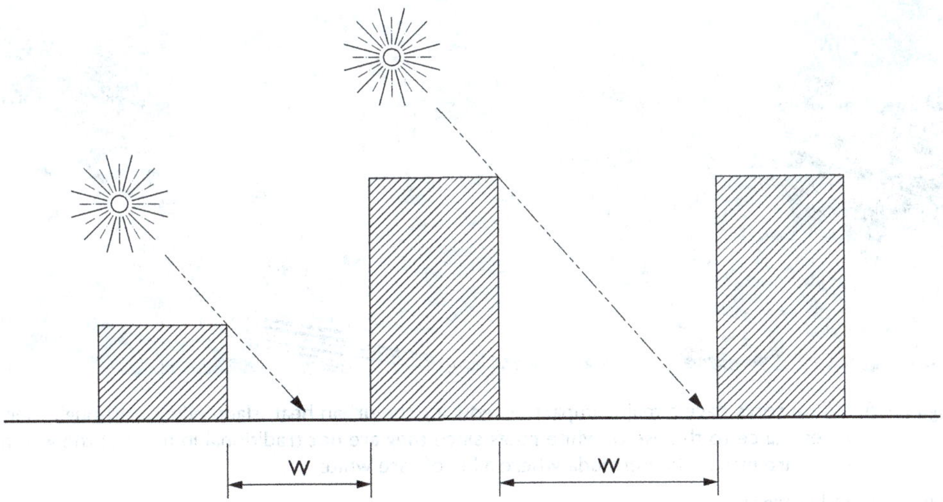

Figure 4.5 To get solar access at the lower floors, streets must be wider as buildings get taller.

Source: Supplied by the author.

The colour of roofs and walls

After orientation, the next lowest fruit we should pick is colour. Because white is the most environmentally sustainable colour, all roofs and most walls should be white. White roofs result in significantly cooler urban heat islands, because much of the incident solar radiation is reflected back into space rather than being absorbed by the roof and turned into heat. The heat gain to a building with a white roof is an incredible 50 per cent of the heat gain through a black roof. The traditional dark roofs will heat the urban air both directly and indirectly. The heat not immediately transferred to urban air will be absorbed and flow down into the building, increasing the cooling load. Electrical energy is then needed to power the air-conditioning equipment to pump the heat back out of the building into the urban air. In addition, the electricity used is also turned into heat, which also ends up in the urban atmosphere.

Although the heat gain through white walls is only about 65 per cent of the heat gain through black walls, there is still a significant benefit to both the building and the urban heat island in using white walls. Furthermore, white walls provide more and better quality daylight. Instead of the normal daylight (i.e. from both the sky and the sun) entering a window from above, more light enters from below, illuminating the ceiling, thereby providing the dual benefit of daylight going further indoors and being of better quality. The white exterior walls also provide more light at night by reflecting some of the street and floodlighting back to the ground. Not only will there be more light, but the light will also be more diffuse, thereby better illuminating what would otherwise be dark and dangerous corners and alleys. Because of the many specific benefits of white roofs and walls to the urban environment, they should be a requirement enforced through zoning and building codes. When he was the United States Secretary of Energy, Steven Chu observed: 'Rapid development of cool materials [white surfaces] represents one of the largest and most cost-effective opportunities we have to counter global warming, improve health, and strengthen security.'

Figure 4.6 White roofs have a major impact in reducing the urban heat island effect. Although there is resistance to the use of white roofs since they are not traditional in most of the world, there are places like Bermuda where all roofs are white.

Source: Supplied by the author.

The most common argument against white, sloped roofs is based on aesthetics. In most cultures, white roofs are rare or non-existent. Since much of aesthetics is based on what is familiar and traditional, people believe that white roofs are ugly because they are used to seeing roofs that are dark in colour. However, when they are confronted with an image of a building with a white, sloped roof, most people acknowledge that white roofs are not ugly. For the readers who doubt that sloped white roofs could become the new norm, it is worth noting that all roofs in Bermuda are white (see Figure 4.6). It cannot be that all people in Bermuda have bad taste or that they suffer physically or psychologically because of all the white roofs. Furthermore, white roofs are becoming quite common along the coastlines of the southern United States. When the aesthetic objection to white roofs is retracted, people change the focus of their objection to the 'fact' that white roofs get dirty and therefore become visually unappealing. This objection is only partly true because dirt also shows up on dark roofs. Furthermore, the most reflective roofs are made of white metal and their smooth surface minimizes the collection of dirt.

There is much less objection to white walls, since they are already common in many cultures. White walls have often been used in very hot climates because of their established benefit for thermal comfort. In cooler climates, white became popular during classical revival periods because of the erroneous belief that classical Greek buildings were all white. In fact, the ancient Greeks painted parts of their buildings in bright colours. In any case, white walls are aesthetically acceptable in much of the world. Although the main objection is based on the belief that white walls cause glare, there are many cities where all walls are white and yet glare is not a problem. For example, the author did not suffer from glare in Casablanca (a city with a preponderance of white houses). For low-rise buildings, the white roof is most helpful, and for high-rise buildings, white walls become the dominant element, since the wall area is much greater than that of the roof. Sustainable cities are white cities.

Fortunately, there is growing international awareness of the importance of white and light-coloured surfaces. The Global Cool Cities Alliance is a non-profit organization working to mitigate the effects of climate change through 'global cooling' (i.e. reflecting solar radiation back into space).

Window size, placement, orientation and numbers

In modern office, institutional, and even residential buildings, the walls are increasingly being replaced by glass, thereby greatly expanding the window area. From a sustainability point of view, all-glass buildings are inferior to buildings with white walls and limited window areas. The reason for this may be better understood by looking at the third-lowest fruit within easy reach: the design of windows in regard to size, orientation and number. For proper window design, it is most important to abandon the aesthetic idea that all façades should look the same. This arbitrary aesthetic concept is most harmful, because each façade experiences a different solar impact, and therefore requires a different window design. As mentioned earlier, ideally the east and west façades would have no windows. Since that is not always possible in reality, the number and size of windows on those orientations should be minimized. Furthermore, any necessary east and west windows should be wide rather than tall, because short windows are easier to shade. For sustainability, most windows should face north and south, but the optimum area of these windows will vary with climate and building type. Only buildings with a need for a substantial amount of daylighting should have a large north-facing glazed area, while all buildings in cold climates should have a large south-facing glazed area to maximize the harvest of the winter sun. In very warm climates, only buildings needing daylighting should have a moderately large south-facing glazed area.

The reasons outlined above for window placement and size indicate that all-glass buildings are much less sustainable than buildings with white walls and optimal glazing areas. The most sustainable buildings would have large areas of white walls on the east and west façades and a smaller amount of white walls on the north and south façades.

The discussion thus far has assumed that buildings will be aligned with the cardinal directions of the compass, but of course many existing cities have street layouts that force buildings to have less than ideal orientations. Because solar geometry is quite complex, and building orientation may vary greatly, it is not possible to impose rules for all of the possible orientations. Fortunately, devices called heliodons can greatly aid in the design of buildings of any orientation. Heliodons will be described in more detail later.

Shading

Because most people on the planet live in hot climates, and because even cooler climates such as Northern Europe and North America have an overheating problem in the summer, shading is an almost universal requirement. Although shading saves energy by rejecting the sun, it is one of the most important solar-responsive design strategies. Outdoor shading is the most effective strategy (see Figure 4.7), indoor shading is the least effective, and shading by means of glazing is somewhere in between. Indoor shading devices should be used primarily as back-up and for controlling the glare of the sun at low angles.

Not only is shading with glazing less effective than outdoor shading, it is also a problem for the urban environment. Since much of the sunlight hitting glazing is reflected off the surface in a specular (mirror-like) manner, the sunlight is reflected onto neighbouring buildings and ground surfaces, greatly increasing their solar load. Thus, a sustainable city would require outdoor shading devices not only to directly reduce the cooling load on buildings, but also to prevent the undesirable solar reflections onto other buildings and the ground. In summer, the reflections off glass may be considered to be sunlight pollution, and therefore a 'good-neighbour' building would reflect whatever sunlight it cannot use back into space.

Figure 4.7 Outdoor shading has a major benefit in reducing the urban heat island effect. Outdoor shading also improves the aesthetics of a city. A building with shading devices is much more interesting than a building with plain glass façades.

Source: Supplied by the author.

Because of the dynamic nature of the environment, dynamic devices are usually better than static ones. For example, south windows should collect the most sunlight in the winter and the least in the summer; east windows need shade in the morning but not in the afternoon; west windows need the reverse; and all windows need shade in the summer on sunny days but not on cloudy days. Only a dynamic shading system will be able to respond to these changing conditions. Before air-conditioning became available, awnings were frequently extended in the summer and retracted or stored in the winter. Automatically or manually operated shading devices are now both readily available and reliable, negating the argument that they are too hard to maintain.

Vegetated roofs and walls

Dynamic shading can be provided not only by moveable awnings but also by deciduous plants that automatically respond to the changing seasons, as may be seen in Figure 4.8 (an apartment house in Rome). Using plants for shading is especially appropriate in urban settings because of the many additional benefits that living plants provide. They cool the air by evapotranspiration, they filter the air of pollutants, they provide extra oxygen, they provide the psychological benefit of pleasing aesthetics and, possibly most important of all, they provide for our genetically hard-wired need for living things ('biophilia'), as suggested by biologist E.O. Wilson.

Urban greenery was traditionally provided mainly by parks and sometimes by trees along streets and avenues. There is now renewed interest in planting urban trees, as demonstrated by the existence of groups such as Cool Communities, which is promoting not only light colours but also urban street trees. Instead of just planting trees and hoping for the best, it is now recognized that newly planted trees must also be provided with good soil and an adequate supply of water, which is possible through the use of permeable paving materials.

Urban greenery is also being provided by vegetated roofs. The phrase 'vegetated roof' is better than 'green roof' because, after all, white is the 'greenest' colour. In dense urban areas, the roofs of high-rise buildings are often covered with cooling towers, and even when they are not, the roof area is much smaller than the area of the walls. Thus, vegetated walls are more valuable than vegetated roofs in dense urban high-rise areas. Since east and west walls should have no or few windows, they are well suited to being mostly covered with plants, such as vines growing on cable trellises. Of course, traditional balconies with planters are also a great method for shading walls and windows with plants (see Figure 4.8).

Cities vary greatly in their use of greenery, and Singapore is a leader among them. Besides all of the standard techniques for providing urban greenery, Singapore's elevated highways and pedestrian overpasses are draped with hanging plants (see Figure 4.9), reminiscent of the Hanging Gardens of Babylon, which were one of the wonders of the ancient world not only because of their technical achievement, but also because such a city was perceived as an ideal. A modern city of hanging gardens is not only technically and economically feasible; it is also desirable from a sustainability point of view.

Providing living plants in an urban setting is an example of solar-responsive urban design. After all, it is the absorption of solar energy that allows plants to grow and thereby provide shade and the many other benefits of urban greenery mentioned earlier.

Figure 4.8 Because the environment is not static, dynamic shading devices such as awnings perform much better than static ones. Deciduous plants are also dynamic on a seasonal basis. Plants also provide a critical amenity to the urban fabric, as shown in this apartment house in Rome.

Source: Supplied by the author.

Figure 4.9 The importance of plants in urban areas cannot be overstated. Besides their many functional benefits, they satisfy a human need for nature called 'biophilia'. This urban street scene is in Singapore.

Source: Supplied by the author.

Passive solar space heating

Before the meaning of the word 'solar' was equated with the concept of photovoltaics in most people's minds, it was most often equated with passive solar space heating. Because photovoltaics are the highest fruit we could pick, this change in perception has not been helpful in promoting either solar energy or sustainability. Since passive solar is essentially free, it should be a major strategy for any city in a climate where the buildings need heating systems. Passive solar is so simple and cost-effective because all south-facing windows are net energy gainers. Over the whole winter season south-facing windows harvest more heat from solar energy during daylight hours than the heat they lose during all winter hours.

When picking the very low-hanging fruits of orientation and window design, passive solar heating is an automatic by-product. However, the energy-collecting performance of south-facing windows can be significantly enhanced through some additional features that are either free or modestly priced. For example, by adding thermal mass, excess solar heat during the day can be stored for night-time use. Since the floor is the ideal location for the mass, concrete floor slabs, so common in cities anyway, provide free thermal mass. The only limitation is to not cover the floor with carpets or wood, which act as insulators. If a floor covering is desired, it must be of a conducting material such as ceramic tiles. However, exposed concrete can be made very attractive by the use of an acid stain. The use of massive walls further increases thermal comfort in the winter. Thermal mass is also essential in certain passive cooling strategies, and it can reduce energy consumption in air-conditioned buildings. Consequently, adding thermal mass to a building is a major sustainability strategy.

Passive solar space heating is also enhanced with moveable indoor insulation covering the windows at night to reduce the heat loss. This same insulation may be used in the summer to keep the heat out, but it does not replace moveable outdoor shading devices that shade more effectively while still providing a view. However, the outdoor shading must be retracted in the winter so that the windows have full access to the sun. Making use of passive solar space heating should be required by code in any city located in a climate where winter heating is required. Of course, allowances must be made for spaces that do not have windows exposed to the winter sun, which is especially likely in existing cities not designed for providing solar access.

Daylighting

Typically, about 40 per cent of the energy used by an office building is for lighting. By its very nature, most office work occurs during daytime hours and could therefore be illuminated

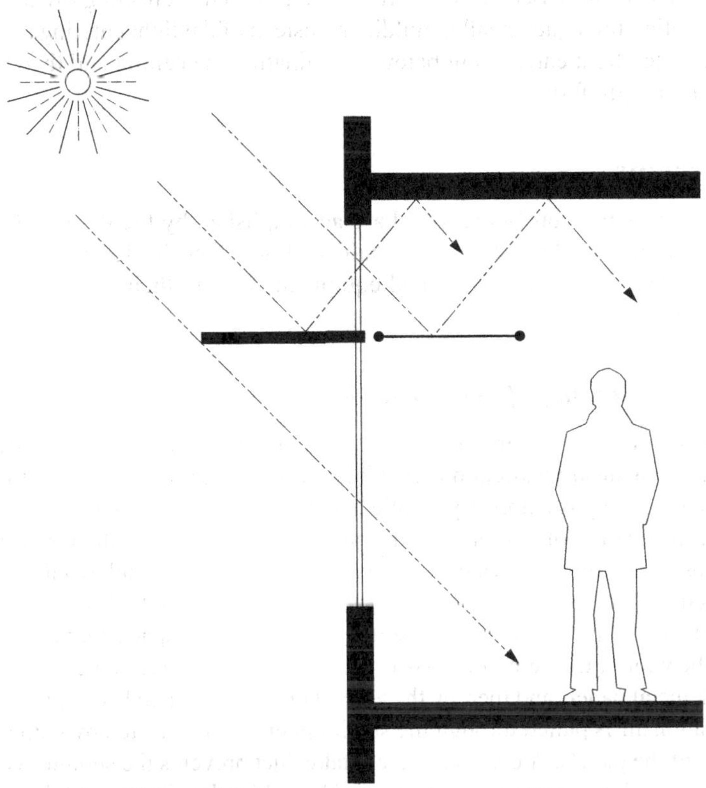

Figure 4.10 Both the quality and quantity of daylighting is improved by the use of light shelves. Besides light shelves, the Singapore Central Library also uses shading devices and white façades to make the building as sustainable as possible.

Source: Supplied by the author.

by daylighting rather than by energy-consuming electric lighting. Consequently, daylighting is another major sustainability strategy to be picked from the solar-responsive design tree.

As with passive solar, when picking the low-hanging fruits of orientation and window design, daylighting is an automatic by-product. As is widely known, large areas of north-facing windows will supply constant and quality daylight for their adjacent spaces. However, south-facing windows will supply not only high-quality light but also supply it in much larger amounts, especially with the additional feature of a light shelf (see Figure 4.10). The beauty of the light shelf is that it performs several valuable functions: it reflects light high up to the ceiling, producing a high-quality reflected light over a large area; it shades the view window, thereby reducing the heat gain and puddles of sunlight; and it eliminates the glare from the direct sun or the bright, hazy sky.

When necessary, east and west windows can also provide daylighting but the quality is poor for part of each day when the sun is low in the sky (i.e. early morning on the east and late afternoon on the west). To minimize these problems, any windows facing east or west should be small and short while the light shelves should be much deeper than those on the south-facing windows.

The quality and amount of daylight is also a function of building space and the colour of the exterior walls, as discussed above. Furthermore, the light reflecting off white roofs can provide daylighting for adjacent taller buildings. Instead of daylight entering a window from above, the reflected light enters from below, illuminating the ceiling and thereby providing a better quality of light further indoors.

Active systems

While the strategies previously described are accomplished by the design of the building itself, all of the following three strategies require additional mechanical and electrical equipment to make them work. This additional equipment adds to their cost and lengthens the payback period.

Active solar preheating of ventilation air

Because infiltration has been largely eliminated in modern, airtight, energy-efficient buildings, positive mechanical ventilation is required. However, preheating ventilation air in the winter with solar energy is extremely simple and relatively inexpensive.

In the solar preheating of ventilation air, the exhaust fan pulls fresh air into the building through a transpiring collector, which consists of dark-coloured (black would be best) metal panels covered with small holes or slits for the air to pass through. The vertical panels are mounted on the non-window parts of the south wall with an air space between the panel and the wall. In the winter, the ventilation air is drawn from this space. When the sun shines, it heats the dark metal panels and thereby the air on both the front and back sides of the panels. As the ventilation air is pulled through the small openings, it is heated by both the front and back surfaces of the panels. A diverter in the intake duct prevents the summer ventilation air from passing through the transpired collectors. The colder the climate and the more ventilation air that is required, the greater the benefit of this system. This system is appropriate for apartment houses, offices, schools and shopping centres. These ventilation air preheaters are especially useful in buildings such as hospitals that require large amounts of fresh outdoor air to minimize the transmission of disease.

Active solar sanitary hot water heating

Because of the very high heat capacity of water and because of the large amount of sanitary (domestic) hot water used for washing and cleaning in homes, apartments, hotels, hospitals, commercial laundries and food services, much energy can be saved by heating water with the sun. Except in the cloudiest climates, active solar hot water collectors are economical and always sustainable. If the payback period is long, it is due to the inappropriately low cost of energy supplied by fossil fuels, and/or the lack of mass production and mass consumption of solar hot water collector technology. When the real cost of fossil energy is considered, then the payback period for active solar hot water systems is quite short. China is a good example of a country where even relatively poor people own solar hot water systems (see Figure 4.11). China is also an example of how mass production and mass installation of systems lowers the price of each unit.

Solar hot water collectors may also be used for space heating, but this function is much less economical than the use of solar for domestic hot water because space heating is only needed for that part of the year when solar energy is least available. Furthermore, because passive solar space heating is much more economical than active solar space heating, passive solar should be used whenever possible.

Active solar hot water collectors should not be confused with photovoltaic (PV) collectors, although that confusion is widespread. Too often solar is rejected because it is too expensive, even though that may only be true for PV, which is the fruit that hangs the highest on the solar-responsive tree. The confusion is partly due to the fact that active solar hot water panels and PV panels look very similar and usually have the same orientation and tilt requirements.

Photovoltaics (PV)

Although photovoltaics (PV) are an ideal renewable energy source, their high cost places them at the top of the solar-responsive design tree. This fruit should be picked, but only after all of the fruit that is within easier reach has been picked. Even if PV is not initially included in a new building, allowance should be made for its later addition. Since the requirements for active solar hot water panels are the same as for PV panels, they can be placed side by side, and any design for including active solar will automatically also be appropriate for PV.

Figure 4.11 China is one of the leading countries in using active solar hot water collectors, as seen here in the city of Dezhou where most buildings are situated along east–west streets. There are few places and climates where solar hot water heating is not appropriate.

Source: Supplied by the author.

In most cases, the ideal orientation for collector panels is due south and the ideal tilt is equal to the latitude of the building site. Thus, sloped roofs should have an angle equal to the latitude so that the panels are part of the roof for both functional and aesthetic reasons. On flat roofs, the mounting hardware is adjusted so that the panels are tilted to the angle of latitude. Since the panels will also provide valuable shade for the roof, they may be a better choice than a vegetated roof. However, vegetated roofs may be a better choice if the roof is used as a garden and if storm water retention is valuable in that particular city.

When the cost of PV has declined sufficiently, it will be practical to cover not only the roof but also the south wall, and then the west wall, and finally the east wall. Since PV panels can be tilted when mounted on shading devices, some PV is already practical on walls over windows.

It is important to remember that the use of PV does not eliminate the need for the passive strategies described above. The cost and mounting area of PV panels is too great for it to provide all of the energy for a building unless the energy consumption of a building has been greatly reduced by various methods, including the lower-hanging fruit on the solar-responsive tree.

Solar zoning

There are no benefits from most solar-responsive design if there is no access to the sun. Thus, solar access zoning must be provided for both new and existing communities. Without such protection, it makes little sense to design a building to harvest sunshine, since some future construction could block it. At the very least, the solar access zoning should guarantee that the roof will have access to the sun for most of each day. Only in low-density areas can solar access be guaranteed for the whole building (including the ground-floor windows) for passive solar heating and daylighting.

Solar access can be provided by several different zoning methods. In conventional zoning, solar access is provided by limiting the height of structures and requiring setbacks from the

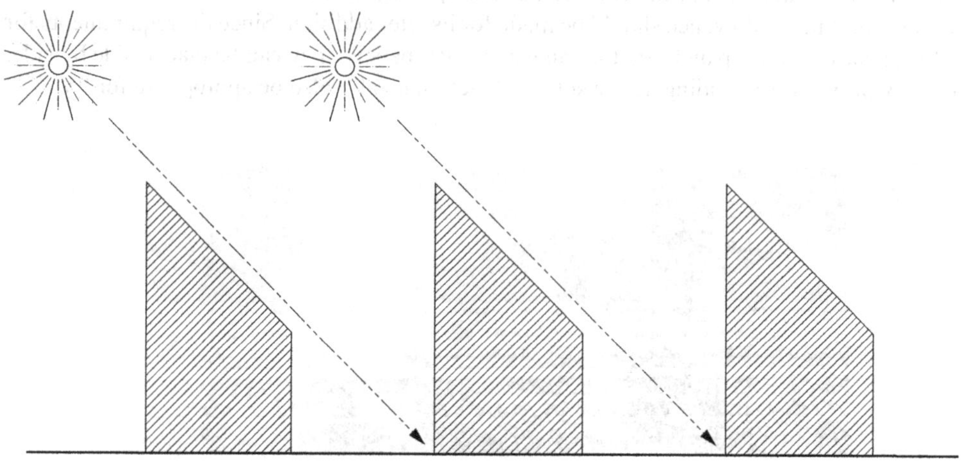

Figure 4.12 Solar zoning ordinances should be used to provide the maximum solar energy to every building in a city. Because the bulk-plane slopes down towards the north end of the property, it provides considerable solar exposure to the neighbouring property.

Source: Supplied by the author.

property line. A better solar zoning approach is to control the height of a building with an inclined plane sloping down towards the north end of the property (see Figure 4.12). The most generous method, which is called 'solar-envelope zoning', provides the most solar access by limiting the building height with four sloping planes to create a pyramid type of building boundary. Both zoning ordinances and knowledgeable urban designers are critical to assure solar access for each building.

Heliodons

The potential of solar-responsive design is greatly underused because of two factors: too many designers and planners are not aware of its great potential, and many of those who use it lack the knowledge to apply it effectively. Both problems are a result of the complexity of solar geometry that is the foundation of solar-responsive design. Fortunately, devices called heliodons make the learning of solar geometry and solar-responsive design both easy and fascinating. Heliodons allow for the simulation of sun angles on a scale model for the variations of latitude, time of year and time of day. Because the conceptually clear types of heliodons simulate our everyday experience of the apparent movement of the sun, they are best for the initial learning of solar-responsive design, including the impact of orientation (see Figure 4.13). More information on heliodons is available at www.cadc.auburn.edu/heliodon.

Conclusion

Although the impact of the sun has been largely ignored in the design and planning of cities, there are some exceptions. An ancient exception was the Greek city of Olynthus (*c.* 400 BCE),

Figure 4.13 Conceptually clear heliodons are powerful tools for promoting, understanding and designing solar-responsive buildings and cities. Such heliodons should be located in every architecture and planning school, every science museum, and the planning departments of large cities.

Source: Supplied by the author.

where the streets were purposely oriented east–west so that all buildings could harvest the winter sun while rejecting the summer sun. A modern exception is the development of Village Homes in Davis, California, where again all houses were built on east–west streets so that they could take advantage of all of the solar-responsive design strategies. More recent examples of good urban design include the Vauban solar development in Freiburg, Germany and the Beddington Zero Energy Development (BedZED) in London, UK.

Unfortunately, even some cities that claimed to be solar, such as the solar city in Linz-Pilching, Austria, failed to utilize that most fundamental and important of all solar strategies, namely orientation.

Designing all new cities and developments to be solar responsive is critical for a more sustainable future because it will significantly reduce energy consumption. Furthermore, as much solar access as possible must be provided in existing cities through zoning ordinances and building codes. Solar-responsive design will be able to save a great deal of energy, but only if urban designers and planners make it possible.

Bibliography

Atkinson, W. 1912, *The Orientation of Buildings or Planning for Sunlight*, John Wiley & Sons, New York.

Bainbridge, D., Corbett, J. and Hofacre, J. 1979, *Village Homes' Solar House Designs: A collection of 43 energy-conscious house designs*, Rodale Press, Emmaus, PA.

Erley, D. and Jaffe, M. 1979, *Site Planning for Solar Access: A guidebook for residential developers and site planners*, U.S. Department of Housing and Urban Development, Washington, DC.

Francis, M. 2003, *Village Homes: A community by design: case studies in land and community design*, Landscape Architectural Foundation, Washington, DC.

Givoni, B. 1998, *Climate Considerations in Building and Urban Design*, Van Nostrand Reinhold, New York.

Jaffe, M.S. and Erley, D. 1979, *Protecting Solar Access for Residential Development: A guidebook for planning officials*, Office of Policy Development and Research, U.S. Department of Housing and Urban Development, Washington, DC.

Knowles, R.L. 1974, *Energy and Form: An ecological approach to urban growth*, MIT Press, Cambridge, MA.

U.S. Department of Energy, Office of Project and Facilities Management 1985, *Site Development Planning for Energy Management*, U.S. Department of Energy, Washington, DC.

U.S. Environmental Protection Agency 1992, *Cooling Our Communities: A guidebook on tree planting and light coloured surfacing*, U.S. Environmental Protection Agency, Washington, DC.

Van der Ryn, S. and Calthorpe, P. 1986, *Sustainable Communities: A new design synthesis for cities, suburbs, and towns*, Sierra Club Books, San Francisco, CA.

Chapter 5

Trends in low carbon transport and urban development in 33 cities, 1995/1996 to 2005/2006

Some prospects for lower carbon transport

Jeff Kenworthy

Summary

There were many changes in land use and transport systems in the world's cities between 1995/1996 and 2005/2006. As cities struggle to become more ecological and liveable, with lower carbon passenger transport, their shape in terms of urban sprawl or more compact development, more car dependence or more reliance on transit, walking and cycling, will be critical factors. If cities are not addressing these fundamental characteristics, lower carbon transport policies will be severely undermined. This chapter examines two snapshots in time for a sample of 33 cities in the USA, Canada, Australia, Western Europe and Asia by assembling a unique set of 2005/2006 land-use, private and public transport service, usage and infrastructure data, as well as factors such as private and public transport energy use, and comparing them to an identical set of 1995/1996 data. The data show many positive changes, such as evidence supporting the idea of 'peak car use' in cities, as well as a renaissance in transit, especially urban rail. Such changes tend towards lower carbon transport. However, there are also warnings in the data, such as continued growth in car ownership and failures to have all the policy strings pulling in the same direction. Some policy implications are drawn from the results of the decadal changes in cities, with a focus on lower carbon urban passenger transport.

Introduction

There is little doubt that cities need to become more ecologically oriented and part of that is making transport less carbon-intensive (Kenworthy 2006). The world's seven billion people are increasingly located in cities, while rising motorization and higher transport energy use are becoming major problems (Dimitriou and Gakenheimer 2010, Newman and Kenworthy 2011a). If cities are to have lower carbon transport systems, then urban mobility patterns and the many physical factors that underpin them must change. We know that the cities of North America, Australia and many other places are already highly automobile dependent, leading to high passenger transport energy use, while a huge proportion of the world is undergoing rapid motorization (China, India, Brazil, etc.), leading to increased carbon footprints from transport.

This chapter focuses on the passenger transport systems in high-income cities and how they are evolving at this critical juncture where 'peak oil', climate change and many other problems threaten to challenge every assumption that has governed urban development since the 1950s. It looks at some of the critical changes that occurred between 1995/1996 and

2005/2006 in 33 cities in the USA, Canada, Australia, Western Europe and Asia by assembling a large array of land-use, private and public transport service, usage and infrastructure data, and transport energy use. Collectively, these data help us to characterize transport patterns in cities and to assess the prospects for lower carbon transport.

Methodology

This chapter presents the first results of a partial update of the *Millennium Cities Database for Sustainable Transport* by Kenworthy and Laube (2001). This database contained 1995/1996 data for a wide range of land-use, transportation, economic and energy/externalities items for 100 cities worldwide, including all the cities mentioned.

The data presented here for 2005/2006 provide a decadal perspective on a number of key land-use and transport variables in the 33 cities cited in Table 5.1. For these cities, a large enough sample of key data has now been consistently obtained so that a meaningful comparison may be made. This is a work in progress, as it takes many years to obtain the full range of data that were collected for all the 100 cities in the original study. All the regional averages reported for each variable for 1995/1996 and 2005/2006 are consistent for the cities in Table 5.1.

To calculate the standardized data for this chapter, 23 primary data items per city were collected (public transport data, referred to here mostly as 'transit', were collected for each mode (e.g. buses, LRT, metro), with up to seven modes per city, but only the total transit variable is included in the count of 23 data items). Data collection was first undertaken from published sources in each city, including national censuses and annual reports of agencies, but mainly these data are not readily available from published sources in the form required for the metropolitan areas defined for this research. Therefore most data were collected via many emails exchanged with hundreds of individuals working in a multitude of different organizations in each city.

For this chapter, a set of 26 standardized variables (first listed in Table 5.2) have been calculated to capture some of the salient changes in land-use and transport-related factors that together provide a substantial insight into the urban passenger transport field.

Table 5.1 The 33 cities surveyed in this update study

USA	*Canada*	*Australia*	*Europe*	*Asia*
Atlanta	Calgary	Brisbane	Berlin	Hong Kong
Chicago	Montreal	Melbourne	Copenhagen	Singapore
Denver	Ottawa	Perth	Frankfurt	
Houston	Toronto	Sydney	Hamburg	
Los Angeles	Vancouver		Helsinki	
New York			London	
Phoenix			Manchester	
San Diego			Munich	
San Francisco			Oslo	
Washington			Stockholm	
			Stuttgart	
			Zurich	

Results and discussion

Tables 5.2 and 5.3 present the results as regional averages for 1995/1996 and 2005/2006 with relevant percentage changes in each factor. Table 5.2 covers US, Canadian and Australian cities. Table 5.3 covers European cities and Singapore/Hong Kong. These latter two cities are well known for their low dependence on the automobile and provide a strong contrast to the other cities.

Urban form

The importance of urban form in helping to explain patterns of urban transport, especially car dependence and transport energy use, has been widely discussed (Cervero 1998; Kenworthy and Laube 1999; Newman and Kenworthy 1989, 1999a). As densities rise in cities, per capita car use and transport energy decline (Newman and Kenworthy 1989, 1999a). Selectively raising urban densities, such as in centres linked to quality transit with good provision for walking and cycling, can help to reshape cities for reduced car use and lower transport energy by improving the viability of transit (Thomson 1977; Newman and Kenworthy 2006). Decentralized concentration of jobs in sub-centres outside the CBD is important in creating more polycentric metropolitan areas that are better served by non-auto modes. There are many studies that have shown how 'salt and pepper' dispersal of jobs has led to radical drops in transit use (Cervero and Landis 1992; Baillieu Knight Frank 1991), as well as large increases in transport energy use and car trip lengths (Alexander 1980, 1981).

Densities

Australian cities remain the lowest-density cities in this sample, followed closely by cities in the US, but with a significant jump upwards in Canadian cities, a fact that is reflected in many of the other factors discussed in this chapter. European cities remain medium density, while Singapore/Hong Kong clearly lead with an average urban density over five times higher than European cities (with commensurate large differences in the factors discussed below).

The data show that urban population density is mostly on the rise or has more or less stabilized. 'Activity density' (population and jobs combined) is even stronger in this positive change. These so-far modest trends of 1 per cent to 8 per cent increases in urban and activity density are significant because they represent a reversal of the history of these cities since the 1950s, from which time densities continuously declined through automobile-based sprawl.

Centralization of jobs

Job centralization is measured as the proportion of metropolitan jobs located in the central business district (CBD). Most transit systems in the world are still focused on the central city, despite a growth in cross-town or circumferential services in many cities in response to job decentralization. The data show that the CBD as a location for work is declining nearly everywhere relative to other parts of the city, even though the absolute number of jobs is still growing in many CBDs.

Private transport infrastructure

Urban form acts in concert with both private and public transport infrastructure to shape travel patterns in cities. Four factors are considered here: car ownership, length of road per person, length of freeway per person and CBD parking spaces.

Car ownership

In all groups of cities except the Canadian cities, car ownership has risen. Singapore, however, increased in car ownership from a meagre 99 cars per 1,000 people in 1995 to only 100 in 2005. Singapore's punitive car ownership costs under its Certificate of Entitlement system, its generally high car-operating costs, the Area Licensing Scheme for entry into the central city and significant development of its transit system have all helped to keep car ownership low and stable in Singapore, while still maintaining a high level of personal mobility.

Length of road per person

In these cities, road provision per person mostly reduced, or at least remained about the same over the decade. Australian cities are the highest overall providers of roads (7.6 metres per person), followed by US (6.0), Canadian (5.4) and European cities (3.2), and Singapore/Hong Kong, a tiny 0.5 metres of road per person.

Length of freeway per person

Of more critical importance to the prospects for lower carbon transport is the level of freeway provision, because freeways are premium road infrastructure and are symbols of a city's priorities in transport infrastructure. Cities choose to build or not to build freeways through their policy orientation. The City of Vancouver chose to dismantle the freeway plans of the 1960s and never built any. It remains one of the world's most liveable cities, while retaining a highly functional, diversified passenger transport system (Schiller *et al.* 2010). Researchers have shown how urban freeways directly facilitate higher car and energy use in cities through longer travel distances, declining transit systems and deteriorating conditions for pedestrians and cyclists since the 1970s (e.g. Watt and Ayres 1974; Newman and Kenworthy 1984, 1988b, 1999b; Schiller *et al.* 2010).

The evidence between 1995/1996 and 2005/2006 is mixed. US cities seem to have stabilized in freeway provision per person, but in practice some continue to build freeways while others do not. Australian cities have declined 4 per cent on average, while Canadian cities appear to be providing significant new freeway infrastructure, growing by 29 per cent, and are now a tiny fraction above the US sample. Australian cities provide little more than half as much freeway length per capita as North American cities (0.083 metres per person). European cities have increased very slightly in this factor, due to a combination of minimal freeway construction and generally small population growth in the 12 cities examined. Interestingly, these European cities had fractionally more per capita freeway provision than Australian cities in 2005/2006. Singapore and Hong Kong, although increasing slightly, are the least freeway-oriented cities in the sample.

Parking spaces per 1,000 CBD jobs

Parking provision in the CBD is an important factor in determining modal split for trips to these space-constrained areas of cities. The availability of parking, much more than price, determines the attractiveness of car commuting to the central city (Newman and Kenworthy 1988a). High levels of parking will encourage the most energy-consuming trips to work, while well-loaded, radial rail systems will use the least energy.

The data here suggest that parking supply in the CBDs of the most auto-oriented cities is generally on the decline, and significantly so. Even Singapore/Hong Kong, the least parking-oriented cities in the sample, declined further in CBD parking. Despite the declines in US cities, they are still much higher than the Canadian and Australian cities in this factor, while European cities are even lower again than these latter two groups. It appears that some cities are actually reducing their combined off-street and on-street CBD parking supply through new development on parking areas, or at least job growth is outstripping parking supply (or there is a combination of these factors involved).

Reducing CBD parking will support transit use to the CBD, including during the off-peak period where excess transit capacity exists. Parking availability in sub-centres also needs to be reduced if transit is to compete better with private car use, while walking and cycling benefit greatly from reduced parking through a better public realm.

Public transport infrastructure

Length of reserved public transport routes per person

One of the critical requirements for transit is for it to be reliable and legible (easily visible in its location and where it goes) and for it to compete in speed with road traffic. Therefore transit routes need to be free from general traffic, and stops and alignments must be easily understood by users and not subject to change. Regular bus systems provide neither, since they suffer traffic congestion, and their stops and routes can literally change overnight. This factor is measured here as the per capita length of reserved transit right-of-way (bus lanes and rail systems operating on separated rights-of-way). The data show that in all groups of cities except the Australian cities, this factor has generally significantly improved.

Ratio of reserved public transport infrastructure versus freeways

Reserved transit routes are the corollary of freeways because both are premium, higher-speed transport infrastructure, but lead to very different results for mobility and carbon footprints. The ratio between these critical transport infrastructures is important, as it measures whether private or public transport is gaining ascendancy in cities. In every group of cities, except in Australia, there is a significantly growing ratio in favour of transit-reserved routes compared to urban freeways. Australian cities, while not only growing in freeway provision, have also failed to construct rail systems and bus lanes (except Perth).

US cities, while growing in this factor, still only have a little over 50 per cent as much reserved transit route as freeways, and Canadian cities are now identical due to very low growth in this factor. European cities grew in this ratio and have some 5.5 times as much reserved transit route as freeway. Singapore and Hong Kong increased by 53 per cent and now have significantly more transit-reserved route than freeway, whereas the reverse was true in 1995. Australian cities on the other hand declined, though they are still much higher than the North American cities.

Public transport service, use and performance

It is not only necessary to have good transit infrastructure, but also to operate high levels of service and to have other attractive performance features to attract many users.

Total public transport vehicle and seat kilometres per person

There are two key measures of transit service examined here: vehicle kilometres per person and seat kilometres per person. The first measures the per capita service run by the vehicles each year (wagons for trains). The latter variable adds the element of capacity by measuring the seat kilometres delivered by the vehicles. This factor is simply the vehicle kilometres multiplied by the average number of seats provided by each type of vehicle. This gets much higher for rail modes because of more seats per vehicle, and the ability to string together up to ten wagons per train in the case of some cities.

In all cities, the level of transit service per capita, as measured by both vehicle and seat kilometres, is going up. US cities remain the lowest in vehicle kilometres, followed some way ahead by Canadian cities, then Australian cities, European cities, and finally Singapore/Hong Kong with over three times more than US cities. Seat kilometres follow the same pattern with Singapore/Hong Kong offering nearly four times as many seat kilometres of service than US cities.

Ratio of public transport system speed to road traffic speed

It is not just transit service levels and infrastructure supply that are important. The speed at which transit operates, and particularly this speed in relation to the car, is very important in determining people's modal choice. Here we measure kerb-to-kerb speed.

The data show a mixed situation. Canadian cities are slightly down, there is no change in US cities and Australian cities are slightly better. However, more importantly, the data show that in US and Canadian cities, average speed of transit is barely above half that of the car, while in Australian cities it is above three-quarters. Average transit speed in the European cities is close to that of the car, and in Singapore/Hong Kong it is still a little lower than the car. Metro and suburban rail generally offer the most competitive speeds. Tram and LRT systems can compete with cars in certain circumstances, but tend to have lower speeds because they are generally installed in denser settings with more intersections and closer stop spacing. Where they operate on reserved rights-of-way and have green-wave traffic signals, they can compete with the car.

Total public transport boardings per person

The litmus test for transit is not how much service is provided, but how much it is used. The number of boardings that occur in the transit system is one way of measuring this, along with passenger kilometres.

While boardings per capita have risen nearly everywhere, the actual increase is small, given that it is over a decade. US cities' residents remain by far the lowest users of transit, followed by those in Australian cities. Canadian cities' residents are significantly higher 'boarders' than both US and Australian cities. Even Sydney, Australia's highest transit-using city, is 10 per cent below the Canadian cities. It appears that the significantly higher density of Canadian cities and better transit-oriented development enables them to perform much better in transit use, despite having lacklustre infrastructure and service levels compared to Australian cities.

Total public transport passenger kilometres per person

This factor also measures transit use except that it represents the actual distances people travel on transit each year, not just the number of trips they make. In each group of cities, there has been a significant increase in this factor. In other words, travel distances on transit have mostly increased more than the actual trip-making. The pattern is very similar to boardings. The only difference is that whereas the per capita transit boardings in Canadian cities were well above Australian cities, the per capita transit passenger kilometres in Australian cities are a little higher than in Canadian cities. This is because Canadian cities are on average nearly twice as dense as Australian cities so shorter trips are more common.

Proportion of total motorized passenger kilometres on public transport

This factor measures transit use against car use. The results show that nearly everywhere transit performs a slightly higher proportion of the motorized passenger task in 2005/2006 as it did in 1995/1996. However, this role is still very small in US cities (3.2 per cent). In Australian cities it is 8.1 per cent and in Canadian cities a little better at 11.3 per cent. By contrast, transit systems in European cities account for 22 per cent of motorized travel, while in Singapore/ Hong Kong it is a massive 65 per cent. If one included motorcycle and taxi passenger kilometres in private transport and took a new percentage, transit's contribution would be a little lower in each case, but not significantly so.

Private transport use

Car vehicle and car passenger kilometres per person

Whether cities are moving towards a lower carbon footprint in transport is ultimately determined by car use. Car use can be measured by car vehicle kilometres travelled (VKT) per capita (the annual distance travelled by cars themselves on a per capita basis) and by car passenger kilometres per person, or the distances travelled by the people in the cars, including the driver. Car passenger kilometres (PKT) are the car vehicle kilometres multiplied by average car occupancy (7 days a week, 24 hours a day).

In both of these factors, the results were not particularly good, but there are some positive signs. US cities have grown only a little in both measures, while Canadian cities have increased in VKT but declined in PKT. Australian cities have increased in both factors. European cities have also increased a little in car use per person. Singapore/Hong Kong increased by only 128 car VKT per person from 1995 to 2005, whereas Australian cities increased by over 1,000 car VKT per person.

The data show that there is a general fall in car occupancy (except in the European cities, where there is a slight rise), not actually less driving. This means more single-occupancy vehicle travel, which results in highly inefficient resource use. It partly reflects the trend towards smaller household sizes in cities and more cars per 1,000 people, so fewer people are travelling in each car.

There is, however, significantly reduced car use 'growth' compared to previous decades, where per capita car use rose significantly. While the 2005 to 2010 trend is yet to be completed, the first signs of a decline in car use may be seen in Figure 5.1, which summarizes the changes in car VKT per capita in prosperous cities over the 45-year period from 1960 to 2005. In these American, Canadian, Australian and European cities, the growth in car use is slowing down and this is likely to continue in the twenty-first century. Newman and Kenworthy (2011b) describe this phenomenon of 'peak car use' in many developed countries and cities as starting in about 2004. The data in this paper up until 2005 are consistent with these later observations.

It is clear, though, that US cities remain dramatically ahead of any other cities in their level of car use. Australian cities are still well below their US counterparts (34 per cent less), while car use in Singapore/Hong Kong is 90 per cent less than in US cities. Interestingly, Canadian cities have for decades been positioned between the more extreme car-using cities in the US and Australia and the more moderate European cities. With European car use going up and car use in major Canadian cities stabilizing or even declining, Canadian cities are closer to European cities than they are to their American and Australian cousins. Other factors such as higher densities and more transit use conform to this positioning. Goldberg and Mercer (1986) showed how very different US and Canadian cities are on several factors, including transport.

Energy and externality factors

Total private passenger transport energy use per person

Of critical interest to the future of cities, lower carbon transport and climate change is the extent to which cities are dependent on liquid fossil fuels. World oil production has either already peaked or will peak within the next few years. Oil prices hit almost US$150 per barrel in mid-2008 and sent shock waves throughout the global economy. This contributed to the sub-prime mortgage meltdown in the US, which ultimately set off the Global Financial Crisis. Oil now hovers around US$110 to $130 per barrel.

All cities in this study, except Australian and European cities, have reduced their per capita demand for energy in private passenger transport – US cities by 11 per cent, Canadian by 5 per cent, and Singapore/Hong Kong by 6 per cent. A combination of improved technologies and stabilizing car use appears to be behind these changes. The US and Canadian cities' vehicle fleets were so inefficient that trimming their fuel consumption over ten years was achieved mainly by downsizing. By contrast, Australian cities grew in car use by 14 per cent and energy use by 16 per cent, suggesting that urban vehicular fuel consumption rates per kilometre actually increased marginally. European cities increased in this factor by only 4 per cent.

Even with this decline in energy use in US cities, they are still the highest transport energy consumers, with over 53,000 MJ per capita in 2005, compared to Australia, which averaged 36,000 MJ. Canadian cities are even lower, at 31,000 MJ. European cities use 16,500 MJ and Singapore/Hong Kong a meagre 6,000 MJ. Such stark differences point to very important lessons that must be learned from the lower-energy cities, such as more compact land use, better transit, and increased walking and cycling.

Total public transport energy use per person

Public compared to private transport energy use is mostly relatively small. In American cities, it is less than 2 per cent of the private figure, Canadian cities are 4 per cent, Australian 3 per cent and European 10 per cent, but Singapore/Hong Kong is 44 per cent. Due to increases in transit services and more energy-consuming features such as air conditioning, transit energy use per person has risen consistently in all cities (14 to 41 per cent increases). Whether this raises carbon output depends on whether private transport energy use declines due to more use of transit and also on where transit energy increases are sourced (many electric systems depend on hydro, nuclear and renewables such as wind (e.g. Calgary runs its LRT system only on wind power)). The important thing is that transit has huge energy-conserving potential. Singapore/Hong Kong uses only 2,691 MJ per person to carry 65.3 per cent of their entire motorized mobility. US cities use 963 MJ per person to carry 3.2 per cent of their motorized mobility. Given changes to city structure and other factors that favour greater use of existing transit services, transit has the potential to cater for a massive amount of any city's mobility needs with comparatively little energy use or carbon. Transit systems also use far less energy per PKT than cars. US transit is the most energy consumptive (1.7 MJ/PKT), but cars still consume 71 per cent more (2.9 MJ/PKT). Transit in Singapore/Hong Kong consumes only 0.71 MJ/PKT compared to cars at 3.1 MJ/PKT (four times more). Overall, an increase in transit energy use can probably be considered beneficial for lower carbon transport.

Transport deaths per 100,000 people

A factor in this study that shows consistent improvement is the annual transport deaths per capita, mostly reductions of around 30 per cent. US cities are still significantly worse than all

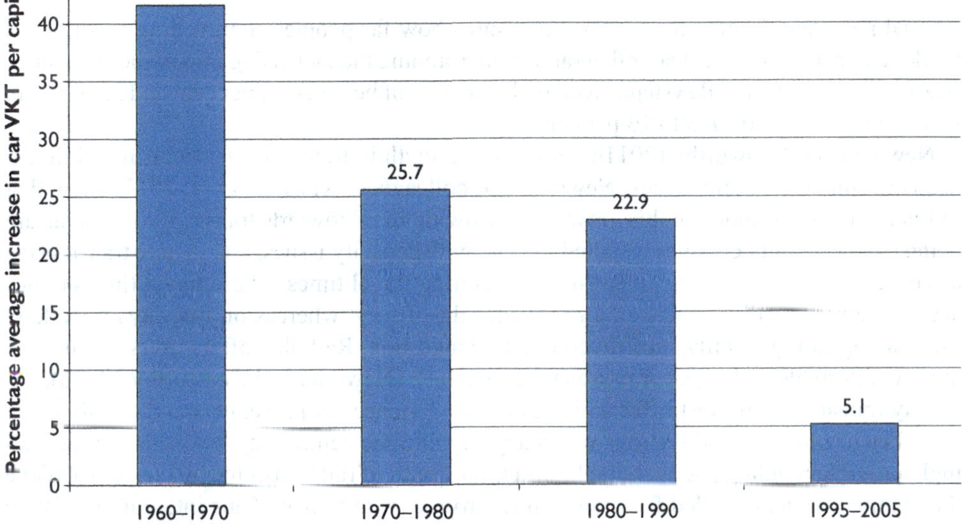

Figure 5.1 Car-use growth trends in prosperous cities from 1960 to 2005.

Source: See Endnote 1 of Newman and Kenworthy (2011b) for details.

the other cities, while European cities are the best. The pattern of transport fatalities per capita seems to largely follow car use.

Urban rail systems and their role in city sustainability

Many of the transit trends are positive. Examination of the data by mode reveals that underpinning this transit renaissance is the strong performance of urban rail modes compared to buses. This chapter now explores three items in Tables 5.2 and 5.3 to show this trend between rail and buses.

Growth in rail service compared to buses

Annual rail seat kilometres per person grew substantially in all groups of cities over the ten years (24 to 33 per cent), with the exception of Australian cities, where it grew by only 4 per cent. On the other hand, bus services expanded by only 4 to 9 per cent in the cities that grew in this factor, while in the Australian and Canadian cities it actually declined by 1.4 per cent and 5.3 per cent respectively.

Growth in rail trip-making compared to buses

Growth in rail trip-making significantly exceeded that of buses. Apart from Australian cities, the other cities increased their rail use by between 10 per cent and 22 per cent (Australian cities 6 per cent). However, bus usage increased by only 2 to 7 per cent in all cities that recorded positive increases, while bus use in Singapore and Hong Kong actually declined by 13 per cent.

Growth in passenger travel by rail

Annual passenger kilometres of travel measures how far people are travelling by different modes and here it is even clearer that rail is underpinning the increasing importance of transit. In all groups of cities, rail systems averaged increases of between 12 per cent and 32 per cent, while buses ranged from 5 to 13 per cent.

Newman and Kenworthy (2011b) show how growth in transit use is occurring while car use is declining in many places. Newman and colleagues (Newman *et al.* 2013) show that rail is the favoured mode in this growing worldwide trend towards transit. Rail systems are being extended and new lines installed to cope with capacity issues, and to give transit users a secure way of overcoming congestion and cutting travel times. These travel time savings are secure where rail operates on a reserved right-of-way, whereas on freeways travellers' time savings are generally short-lived due to congestion. Rail also offers users a generally more comfortable and superior passenger experience (Newman and Kenworthy 1999b).

Any transit system must offer an integrated experience for passengers and the ability to interchange and change directions without going into the central city. Systems are increasingly providing high-capacity, reliable structural spines of rail transit fed by buses to achieve the competitiveness needed for modal shifts from cars to transit. Thus integration of both modes is needed, but rail is generally critical to reversing car dependence in cities (Henry and Litman 2011).

Policy implications

A number of policies to reduce car use and carbon become evident.

Urban form, centres and parking

Densities are generally showing an upward or largely stable trend in most cities, though the upward trend is small on a metropolitan scale and there are some exceptions (e.g. minor declines in Canadian and European cities). There are many reasons to continue increasing densities, especially in focused centres, which create a more transit-oriented urban form (polycentrism, decentralized concentration). This should lead to a greater modal share for non-auto modes, shorter trip distances and lower energy use (Newman and Kenworthy 2006).

The sub-centres or CBDs of cities should be a focal point for employment growth and an increasingly important site for population increases so that centres also become better neighbourhoods. Such centres are best served by rail with feeder buses or trams and it is very important to have high-quality public realms to promote both greater transit and non-motorized trips to and within each centre. Dispersal of work into locations that are only easily accessible by car needs to be limited.

Parking supply in CBDs is generally on a downward trend. Where surface parking is replaced by public facilities, housing and other uses, the liveability and human attractiveness of the city also rises. For example, central and inner Portland (Oregon) has changed into a vital, colourful and economically successful hub over the space of 35 years through a wide variety of rejuvenation projects focused on alternative transport, pedestrian space and superb urban design.

Reduced CBD parking also strengthens the possibility of residential and job growth, since parking can occupy vast amounts of land. CBDs need to continue growing in jobs but not in parking supply, and surface commuter parking areas need to be built out with new housing (Newman and Kenworthy 1988a). It should also be feasible to replace parking structures with workplaces, better public environments and housing.

Private transport infrastructure

Car ownership in cities is still on the rise, except for Canadian cities where there has been a small decline. This has many implications, including the potential for more car use as well as extra land consumption because each new car has multiple parking places located at shopping areas, workplaces, hospitals, etc. The economic incentives and disincentives for cars need to be closely examined, for example, how cars are treated through taxation and free work parking. Cities should examine successful schemes for limiting the growth in car ownership such as those in Singapore and Shanghai, which seek to match new car registrations with the attrition rate of old cars through an auctioning system for the right to buy a car. Car sharing should be more widely promoted and supported by government and new ways of controlling congestion explored, for example, through congestion offsets (Bradley and Kenworthy 2012).

Road provision in cities needs to continue its general stabilization/decline by a combination of minimal new construction and reduced local roads through more compact land use. In particular, new freeways are still a major issue in some cities (e.g. Calgary). New freeways encourage car use and limit lower carbon transport, especially in cities in emerging

Table 5.2 Trends in 26 key transport-related variables in US, Canadian and Australian cities, 1995/1996 to 2005/2006

Variable	*US cities*			*Canadian cities*			*Australian cities*		
	1995	*2005*	*% Change*	*1996*	*2006*	*% Change*	*1996*	*2006*	*% Change*
Urban form factors									
Urban density (persons per ha)	14.9	15.4	**3.4%**	26.2	25.8	**−1.5%**	13.3	14.0	**5.3%**
Activity density (persons + jobs per ha)	22.4	23.6	**5.4%**	38.8	39.9	**2.8%**	18.8	20.3	**8.0%**
Proportion of jobs in CBD (%)	9.2	8.2	**−10.9%**	15.7	15.0	**−4.5%**	13.3	12.7	**−4.5%**
Private transport infrastructure factors									
Length of road per person (metres)	6.5	6.0	**−7.7%**	5.3	5.4	**1.9%**	8.6	7.6	**−11.6%**
Length of freeway per person (metres)	0.156	0.156	**0.0%**	0.122	0.157	**28.7%**	0.086	0.083	**−3.7%**
Parking spaces per 1,000 CBD jobs	555	487	**−12.3%**	390	319	**−18.2%**	367	298	**−18.9%**
Cars per 1,000 persons	587	640	**9.0%**	530	522	**−1.5%**	591	647	**9.4%**
Public transport infrastructure factors									
Total length of reserved public transport route per person (m/1,000 persons)	48.7	71.7	**47.2%**	56.3	66.8	**18.7%**	169.7	160.0	**−5.7%**
Ratio of reserved public transport infrastructure versus freeways	0.41	0.56	**36.6%**	0.55	0.56	**1.8%**	2.18	1.98	**−9.2%**
Private transport use factors									
Car vehicle kilometres per person	12,847	13,100	**2.0%**	6,386	6,519	**2.1%**	7,652	8,698	**13.7%**
Car passenger kilometres per person	18,155	18,703	**3.0%**	8,645	8,495	**−1.7%**	12,114	12,447	**2.7%**
Public transport service, use and performance factors									
Total public transport vehicle kilometres per person	31.6	39.2	**24.1%**	47.9	52.1	**8.8%**	57.4	58.9	**2.7%**
Total public transport seat kilometres of service per person	1,560	1,874	**20.1%**	2,290	2,368	**3.4%**	3,997	4,077	**2.0%**
Total rail seat kilometres per person	756	1,006	**33.1%**	676	841	**24.4%**	2,668	2,763	**3.6%**
Total bus seat kilometres per person	804	855	**6.3%**	1,607	1,522	**−5.3%**	1,283	1,265	**−1.4%**
Total public transport boardings per person	60	67	**11.7%**	140	151	**7.9%**	90	96	**5.8%**
Total rail boardings per person	23	28	**21.7%**	45	53	**17.8%**	47	50	**6.4%**
Total bus boardings per person	37	38	**2.7%**	95	97	**2.1%**	42	44	**4.8%**

Total public transport passenger kilometres per person	492	571	**16.1%**	917	1,031	**12.4%**	966	1,075	**11.3%**
Total rail passenger kilometres per person	274	341	**24.5%**	339	407	**20.2%**	638	713	**11.8%**
Total bus passenger kilometres per person	217	228	**5.3%**	577	620	**7.5%**	319	349	**9.3%**
Proportion of total motorized passenger kilometres on public transport (%)	2.9	3.2	**10.3%**	9.9	11.3	**14.1%**	7.6	8.1	**6.6%**
Ratio of public transport system speed to road traffic speed	0.52	0.52	**0.0%**	0.60	0.59	**–1.7%**	0.75	0.78	**4.0%**
Energy and externality factors									
Total private transport passenger energy use per person (MJ)	60,034	53,441	**–11.0%**	32,519	30,804	**–5.3%**	31,044	35,972	**15.9%**
Total public transport energy use per person (MJ)	811	963	**18.7%**	1,044	1,190	**14.0%**	876	1,036	**18.3%**
Total transport deaths per 100,000 people	12.7	9.5	**–25.2%**	6.5	6.3	**–3.1%**	9.1	6.2	**–31.4%**

Table 5.3 Trends in 26 key transport-related variables in European cities and Singapore/Hong Kong 1995 to 2005

Variable	*European cities*			*Singapore/Hong Kong*		
	1995	*2005*	*% Change*	*1995*	*2005*	*% Change*
Urban form factors						
Urban density (persons per ha)	43.8	42.5	**–3.0%**	215.4	217.3	**0.9%**
Activity density (persons + jobs per ha)	69.3	69.3	**0.0%**	317.7	330.6	**4.1%**
Proportion of jobs in CBD (%)	19.6	17.5	**–10.7%**	11.4	9.2	**–19.3%**
Private transport infrastructure factors						
Length of road per person (metres)	3.2	3.2	**0.0%**	0.6	0.5	**–16.7%**
Length of freeway per person (metres)	0.086	0.089	**3.5%**	0.025	0.027	**8.0%**
Parking spaces per 1,000 CBD jobs	224	241	**7.6%**	136	121	**–11.0%**
Cars per 1,000 persons	394	451	**14.5%**	73	78	**6.8%**
Public transport infrastructure factors						
Total length of reserved public transport route per person (m/1,000 persons)	240.5	300.2	**24.8%**	18.3	33.8	**84.7%**
Ratio of reserved public transport infrastructure versus freeways	4.73	5.44	**15.0%**	0.93	1.42	**52.7%**
Private transport use factors						
Car vehicle kilometres per person	4,759	5,130	**7.8%**	1,206	1,334	**10.6%**
Car passenger kilometres per person	6,737	7,232	**7.3%**	1,978	1,975	**–0.2%**
Public transport service, use and performance factors						
Total public transport vehicle kilometres per person	99.4	113.2	**13.9%**	119.9	134.5	**12.2%**
Total public transport seat kilometres of service per person	5,599	6,893	**23.1%**	6,882	7,267	**5.6%**
Total rail seat kilometres per person	3,594	4,718	**31.3%**	1,146	1,480	**29.1%**
Total bus seat kilometres per person	1,972	2,131	**8.1%**	5,468	5,691	**4.1%**
Total public transport boardings per person	304	330	**8.6%**	477	450	**–5.7%**
Total rail boardings per person	183	202	**10.4%**	152	169	**11.2%**
Total bus boardings per person	120	127	**5.8%**	319	277	**–13.2%**
Total public transport passenger kilometres per person	1,744	2,021	**15.9%**	3,169	3,786	**19.5%**
Total rail passenger kilometres per person	1,204	1,445	**20.0%**	1,286	1,704	**32.5%**
Total bus passenger kilometres per person	533	570	**7.0%**	1,826	2,055	**12.6%**
Proportion of total motorized passenger kilometres on public transport (%)	20.6	21.9	**6.3%**	63.3	65.3	**3.2%**
Ratio of public transport system speed to road traffic speed	0.86	0.95	**10.5%**	0.77	0.86	**11.7%**
Energy and externality factors						
Total private transport passenger energy use per person (MJ)	15,860	16,530	**4.2%**	6,447	6,077	**–5.7%**
Total public transport energy use per person (MJ)	1,264	1,694	**34.0%**	1,905	2,691	**41.3%**
Total transport deaths per 100,000 people	5.2	3.3	**–36.5%**	5.3	3.8	**–28.3%**

economies (Song 2013). Vancouver is a leader in transit-oriented design today partly because it built no freeways and it anchors a huge amount of new development around an ever-expanding rail system that provides the fastest travel option in the region for many trips (Schiller *et al.* 2010).

Public transport infrastructure, service and use

Apart from Australian cities, most cities are doing well in providing new, dedicated rights-of-way for transit, and in US, Canadian and European cities, as well as in Singapore and Hong Kong, there is an improving ratio between the provision of high-quality transit and freeways. For this to continue, not only must new transit systems, especially rail systems, be constructed but freeways need to be removed in strategic areas (e.g. in Seoul, Portland, San Francisco and Milwaukee). Such removals not only assist the competitive position of non-auto modes, but they generally radically improve the urban design and human qualities in affected areas. Every city has its opportunities for such removals, but the politics is often torturous, as the example of the Turcot Interchange in Montreal shows (Kenworthy and Townsend 2009).

The picture for transit service supply is a rather positive one, with all groups of cities expanding their service, in particular the amount of rail. All transit systems need to have non-timetable frequencies of ten minutes or less and the connection and integration between transit systems needs to improve. This integration needs to be in people getting between services conveniently and pleasantly, with maximized opportunities for linking trips (e.g. making shopping, business and social visits), as well as better fare integration, which provides attractive prices compared to the car, including rewarding committed users with big discounts. Worldwide, successful transit is built on transfers between modes and operators and on increasing committed users, generally with prepaid tickets. Physical connections between transit modes are best achieved in centres that operate on a time-pulse transfer system or that have saturation levels of service that make timetables irrelevant. The critical thing is the quality of transit services and how they relate to the urban structure of cities, as seen in Canadian cities where usage is high relative to infrastructure and service provision.

Transit use has, on average, headed in an almost universally positive direction in all groups of cities. In particular, rail modes have improved universally in both their trip-making rate as well as in passenger kilometres, while buses have fared worse. While transit has universally gained marginally over the car in its share of total motorized passenger kilometres, the actual increases are modest. Even in the transit-oriented European cities, the transit share of motorized passenger travel is still only 22 per cent. It is only in the dense Asian cities that transit mobility actually accounts for almost two-thirds of all motorized movement, an achievement that offers something to aim for. Policies to increase transit quality and choice need to be supported everywhere with complementary land-use policies to better integrate development, especially in dense centres, revised parking policies, a cessation of freeway building and charging the true cost of urban motoring, all of which are strongly related to each other as policy levers that need to be pulled in the same direction.

Private transport use

Measures are needed to curb the still growing (but slowing) car use in most cities. In Perth, for example, despite new rail lines and other positive directions, total annual car use per person over the ten years grew by over 1,300 km. In common with many cities, Perth suffers from some policy conflicts, for example, improving transit while at the same time building new major roads. Car use can be diminished with the planning-oriented policies already outlined, but there is also much that can be done through economic and fiscal changes, despite these being politically unpopular. This includes charging the 'right prices' for car use through congestion charges as in London, road pricing schemes, the price of fuel, and other 'carrots and sticks' such as the Certificate of Entitlements in Shanghai, and Singapore and Beijing's lottery for the right to purchase a car (Song 2013). Based on the many positive trends revealed here, it could be argued that there should have been greater increases in transit use and bigger reductions in car use. It could be postulated that 'road pricing' or more widespread 'congestion control' (Bradley and Kenworthy 2012) is the missing link in all cities to make more of the otherwise positive achievements in transport.

Energy and externalities

Australian cities and, to a much lesser extent, European cities, stand out as the only cities studied where private transport energy use is still growing. In all the other groups of cities, this use has declined. Gains in the fuel efficiency of cars are capable of delivering reductions in energy use, as witnessed in US cities. Here, the car fleet in 1995 was so profligate in fuel use (Kenworthy and Laube 2001) that when combined with almost stable car use, energy use per capita had declined by a significant 11 per cent by 2005. The same may be said for the Canadian car fleet, and to a lesser degree the Australian car fleet.

The evidence is that technological changes, which produce more fuel-efficient cars, can easily be exceeded by increases in car use (Newman and Kenworthy 1988b). Policies to limit car use and concurrently provide better propulsion systems and other energy-efficient design changes in cars are needed to promote lower carbon transport. Policies and technologies that promote renewable energy in transport need to be pursued, but they should not be seen as a panacea to the transport energy problem. For example, 'electro-mobility' in Europe is presented by engineers and technologists as a problem of replacing all existing car VKT with the same VKT driven by electric vehicles. Electric cars make more sense in a scenario of significantly lower demand for car mobility, even if those cars are more 'environmentally friendly'. Substituting the current massive demand for private mobility based on liquid fossil fuels is not possible, but it may be possible to meet the energy demands of significantly reduced car use per se (Kenworthy 2011). Transit systems have huge untapped capacity for reducing energy use and carbon emissions, especially through electric systems running on renewables.

Transport deaths are on a heartening downward trend in all the prosperous cities examined in this study. However, what may be behind this fairly consistent trend is open to interpretation and may have an element of risk avoidance or reduced exposure to accidents (Hillman 1992). For example, parents may be restricting the independent mobility of their children in response to traffic danger.

Conclusions

In 11 of the 26 variables (42 per cent), the trends are consistently positive for the groups of cities as a whole, with naturally some within-group variability. Of these 11 variables, nine are related to aspects of the public transport system, such as growing service supply, increased use and so on. Increasing activity density and a reduction in transport deaths per 1,000 persons are the other two consistently positive stories to come out of the research.

Furthermore, another ten variables achieve a mostly positive change, with generally only one group of cities diverging from that trend. Six of these mostly positive variables are related to transport infrastructure items, both private and public transport infrastructure. Thus, of the variables examined, 21 out of the 26 (81 per cent) demonstrate a generally positive outlook for lower carbon transport.

Only two variables demonstrate a consistently negative trend: car usage (vehicle kilometres) and the proportion of jobs located in the CBD. Only one further variable exhibits a mostly negative trend: car ownership. On the positive side, car use had already begun to show some positive signs by 2005 through the slowing of the growth rate, which appears to have continued after 2005 (Newman and Kenworthy 2011b; Puentes and Tomer 2009; Millard-Ball and Schipper 2010). The decline in the proportion of jobs in CBDs does not necessarily mean a decline in the absolute number of jobs located there. If non-CBD jobs are shifting to TOD-based sub-centres, then non-auto modes can still service them, but scattered jobs will be car-based. The two mixed-trend variables are car passenger kilometres, where the increases are very small and two groups show declines, and bus seat kilometres.

In summary, for these 33 cities, and possibly for many others, there is some turnaround in factors that have historically shown trends that work against lower carbon transport. However, it appears that cities will need to impose citywide congestion controls for low carbon urban transport to be given a major boost.

References

Alexander, I. 1980, 'Office dispersal in metropolitan areas II: case study results and conclusions', *Geoforum*, vol. 11, pp. 249–275.

—— 1981, 'Employment dispersal in metropolitan areas: equitable and energy saving?', *in Proceedings of 51st ANZAAS Conference*, Architecture and Planning Section, May.

Baillieu Knight Frank 1991, 'The wasting of the CBD: a paper on infrastructure use and employment in Melbourne', Baillieu Knight Frank Melbourne, Melbourne, Vic.

Bradley, M. and Kenworthy, J. 2012, 'Congestion offsets: transforming cities by letting buses compete', *World Transport Policy and Practice*, vol. 18, no. 4, pp. 46–69.

Cervero, R. 1998, *The Transit Metropolis: A Global Inquiry*, Island Press, Washington, DC.

Cervero, R. and Landis, J. 1992, 'Suburbanisation of jobs and the journey to work: a submarket analysis of commuting in the San Francisco Bay area', *Journal of Advanced Transportation*, vol. 26, no. 3, pp. 275–297.

Dimitriou, H. and Gackenheimer, R. (eds) 2010, *Transport Policy Making and Planning for Cities of the Developing World*, Edward Elgar, London.

Goldberg, M. and Mercer, J. 1986, *The Myth of the North-American City: Continentalism challenged*, University of British Columbia Press, Vancouver, BC.

Henry, L. and Litman, T. 2011, *Evaluating New Start Transit Program: Comparing rail and bus*, www.vtpi.org/bus_rail.pdf, viewed 6 March 2012.

Hillman, M. 1992, *Uses and Abuses of Transport and Road Safety Statistics in Policy Formation*, www.mayerhillman.com/Articles/EntryId/31/Uses-and-abuses-of-transport-and-road-safety-statistics-in-policy-formation.aspx, viewed 6 March 2012.

Kenworthy, J. 2006, 'The eco-city: ten key transport and planning dimensions for sustainable city development', *Environment and Urbanisation*, vol. 18, no. 1, pp. 67–85.

—— 2011, *International Benchmarking and Best Practice in Adapting to a Future of Electric Mobility in Germany: Sustainable Transport or Just Electric Cars?* Electric mobility in Hessen, Germany,

Consulting project to Land Hessen through the University of Applied Sciences, Frankfurt am Main, Germany.

Kenworthy, J. and Laube, F.B. 1999, *An International Sourcebook of Automobile Dependence in Cities, 1960–1990*, University Press of Colorado, Boulder, CO.

—— 2001, *The Millennium Cities Database for Sustainable Transport*, CD ROM, International Union of Public Transport (UITP), Brussels and Institute for Sustainability and Technology Policy (ISTP), Perth.

Kenworthy, J. and Townsend, C. 2009, 'Montreal's dualistic transport character: why Montreal needs upgraded transit and not more high capacity roads', in P. Gauthier, J. Jaeger and J. Prince (eds), *Montreal at the Crossroads: Superhighways, the Turcot and the environment*, Black Rose Books, Montreal, QC.

Millard-Ball, A. and Schipper, L. 2010, 'Are we reaching peak travel? Trends in passenger transport in eight industrialized countries', *Transport Reviews*, vol. 31, no. 3, pp. 1–22.

Newman, P.W.G. and Kenworthy, J.R. 1984, 'The use and abuse of driving cycle research: clarifying the relationship between traffic congestion, energy and emissions', *Transportation Quarterly*, vol. 38, no. 4, pp. 615–635.

—— 1988a, 'Parking and city centre vitality', paper presented to International Parking Conference, Perth, October.

—— 1988b, 'The transport energy trade-off: fuel-efficient traffic versus fuel-efficient cities', *Transportation Research*, vol. 22A, no. 3, pp. 163–174.

—— 1989, *Cities and Automobile Dependence: An international sourcebook*, Gower, Aldershot, UK.

—— 1999a, *Sustainability and Cities: Overcoming automobile dependence*, Island Press, Washington, DC.

—— 1999b, '"Relative speed" not "time savings": a new indicator for sustainable transport', in: *Papers of the 23rd Australasian Transport Research Forum*, Perth, WA, vol. 23, pt. 1, pp. 425–440.

—— 2006, 'Urban design to reduce automobile dependence', *Opolis*, vol. 2, no. 1, pp. 35–52.

—— 2011a, 'Evaluating the transport sector's contribution to greenhouse gas emissions and energy consumption', in R. Salter, S. Dhar and P. Newman (eds), *Technologies for Climate Change Mitigation: Transport sector*, UNEP RISØ Centre on Energy, Climate and Sustainable Development and Risø DTU National Laboratory for Sustainable Energy, Roskilde, Denmark.

—— 2011b, '"Peak car use": understanding the demise of automobile dependence', *World Transport Policy and Practice*, vol. 17, no. 2, pp. 31–42.

Puentes, R. and Tomer, A. 2009, *The Road Less Traveled: An analysis of vehicle miles traveled trends in the U.S.*, Brookings Institution, Washington, DC.

Schiller, P., Bruun, E. and Kenworthy, J. 2010, *An Introduction to Sustainable Transportation: Policy, planning and implementation*, Earthscan, London.

Song, Z. 2013, 'Beijing: transition to a transit city', in Institute for Mobility Research (ed.), *Megacity Mobility Culture: How cities move on in a diverse world*, Springer Verlag, Berlin.

Thomson, J.M. 1977, *Great Cities and Their Traffic*, Penguin Books, Middlesex, UK.

Watt, K.E.F. and Ayres, C. 1974, 'Urban land use patterns and transportation energy cost', in *Proceedings of the Annual Meeting of the American Association for the Advancement of Science*, San Francisco, CA.

Chapter 6

How city dwellers affect their resource hinterland

A spatial impact study of Australian households[1]

Manfred Lenzen and Greg M. Peters

Summary

This work links databases on household consumption, industrial production, economic turnover, employment, water use and greenhouse gas emissions in Australia into a spatially explicit model. The causal sequence starts with households demanding a certain consumer basket. This demand requires production in a complex supply-chain network of interdependent industry sectors. Even though the household may be confined to a particular geographical location, say, a dwelling in a city, the industries producing the life-cycle inputs for the commodities that the household demands will be spread at least all over Australia, if not beyond. Industrial production represents local points of economic activity, employment, water use and emissions, which have local economic, social and environmental impacts. This work follows the consumer basket of a typical household in Australia's two largest cities – Sydney, New South Wales, and Melbourne, Victoria – along its upstream supply chains, and to the numerous locations of production within Australia. The spatial spread is described using a detailed regional inter-industry model. Using industry-specific emissions profiles, industrial production is then translated into local impacts. We show that a typical household causes the emission of around 80 tonnes of greenhouse gas emissions, and the use of around three million litres of water, stimulates about AUD 140,000 (EUR 98,730) to circulate in the wider economy, and provides labour worth just under three full-time employment years. We also show maps that visually demonstrate how a local household affects the environment across an entire continent. Our model is unprecedented in its spatial and sectoral detail, at least for Australia.

Introduction

With environmental issues becoming more pressing every year, there is an abundance of databases that relate environmental variables to spatial entities. Examples of such databases are maps that reflect the spatial distribution of variables such as land use, species density, emissions of toxic compounds to land, air and water, changes in mean temperatures, rainfall, water availability, and many more. In Australia, climate change, land use and water management receive considerable attention due to the fact that Australia is on the one hand characterized by thin and poor topsoils as well as a dry and unpredictable climate, but on the other hand has agricultural industries that are significant exporters. As a response to an increased need for information supporting land and water management, Australian government agencies have produced, for example, the National Land and Water Resources Audit (National Land and Water Resources Audit 2008).

While such reference works provide a valuable overview of environmental pressures across an entire region at a given point in time, there is generally no information about the producing industries causing these environmental pressures. By contrast, databases linking industry sectors with environmental information, such as the National Greenhouse Gas Inventory (AGO 2008) and the Australian Water Accounts (ABS 2006; Vardon *et al.* 2006), generally do not contain any spatial information.

One consequence of this lack of information for the disciplines of industrial ecology and sustainable consumption is that widely applied input–output-based impact analyses of consumption cannot be undertaken in spatial detail. In Australia, this situation has improved recently with: (1) the construction of a Multi-Region Input–Output (MRIO) database for Australia at the state level (Gallego and Lenzen 2009); (2) the publication of an integrated regional database on land use (Australian Collaborative Land Use Mapping Program 2008); and (3) spatially explicit Australian Business Registers (ABS 2007c), labour surveys (ABS 2008) and National Pollutant Inventories (National Pollutant Inventory 2008), all of which contain information on production characteristics for statistical subdivisions (SSDs) and statistical local areas (SLAs).

The main aim of this chapter is to show how combining an MRIO of Australia with spatial databases on industrial production can enable input–output-type modelling of impacts at a detailed spatial disaggregation. Further, we show that because of its consistent treatment of system boundaries, input–output modelling is ideally suited for simultaneously quantifying environmental, social and economic impacts across entire product life cycles, and thus allowing valid comparisons and trade-offs across sustainability objectives.

In the following pages, we describe how consumption at one location causes economic, environmental and social impacts at other locations across Australia. In short, demand for an initial bundle of commodities at a particular place drives regional industries, which, in turn, engage in productive activity and exchange a wide range of intermediate commodities through a sectoral and spatial network of intra- and inter-regional supply chains. Summing up all industrial activity yields total levels of regional sectoral turnover, which can be translated into regional impacts. Thus, we are able to add to the existing notions of 'carbon footprint' and 'virtual water', an economic and social footprint measured in terms of turnover and employment, all at fine spatial detail.

This approach is in line with innovations in life cycle assessment (LCA) over the past decade, where researchers have moved to site- and receptor-dependent impact assessment (Potting *et al.* 1998; Sadamichi and Kato 2006). Examples of this phenomenon include the Japanese SAME (Spatial Area of iMpact Equivalency) approach, which overlays spatial emissions distributions with population density distributions in order to arrive at a measure of vulnerability (Nansai 2005), and efforts in Europe and North America to elaborate regional differences in the effect of eutrophying and acidifying emissions (Huijbregts *et al.* 2000; Bare *et al.* 2003).

The objective of the current analysis is to apply MRIO modelling to enumerate the impacts of typical households in Australia's two largest cities – Sydney and Melbourne – in terms of two environmental indicators (greenhouse gas emissions and water use), one economic indicator (turnover) and one social indicator (employment). The remainder of the chapter is organized as follows. Section 2 reviews the literature on subnational MRIOs and briefly describes how we couple our regional inter-industry model to new spatial databases. In Sections 3 and 4, the model is applied to our two case studies. The environmental, economic and social impacts caused by the consumption of these families are cast in terms of greenhouse gas emissions, water use, gross turnover and employment. Section 5 draws conclusions from these results.

Literature review

The literature on sub-national MRIOs provides abundant evidence for the fact that the integration of physical and economic data at this spatial level is generally fraught with difficulties because physical and financial industry statistics exist at different levels of industry sector detail. For example, greenhouse gas statistics may contain disaggregated information only on large fuel users such as electricity generation, metals manufacturing and transport, water-use statistics may focus only on water-intensive industries such as rice and cotton growing, and financial and employment statistics may disaggregate service sectors with high turnover. To integrate physical and monetary databases into one framework, researchers have used the strategy of finding the largest set of industries common to all statistics. Those industries that are disaggregated in one dataset but not in another have to be aggregated. This procedure often leads to a high degree of aggregation and a consequent loss of detailed information. Instead, in this work we disaggregate aggregated data sources by prorating or other established non-survey methods such as gravity methods (Gallego and Lenzen 2009). As a result, our MRIO – like virtually every input–output table – is based in part on survey data (the more important entries measured by data sources), and in part on non-survey data (modelled by the MRIO matrix-balancing procedures; see Lenzen *et al.* 2009). The non-survey data are naturally more uncertain, but they are also the less important entries. Under these circumstances, even a large number of uncertain but small elements in an input–output table do not necessarily pose a problem; Hewings and colleagues (Hewings *et al.* 1988) have shown conclusively that such frequently occurring major uncertainties hardly affect the magnitude of multipliers that form the core of our results. Jensen and West (1980) reported that a surprisingly large number of smaller elements can even be removed before output multipliers show a significant change.

The issues surrounding MRIO entry size and reliability are summarized in the work of Jensen (1980) on regional input–output tables, which introduced the concept of 'holistic accuracy'. According to Jensen, 'partitive', or table, accuracy represents the conventional understanding of a measure of the deviation of all estimated matrix elements from their 'true' values, whereas 'holistic', or operational, accuracy is concerned with the representativeness of a table of the synergistic characteristics of a regional economy. From this perspective, the accuracy of single elements may be unimportant, as long as the results of modelling exercises yield a realistic picture for the analyst or decision-maker's purposes. This is exactly the case in this work; we report multipliers and total impacts that are relatively insensitive to partitive accuracy (Gallego and Lenzen 2009).

A practical example highlights the importance of working at the highest possible detail. Consider a model application concerned with the embodied water passed on from wool growing to wool textiles manufacturing. Assume that in water data sources wool growing and cotton growing are disaggregated, and that cotton uses much more water than sheep. Assume further that other applications of the model could be about employment in manufacturing sectors, and that employment figures are only available for agriculture as a whole. Assume finally that in all data sources wool and cotton textiles are aggregated in one 'textiles manufacturing' sector, and that financial gross output exists at the disaggregated level for cotton growing. Our strategy would be to:

1 estimate employment for wool and cotton by prorating aggregate employment in agriculture across subsectors according to gross financial output;
2 split the 'textiles manufacturing' sector into two subsectors called 'wool textiles' and 'cotton textiles';
3 estimate gross financial output of wool and cotton textiles and split all inputs into the two textiles subsectors according to gross financial output, except set to zero the transactions 'wool into cotton textiles' and 'cotton into wool textiles'.

Subsequent analyses, for example, of effects on water use as a consequence of demand shifts for wool and cotton textiles, would make use of the fact that disaggregated water-use figures were available for both wool and cotton growing, which are the main supplier of wool and cotton textiles, respectively. Even though all other (financial, employment, etc.) quantities for these subsectors are not based on specific data but prorated according to the gross output proxy, this would not matter, since water use is the quantity of interest. Hence, and summarizing, there is no penalty in generating a large number of uncertain but small estimates in an inter-industry model, as long as important (large) data items are sufficiently accurate.

Methodology

Our MRIO is essentially a collation of regional transaction tables connecting supply industries with using industries. The economic inter-industry system is accompanied by matching physical data on greenhouse gas emissions, water use, employment and 1,273 other indicators. The methodology is explained in detail in Lenzen and Peters (2010) and further information may be found in Gallego and Lenzen (2009).

Results

Overall results

Two family archetypes, one from Sydney and one from Melbourne, were selected to represent typical high-income, inner-city-dwelling, well-educated, possibly environmentally aware, busy professionals with no children or only one child (see Table 6.1). Socioeconomic and demographic data on these family types were taken from the Australian Household Expenditure Survey (ABS 2007b) and the census (ABS 2007a). The families both belong to

Table 6.1 Characteristics of the household groups selected for Sydney and Melbourne

	Sydney	*Melbourne*
Average weekly dispensable household income	$ 1,174.20	$ 1,067.89
Percentage of persons under 18 years	24%	24%
Average number of persons in the household	2.7	2.7
Tenure type (% owners with or without a mortgage)	69%	74%
Percentage of employed persons in household	75%	73%
House type (% of persons living in a separate house)	72%	82%
Education (% of persons with a degree)	43%	44%
Percentage of persons travelling to work by car	68%	80%
Car ownership (vehicles per household)	1.41	1.57

the upper income brackets; annual household income is in excess of AUD 50,000 (EUR 35,000) after tax. Such families usually own a detached house, the adults are in full-time, paid employment, and they generally use one or even two cars to drive to work.

Note that differences between the impacts we associate with the families in the two cities are partially due to differences in their expenditure patterns as reported in the Household Expenditure Survey (ABS 2007b). Neither the two survey samples nor the selection of families by the authors are representative for the city or the state. Therefore, the following results must not be taken as a general comparison.

The families' nationwide impacts in terms of greenhouse gas emissions, water use, economic turnover and employment are summarized in Table 6.2: the households cause around 80 tonnes of greenhouse gas emissions, and the use of around three million litres of water, stimulate about AUD 140,000 (EUR 98,197) to circulate in the wider economy, and provide labour worth just under three employment years at full-time equivalent. The differences between the two households may partly be explained by Melbourne's colder climate and higher rainfall (and correspondingly fewer water restrictions and the more agricultural irrigation) and more emissions-intensive brown-coal use in Victoria, compared to New South Wales. These impacts are both direct and indirect, with indirect supply-chain effects exceeding direct effects for indicators except employment. In other words, the families cause significantly more greenhouse gas emissions and the use of more water, and stimulate more money to circulate around the wider economy through their consumption of goods and services, than directly by driving cars, burning household gas, running their taps or spending money at the shops. This result highlights the importance of indirect effects for decision-making, be it at the household or at the government policy level (Lenzen 2001). These families do, however, provide more employment directly than they create elsewhere through their consumption; the situation is reversed for low-income households, which create more jobs indirectly than they contribute themselves.

Direct emissions exclude those from electricity use and air travel (because these are indirect emissions caused by the power plants and airlines), but they include private car travel and household gas use. Direct water use is metered water drawn from household taps. Direct

Table 6.2 Total annual impacts per household for Sydney and Melbourne in terms of four indicators

		Sydney	*Melbourne*
Total	Greenhouse gas emissions (t)	77.4	91.3
	Water use (ML)	2.3	3.4
	Economic turnover ($)	130,549	150,735
	Employment (FTE)	2.85	2.90
Direct	Greenhouse gas emissions (t)	9.3	13.0
	Water use (ML)	0.5	0.7
	Economic turnover ($)	61,058	55,530
	Employment (FTE)	1.88	1.84
Indirect	Greenhouse gas emissions (t)	68.2	78.3
	Water use (ML)	1.9	2.7
	Economic turnover ($)	69,490	95,205
	Employment (FTE)	0.97	1.07

Note: t = tonnes, $ = Australian dollars, ML = 1 megalitre, FTE = full-time equivalent.

turnover is the money spent by household members. Direct employment is the labour provided by the household members themselves. Note again that the values in Table 6.2 are not representative city-wide or state-wide averages and therefore cannot be used to conclude comparative performance of the cities.

Spatial results

The maps presented in the following pages distinguish 1,406 statistical local areas (SLAs) across the entire Australian continent. The size of SLAs is determined by their population. As a consequence, SLAs in the arid interior of Australia are far larger than inner-city SLAs.

Greenhouse gas emissions

Consumption by the Sydney family causes greenhouse gases to be emitted all over Australia (see Figure 6.1). Prominent contributions to the family's greenhouse footprint[2] are emissions from large power plants in New South Wales (NSW), two around Muswellbrook in the Hunter Valley, one major plant near Lake Macquarie, and two near Lithgow. These power plants are all co-located with coal deposits and energy-intensive industries. Our Sydney household also draws indirectly on inter-state power from the Latrobe SLA in Victoria (VIC), from Queensland (Nanango, Banana and Gladstone SLAs), and (despite the absence of a direct power connection) the Collie SLA in Western Australia (WA). Furthermore, significant emissions caused by our Sydney household occur as follows:

- in iron and steel semi-manufacturing in Port Kembla south of Sydney;
- through land clearing for, and enteric fermentation in, beef cattle in NSW (Carrathool, Balranald, Wentworth, Central Darling, Brewarrina, Walgett, Gunnedah);
- during petrol refining at Kurnell on Botany Bay, and at Parramatta, NSW;
- from softwood timber felled around Tumut, Oberon and Bombala, NSW;
- through enteric fermentation in sheep for wool in NSW (Bourke, Cobar);
- during cement and lime production (Wingecarribee, NSW and Kalgoorlie, WA);
- from natural gas operations near Roebourne and Ashburton, WA.

Our Melbourne household's carbon footprint extends less beyond state borders (see Figure 6.2). This is partly because Victoria produces more electricity than it consumes, with the remainder being sent to South Australia (SA) and NSW. Moreover, Victoria includes a major part of the Murray-Darling Basin where most of Australia's food, especially dairy products, is produced. As a consequence, Victoria is relatively self-sufficient in terms of emissions-intensive products. Major contributions to the family's greenhouse gas budget are as follows:

- electricity (Latrobe, VIC);
- iron and steel (Wyndham, VIC);
- natural gas (Rosedale, VIC);
- beef cattle (Mildura, Swan Hill, Towong, all VIC);
- basic chemicals (Altona and Maribyrnong, VIC);
- petrol refining (Corio and Altona, VIC);
- air transport (Craigieburn, VIC);
- softwood timber (Glenelg and Rosedale, VIC);

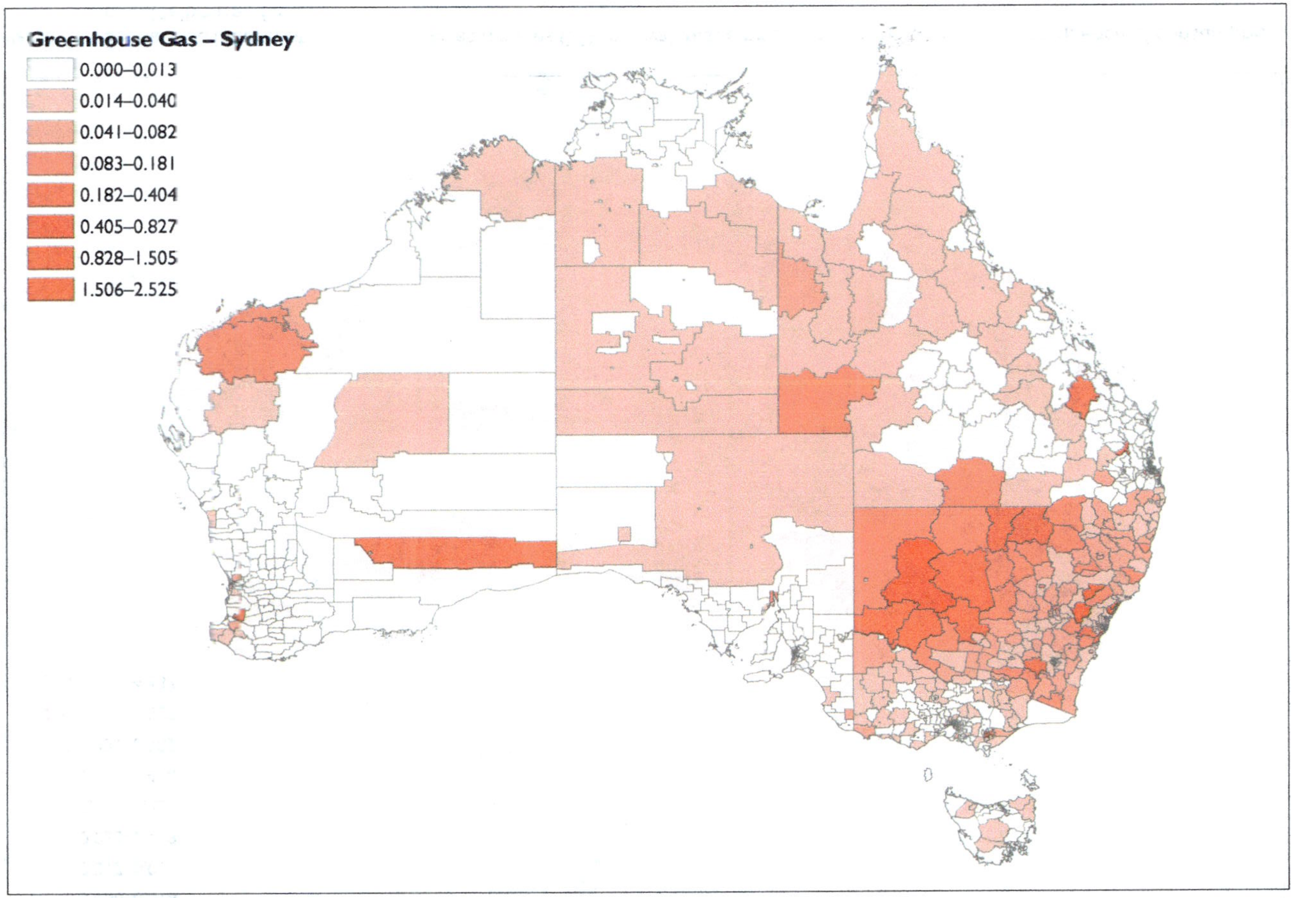

Figure 6.1 Spatial distribution of greenhouse gases (tonnes CO_2 equivalents) emitted across Australia as a consequence of consumption by a Sydney family. One tonne (t) = 103 kilograms (kg, SI) × 1.102 short tons.

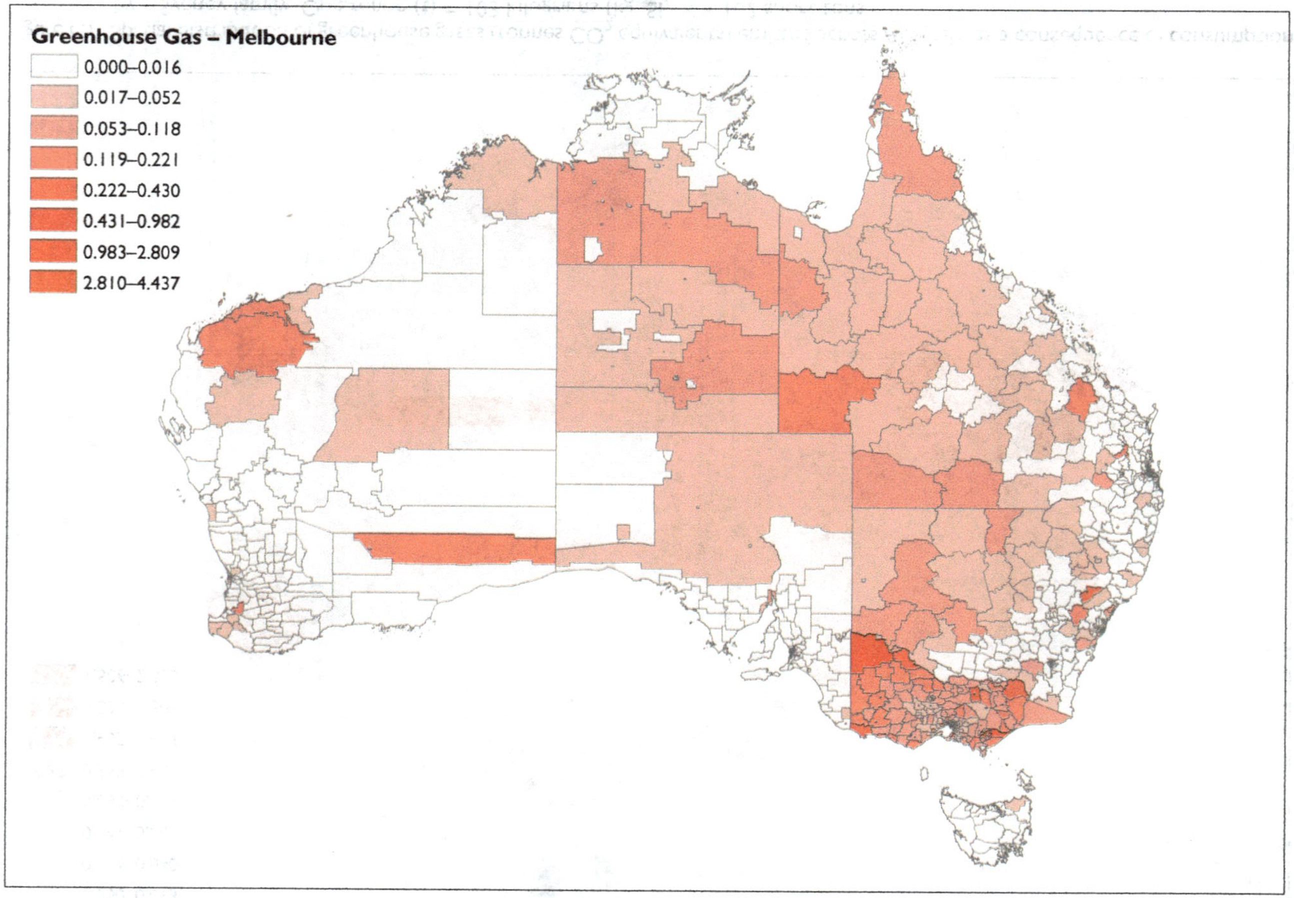

Figure 6.2 Spatial distribution of greenhouse gases (tonnes CO_2 equivalents) emitted across Australia as a consequence of consumption by a Melbourne family.

- cars (Broadmeadow, VIC);
- natural gas (Roebourne and Ashburton, WA);
- sugar (Maribyrnong, VIC).

Water use

The water-use maps (see Figures 6.3 and 6.4) show water-use points largely confined to the household's resident state, which is a reflection of irrigated agricultural commodities not extensively crossing state boundaries. This is especially true, for example, for untreated milk, which appears in the Sydney household's water budget as produced around Bega, Taree, Wollondilly and Singleton (all NSW), and in the Melbourne household's budget as produced around Loddon, Shepparton, Gannawarra and Campaspe (all VIC). Further important items are fruit (Griffith and Leeton, NSW, and Mildura and Swan Hill, VIC), rice (Griffith and Murrumbidgee, NSW) and vegetables (Hay, Narromine and Cowra, NSW, and Casey and Baw Baw, VIC).

For the Sydney household, the top inter-state water-use points are untreated milk (Loddon, Shepparton, Gannawarra and Campaspe, VIC), and grapes for wine from Mildura (VIC). For the Melbourne household, the top inter-state water-use points are unginned cotton (Moree, Narrabri, Walgett and Warren, NSW), rice (Griffith and Murrumbidgee, NSW), grapes for wine (Gawler, SA), and beef cattle (Berrigan, NSW and Johnstone, QLD).

Economic turnover

In contrast to the two environmental indicators – greenhouse gas emissions and water use – economic turnover consequences of our households' consumption occur much more within the population centres, and in very different sectors (see Figures 6.5 and 6.6). The payment of rents and mortgages for private dwellings creates considerable turnover in the wealthy residential areas of inner Sydney and Melbourne, Woollahra, Ku-ring-gai and North Sydney (NSW), and St Kilda, Brighton and Caulfield (VIC). Hotels, clubs, restaurants and cafés (inner Sydney and Marrickville, NSW, and inner Melbourne and Prahran, VIC) also form a large part of the households' economic footprint. Further important components are large city retail malls (Penrith, Rockdale, Marrickville and Warringah, NSW, and Ringwood, Dandenong and Frankston, VIC), followed by banking, clothing, telecommunications and other city-based businesses.

Prominent agricultural products are, for example, wine (Cessnock, NSW), poultry (Casey, VIC), fish (Bega, NSW) and fresh meat (Melton, VIC). However, these are rare and generally low-ranking occurrences in the components list, which goes to show that although agricultural commodities cause substantial environmental and resource impacts, they do not fetch high prices and are hence not endowed with large economic turnover.

Significant inter-state turnover is generated, for example, for trade advertisers in Flynn (ACT), and LPG and LNG producers in Toodyay (WA).

Employment

Employment impacts of our households' consumption are distributed across a wide range of mostly service industries such as banking, market research, business management, doctors, dentists, optometrists, clothing, retail trade, legal services, hotels, clubs, restaurants, cafés,

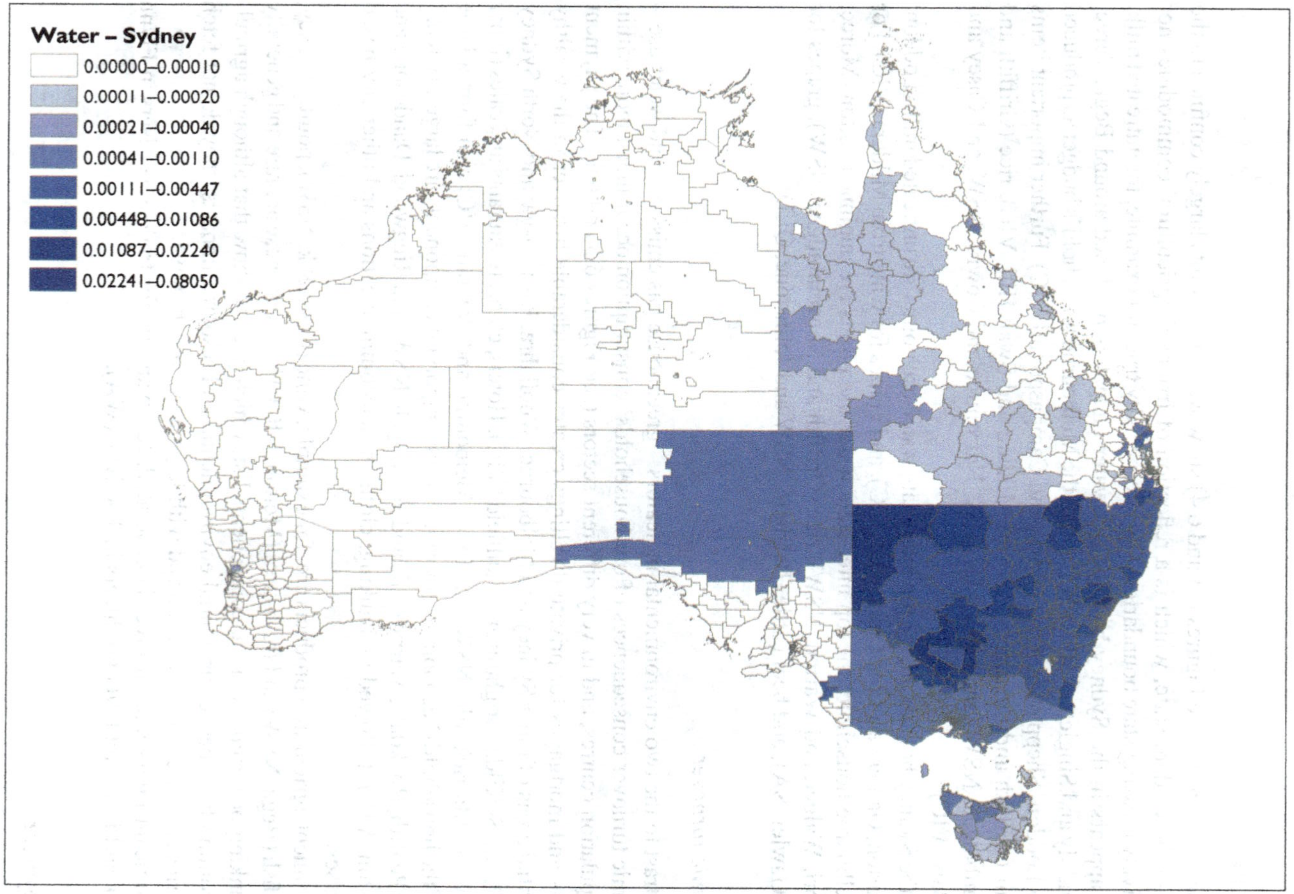

Figure 6.3 Spatial distribution of water (ML) consumed across Australia as a consequence of consumption by a Sydney family.

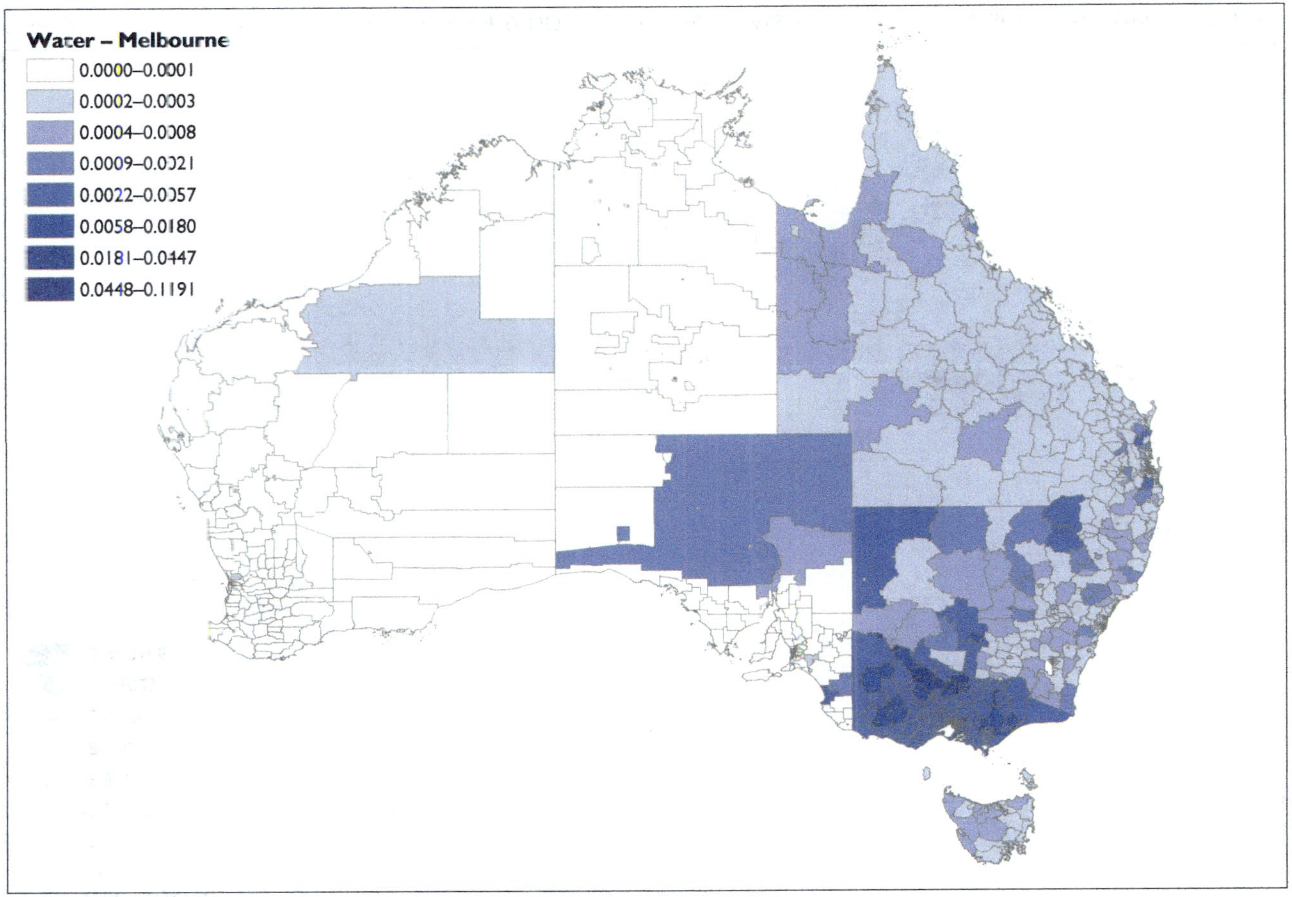

Figure 6.4 Spatial distribution of water (ML) consumed across Australia as a consequence of consumption by a Melbourne family.

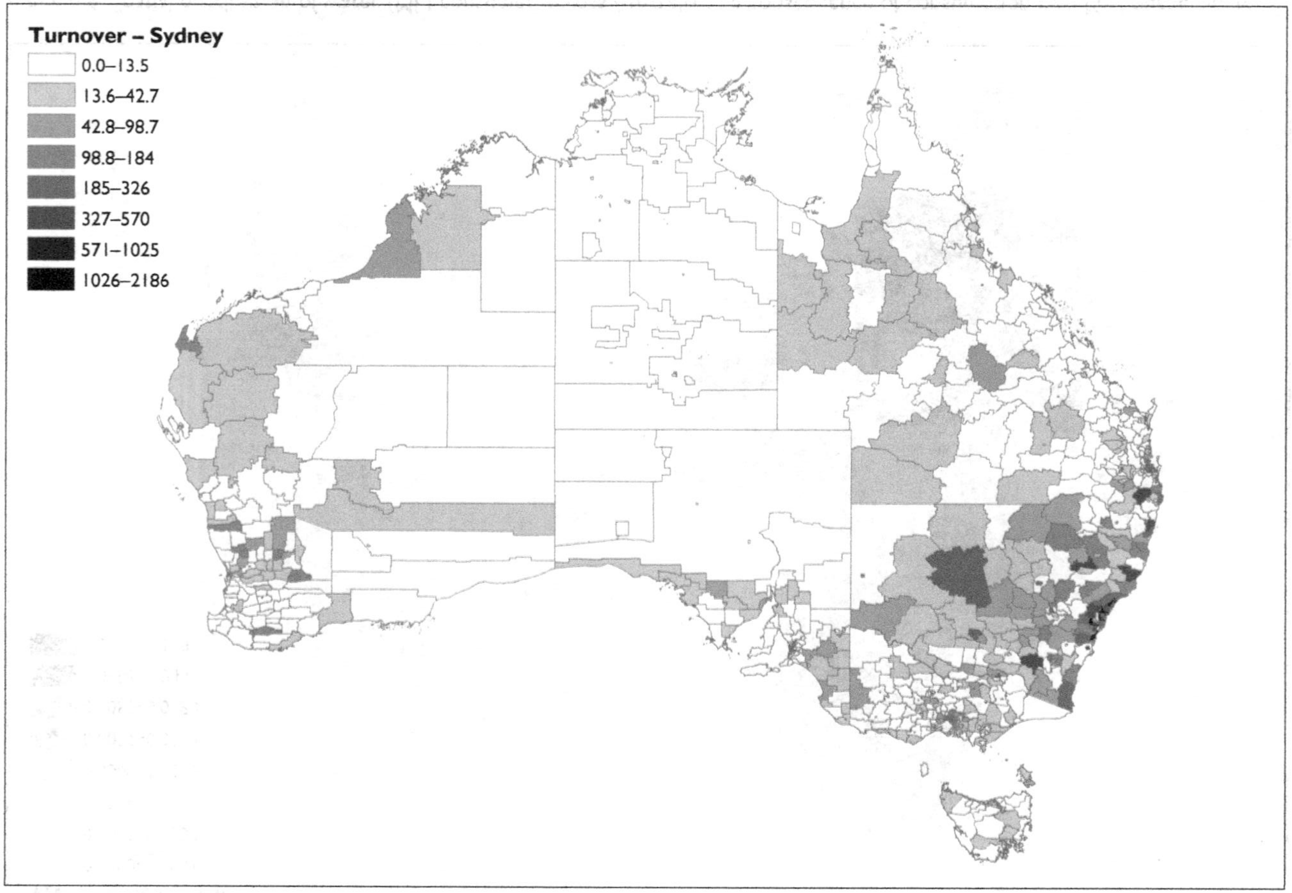

Figure 6.5 Spatial distribution of economic turnover (AUD) stimulated across Australia as a consequence of consumption by a Sydney family.

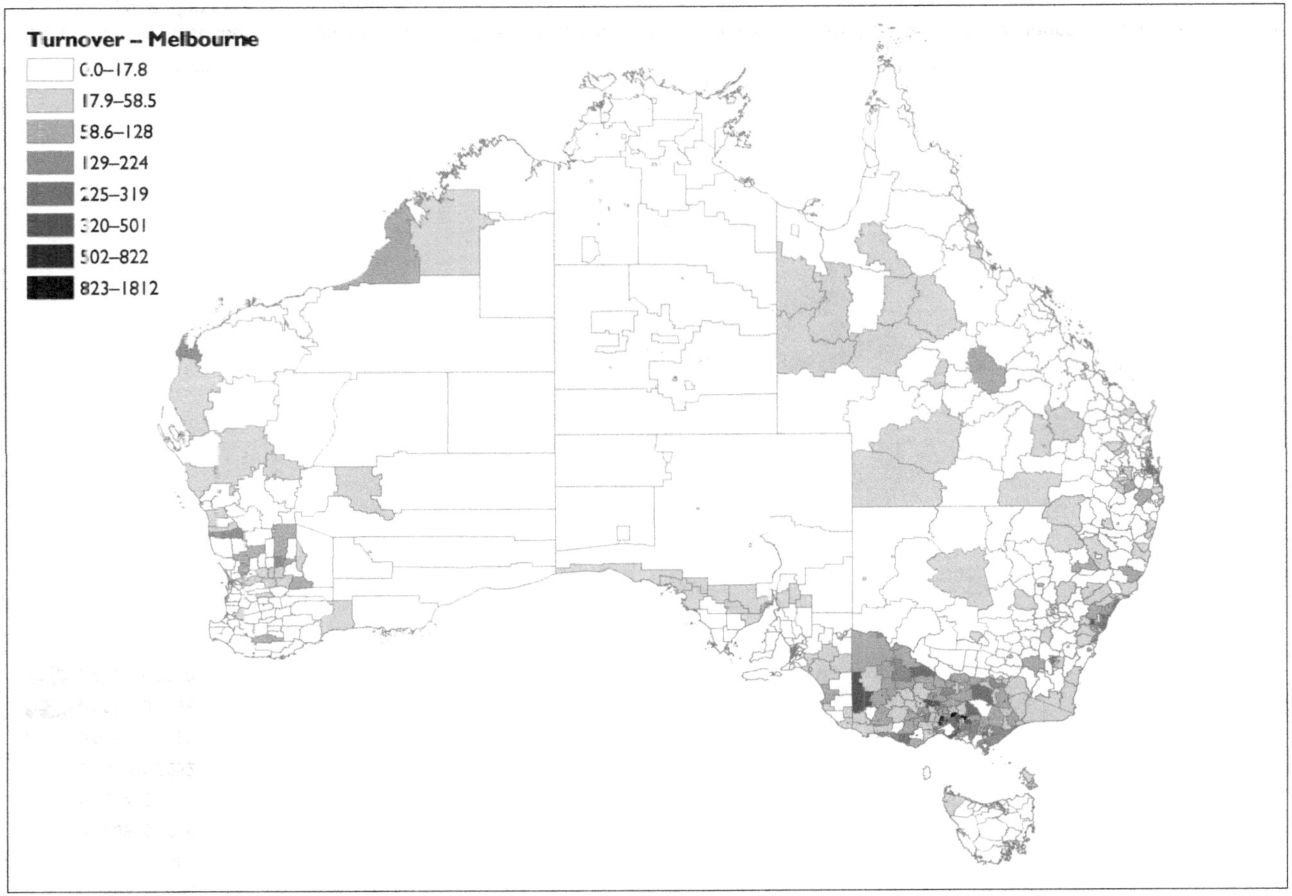

Figure 6.6 Spatial distribution of economic turnover (AUD) stimulated across Australia as a consequence of consumption by a Melbourne family.

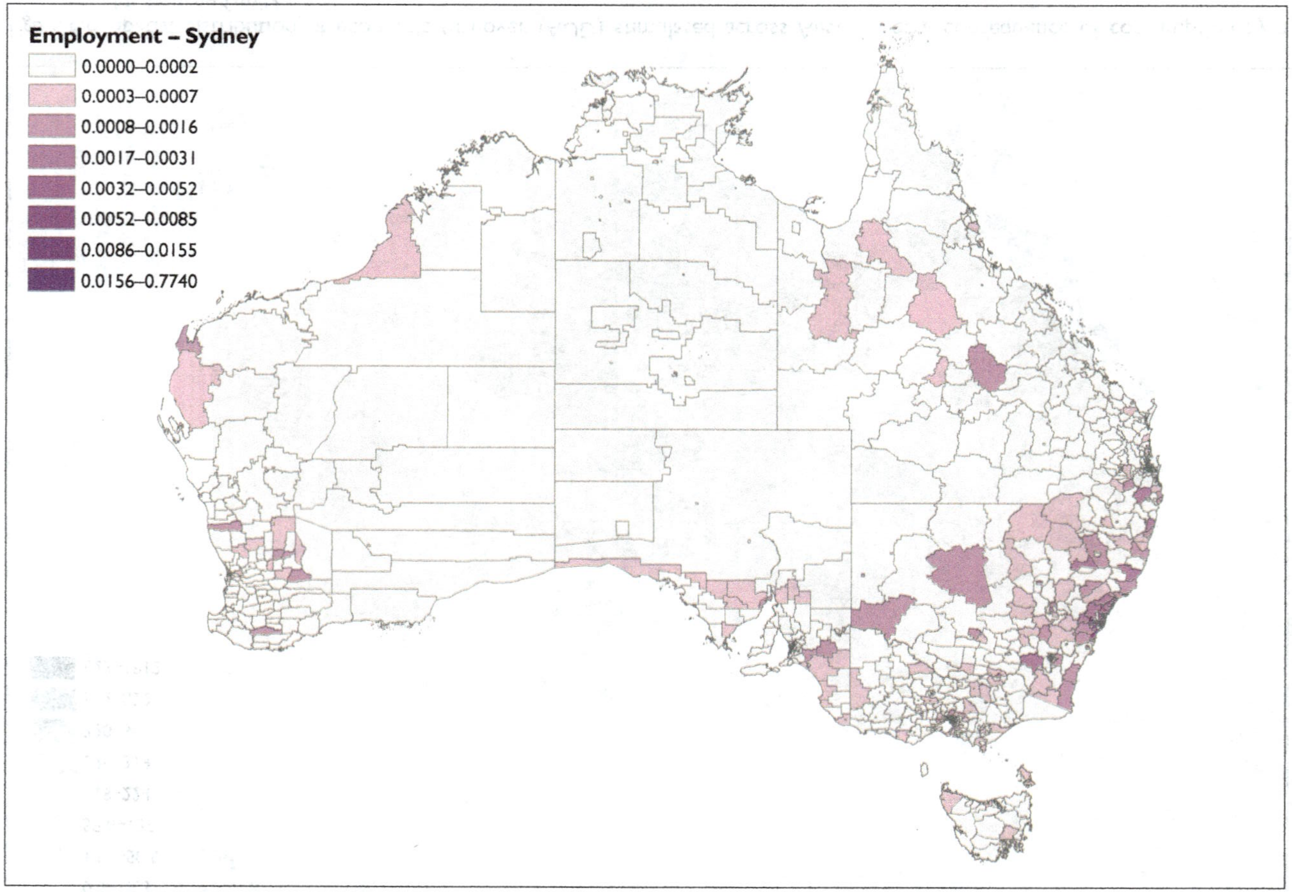

Figure 6.7 Spatial distribution of employment (full-time equivalent) generated across Australia as a consequence of consumption by a Sydney family.

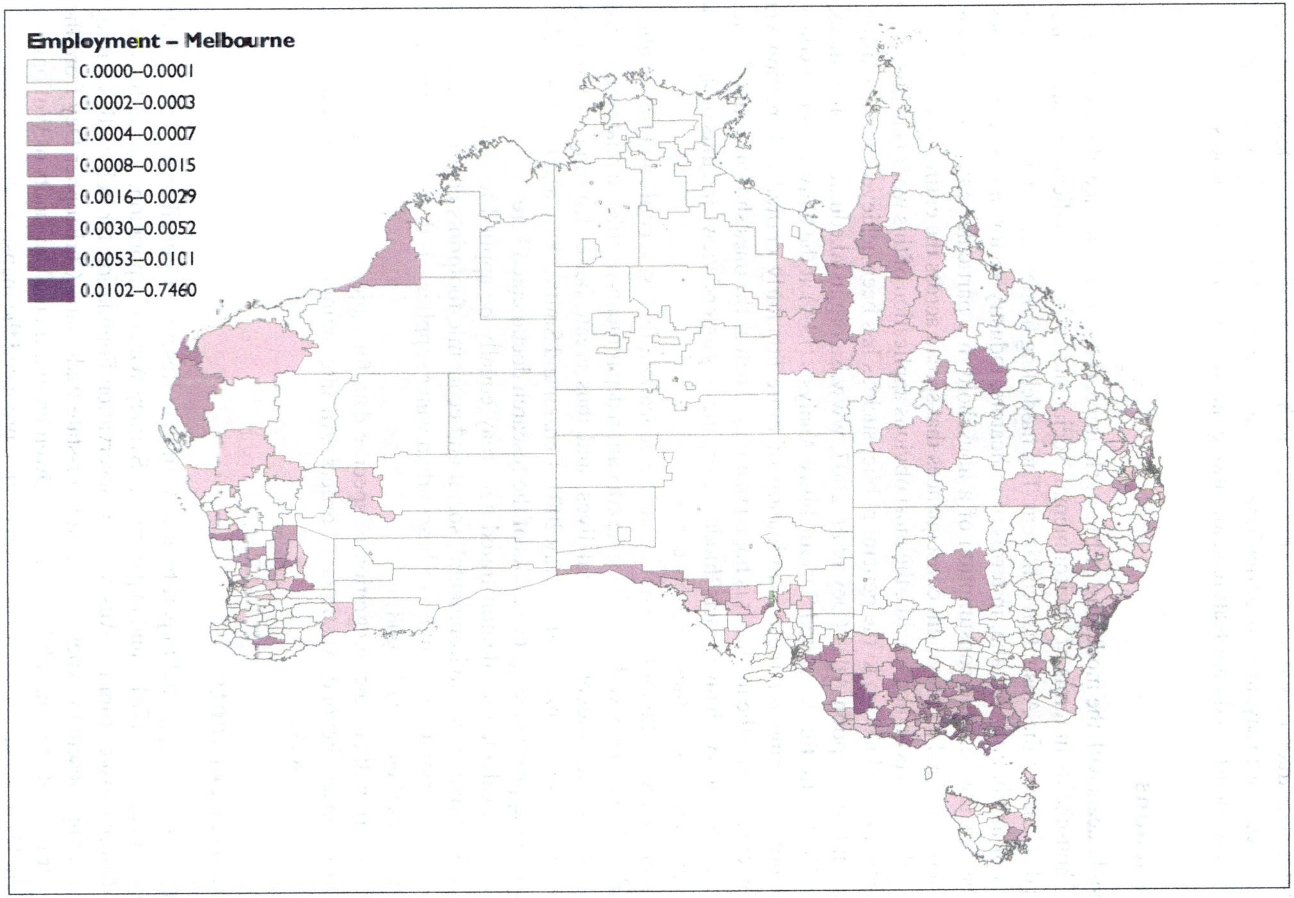

Figure 6.3 Spatial distribution of employment (full-time equivalent) generated across Australia as a consequence of consumption by a Melbourne family.

property services, education, insurance and community services. Since these industries are mainly located in urban centres, the main employment impacts also occur there (see Figures 6.7 and 6.8). Since these service industries often require personal interaction, there is very little inter-state trade of service commodities, and hence very few inter-state employment impacts. Needless to say, modern agricultural and resource industries are highly mechanized, so that these industries do not feature at all among the top employment impacts.

Conclusions

This study deals with the modelling of life-cycle impacts as a consequence of the consumption of households in terms of environmental, social and economic indicators, and at high spatial detail across the Australian continent. The modelling framework is at present the most detailed of its kind in Australia, in terms of indicators, industry sectors and spatial entities. Our results show that the consumption of a household in a particular city has emissions, water, monetary and employment consequences that spread across the entire country.

Since the focus is on households, the most obvious application of this work is in providing information and education to the general public, such as has been achieved by the existing Australian Consumption Atlas (Dey *et al.* 2007, www.isa.org.usyd.edu.au/research/index.shtml). The Centre for Integrated Sustainability Analysis (ISA) at the University of Sydney is currently preparing for the dissemination of more popularly crafted information via museum exhibits and the media. It is hoped that maps such as the ones shown in this chapter will unveil the supply-chain impacts 'hidden' from the consumers' view, and thus bring home the sustainability issues these impacts entail.

A second outlet of this information is public education; that is, schools and universities. Similarly, ISA has been working with state education departments in order to relate embodied emissions more to students' personal lives, and thus create better learning experiences (Lenzen *et al.* 2002; Lenzen and Murray 2001; Lenzen and Smith 2000), and it is planned to replicate this approach using the results of the research documented here.

Finally, the calculations demonstrated here may equally be applied to companies, and maps of the companies' environmental, social and economic footprints can be created. An example of a corporate case study – a large urban water supplier – is documented by Peters and colleagues (Peters *et al.* 2008).

Future work will see the integration of specific data for major trading partners, and the tracing of impacts beyond Australia's borders. An international MRIO seems a natural extension of this work.

Acknowledgements

This research was supported in part by the Australian Research Council (ARC) through its Linkage Project LP0347812, and in part by the Sydney Water Corporation. The authors also thank Rowena Joske from the Australian Conservation Foundation for help with sourcing and screening household expenditure and inter-state trade data. Christopher Dey from the Centre for Integrated Sustainability Analysis, Owen Price from the University of Wollongong and Craig Davis from Sydney Water Corporation produced the maps.

Notes

1 This chapter was originally published as M. Lenzen and G.M. Peters. 2010, 'How city dwellers affect their resource hinterland: a spatial impact study of Australian households', *Journal of Industrial Ecology*, vol. 14, no. 1, pp. 73–90. © 2009 by Yale University, http://dx.doi.org/10.1111/j.1530-9290.2009.00190.x, and is reproduced by permission.
2 Our figures go beyond the carbon footprint because they include CH_4 and N_2O.

References

Australian Bureau of Statistics (ABS) 2006, *Water Accounts for Australia 2004–05*, cat. no. 4610.0, Australian Bureau of Statistics, Canberra, ACT.

—— 2007a, *2001 Census Data by Location, Community Profiles*, Australian Bureau of Statistics, www.abs.gov.au/websitedbs/d3310114.nsf/home/Census+data, viewed 1 April 2013.

—— 2007b, *Household Expenditure Survey: Detailed expenditure items*, cat. no. 6535.0, Australian Bureau of Statistics, Canberra, ACT.

—— 2007c, *Counts of Australian Businesses, Including Entries and Exits*, cat. no. 8165.0, Australian Bureau of Statistics, Canberra, ACT.

—— 2008, *Labour Force, Australia, Detailed*, cat. no. 6291.0.55.001, Australian Bureau of Statistics, Canberra, ACT.

Australian Collaborative Land Use Mapping Program 2008, *Catchment Scale Land Use Mapping for Australia*, Bureau of Rural Sciences, Canberra, ACT.

Australian Greenhouse Office (AGO) 2008, *National Greenhouse Gas Inventory*, Department of the Environment and Water Resources, Canberra, ACT.

Bare, J.C., Norris, G.A., Pennington, D.W. and McKone, T. 2003, 'TRACI: the Tool for the Reduction and Assessment of Chemical and other Environmental Impacts', *Journal of Industrial Ecology*, vol. 6, nos 3–4, pp. 49–78.

Dey, C., Berger, C., Foran, B., Foran, M., Joske, R., Lenzen, M. and Wood, R. 2007, 'Household environmental pressure from consumption: an Australian environmental atlas', in G Birch (ed.), *Water Wind Art and Debate: How environmental concerns impact on disciplinary research*, Sydney University Press, Sydney, NSW.

Gallego, B. and Lenzen, M. 2009, 'Estimating generalised regional input–output systems: a case study of Australia', in M. Ruth and B. Davíðsdóttir (eds), *The Dynamics of Regions and Networks in Industrial Ecosystems*, Edward Elgar Publishing, Boston, MA.

Hewings, G.J.D., Sonis, M. and Jensen, R.C. 1988, 'Fields of influence of technological change in input–output models', *Papers of the Regional Science Association*, vol. 64, pp. 25–36.

Huijbregts, M.A.J., Schöpp, W., Verkuijlen, E., Heijungs, R. and Reijnders, L. 2000, 'Spatially explicit characterization of acidifying and eutrophying air pollution in Life-Cycle Assessment', *Journal of Industrial Ecology*, vol. 4, no. 3, pp. 75–92.

Jensen, R.C. 1980, 'The concept of accuracy in regional input–output models', *International Regional Science Review*, vol. 5, pp. 139–154.

Jensen, R.C. and West, R.G. 1980, 'The effect of relative coefficient size on input–output multipliers', *Environment and Planning A*, vol. 12, pp. 659–670.

Lenzen, M. 2001, 'The importance of goods and services consumption in household greenhouse gas emissions calculators', *Ambio*, vol. 30, no. 7, pp. 439–442.

Lenzen, M. and Smith, S. 2000, 'Teaching responsibility for climate change: three neglected issues', *Australian Journal of Environmental Education*, vol. 15/16, pp. 69–78.

Lenzen, M. and Murray, J. 2001, 'The role of equity and lifestyles in education about climate change: experiences from a large-scale teacher development program', *Canadian Journal of Environmental Education*, vol. 6, pp. 32–51.

Lenzen, M. and Peters, G. 2010, 'How city dwellers affect their resource hinterland: a spatial impact study of Australian households', *Journal of Industrial Ecology*, vol. 14, no. 1, pp. 73–90.

Lenzen, M., Dey, C.J. and Murray, J. 2002, 'A personal approach to teaching about climate change', *Australian Journal of Environmental Education*, vol. 18, pp. 35–45.

Nansai, K. 2005, 'Site-dependent Life-Cycle Analysis by the SAME approach: its concept, usefulness, and application to the calculation of embodied impact intensity by means of an input–output analysis', *Environmental Science & Technology*, vol. 39, no. 18, pp. 7318–7328.

National Land and Water Resources Audit 2008, Atlas and Data Library.

National Pollutant Inventory 2008, *2006–07 NPI Reports and Maps*, Department of the Environment, Water, Heritage and the Arts, Canberra, www.npi.gov.au/overview/view.html, viewed 1 April 2013.

Peters, G., Sack, F., Lenzen, M., Lundie, S. and Gallego, B. 2008, 'Towards a deeper and broader ecological footprint', *Engineering Sustainability*, vol. 161, no. 1, pp. 31–37.

Potting, J., Schöpp, W., Blok, K. and Hauschild, M. 1998, 'Site-dependent life-cycle impact assessment of acidification', *Journal of Industrial Ecology*, vol. 2, no. 2, pp. 63–87.

Sadamichi, Y. and Kato, S. 2006, 'Life cycle impact assessment of fuel procuring and electricity generating processes in Japan by using an "LCA-NETS" scheme', *International Journal of Emerging Electric Power Systems*, vol. 7, no. 1, p. 4.

Vardon, M., Lenzen, N., Peevor, S. and Creaser, M. 2006, 'Water accounting in Australia', *Ecological Economics*, vol. 61, no. 4, pp. 650–659.

Part II

New directions

Designing for low carbon cities

Introduction to Part II

Part II provides case studies of cities in transition to low carbon urbanity and lifestyles. Green buildings and low carbon precincts minimize pollution and environmental damage by respecting their surroundings. They reuse rainwater and treat and recycle wastewater on site. They do not dump waste materials off site but reuse them as much as possible. They reduce their energy needs by using the maximum low-embodied-energy insulation possible. Low carbon precincts of green buildings have good natural ventilation, use low-energy lighting and electrical appliances, and make use of passive and active solar power. Low carbon precincts have reduced embodied energy by using mainly locally sourced low energy materials from sustainably managed, renewable or recycled sources, with minimal use of imported materials.

Information technologies and green infrastructure are likely to be critical in any urban transformation, as in the future precincts will harvest energy, with small distributed technologies (woven seamlessly into the urban fabric) that will generate energy within the city, store it effectively and distribute it wisely. It is likely that such decentralized technologies are inherently more amenable to innovation than large, centralized and capital-intensive infrastructure.

The chapters in Part II foreground urban and architectural design as the most powerful tool for connecting people and place in a low carbon urban future. Mitchell Joachim challenges designers to rethink scale in order to imagine a future of 'urbaneering', illustrating his idea in relation to the city of New York, with a particular emphasis on Brooklyn. Nico Tillie, Andy van den Dobbelsteen and Sebastian Carney explain how they have combined various tools and methods in Rotterdam, such as 'exergy' (balancing energy needs and reusing waste heat) to develop a systematic process for shifting the city to a low carbon future. Steffen Lehmann describes the possibilities of green districts and carbon engineering as ways to reduce urban heat islands and generate energy, and identifies an urban renewal project in Sydney as a possible exemplar. Christoph Ingenhoven, Martin Reuter and Ben Dieckmann report how practising architects are contributing to low carbon cities through their designs for a university precinct in Dublin and an innovative office tower in Sydney. Robert Cervero documents the low carbon changes underway in Seoul and draws attention to the importance of leadership and governance when taking new directions in urban development. Shipra Narang Suri notes the enormous challenge of making India's cities more ecologically and socially viable, reviews government policies on sustainability that relate to urbanity and suggests that the low carbon goal may be both more feasible and helpful in policy and practice.

Chapter 7

Future ecological design as urbaneering

New positions on city-making without scale

Mitchell Joachim

Summary

Whose job is it to create a city? Our intention is to jumpstart a new profession that can reinvent and negotiate the complex mix that encompasses a city. We have defined a radical new occupation to regenerate, pioneer and sustain the future urban realm. These innovative multidisciplinary advocates are called 'urbaneers'. Their immense task is to manifest and

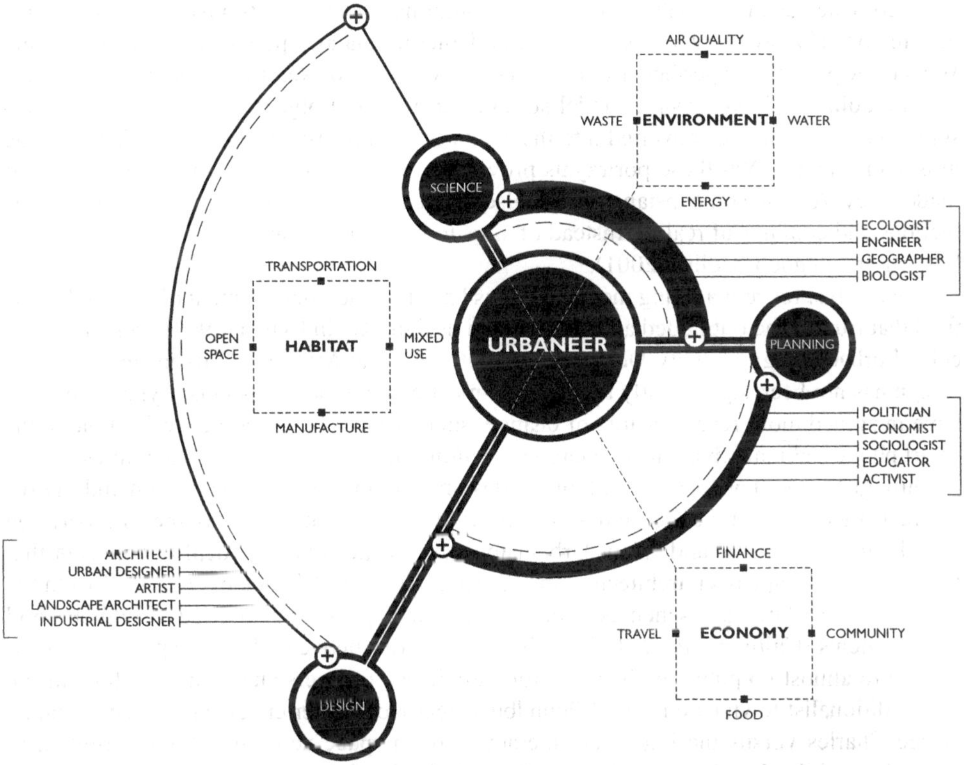

Figure 7.1 The field of urbaneering.

Credits: Mitchell Joachim, Maria Aiolova, Melanie Fessel, Caleb Lowery.

facilitate the City 2.0 across the globe. Each urbaneer is an individual with a different set of versatile abilities that merge previously disparate occupations. They range from combined ecological architects and engineers to action-based urban planners and developers. Almost any recombined professional activities will work, so long as they meet the constantly changing needs of urbanization. Urbaneers perform in a role akin to Jane Jacobs, but at the magnitude and accomplishment of Robert Moses. An excellent historical example of an urbaneer is Frederick Law Olmsted.

Urbaneering utopia: a new profession for the design of cities

Who is the primary authority in the making of utopia or any extraordinary future city? Urbaneering is a burgeoning discipline based on urban design that can negotiate the complex mix of technology, theory and practice that embraces the re-invention of the city to exceed the needs of the planet. Today, this nascent interdisciplinary field is in a state of radical development.

Throughout human history, sparks of utopian reflection have been indispensable in evolved societies (More 1516). Utopias, for the most part, are a necessary paradigm. Utopias display maximal solutions to existing real-world problems. They tackle upheaval with orderly redistribution. In nearly all variations, utopias are deliberately excessive. They overshoot the answer to a crisis to accentuate the problem.

To convince us of the viability of utopian solutions, society needs a psychological frame of reference (Lasswell 1930). It is helpful to depict utopia as a personal gym membership. What is the picture of a perfect physique? Why do we need to exercise? Aspirations to appear like an idolized athlete or supermodel sustain our work-out objectives. Many of us understand that we cannot be converted into the unattainable airbrushed automatons that decorate glossy magazines. Yet these portrayals provide a common measurement to reflect on our inadequacy. And, in some instances, this false imagery makes us admire the precious imperfections and confines of reality. Instead of aims to be grandiose and ideal, the goal shifts to be earnest and good (Collins 2001).

One catalyst for reimagining the good city is based on the establishment of a patently new field that trumps the outmoded agenda of urban design. Kevin Lynch (1960) originally conceived urban design as 'City Design' in the late 1950s at MIT. Since its inception, urban design has not been significantly upgraded to meet contemporary insights. Lynch could not have imagined more recent factors of change, such as Google, social networks, ubiquitous smart phones, climate dynamism, energy addictions and global economic calamities.

Throughout the developed world, urban design is at an impasse, unable to mend the rift between theory and practice, and stuck in cryptic arguments, such as the one between 'Landscape Urbanism' and 'New Urbanism'. Landscape urbanists deploy concepts that favour the landscape over architecture in order to plan a city (Waldheim *et al.* 1997). On the contrary, New Urbanist schemes promote historical pedestrian-centred neighbourhood developments (Calthorpe *et al.* 1991). The dissimilarity between the two approaches is of interest to almost no public body apart from architects, planners and their ilk. Think of the neo-traditionalist town extension of Poundbury near Dorchester endorsed by New Urbanist Prince Charles versus the high-tech interactive open space zone of Schouwburgplein in Rotterdam of the Landscape Urbanists West 8. Both of these urban factions have their merits, yet they fail to agree on a holistic idea of the future city. Moreover, the public – citizens who occupy these spaces – invest little time in understanding the minutiae. For them, it is an archaic clash between tradition and novelty. Urbaneering embraces both and envisages much more.

Urbaneering undertakes a diverse range of projects as a method for maximal design. It practices totalized schemes that rethink all scales of involvement from the doorknob to the state. Its projects can range from new materials, transportation systems and open spaces, to buildings, cities and surrounding regions. Recently, a few urbaneers have shaped phyto-remediation ponds, living woody plant structures, roof-top farms, soft cars and buses, urban junk space and city-wide action plans. To inspire interdisciplinary innovation, urbaneers encourage people to switch roles; architects must design cars, automotive engineers must devise eco-systems, and ecologists must draw up buildings.

At the core of urbaneering is a variety of utopian agitation that dispels the defunct myths of modernism with equitable objectives. An urbaneer replaces implausible rules and master planning with suggestive memes and polemical models. It is hard to argue with amorphous memes like Burnham and Bennett's 'city beautiful', Ebenezer Howard's 'garden city' or 'smart growth' (Calthorpe and Kelbaugh 1989). The public can rally around these open-ended symbolic gestures and phrases. Since the meme is not fully explicit, the concept leaves room for broad cultural interpretations. It is almost exactly what communities yearn for – freedom to define their own urban spaces.

The urbaneers' aim is to support people to become part of an advanced intellectual initiative framed on the recalibration of the city. Projects such as London's Canary Wharf, Berlin's Potsdamer Platz, the New York City High Line, Masdar City in the United Arab Emirates and the Sino-Singapore Eco-City in Tianjin, China already demand fresh directives. The new profession of urbaneering provides them in astonishing and collaborative ways.

In the past, urban design has been largely interdisciplinary, but it has not been revamped since its formal inception. Urbaneering will involve city design in a huge range of new ideas, including crowdsourcing, DIY projects, localized energy, shared transport, e-government, high-throughput computation, biotechnology and ecology. Urbaneers focus as much on cities' ecosystems and infrastructure – areas ripe for improvement – as on more conventional subjects such as buildings and parks.

What does urbaneering look like? Form follows anything as long as no shape is unmotivated. Diagrams that combine ecology and urbanity, called 'ecograms', prioritize design directives. Light and air are among the causal factors that respond to context in such ecograms. Equally, encapsulation of programmes is correctly orchestrated via any low-energy embodied volumes or dense generic loft spaces. These highly malleable spaces support growth, with elaboration and magnification of character at their limits. Ecograms are spaces of rampant plurality in which lifestyle is elective and fluid. Depending on the nerve that the ecogram presses, the reinvention and re-privatization of consequences could prove to be illimitable.

Beyond utilizing the ecogram, urbaneers have a set of proficiencies that merge previously disparate occupations. The discipline is home to almost any recombined professional activities, so long as they meet the constantly changing needs of urbanization. An excellent historical example of someone who today would be an urbaneer is Frederick Law Olmsted (1822–1903), the nineteenth-century activist who combined journalism, social action and landscape architecture to a single political end (Beveridge and Rocheleau 1998).

The next city needs a new breed of communicator, a person skilled in the art of cities beyond the typical planners, civil engineers and architects of the present day. These fields need a multifaceted filter of reason to incorporate a profound knowledge of place. An urbaneer posits the solutions to municipal problems that normally take multiple disciplines to solve. Urbaneers look to merge the edification and expertise needed to reform the city of today for the utopia of tomorrow.

Urbaneering Brooklyn

What is the key objective for ecological cities? A primary aim for the next incarnation of the 'good city' is that all necessities are provided inside its accessible physical borders. In this intensified version of urbanity, all vital needs are supplied for its population. In this city, food, water, air quality, energy, waste, mobility and shelter are radically restructured to support life in every form. Infrastructure is celebrated as the new centre.

The strategy includes the replacement of dilapidated structures with vertical agriculture and housing merged with road networks. Former streets become snaking arteries of liveable spaces embedded with renewable energy sources, soft cushion-based vehicles for transportation and productive green rooms. The plan uses the former street grid as the foundation for up-to-the-minute networks. By re-engineering the obsolete streets, we can install radically robust and ecologically active smart pathways. These operations are not just about a comprehensive model of tomorrow's city but an initial platform for discussion. Urbaneers expect that the future will necessitate marvellous dwellings coupled with a massive cyclical resource net.

Figure 7.2 A spectacle of smart infrastructure highlights the new downtown.

Figures 7.2 to 7.6 Urbaneering Brooklyn 2110, city of the future.

Credits: Mitchell Joachim, Maria Aiolova, Melanie Fessel, Dan O'Connor, Celina Yee, Alpna Gupta, Sishir Varghese, Aaron Lim, Greg Mulholland, Derek Ziemer, Thilani Rajarathna, John Nelson, Natalie DeLuca.

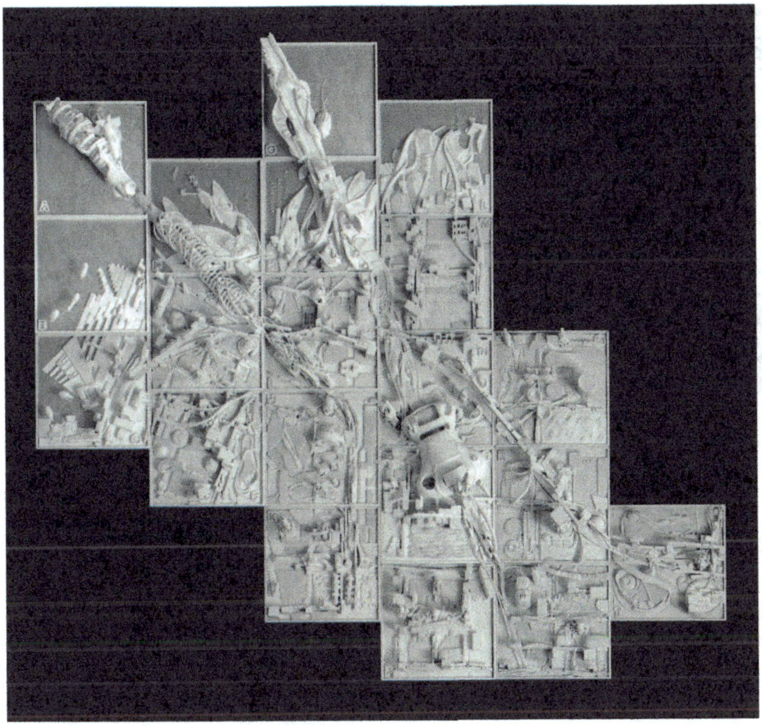

Figure 7.3
All necessities are provided inside the city boundaries.

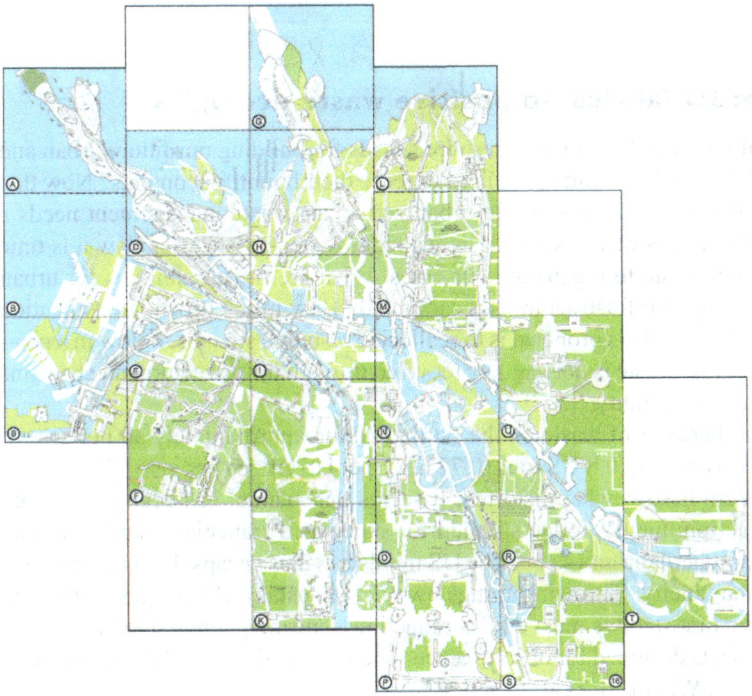

Figure 7.4
Site plan includes quadrants for energy, food, water and waste.

Figure 7.5 View of Brooklyn Bridge into new downtown.

Figure 7.6 Green brain park space used for algae production.

Rapid Re(f)use: 3D fabricated positive waste ecologies

Imagine our colossal municipal landfills as resource sheds for building our future urban and peri-urban spaces. What kind of effort is required to reuse their bountiful contents? Now that the bulk of humanity has chosen to settle in urbanized areas, waste management needs a radical revision. For hundreds of years, we designed cities to generate waste. Now it is time that we begin to design waste to regenerate our cities. What are the possibilities for urban environments after our aged infrastructure is recalibrated? How might urban intensification and waste mix? Terreform ONE's proposal is to reallocate resource streams to flow in a positive direction.[1] In this case, waste is not vaguely recycled through infrastructural mechanisms but instead up-cycled in perpetuity.

The USA is the lead creator of waste on the Earth, making approximately 30 per cent of the world's trash and tossing out 0.8 US tons (0.72 tons) per US citizen per year (EPA 2008). America's value system is somewhat distorted. It seems that 'value' has degenerated into rampant waste production: mega-products scaled for supersized franchise brands, big-box retail, XXL jumbo paraphernalia and so on. The US mindset is thus encapsulating a joint race for ubiquity and instantaneity. Throwing things away is unsustainable (Rogers 2006, pp. 54–67, 104–132). The first step we must take is reduction – meaning a massive discontinuation of producing objects designed for obsolescence. Then we need a radical reuse plan. Our waste crisis is immense. What is our call to action?

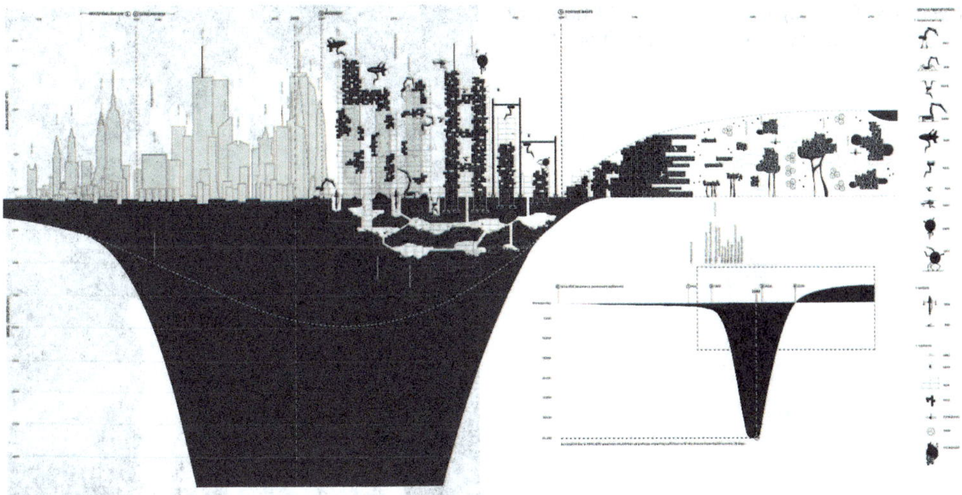

Figure 7.7 History and future of garbage in New York.

Figures 7.7 to 7.9 Rapid Re(f)use: waste to resource city 2120.

Credits: Mitchell Joachim, Maria Aiolova, Melanie Fessel, Emily Johnson, Ian Slover, Philip Weller, Zachary Aders, Webb Allen, Niloufar Karimzadegan, Lauren Sarafan.

Figure 7.8 Every hour, New York City produces enough waste to fill the Statue of Liberty.

Figure 7.9 One-day tower made of 24 hours of compacted waste in New York City.

One such dilemma lurks in New York. New York City is currently disposing of 36,200 US tons (32,840 tons) of waste per day (Cohen 2008). Previously, most of this discarded material ended up in Fresh Kills on Staten Island, before operations were blocked. Manhattan's inhabitants discard enough paper products to fill a volume the size of the Empire State Building every two weeks. Terreform ONE's Rapid Re(f)use and Homeway projects strive to capture, reduce and redesign New York's refuse infrastructure. The initiative supposes an extended city reconstituted from its own junked materials. The concept remakes the city by utilizing all the trash entombed in the Fresh Kills landfill. Theoretically, the method should produce, at minimum, seven entirely new Manhattan Islands at full scale. New York City's premier landfill was started by Robert Moses and driven by apathetic workers and machines. Now, guided by a prudent community with smart equipment, we must reshape it.

How could this work? Outsized automated 3D printers could be modified to rapidly process trash and to complete the task within decades. These potential automatons would be entirely based on existing techniques commonly used in industrial waste compaction devices.

To accomplish this job, nothing drastically new needs to be invented. Most of the technologies needed are available off the shelf. Instead of machines that crush objects into cubes, compaction devices could benefit from adjustable jaws that would craft simple shapes into smart 'puzzle blocks' for assembly. The blocks of waste material could be predetermined, using computational geometries, in order to fit domes, archways, lattices, windows, or whatever patterns are required. Different materials could serve specified purposes: transparent plastic for fenestration, organic compounds for temporary decomposable scaffolds, metals for primary structures and so on. Eventually, the future city would make no distinction between waste and supply. Admittedly, this meta-design theme is not entirely novel. At approximately the same time that Rapid Re(f)use was initiated, the feature film *WALL·E* (Stanton 2008) was conceptualized. *WALL·E* profoundly influenced Terreform ONE's research agenda.

Excursion to Disneyland

Inspired by an interest in creations of the future such as Disney's Tomorrowland, Terreform ONE visited the Walt Disney Imagineering (WDI) headquarters in Glendale, Southern California. The group had prepared a presentation that would unpack a comprehensive view of its version of the future: a world free of carbon loading in the atmosphere and abundant in self-sufficient lifestyles. As architects invested in an ecologically sustainable future, the team had meticulously crafted cities within the rubric of a socio-ecological domain – rethinking the design of entire systems, from doorknobs to democracies.

When Ben Schwegler, mastermind and chief imagineer, pulled back the proverbial curtain to reveal WALL·E, the group was crestfallen. Disney had beaten them to it. WALL·E was perfect – almost: a tightly packaged, solar-powered, curious, obedient, evolved, robotic trash compaction and distribution device. His name is an acronym: Waste Allocation Load Lifter Earth class. Left behind by mankind, he toils with trillions of tons of non-recycled inner-city trash. Not only is WALL·E a highly advanced rubbish manager, he is also a mechanized and inventive architect. He accomplishes his immense tasks while remaining completely adorable. Not easy to do. WALL·E's life is a tale of an ultramodern trash compactor in love. Ceaselessly, he configures mountains of discarded material. Why pyramids of trash? WALL·E's daily perpetual feats seem almost futile. Disney's film omits exactly why he is programmed to pile refuse; and there is the shortcoming.

Future waste and past cities

Collaborators at Terreform ONE were interested in exploring a deeper motivation for stacking refuse. What if the refuse was refabricated to become real urban spaces or buildings? If it is plausible to adapt current machinery, how much material is there? At first sight, any sanitary landfill may be viewed as an ample supply of building nutrients. Heavy industrial technologies to compact cars into lumber or to automatically sort out garbage are readily available. Other technologies, which would make possible the articulation of specific forms, are also available if scaled in larger sizes. 3D printing has exhaustive capabilities if adjusted to larger scales. This is where Terreform ONE's city began.

The envisioned city would be derived from trash – not ordinary trash, but 'smart refuse'. A significant factor of the city created from smart refuse is 'post-tuning'. Unitized devices would not immediately adapt. Integration into the city texture would be a learning process.

In time, the responses would eventually become more attuned to the needs of the urban dweller. This city is envisioned from trash, but each individual component would be enhanced with a modicum of CPU power. Brief durational events would endow these 'smart units' with experiences needed for their evolution.

The main objective for the city of Rapid Re(f)use is to establish a smart, self-sufficient, perpetual-motion urbanism. It has been argued that perpetual motion cannot exist. Perpetual motion defies the laws of thermodynamics and energy conservation, since it would necessitate a machine that produces more energy than it consumes. Unlike machines, cities are similar to a complex ecology. Lynch (1984, pp. 26–27) explains that cities are 'the product of beings capable of learning. Culture can stabilize or alter the habitat system, and it is not clear whether we wish it to be otherwise.' Ecology is capable of achieving a continuous harmonious state, or even a positive intensification. If ecological models are productively everlasting, urban models can logically follow their example.

Architects have ruminated over improbable instruments of physics since the Middle Ages. In the thirteenth century, evidence of the *perpetuum mobile* was uncovered in the sketchbooks of French architect Villard de Honnecourt (Bowie 2006, pp. 32–49). What if the Rapid Re(f)use city was like an instrument that produces more energy from renewable sources than the energy it consumes? In this case, 'nothing can be thrown away'. Every bit would be a vital piece of stored energy, poised to be reused in a cyclical nutrient stream (McDonough 1998). Rapid Re(f)use is imagined as a city without a tail pipe; a city that not only has zero impact, but also makes a positive contribution towards its natural surroundings. This is not only about solving our ecological issues, but also about returning to a system of perpetuity. This is the only possible future for a truly breathing, interconnected, metabolic urbanism. Cities have passed the age of industrialization and entered the age of recovery. After this great cleansing, we may transition into a greater order: 'positive waste'. This is an order that captures our socio-ecological needs: not utopia, but a place where everything is precious and nothing is disposed.

Envisioning ecological cities

How should urban design imagine new instrumentalist technologies for cities? For 150 years, the innovation of the elevator has done more to influence urban design than most urban designers. Elevator systems had incredible success in creating compact and greener cities. Imagine what the advent of the jet pack will do for cities. Urban design is greatly altered by such devices. For instance, automobiles have defined limits in cities for almost a century. Unlike the elevator however, the car has arguably caused more problems than it has solved. Perhaps it is time for urban design to rethink technologies to fit cities, not constrain them. As a wide-ranging discipline, it can effortlessly illuminate the technological potentials for cities. Urban design will successfully situate itself through the production of future macro-scaled scenarios predicated on innovative devices.

Physicist and polymath Freeman Dyson has said that the best way to comprehend our near urban future is to examine science fiction, not economic forecasts. In his experience, sci-fi is good for decades of technological fulfilment. Unfortunately, economic forecasts are only accurate for five to ten years. Most predictive economic models are quantitatively based and find it difficult to extrapolate the qualifiers associated with creativity. Sci-fi is a phenomenal way to explore our plausible urban future that should not be dismissed by urban designers. Dyson is certain that the urban era of information will soon transition into 'the age

Figure 7.10 Blimp bumper bus.

Figures 7.10 to 7.13 SOFT mobility.

Credit: Mitchell Joachim.

Figure 7.11 The hug-and-kiss sheep car has zero accident victims.

Figure 7.12 Urban stackable cars with solar recharge port.

Figure 7.13 Homeway: the great suburban exodus.

Credits: Mitchell Joachim, Maria Aiolova, Melanie Fessel, Philip Weller, Ian Slover, Landon Young, Cecil Howell, Andrea Michalski, Sofie Bamberg, Alex Colard, Zachary Aders.

of domesticated biotechnology' (Dyson 1998). In his lectures, *Infinite in All Directions*, he states: 'Bio-tech offers us the chance to imitate nature's speed and flexibility.' He envisions a realm of functional objects and art that humans will 'grow' for personal use. According to a *New York Times* article on Dyson, he also believes that the impact of climate change is profoundly mis-stated, speculating that 'if CO_2 levels soared too high, they could be soothed by the mass cultivation of specially bred "carbon-eating trees"' (Dawidoff 2009). He is not concerned with predicting the future but rather with expressing its possibilities. These expressions are founded along societal desire lines as a kind of relevant optimism. Dyson assesses whatever our civilization wants and advances our expectations.

At some level, urban design engages this position that promises a better tomorrow. Many practitioners and urbanists suffer, to some extent, from this search for direction and clairvoyance. Alex Krieger (Krieger and Saunders 2009) strongly asserts that the broadly defined vocation is more of a scrupulous sensibility than an exclusive authority. The profession is torn between several incompatible agendas, weighty theories and oversimplified applications, ivory towers and new urbanism, developers' brands and radical ecologies, and vernacular forms and futurology. One of my research group's chief aims is to create projects at the intersection of technology and urbanism, under the rubric of 'ecology'. Our projects range from highlighting the possible effects of self-sufficient cities to studying flocks of jet packs. These ideations keep us thriving as urban design researchers. It is our supposition that the prospective ecological city is about extreme solutions to an extreme predicament. Our future fundamentally depends on the immensity our solutions envision.

Envisioning is by definition a view or concept that evolves beyond existing boundaries. This notion of foresight may be interpreted in many different ways, each foregrounding particular ideations and processes describing the next event. Here in the US, we need these radical new visions to help solve our current global calamity. The Earth's climate is currently in an unremitting state of trauma. We seek precise prescriptions that cover a wide scope to alter this massive dilemma. John F. Kennedy once said: 'If man created problems, man can solve them.' This future vision unfolds a truly breathing, interconnected metabolic urbanism. How does it reify from statistics to architectural form? What does the future look like for US cities? How do technological devices affect these functions?

For a popular audience, the film *WALL·E* enabled society to contemplate one conceivable future. The film is set in a generic city that is completely buried in trash. Humans have abandoned life on Earth for off-world dwelling, leaving behind one lone solar-powered robot to clear the rubbish. Part of the message of the film is that technology alone cannot solve humanity's 'affluenza', yet the film's powerful computer-generated visuals encourage us to confront our colossal wastefulness and rethink the city.

Urbaneers foresee strategies for people to fit symbiotically into their natural surrounds. To achieve this, all things are considered possible. We design the scooters, cars, trains and blimps, as well as the streets, parks, open spaces, cultural districts, civic centres and business hubs that comprise the future metropolis. For centuries, cities have been designed to accommodate the theatre of our human desire. Urbaneers have joined the ranks of those delivering a new sense of the city, one that privileges the play of nature over anthropocentric whims. We are constantly vying for a profound clairvoyant perspective. We desire to preview a likeness of our collective future yet untold.

Our foresight of ecological design is not only a philosophy that inspires visions of sustainability but also a focused scientific endeavour. The mission is to ascertain the consequences of fitting a project within our natural environment. Solutions are derived from numerous examples: living material habitats, climatic tall building clusters and mobility technologies. These design iterations succeed as having activated ecology both as a productive symbol and an evolved artefact. Our current research attempts to establish new forms of design knowledge and new processes of practice at the interface of design, computer science, structural engineering and biology.

The necessity of all scales: planetary design in the age of globality

Figure 7.14 View of a clean-tech industrial water park.

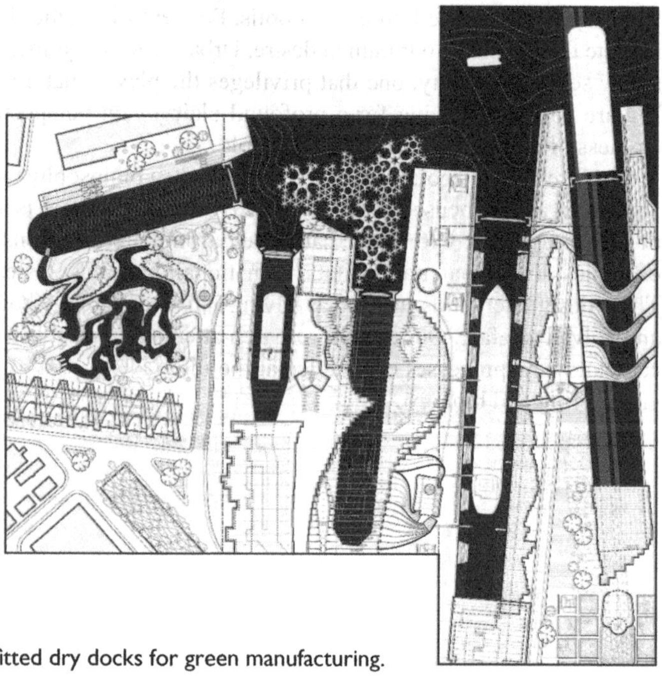

Figure 7.15 Site with five retrofitted dry docks for green manufacturing.

Figure 7.16 Super-sheds over docks merge landscapes.

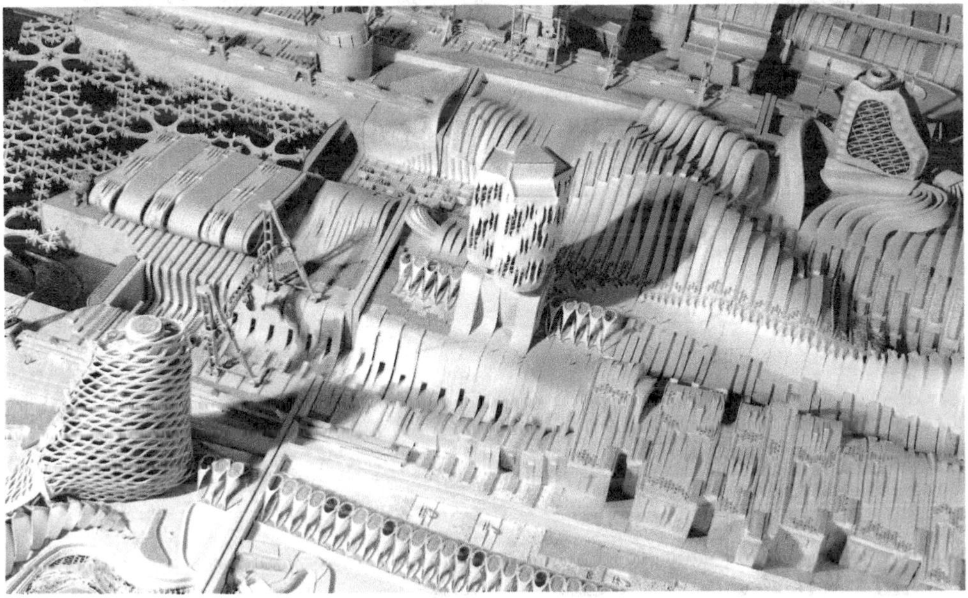

Figure 7.17 Clean-tech office towers.

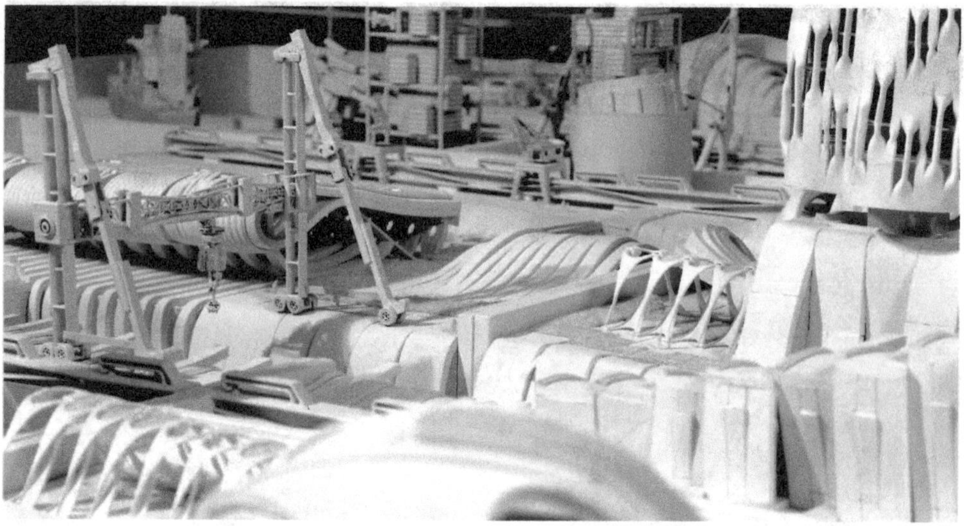

Figure 7.18 Large-scale 3D printer over dry docks.

Figure 7.19 Restorative ecology with phytoremediation systems in surfaces.

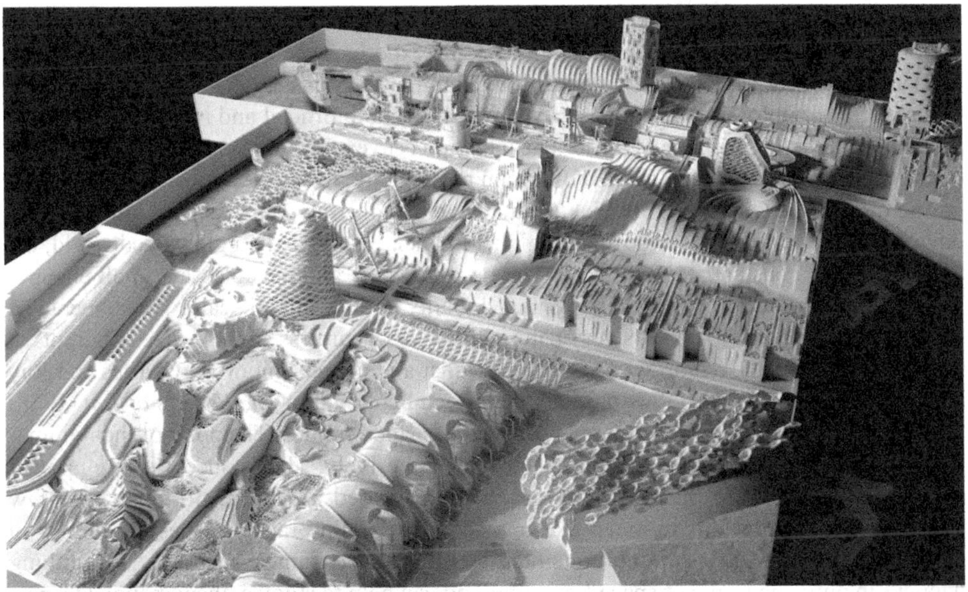

Figure 7.20 Overall view with converted warehouses and sewer management.

Figures 7.14 to 7.20 Super docking, Brooklyn Navy yard.

Credits: Terreform ONE + Planetary ONE, Mitchell Joachim, Maria Aiolova, Nurhan Gokturk, David Maestres, Jason Vigneri Beane; Design Team: Carlos Barrios, Alex Felson, Walter Meyer, Melanie Fessel, Zafirah Bacchus, Ivy Chan, Courtney Chin, Adrian De Silva, Julianne Geary, Francisco Gill, Shima Ghafouri, Jacqueline Hall, Kelly Kim, Florian Lorenz, Bart Mangold, Dustin Mattiza, Chema Perez, Alsira Raxhimi, Daniel Russoniello, Melody Song, Allison Shockley, Katherine Sullivan.

Globalization has a final outcome; it is not unending. Humanity is in the age of globality, the final state of the globalization process. Globality is the endgame, an all-inclusive terrestrial status. In this fully connected world, populations will compete with everyone, everywhere, for everything, at all times and in all scales. Vast networked communication linkages will reverse developing world economies and cause developed nations to partially corrode if not yield. A conceivable state of planetary equilibrium ought to influence the major sectors of industry, commerce, resource management, infrastructure, technology, energy and governance, resulting in profound transformations. What is design like in this condition of globality?

The American economist Jeremy Rifkin (2000) suggests that the modern age has been characterized by a 'Promethean spirit', a restless energy that preys on speed records and rapid solutions, unmindful of the past, uncaring of the future, existing only for the moment and the quick fix. The earthly rhythms that characterize a more pastoral way of life have been shunted aside to make room for the fast track of an urbanized existence. Lost in a sea of perpetual technological transition, people find themselves increasingly alienated from the ecological choreography of the planet. Humans expect jet airliners to collapse the space of a continent in a matter of hours. We have partly lost our sense of scale, time and distance.

Scale also determines our profound connection to place. Yi Fu Tuan's (1974) environmental conjecture conveys knowledge and love of place with descriptions, maps and itineraries enabling disconnected contemporary humans to appreciate their location as pre-modern people did or animals do. Such literature also enlivens the experience of everyday places with new facts and rhetorical devices that can recalibrate familiar environs to keep alive a sense of the undiscovered country of the nearby. These texts direct both official and intuitive knowledge towards 'topophilia', the love of place. The intention is to expand our sense of the land, not to confine it to one size. We are in need of terrain propinquity, not dislocation.

The disasters at Chernobyl and Fukushima have demonstrated far-reaching influences on environmental health worldwide and on global financial markets. In a globalized world, the effects of scale constantly transfer between the irreducible and the colossal. Small changes ramify into massive results and vice versa. Globality operates in all scales simultaneously, without privileging one form. Its very nature implies measureless shifts in size. Attempts at thinking inside strict categories of scale are defunct and counter-intuitive. Charles and Ray Eames provided the perfect illustration in their documentary film *The Powers of Ten* (1968) where scale is conveniently defined in neat square-shaped frames. The point of their animation is to bridge the different perceptions of scale cohesively. Their concept is to empower individuals to visualize the ranges of observation melded as one. Unfortunately, a few viewers also interpret this to mean that one actually should bracket places and things in specific scales. That is a common misinterpretation of the Eames' message, but nothing happens in only one frame of space/time. Framing can help to study a phenomenon at a particular moment but things always shift. Moreover, artificially binding a place to a numerical scale is to some extent random and arbitrary. What law declares that measures must be in units divisible by ten? In his lectures, Jamer Hunt (2011) points this out nicely by making a reference to the film *This Is Spinal Tap*. One scene shows the actor Christopher Haden-Guest referring to his guitar amplification equipment with a volume knob that goes to 11. Within the central idea of this farcical narrative, scale is portrayed as being truly capricious.

Rem Koolhaas and Bruce Mau intended an emphasis on scale in their 1995 book, *S, M, L, XL*, yet they also alluded to the in-between thresholds and differentiation of projects. A false reading of *S, M, L, XL* is to suppose that the Office for Metropolitan Architecture's (OMA) works fit cleanly into categories measurable by an orderly unit. What exactly is designated a 'small' project: a door hinge detail, an elevator, a wooden deck addition? More notably, what 'small' project does not have significant impacts on an extra-large scale? Everything has consequent ramifications. Koolhaas and colleagues are of course fully aware of this, but may have failed to communicate their message in the title. Like *The Powers of Ten*, *S, M, L, XL* underscores an explicit simplification in degrees of change. Urbanism cannot be compiled into tidy categories defined by size. Design needs to break out of the emblematic question of size, especially as it relates to one planet. One size does not fit all; rather *all sizes fit one*.

Visualizing in one scale is markedly problematic. An analysis comparing the views of Johann Wolfgang von Goethe and Isaac Newton best demonstrates this assertion. In 1810, Goethe (1970) published a comprehensive, lucid description of colour within a perceptual human context. He argued ardently that colours were defined by an inseparable relationship called *Zur Farbenlehre* (*Theory of Colours*). One colour cannot be reduced to a single element. It requires a setting of other colours and circumstances to be fully perceived as an observable phenomenon by humans. He proclaimed that to identify blue you need to, at some level, recognize red, orange, yellow, and the entire spectrum. Goethe was responding to Newton's *Opticks*, published in 1704. Newton had treated colour analytically and saw each colour as discrete wavelengths in an optical spectrum (Newton 2012[1730]). Newton surmised that it is absolutely possible to observe one colour disconnected from its domain.

Every colour has a particular frequency and may be defined as such. While both thinkers are correct, the problem that design professionals have with scale is identical. Scale may be Newtonian and viewable at one frame of unitized reference. Like Goethe's view of colour, however, scale is best perceived holistically, in relationship to other scales, in order to understand its true phenomenon.

Most archetypal designers tolerate a proclivity to divide concepts into specific units to help comprehend or visualize the problem better. However, this is inherently misguided. Design is a prescience that affects all disciplines and cannot be conceptualized as a bracketed or contained field. The condition of globality and scale means that the design professions as we know them need to be restructured. A new breed of designers who can speculate and produce at all levels from the nano-scale up to the planetary are required. These thinkers may be referred to as planetary designers.

As a globalized, meta-Pangaea community, design is obliged to be pervasive. It simply cannot limit itself to any one scale or project scope. If so, its relevance and instrumentality are greatly diminished. The principal operations of scale and the systems that deploy it restrain and confuse the complicated reality of design problems. The more scale is used, the more designers fail to envision the whole picture. Computer-aided design is part of this dilemma, both a solution and a quandary. Most design software unwittingly forces designers into forms of measure. As soon as the file opens, designers are asked to define the units of measure, view points and scale. The same is not true when an individual picks up a pencil. Freedom to draw and therefore conceptualize without boundaries is practically inestimable. To quote Frank Stella, 'artists don't think in units'.

The processing capacities of computers distort the implications of measure. Software allows for seemingly limitless flexibility. Operators can shift from the smallest possible detail to the largest components. It is possible to zoom endlessly outwards into entire regions. On one level, this is a tremendously effective visualization capacity. Consider the design of a rail spike and its connection to the track; zoom onto the tracks themselves and further out to all the trains on those tracks and advance past Penn Station and finally out of the entire New York City metropolitan region. This is an acutely impressive tool. However, the software does not make any distinctions about the conventions of physics and natural forces that govern each zoomed layer. In many cases, the current limitations of memory also fail to provide all the essential detail within each consecutive zone. Furthermore, the material and chemical behaviours of the objects and places are not described in relationship to each other beyond geometric location. Admittedly, the current trend is to restructure computation to account for these missing characteristics. Envision a computer program that fully simulates the ecosystem of the Earth and all the associative quantum mechanics. It would become a Jorge Luis Borges map of the world in such vast specificity that it would correlate to the exact size of the world at 1:1 (Borges 1975).

Additional scenarios of scale in relation to the planet are depicted in *The End of Nature* by Bill McKibben (1999). He marshalls the latest scientific evidence about the greenhouse effect, the depletion of the ozone layer and a harrowing array of other ecological ills, and explains the frightening implications of the destruction cities have wrought on our planet. McKibben examines the meaning of these changes, and considers the wretchedness of life where there is no escaping mankind. Although human civilization has pillaged and polluted the Earth for centuries, in the past those aggressions were relatively localized; now, with the globality shifts caused by greenhouse gases and ozone depletion, humans and their cities have altered the most elemental processes of life everywhere. Nature itself has been tainted, becoming the equivalent of a vast heated room. By turning nature into an artefact or b(u)y-product of economic development, society has lost something of profound importance: nature as a quasi-celestial source of ultimate meaning and value. It is this loss that McKibben refers to as an apocalyptic calamity. The ending of nature is something independent of, larger than and uncontrolled by man. In this nihilistic world, every measurable space has been touched by human interference. Here even the right of all life to breathe has been impacted.

Many designers and planners have been concerned in recent years with revealing 'truth windows' into nature to avoid its end. They highlight ecological processes in their designs so that the users of the environment may experience, comprehend and appreciate the scales of those processes aesthetically. In practice, revelation of ecological process has meant everything, from capturing stormwater on the surface of the land before it drains away into sewers, to planting a row of trees in an urban plaza where a creek once existed. In addition, the ecological processes that are revealed may themselves be truly natural, in the sense that they could continue to subsist without the management of society, or they may be deeply artificial, engineered systems that need relentless supervision if they are to persist in an urbanized context. Ultimately, the intention is to make the scales of ecology visible and thereby expose an alleged spectacle of beauty otherwise unseen.

Our future in architecture recognizes that there is an immeasurable ecological quality that goes well beyond the borders of the building site. It is clear that ecological principles comprise a web of interconnected concepts, and a breathing, healthy metabolism has to fit inside. Thus architecture in the future, or even now, must be understood without a solitary scale. It must be planetary, on a scale that encompasses the outer edges of the atmosphere all the way into the deep regions of inner space.

The new field of geo-engineering, for instance, exemplifies the lack of scale comprehension. In geo-engineering, exertion is made on the scope of an entire continent. Geo-engineers produce efforts equivalent to digging the Panama Canal as everyday feats. When looking at consequences and devices that can cause change at the regional level, we must look at not just the region but the entire hemisphere and its various megalopolises, as well as the smallest biological system. We must realize that there is a kind of hubris, an unlimited bravery and power in conquering nature. Designing modifications on a continental scale does suggest that accidents will be unmatchable.

The issue of scale seems to be rampant in the polemical discussions of ecological cities. Scale is a pervasive term in engineering, architecture, urban studies and design. Although not without limitations, it serves as a constant and definitive point of reference to help elucidate a given project. We have separated our professional disciplines, project scopes and programmatic language in terms of size; but this is a redundant supposition in the age of globality. How designers can play a significant role in this expansive territory and live up to our prophetic tasks is worthy of further examination. Designers' responsibility is to re-tool the middle ground, the in-between and the nexus points. In this case, infrastructure with applied innovative ecological directives becomes the penultimate goal, before completely reforming the world. Operations of infrastructure shape a broad range of circumstances. Infrastructure – in all its dimensions and extents – is the actual frontier. When considered alongside technological, social and ecological solutions implemented from other disciplines, design innovations are rapidly being diminished, if not surpassed, by other competitive fields. We designers must take action and modify our stance on all scales and morphologies in order to have a positive effect on the global community. Our hypothesis is first and foremost based on one succinct predicate: the end of scale.

Note

1 Terreform ONE (Open Network Ecology) is a 501c3 non-profit experimental green design group that promotes new concepts for the advancement of cities. As a unique interdisciplinary laboratory for industrial design, synthetic biology, landscape ecology, art, architecture and master planning, we work to explore the theoretical and practical framework of ecological design. Through innovative projects and community outreach efforts, we aim to integrate the ecological possibilities of New York City and inspire like-minded projects in urban centres worldwide. The members of Terreform ONE are Mitchell Joachim, Maria Aiolova, Melanie Fessel, Philip Weller, Ian Slover, Emily Johnson, Landon Young, Cecil Howell, Andrea Michalski, Sofie Bamberg, Alex Colard and Zachary Aders.

References

Beveridge, C.E. and Rocheleau, P. 1998, *Frederick Law Olmsted: Designing the American landscape*, Universe Publishing, New York.

Borges, J.L. 1975, 'A universal history of infamy', in *On Exactitude in Science*, Penguin Books, London.

Bowie, T. 2006, *The Medieval Sketchbook of Villard de Honnecourt*, Dover Publications, New York.

Calthorpe, P. and Kelbaugh, D. (eds) 1989, *The Pedestrian Pocket Book*, Princeton Architectural Press with the University of Washington, New York.

Calthorpe, P., Duany, A. and Plater-Zyberk, E. 1991, *Towns and Town-Making Principles*, Harvard University GSD, Cambridge, MA.

Cohen, S. 2008, 'Wasted: New York City's garbage problem', *New York Observer*, 3 April.

Collins J.C. 2001, *Good to Great: Why some companies make the leap . . . and others don't*, HarperBusiness, New York.

Dawidoff, N. 2009, 'The civil heretic', *New York Times*, 25 March.

Dyson, F. 1988, *Infinite in All Directions*, HarperCollins, New York.

Eames, R. and Eames, C. (dirs) 1968, *The Powers of Ten.*

Environmental Protection Agency (EPA) 2008, *Report on the Environment: Highlights of national trends*, www.epa.gov/roehd/pdf/roe_hd_layout_508.pdf, viewed 11 October 2013.

Goethe, J .W. von 1810[1970], *Theory of Colors*, MIT Press, Boston, MA.

Hunt, J. 2011, 'Attraction and repulsion in critical social design', Lecture at University of Applied Arts, Vienna, 10 November.

Koolhaas, R. and Mau, B. 1995, *S, M, L, XL,* Monacelli Press, New York.

Krieger, A. and Saunders, W.S. (eds), 2009, *Urban Design*, University of Minnesota Press, Minneapolis, MN.

Lasswell, H.D. 1930, *Psychopathology and Politics*, University of Chicago Press, Chicago, IL.

Lynch, K. 1960, *The Image of the City*, MIT Press, Cambridge, MA.

—— 1984, *Good City Form*, MIT Press, Cambridge, MA.

McDonough, W. 1998, 'Waste equals food: our future and the making of things', in J. Laddon, T. Atlee and L. Shook (eds), *Awakening: The upside of Y2K*, Printed Word, Spokane, WA.

McKibben, B. 1999, *The End of Nature*, Anchor, New York.

More, T. 1516, *Utopia*, http://ebooks.adelaide.edu.au/m/more/m83u/book1.html, viewed 1 December 2013.

Newton, I. 1730[2012], *Opticks: Or a treatise of the reflections, refractions, inflections and colours of light, based on the fourth edition London, 1730*, Dover Publications, Mineola, NY.

Rifkin, J. 2000, *The Age of Access: The new culture of hypercapitalism, where all of life is a paid-for experience*, Putnam Publishing Group, New York.

Rogers, H. 2006, *Gone Tomorrow: The hidden life of garbage*, The New Press, New York.

Stanton, A. (dir.) 2008, *WALL·E*, Disney/Pixar Animation Studios.

Tuan, Y-F. 1974, *Topophilia: A study of environmental perception, attitudes, and values*, Prentice-Hall, Englewood Cliffs, NJ.

Waldheim, C., Mostafavi, M., Corner, J. and Geuze, A. 1997, *Landscape Urbanism*, a conference event with proceedings held at Harvard University GSD, Cambridge, MA.

Chapter 8

A planning approach for the transition to low carbon cities

Nico Tillie, Andy van den Dobbelsteen and Sebastian Carney

Summary

Worldwide, cities are responsible for approximately 75 per cent of carbon dioxide equivalent emissions. With few systematic urban planning models to reduce carbon emissions in existence, it is difficult to take effective steps. The Rotterdam Energy Approach and Planning (REAP) tool was developed to support low carbon urban planning in a structured, incremental way. The approach has been used in several studies and applied in new developments. The second step of REAP, regarding tuning, exchanging and cascading of energy at various scales, is relatively new, offering cities additional opportunities. In an urban redevelopment for Rotterdam city ports, novel techniques for the exchange of different temperature levels of heat were designed. These involved a spatial plan based on the energy system chosen. Linking spatial planning to the exchange of waste heat within buildings, between buildings (such as supermarkets, greenhouses and apartments), between neighbourhoods and at city level with district heating provides urban planners with a new perspective on planning: a smart grid for heating and cooling! To build reduction scenario pathways, changing the energy mixes of existing cities into low carbon energy mixes is crucial. Carbon dioxide (CO_2) mapping, energy potential mapping and the REAP principles will help people to choose effective solutions. They were used to build detailed stakeholder-based CO_2 reduction pathways with the GRIP (Greenhouse gas Resinal Inventory Protocol) scenario tool. This chapter discusses REAP, CO_2 mapping, energy potential mapping and a method to build structured reduction pathways and possible spatial consequences.

Introduction

Background

Cities are responsible for approximately 75 per cent of global CO_{2eq}[1] emissions. The majority of these emissions are linked to energy. Currently there are limited approaches available to urban planners and policy makers to reduce carbon emissions in a systemic way, making it difficult to devise long-term strategies, take effective steps and produce focused initiatives. This leaves us with a question: 'How can CO_2 and energy become an integral part of urban planning?'

This chapter explores and presents the results of a practical approach for assessing and managing the transition to a low carbon city. It utilizes a combination of: the New Stepped Strategy (NSS) (Dobbelsteen 2008), the Rotterdam Energy Approach and Planning (REAP) model (Tillie *et al.* 2009a), Energy Potential Mapping (EPM) (Broersma *et al.* 2013) and a stakeholder-based energy scenario planning process, the Greenhouse gas Regional Inventory Protocol (GRIP) (Carney and Shackley 2009). The results and analysis provided are taken from a case study of the City of Rotterdam.

The value of connecting urban planning, energy and CO_2 is apparent from the results of the studies undertaken. The REAP study showed the energy-saving opportunities from exchanging waste heat within and between buildings, neighbourhoods and at the city scale, through a smart district heating grid. We used EPM to present the generation capacity of Rotterdam in a user-friendly format for a wide range of stakeholders. In preparing a stakeholder-based vision of low carbon heating and cooling for Rotterdam, energy potential maps were used with Geographical Information Systems (GIS) and fed into the GRIP scenario process. This process gave urban planners, policy makers and other stakeholders direct insight into the potential future energy mix, which would reduce carbon emissions, and the spatial consequences of such a future.

This chapter is divided into four sections. In the first section, we introduce the problem and discuss the heating energy requirements of the city. In the second section, we examine the NSS and REAP. In the third section, we present EPM in the context of Rotterdam. In the fourth section, we present GRIP and its results in the city. Finally, we present our conclusions.

The significance of heat exchange in cities

An often overlooked opportunity for urban energy efficiency is the potential to reuse waste heat. Buildings in the Netherlands, for example, use approximately 30 to 40 per cent of the total amount of energy used nationally. However, in a Westernized city where heavy industries are often absent, this contribution may be 50 to 75 per cent. Of this, the component used for space and water heating is approximately 70 to 80 per cent (SenterNovem 2008; Broersma *et al.* 2011). In the Netherlands, the predominant source of this heat is supplied by natural gas (producing hot water of 60 to 90°C), often by means of on-site boilers that are energy efficient but have an exergetic efficiency of less than 10 per cent. In other cases, district heating (urban heat grids) provides this heat through transported hot water (at 90 to 130°C), powered by a power plant or industrial site fired by fossil fuels. This heat is seldom transported over distances exceeding 10 km. Thus, while the reuse of waste heat that would otherwise have been wasted may essentially be good, the current system is not very efficient (Dobbelsteen *et al.* 2012). Studies indicate that 40 to 50 per cent of the heat consumption in cities can be reduced through more efficient use of the residual heat within it (Tillie *et al.* 2009a; Kürschner *et al.* 2011).

Exergetic inefficiency

Buildings in cities currently consume energy with a higher thermal content than required. As stated above, temperatures of around 90°C are used to create an indoor temperature of around 20°C. While this may remain necessary for old, unretrofitted buildings with traditional radiators (which cannot be energetically renovated), in modern, well-insulated houses with a water-based floor, heating system temperatures of 30 to 40°C suffice. In passive houses, the source heating requirement is 25 to 30°C. This means that the same amount of thermal comfort can be provided while reducing the energy used to provide it (based on natural gas consumption) by up to 70 per cent (Dobbelsteen *et al.* 2011, 2012).

A city relying on gas or heat from power plants and/or industrial sites may have a serious problem with meeting its energy supply requirements in the near future. A low carbon future will need this heat to be supplied by sources such as geothermal, inter-seasonally stored solar heat and biomass-fired power plants, which are not abundant enough to meet the world's urban areas. If a lower input temperature requirement could be met while maintaining thermal comfort, and, in turn, this requirement is served by waste heat from other buildings, then the overall input energy requirements may be further reduced. This is the 'essence' of exchanging/cascading heat within neighbourhoods or between buildings. Depending on the number of connections or cascading steps and the efficiency of the exchange, the input of primary energy in an urban system could be further reduced, by at least a factor of 2 (Dobbelsteen *et al.* 2012).

Methods to support low carbon urban developments

In this section we focus mainly on heating and cooling the built environment.

The New Stepped Strategy

The New Stepped Strategy (NSS) was developed by Dobbelsteen (2008), based on the Trias Energetica (Lysen 1996), to improve the energy efficiency of buildings. The NSS adds a step between consumption reduction and the deployment of sustainable sources. Furthermore, it includes a strategy for incorporating waste flows. The NSS's four steps are as follows:

1 **Reduce** energy consumption (using smart and bioclimatic design).
2 **Reuse** waste energy.
3 **Produce** renewable energy.
4 **Supply** the remaining demand cleanly and efficiently.

The second step of the NSS, namely reuse waste energy, optimizes the use of waste heat and water both within and between buildings, possibly on a city-wide scale. For example, wastewater can be purified and the silt fermented to form biogas that can be reused in the energy chain. Step 4 will eventually no longer be possible or desired but it will be necessary in the near future, albeit to an ever decreasing extent (Tillie *et al.* 2009a).

The REAP methodology

Within a city, the energy consumption of buildings can, and must, be reduced to meet efficiency and carbon targets. After this reduction has occurred, it is useful to determine whether waste flows from buildings can be reused. This is already being done to some extent, for example, by recycling heat from ventilated air and wastewater from showers. It is much more difficult to purify wastewater to reclaim biogas at the individual building scale than at precinct level. After applying the NSS's second step (reuse waste energy) to individual buildings, a requirement for energy remains that must be met with renewable energy sources. If we consider the city as a system, we must establish how to scale up from the building level to the level of the neighbourhood, district or city/region (Tillie *et al.* 2009b).

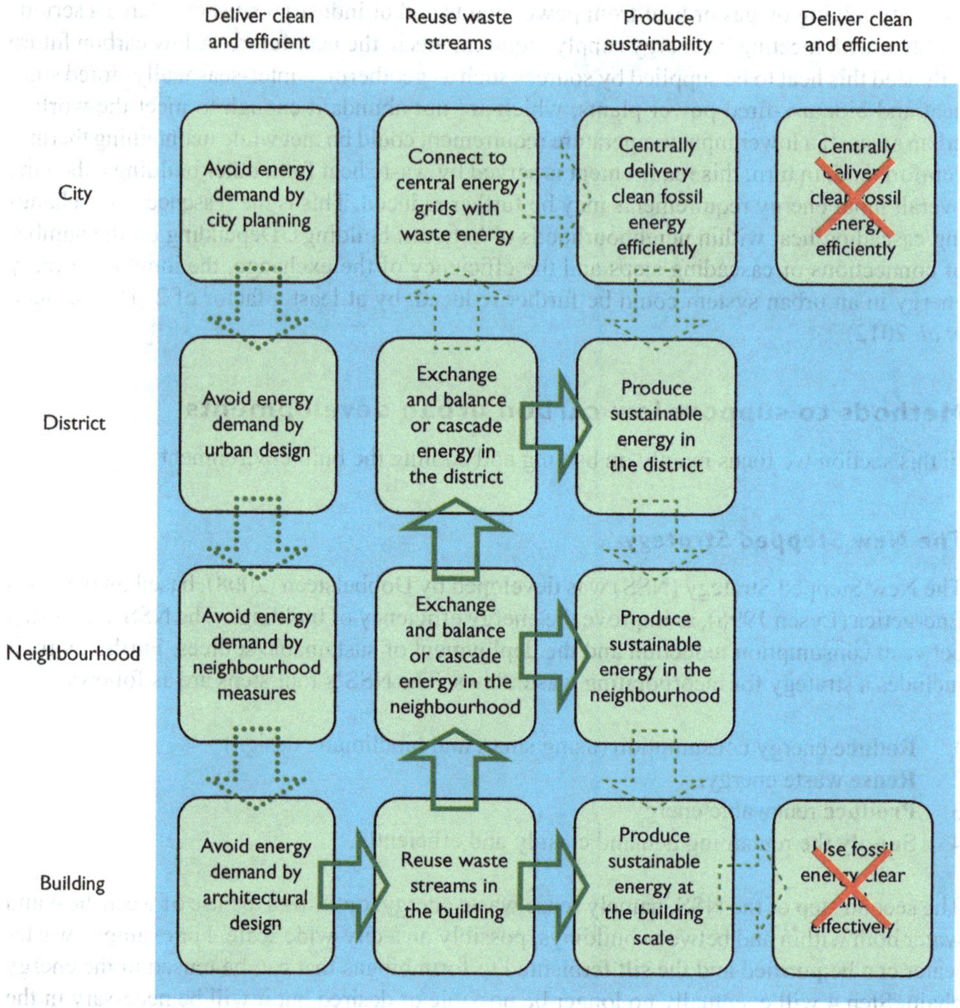

Figure 8.1 The REAP methodology.

Source: Image by Andy van den Dobbelsteen, TU Delft, supplied by the authors.

From building to neighbourhood

There are different ways that buildings' energy can be exchanged within their neighbourhood. In Western Europe, modern offices start cooling down as soon as outdoor temperatures exceed 12°C. By contrast, at these temperatures residences still require heating. This provides opportunities for heat exchange between buildings, particularly during spring and autumn. A particular opportunity exists here to combine the energy requirements of supermarkets (which always require cooling for refrigeration) with residences (which predominantly demand heating). Heating and/or cooling are only in excess when, overall, there is a lower demand for them (i.e. at a particular point in time). There is the potential to optimize these energy flows by storing them and exchanging them as the seasons dictate. A greenhouse captures a considerable amount of passive solar energy, which is usually released into the air through openable glass roof panes. An air-to-water heat exchanger could enable capture of this waste heat (30 to 40°C under sunny conditions) to be used (or intermediately stored) to heat homes, provided that these are well insulated and make use of a low-temperature heating system. This is an example of heat cascading (Tillie *et al.* 2009a).

If all waste flows at cluster or neighbourhood level are being used optimally, the remaining energy requirements need to be generated through renewable technologies. Some technologies are feasible at the individual scale (e.g. PV panels and solar collectors); other forms of generation are potentially more feasible at the neighbourhood level (e.g. ground-source heat pumps and wind) (Tillie *et al.* 2009b).

From neighbourhood to district

If energy planning can be tackled at the level of the urban district, the remaining discrepancies in the energy balance at the neighbourhood level (for example, excess demand for waste heat or cold) may be solved. At the district level, it is likely that there will be a wider mix of functions available, with different patterns of energy demand, and therefore supply. It is also possible to exchange, store and cascade energy at the district level. Larger amenities, such as shopping centres, swimming pools and concert halls, have a very specific energy pattern. Combining a number of these different amenities may achieve a higher level of energy efficiency and waste reduction (Tillie *et al.* 2009b).

In addition to exchange, storage and cascading, another option is possible. Useful functions may be added to complete missing links in the energy balance in an area. Once the existing amenities in the area have been optimized, we can consider any residual demand for heating or cooling. In this case, it is necessary to look for an amenity that requires extra heat on a yearly basis (for example, a swimming pool) or one that requires cold (for example, an ice-skating rink). The provision of renewable energy may then be tackled at the district level. As mentioned above, some sustainable measures may be implemented at building or neighbourhood level, but other, more capital-intensive projects are more appropriate at the district level. An example of this is a biogas fermentation plant that recycles biogas from wastewater and uses co-generation for heat and electricity. Geothermal energy plants are usually only feasible on a grand scale (Tillie *et al.* 2009a).

From district to city and beyond

Amenities in many Northern European cities are regulated centrally, and many have a heat grid (fed with waste heat from conventional power plants, co-generation plants, waste incineration plants or industries). These heat grids provide temperatures of between 90 and 130°C. This temperature level is necessary for poorly insulated buildings that contain central heating systems based on such temperatures. However, new housing projects are more effectively served with lower temperatures – providing the same levels of thermal comfort.

Once connected to the high-caloric heat grid, the whole exercise of exchange, storage and cascading at neighbourhood and district level is no longer logical. In this situation, the heat grid takes care of the heating and potentially also the cooling (via absorption cooling). The problem with this is that the local waste heat can no longer be usefully used and disappears into the environment – the urban surroundings. The lower caloric heat from buildings such as offices, supermarkets, swimming pools and dwellings is abundant and seldom utilized

(Tillie *et al.* 2009b). This has two negative effects: (1) the demand for energy is higher than necessary and energy is lost; and (2) the city is heated up, which in summer can aggravate the urban heat island effect (Dobbelsteen *et al.* 2012), and which is documented as leading to higher mortality rates (Huynen *et al.* 2001).

The REAP methodology aims to solve the problems of matching energy demand and supply starting at the lowest scale (buildings), and moving upward to building clusters and neighbourhoods. At the higher scale, 'help' may be called in from higher levels which – depending on the starting point – can be clusters, neighbourhoods, districts, cities and/or regions. In addition, the heat grid fulfils a useful role as a back-up system, or as a loading and unloading system for too much or too little heat in a district or neighbourhood. REAP can help make an existing neighbourhood sustainable, without requiring drastic urban planning measures. The following discussion of two case studies in Rotterdam will show how this can lead to low carbon neighbourhoods and outline how it may be done (Tillie *et al.* 2009b).

Applying REAP: case studies of the Hart van Zuid area, Rotterdam

The REAP methodology may be used to identify the key decisions that need to be made in the socio-economic and political aspects of urban development and architecture to meet the desired reductions in CO_2 emissions. Following these, the consequences for a city's buildings (and the open spaces in between them) can begin to be identified. In addition, combinations of measures for CO_2 reduction can mix with sustainable development by means of combining functions, social integration and integration of food production in the urban landscape (urban and peri-urban agriculture).

In Rotterdam's Hart van Zuid district, plans for the extensions of the Ikazia Hospital (new situation 33,013 existing m^2) and the Zuidplein shopping mall (new situation 71,050 m^2) provided theoretical test cases for the REAP methodology. The Hart van Zuid district is a major square and transport hub; every year, 33 million people move through the area and 12 million visit the Zudplein shopping mall. The Ikazia Hospital is a major regional hospital that employs approximately 1,300 people. The district is a mixture of 1960s urban development and 1980s architecture (Tillie *et al.* 2009a).

To apply the REAP methodology to the Zuidplein shopping mall, first an inventory of the current energy consumption was made. Then NSS Step 1 (reduce energy consumption) was followed by calculating the potential impact of installing insulation and other possible measures to bring down the consumption for heating and cooling. For example, the Zuidplein Theatre will be renewed and better insulation of the existing shopping mall significantly improves the situation. Also in this step, the demand for heat and cold were balanced as much as possible. For example, an additional 6,000 m^2 of supermarket will join the existing situation. Here useful functions may be added to complete missing links in the energy balance in this area. From an energy point of view, this can be balanced by heat demanding functions such as apartments. As a result (see Table 8.1), the total new energy demand is: heat 8,654 GJ, cold 7,231 GJ, electricity 14,629 GJ. The heat–cold ratio is 1:0.8.

Following NSS Step 2, we looked for opportunities to reuse waste energy. The new supermarket generates sufficient waste heat that 1 m^2 of supermarket can heat approximately 7 m^2 of housing! Adding other small waste heat flows, this means that 45,550 m^2 or 650 apartments can be planned (assuming that excess heat and cold are seasonally stored for later use). The thermal storage covering heating demand is 7,231 GJ. The remaining heating demand is 1,423 GJ.

Following NSS Step 3 (generate renewable energy), we noted that the remaining demand for heat can be solved by the addition of greenhouses on the roofs. A 1 m^2 greenhouse on top of the roof can potentially heat 4 m^2 of housing (extra electricity demand 142 GJ); these could be used for food production and/or as public areas. Another option is the addition of PV panels. PV panels could also be installed on the remaining part of the roof to supply electricity for the shopping centre (2,547 GJ). The total remaining electricity demand for the whole area (12,224 GJ) could be sustainably generated at the district or city scale (Tillie *et al.* 2009a).

The Ikazia Hospital cluster is to be extended (30,030 m^2 existing; 3,000 m^2 new). The hospital consumes a lot of energy at a relatively constant level seven days a week. The possibilities for creating an improved heat–cold balance by adding other functions are limited. As with the shopping mall, an inventory of the current energy consumption was made. Next,

Table 8.1 Heat, cold and electricity usage data for Rotterdam's Zuidplein shopping mall in the Hart van Zuid district

Step 1	*Step 2*	*Step 3*	*Energy*
Reduce energy demand and through insulation + heat cold demanding programme in balance	H:C balance total cluster Heat-Cold ratio H:C 1:0,8	Resulting heating demand sustainably generated by greenhouse and renewable energy generation	contribution PV panels: 2,547 GJ resulting energy demand: 12,224 GJ
total demand cluster: H: 8,654 GJ C: 7,231 GJ E: 14,629 GJ	thermal storage covered heating demand: 7,231 GJ resulting heating demand: 1,423 GJ	H:C (dis)balance contribution greenhouse: 1,423 GJ extra electricity demand due to usage greenhouse: 1,42 GJ	demand: H: 0 GJ C: 0 GJ E: 12,224 GJ

Table 8.2 Heat, cold and electricity usage data for Rotterdam's Ikazia Hospital

Step 1	*Step 2*	*Step 3*	*Energy*
Reduce energy demand and through insulation + heat cold demanding programme in balance	H:C balance total cluster Heat-Cold ratio W:K 1:0.24	Resulting heating demand facade sustainably generated by climate and asphalt collectors	contribution PV panels: 0 GJ resulting energy demand: 6,675 GJ
total demand cluster: H: 32,362 GJ C: 7,918 GJ E: 5,676 GJ	thermal storage covered heating demand: 7,918 GJ resulting heating demand: 24,443 GJ	H:C (dis)balance contribution climate facade contribution : 9,990 GJ aspahalt collector: 8,113 GJ contribution city heating: + 6,340 GJ 24,443 GJ extra electricity demand due to usage climate facade: 999 GJ	demand: H: 0 GJ C: 0 GJ E: 6,675 GJ

NSS Step 1 (reduce energy consumption) was followed by calculating the impact of installing insulation and trying to bring the demand for heating and cooling into balance as much as possible.

The hospital can be optimally insulated by adding a climate façade. A new entrance will provide an improved link to the public areas. 'Pyjama gardens' on the green roofs that incorporate orchards and vegetable gardens with a greenhouse will create a healing environment that will contribute to patients recovering more quickly. As a result, the cluster's total energy demand is: heat 32,362 GJ, cold 7,918 GJ, electricity 5,676 GJ (see Table 8.2). The cluster can reuse waste energy flows by recycling heat from waste air, and water can be applied using heat and cold storage. The resulting heat–cold ratio is 1:0.24.

The demand for heat can be sustainably satisfied using the climate façade and asphalt heat collectors at the new Zuidplein shopping mall. Any remaining energy requirements may be satisfied by renewable energy generation at a district or city scale (Tillie *et al.* 2009a).

Case study REAP2: smart grid for heating and cooling Rotterdam's Merwe-Vierhavens district

The REAP approach was extended and tested in the Merwe-Vierhavens case study, which led to the development of new technical concepts for energy exchange. As in most other urban areas, Merwe-Vierhavens, an inner-city port district in Rotterdam, is confronted by the need for a gradual transition to energy-neutral buildings. At present the area contains warehouses, industrial buildings, commercial offices and a small power plant, many of which will continue operations for several decades, whereas others will be demolished, creating space for new, sustainable premises. This means that the energy demand and supply, of heat in particular, will be very diverse and that the technology and infrastructure (both existing and to be developed) need to be multifunctional, able to serve variable demands now as well as all kinds of demands in the future (Dobbelsteen *et al.* 2012).

The core focus of the project was the elaboration of technical concepts for tuning the demand and supply of heat and cold (as per Step 2 of the NSS: exchange of waste flows). There are four principles for balancing heat and cold:

1 matching supply and demand directly;
2 (inter-)exchanging and storing residual heat and cold;
3 cascading waste heat;
4 storing temporal differences in supply and demand.

When these principles were applied at four different scales – buildings, neighbourhoods (quays), district (the total area of Merwe-Vierhavens) and city (Rotterdam) – five strategies emerged (see Figure 8.2) (Dobbelsteen *et al.* 2012).

The five strategies may be translated into technical solutions that use the available utilities but combine and connect them in a novel way. Figure 8.3 shows the principle of a heat-cascading grid, which uses the return flow of one neighbourhood as the main supply for another.

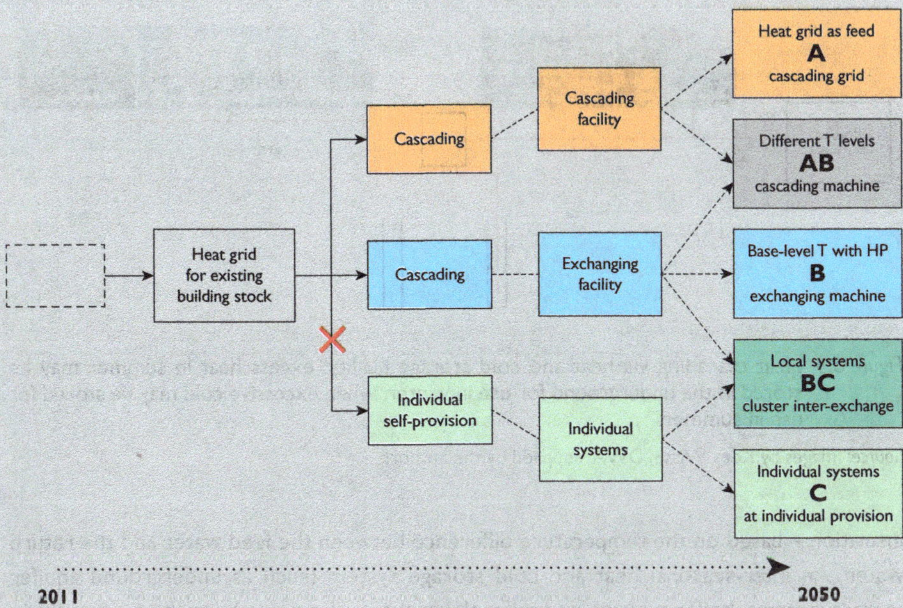

Figure 8.2 The five energy exchange strategies that emerged from REAP2: two are based on connection to the central high-temperature heat grid: A. heat cascading grid, and AB. cascading machine; one introducing a neighbourhood facility: B. exchanging machine; and two disconnected from centralized energy infrastructure: BC. inter-exchange within a cluster and C. individual self-provision.

Source: Image by Andy van den Dobbelsteen, TU Delft, supplied by the authors.

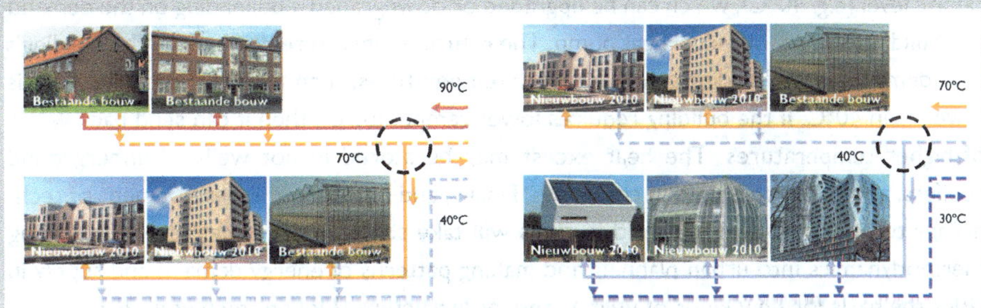

Figure 8.3 Heat-cascading grid, connecting an old neighbourhood with a newer one (left) and connecting a newer neighbourhood with a very well-insulated one (right); the latter grid can be an extension of the first.

Source: Images by Kees Wisse, DWA, supplied by the authors.

Cascading can also be established by a 'cascading machine' connected to the central provision of high-caloric heat. This concept could provide water at different temperatures for buildings with different demands – depending on their heat system and thermal

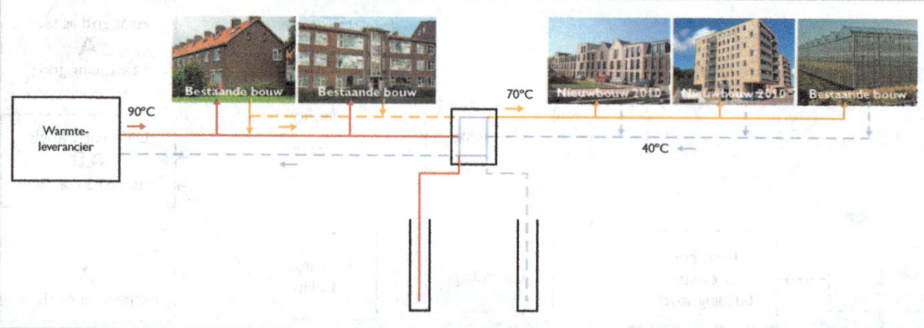

Figure 8.4 Heat cascading via heat and cold storage facility: excess heat in summer may be stored in the underground for use in winter, when excessive cold may be stored for use in summer.

Source: Images by Kees Wisse, DWA, supplied by the authors.

insulation – based on the temperature difference between the feed water and the return water. An inter-seasonal heat and cold storage system (such as underground aquifer storage systems, boiler systems or aquifer thermal energy systems) can shave off peaks in supply and demand, as Figure 8.4 shows.

The five strategies have specific spatial, social, legal and institutional implications. Some of these implications became apparent when they were clarified by sketches of the technical principles and their respective spatial influences, as is shown in Figure 8.5.

The heat-exchanging machine is still a theoretical concept but could be ground-breaking for the exchange of heat and cold in cities. The exchanging machine provides water at a low caloric level (e.g. 40°C), which can be upgraded or downgraded – depending on the demand of a building – by an individual heat pump. The return temperature depends on the building's heat demand. If the building requires higher temperatures, then the return temperature is lower than 40°C. If the building requires lower temperatures, then it can send back water of higher temperatures. The heat excess may be stored in hot wells of underground aquifers, and the excess cold in cold wells. Finally, and not least, the development of technology to exchange heat and cold in cities will take science to a higher level, integrating thermodynamics into urban planning and making patterns of energy demand and supply in cities the basis for new kinds of utilities and infrastructure (Dobbelsteen et *al.* 2012).

The storage of heat and cold will be a valuable asset to a sustainable energy system as, in the future, the availability of renewable energy will fluctuate partly due to demand changes through the seasons. In the case of a cascading machine, a multi-temperature storage system may be welcomed, because it will deliver the possibility of storing heat at different temperatures and avoid greater exergy (quality of energy) losses. However, this multi-temperature system does not exist as yet.

One important aspect that should have been considered in this case study is the economic consequences of the different energy solutions proposed. The introduction of vastly

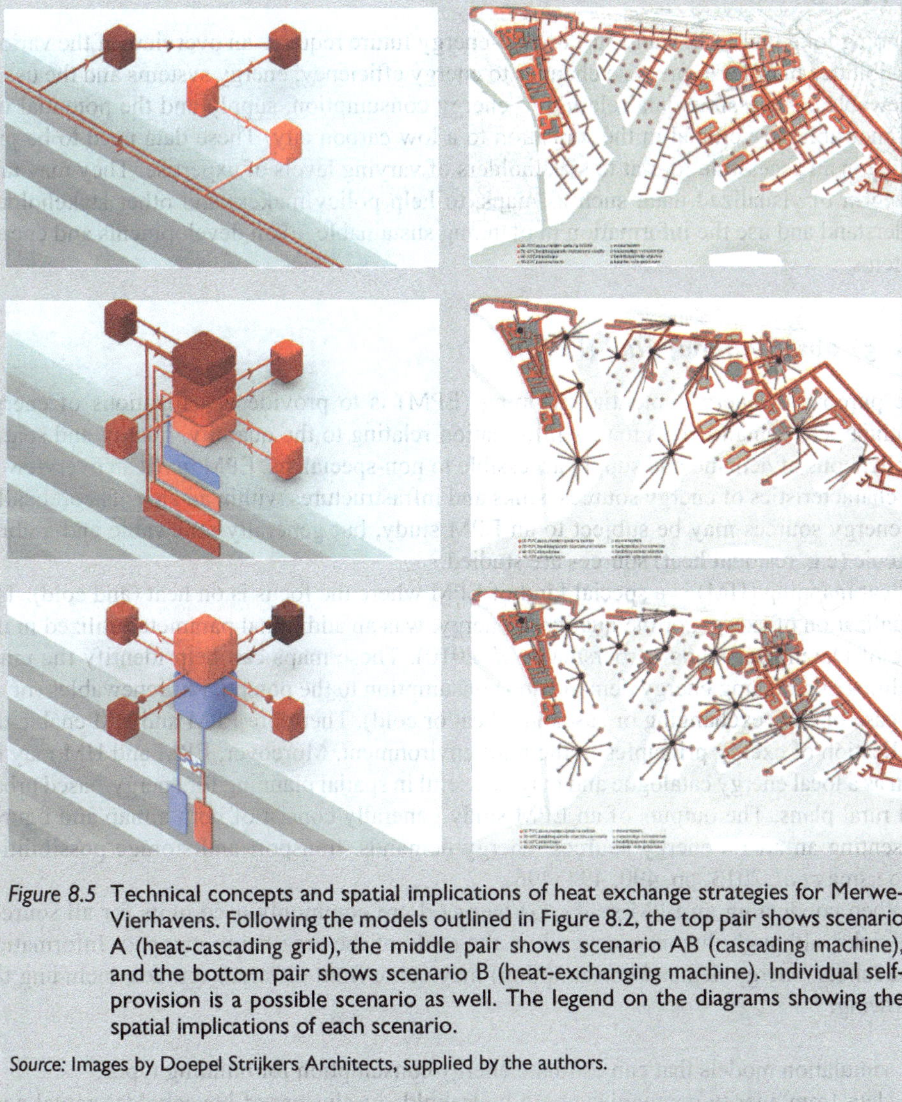

Figure 8.5 Technical concepts and spatial implications of heat exchange strategies for Merwe-Vierhavens. Following the models outlined in Figure 8.2, the top pair shows scenario A (heat-cascading grid), the middle pair shows scenario AB (cascading machine), and the bottom pair shows scenario B (heat-exchanging machine). Individual self-provision is a possible scenario as well. The legend on the diagrams showing the spatial implications of each scenario.

Source: Images by Doepel Strijkers Architects, supplied by the authors.

different and extensive new infrastructures for this kind of heat exchange in cities may be uneconomical when compared to traditional systems. This is, however, dependent on the way things are solved and exactly the reason why low-temperature systems should be tackled at neighbourhood level (typically within a radius of not more than 300 m).

These approaches focus on the development of heat and cold grids in addition to the traditional industry-fed high-caloric heating grids. The local exchange of (low-caloric) heat within and between buildings in an urban neighbourhood or district is not very common. Nevertheless, it has great potential as a way to meet cities' energy needs in the future (Dobbelsteen et *al.* 2012).

Mapping methods

Planning for a low carbon and sustainable-energy future requires an overview of the various possibilities in order to propose changes to energy efficiency, energy systems and the use of renewables. Accessible data relating to energy consumption, supply and the potential use of renewables are crucial in the transition to a low carbon city. These data need to be presented in an accessible format to stakeholders of varying levels of expertise. They may take the form of visualized data, such as maps, to help policy makers and other stakeholders understand and use the information in planning sustainable urban developments and energy systems.

Energy potential mapping (EPM)

The purpose of energy potential mapping (EPM) is to provide visualizations of energy potential and demand. This makes information relating to the quantity, quality and spatial implications of demand and supply accessible to non-specialists. EPM gives an overview of the characteristics of energy sources, sinks and infrastructures within an area. Theoretically, all energy sources may be subject to an EPM study, but generally renewable and anthropogenic (e.g. residual heat) sources are studied.

Heat mapping (HM) is a special kind of EPM where the focus is on heat (and cold). The visualization of exergy, or the quality of energy, was an additional parameter utilized in the case of Dutch heat maps (Broersma *et al.* 2010). These maps can help identify the practicalities of matching energy demand and consumption to the potential of renewables (or, in the case of HM, exchanging or cascading heat or cold). Therefore EPM and HM enable the application of exergy principles in the built environment. Moreover, EPM and HM may be seen as a local energy catalogue and may be useful in spatial planning for energy-based urban and rural plans. The outputs of an EPM study generally consist of both a map and figures presenting an area's energy sources, energy demands, transport and storage possibilities (Broersma *et al.* 2013, pp. 490, 494–496).

When conducting an EPM study, GJ/ha or GJ are commonly used units for all sources and sinks, although this may depend on the scale of the area being mapped. Information about sinks (energy demand/consumption) may come from different sources, including the following:

- simulation models that can estimate energy consumption for building types;
- data from energy companies about households or clusters of households' actual consumption;
- national or international databases, like the International Energy Agency and Eurostat.

When considering possible sources, the potential contributions of both renewable energy (originating from the sun, wind, water, soil and biomass) and waste flows from the built environment need to be calculated (Broersma *et al.* 2013, pp. 496, 497).

In the Netherlands, energy consumption data are available for some 30 different housing typologies for which both average energy consumption and measures to improve energy efficiency are known (SenterNovem 2008). Capacity has been built in to enable this data to be cross-referenced with local datasets and altered if necessary. Consumption may be divided into sectors depending on the data available and how they are organized. The sectors generally used are: residential, services, transport, industry and energy industry (i.e. refineries). So that all data can be mapped and connected to Geographical Information Systems (GIS), data on consumers that have a geographical location such as factories, residential areas and networks are all included. This is especially useful when compiling data to support decision making.

Basic information on climate, topography, geography, land use, underground and built environment is required. For instance, at a basic level, in order to calculate solar potential, the annual radiation per square metre that reaches the area's surface from the sun is considered. For wind, the average annual wind speed is a common parameter. Water energy is considered in different ways; for hydroelectric energy, heights and annual quantities of water flow rates or average river speeds can serve as the required data inputs. In order to define the amount of heat available in geothermal sources, temperatures and volumes of existing underground aquifers are data inputs. For biomass, amounts of waste flows such as from inhabitants, livestock and land use must be determined. Anthropogenic sources (e.g. residual heat from industries) are measurable (Broersma *et al.* 2013, p. 497).

When the theoretical potentials have been calculated, they need to be corrected for what can realistically be harvested taking at least technical efficiencies into account. For some sources, detailed modules (in GIS) are available. For solar energy, for instance, shading, roof angles, weather patterns, cloud coverage, etc. make it possible to obtain realistic harvestable data. Restrictions such as politics, costs and public opinion may be considered and debated during the decision-making process.

Storage is an essential component of an EPM study. Because some renewable resources may have a fluctuating delivery potential, energy storage systems are needed to make their energy available at times other than when it was harvested or produced. As with the energy sources themselves, storage systems deal with geographical and quantitative (as well as qualitative and temporal) input that may be mapped. Heat and cold storage systems have a specific theoretical capacity to store either hot and/or cold water. Water basins or salt caverns can act as storage systems for potential energy to be converted into electricity. Input data required here are volume, height and pressure (as this relates to energy storage of compressed air).

Information about energy infrastructures is also required. Infrastructures have geographical, quantitative and qualitative characteristics (e.g. heat or electricity networks) (Broersma *et al.* 2013, p. 497).

Heat mapping (HM)

In heat mapping, the heat demand is depicted as a series of elevated contours; the heat supply potential present subsequently fills the resulting sink. (Note that, although this may be compared to filling a bucket, height rather than volume indicates the amount of energy, as this makes it possible to assess a collection of areas of different sizes.) When all these individual 3D maps are stacked, the resulting 3D landscape indicates where demand may potentially outstrip supply, and vice versa (see Figure 8.6). Large energy users, such as supermarkets, swimming pools, industries and power plants, are depicted by cylinders, where the colour denotes heating or cooling load and hollow cylinders represent the net demand for an installation. Larger roads may be equipped with integrated pipes to become giant solar collectors and are depicted as yellow bands. When present, a district heating grid providing heat transport opportunities is hatched in red (Broersma *et al.* 2013, pp. 499–500; Broersma *et al.* 2010).

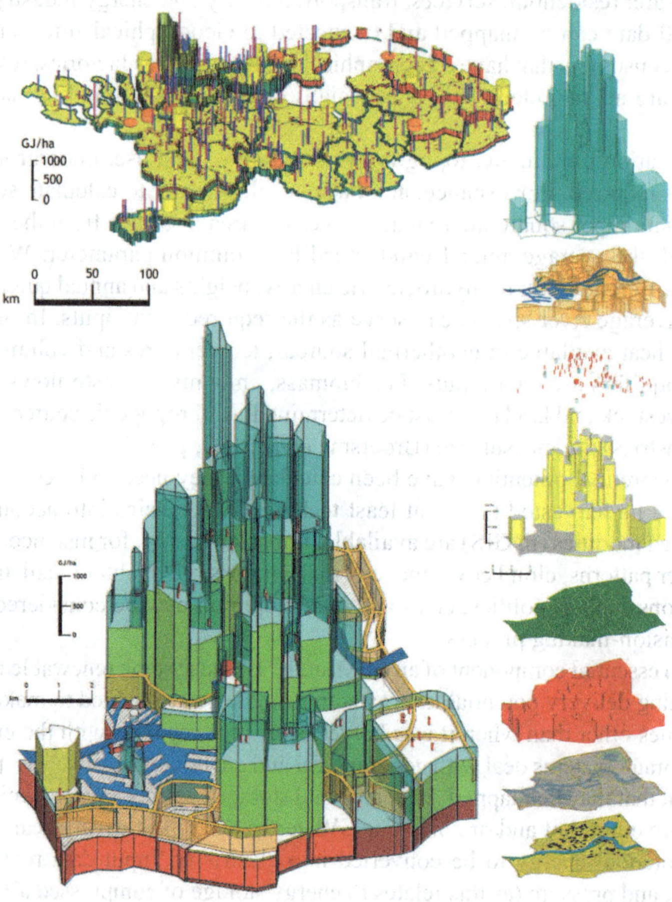

Figure 8.6 Heat maps of The Netherlands (upper left) and the city of Rotterdam (bottom left; SketchUp 3D modelling software was used for the graphs). The graphs on the right show the stacked annual supply potentials (and demand) of the middle graph, from 1 to 8 representing: (1) heat demand of dwellings, (2) heat potential from road collectors, (3) point sources and sinks of heat and cold (supermarkets, hospitals, power plants, etc.), (4) heat potential from roof collectors, (5) heat potential from biomass, (6) geothermal heat potential, (7) map of the area, (8) soil suitability for underground vertical heat exchangers.

Source: Image by Michiel Fremouw, TU Delft, supplied by the authors.

The Rotterdam heat map shows the discrepancy between the heat consumption of the densely populated city centre and of the much more sparsely populated periphery. This presents an opportunity for diverting excess geothermal flux at the latter areas to the former by means of district heating. Also clearly visible is that the demand for heat in the city is generally twice as high as the potential yield from the sun and geothermal sources, an insight supporting the idea that energy savings remain an important means to establish a low carbon future (Broersma *et al.* 2013, p. 500).

Energy scenario planning with GRIP

The previous sections have presented a step-by-step approach to integrating energy (mainly heating and cooling) into spatial planning and making energy data about consumption, storage and supply easily accessible via energy potential mapping. This section discusses the combined use of all these aspects to plan for a low carbon future.

The energy scenario planning process may be run using bottom-up or top-down data. However, the internal consistency of the resulting pathways, and their value for stakeholders, will be much higher if local stakeholders participate and local data are used to determine potential outcomes, as described below.

The Greenhouse gas Regional Inventory Protocol (GRIP) is a three-stage process for forming energy action plans at the local and regional level. The first stage is compiling a greenhouse gas emissions (CO_{2eq}) inventory and energy baseline, the second stage is producing consensus-driven low carbon energy scenarios aided by a scenario tool, and the third stage is using the results and experiences to inform energy plans.

The GRIP scenario process, aided by a user-friendly tool, brings together stakeholders representing different sectors into an iterative exercise. The process is based on the principle of 'to hear is to forget, to see is to remember and to do is to understand' with the design of the exercise based on the learning methods postulated by Bruner (1960, 1961, 1962) and Vygotsky (1987, 1992). The process, facilitated by an energy expert, combines qualitative and quantitative aspects to deliver a whole-system energy scenario that is a consensus-based synthesis of participating stakeholders' views. The results include a breakdown of changes to energy consumption, the fuel mix and energy generation, together with the learning that takes place during the exercise. As the outcomes are consensus-driven, they reflect the vision of the participating stakeholders rather than the interpretations of a researcher or a consultant (Carney and Shackley 2009; Carney 2006; Upham *et al.* 2014).

GRIP has been applied in cities and regions in 15 countries. The results of the scenario process regularly fall short of the set targets (for example, a reduction in emissions of 80 per cent). This is an important finding, as it shows the discrepancy between what stakeholders envisage being delivered and the CO_2 targets being sought. The stakeholders who have partaken in the process all work on energy and climate change; however, feedback from them suggests that just 3 per cent of participants have previously tried to quantify the CO_2 implications of a low energy future. This means that the low carbon energy futures they are working towards may fall substantially short of the CO_2 reductions targeted. This suggests an important gap in experience that needs to be filled. It should also be a concern for policy makers working at regional, national and international levels that there is insufficient experience 'at the coalface', where energy policy is being devised and delivered, to meet emissions reduction targets and associated climate goals.

GRIP was applied in Rotterdam, where four scenario sessions were conducted. These have been used to shape policy. Table 8.3 and Figure 8.7 contain the results of the inventory for Rotterdam; they show the fairly unique situation (for the Western world) of a city where emissions from industry (traditional and energy) account for three-quarters of the city's emissions.

The emissions reductions in the scenarios were 74 per cent, 50 per cent, 83 per cent and 82 per cent. These figures do not include international maritime or aviation-derived emissions. In the two scenarios that failed to meet the emissions reduction target, the limiting factors were the perceived complexities of reducing emissions from industry and, in two of the sessions, a belief that Rotterdam will continue as a location for petroleum refining.

The stakeholders involved in the process included planners, policy makers, academics and local businesspeople. As a consequence of this work, a memorandum of understanding was formed between the city and local businesses for extending the heat–cold grid in the city. This supportive network is now being used as a conduit to form and implement policy locally. Follow-up work is being conducted with Dutch government agencies from all levels of government.

GRIP is being made available through the International Coalition for Local Environmental Initiatives (ICLEI) to aid and inform low carbon energy plans in up to 84 countries.

Table 8.3 Results of the Greenhouse gas Regional Inventory Protocol (GRIP) for Rotterdam

	CO_2	CO_2eqv
Residential	1,259	1,287
Services	1,586	1,591
Industry	5,640	5,646
Energy industry	10,553	10,576
Transport	1,249	1,262
Fugitive	275	275
Total energy	20,562	20,637
Industrial processes	44	148
Waste	0	280
Agriculture	0	132
Total	**20,606**	**21,197**

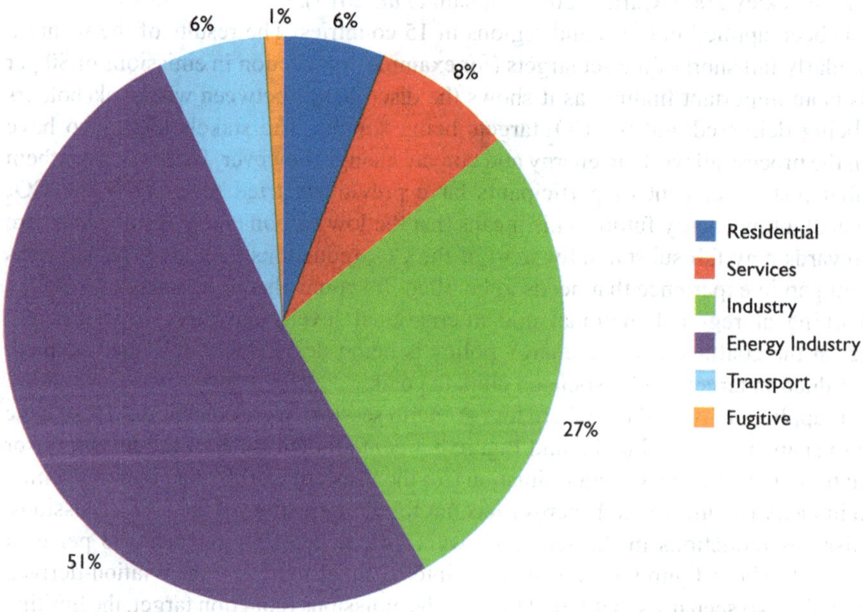

Figure 8.7 Results of the inventory for Rotterdam.

Conclusions, discussion and outlook

Although the NSS and the REAP studies started off as theoretical exercises, over the years they have developed into a practical way to link buildings' energy efficiency and urban planning. The fairly abstract fields of energy use and CO_2 have become part of urban planning. The idea of using waste flows of energy at different scales has led to the need to design smart grids for heating and cooling in many urban scenarios. This has been picked up in several follow-up studies, including the European Celsius Cities project.

Energy potential mapping has proven to be a structured methodology, which is currently evolving into a generic model that integrates energy databases in order to calculate energy potentials for a specific location or area by means of a web tool (Broersma *et al.* 2013, p. 503). Gathering the knowledge of stakeholders in the building process via smart phones can also accelerate data collection and thus facilitate decision making early on in the process. When linking EPM and GRIP to GIS, even more datasets can be used and all energy data can be matched with present socio-economic and urban data. At the moment, a GIS-based transilience approach for cities and neighbourhoods is used to facilitate stakeholders for accelerating the transition into liveable low carbon cities (Tillie 2013). A scalable GIS 'City Scanner' for neighbourhoods, areas or cities with socio-economic and environmental indicators feeds this approach, which allows it to be used as a GIS 'Urban Transition Planner'. Energy and CO_2 are part of this.

The energy scenario planning tool GRIP has recently been combined with REAP, EPM and GIS. Apart from speeding up the process of data collection, it has also made developed scenario paths more valuable as direct links can be made to local situations, potentials and policy making. Scenario paths can easily be altered, making it an easy tool for effecting changes in existing policies. Overall, the combination of these approaches for the transition into low carbon cities has proved to be very valuable. A road map for low carbon cities has become available!

Note

1 CO_2 equivalents: all greenhouse gases converted to carbon dioxide, according to the relative effect on climate change.

References

Broersma, S., Fremouw, M., Dobbelsteen, A. van den and Rovers, R. 2010, 'Warmtekaarten: Nederlandse warmtekarakteristieken in kaart gebracht', Faculteit Bouwkunde TU Delft, Delft.

Broersma, S., Fremouw, M. and Dobbelsteen, A. van den, 2011, *Synergie tussen Regionale Planning en Exergie: SREX*, Publikatieburo Bouwkunde & Delft Digital Press, Delft.

—— 2013, 'Energy potential mapping: visualising energy characteristics for the exergetic optimisation of the built environment', *Entropy*, vol. 15, no. 2, pp. 490–506.

Bruner, J.S. 1960, *The Process of Education*, Harvard University Press, Cambridge, MA.

—— 1961, 'The act of discovery', *Harvard Education Review*, vol. 31, pp. 21–32.

—— 1962, *On Knowing: Essays for the left hand*, Harvard University Press, Cambridge, MA.

Carney, S. 2006, 'The development and application of a stakeholder-led approach for both estimating and exploring the potential for greenhouse gas emissions mitigation on the government office regional scale', doctoral thesis, University of Manchester, Manchester.

Carney, S. and Shackley, S. 2009, 'The Greenhouse gas Regional Inventory Project (GRIP): designing and employing a regional greenhouse gas measurement tool for stakeholder use', *Energy Policy*, vol. 37, pp. 4293–4302.

Dobbelsteen, A. van den 2008, 'Towards closed cycles: new strategy steps inspired by the cradle to cradle approach', in *Proceedings PLEA 2008: 25th Conference on Passive and Low Energy Architecture*, UCD, Dublin.

Dobbelsteen, A. van den, Tillie, N., Fremouw, M., Wisse, K., Doepel, D., Genten, L., Dorst, M. van, Hobma F. and Daamen, T. 2011, REAP2 —Rotterdamse EnergieAanpak & –Planning 2: technische, ruimtelijke, sociale, juridische en strategische uitwerking van het REAP-model, toegepast in de Merwe-Vierhavens; Faculteit Bouwkunde, TU Delft, Delft.

Dobbelsteen, A. van den, Wisse, K., Doepel, D. and Tillie, N. 2012, 'REAP2: new concepts for the exchange of heat in cities', in *Proceedings of Sasbe 2012*, São Paulo, Brazil.

Huynen, M.M., Martens, P., Schram, D., Weijenberg, M.P. and Kunst, A.E. 2001, 'The impact of heat waves and cold spells on mortality rates in the Dutch population', *Environmental Health Perspectives*, vol. 109, no. 5, pp. 463–470.

Kürschner, J., Hakfoort, L., Mantel, B., Dobbelsteen, A. van den and Tillie, N. 2011, 'Leidraad Energetische Stedenbouw', Dienst Ruimtelijke Ordening, Gemeente Amsterdam.

Lysen, E.H. 1996, 'The Trias Energetica: solar energy strategies for developing countries', in *Proceedings of the Eurosun Conference*, Freiburg, Germany.

SenterNovem 2008, *Cijfers en tabellen 2007*, SenterNovem, Utrecht.

Tillie, N. 2013, 'Towards a more resilient Rotterdam: using the transilience approach', ICLEI resilient cities conference, Bonn, Germany.

Tillie, N., Dobbelsteen, A. van den, Doepel, D., Jager, W. de, Joubert, M. and Mayenburg, D. 2009a, REAP: Rotterdam Energy Approach and Planning, Rotterdam Climate Initiative, Rotterdam.

—— 2009b, 'Towards CO_2 neutral urban planning: introducing the Rotterdam Energy Approach and Planning (REAP)', *Journal of Green Building*, vol. 4, no. 3, pp. 103–112.

Upham, P., Carney, S. and Klapper, R. 2014, 'Scaffolding, software and scenarios: applying Bruner's learning theory to energy scenario development with the public', *Technological Forecasting and Social Change*, vol. 81, pp. 131–142.

Vygotsky, L.S. 1987, 'An experimental study of concept development', in R.W. Rieber and A.S. Carton (eds), *The Collected Works of L.S. Vygotsky*, vol. 1, *Problems of General Psychology*, Plenum, New York, pp. 121–166.

Vygotsky, L.S. 1992, *Linguagem, desenvolvimento e aprendizagem*, Editions Ícone, São Paulo, Brazil.

Chapter 9

Green districts and carbon engineering

Increasing greenery, reducing heat island effects and generating energy

Steffen Lehmann

Summary

A search for the ideal urban form for tomorrow's sustainable cities is going on worldwide. Tomorrow's urban precincts will have to generate at least half of their power themselves, while offering new forms of public green spaces, both for recreation and to mitigate a warmer urban climate. In addition, future urban precincts will be test beds for new low carbon construction systems. None of this will be possible without zero-energy and even plus-energy buildings. To create integrated urban developments with a focus on energy, construction systems, water, greenery and the urban microclimate, urban designers and planners will need to engage with policy makers in order to drastically reduce our cities' consumption of energy and resources. This chapter introduces several important aspects of low carbon cities, including the following:

- precincts that generate more on-site energy than they consume, so-called plus-energy districts;
- precincts that are transformed towards decentralized infrastructure and carefully densified;
- prefabricated low carbon construction systems using engineered timber, for a new generation of carbon-engineered, high-performance buildings;
- the reintroduction of urban greenery into the city, using green roofs and façades.

New design concepts will guide the reintroduction of greenery and biodiversity into the urban built environment. The Barangaroo waterfront development at East Darling Harbour in Sydney is a large urban renewal project and a suitable case study. At Barangaroo, all roofscapes will be green roofs, contributing to the mitigation of the urban heat island effect and collecting rainwater; this inner-city precinct is setting new standards for Australian urban renewal, including solutions for environmental, energy and roofscape issues. Here, the 'low carbon precinct' concept is being developed further into a plus-energy city district that is compact, mixed use and well connected to public transport.

Introduction: cities and climate change

What will the energy and material implications of future urbanization be? As the world continues to urbanize, we find shrinking, undynamic and insufficiently developed urban precincts with limited investment and obsolete infrastructure alongside fast-growing, dynamically changing precincts. We need innovative and comprehensive strategies that enable us to manage the coming demographic and structural changes more effectively. But can the city as we know it today ever be sustainable? We need to pursue integrated urban development with a focus on energy, water and the urban microclimate, and policy makers will have to engage with it in order to drastically reduce our cities' consumption of energy, water and resources (ICLEI 2007).

Over the past decade, the link between cities and climate change has been firmly established. Transforming our cities to be more compact and use less carbon not only helps mitigate the impacts of climate change, but also brings about improvements in liveability and well-being for city dwellers.

There is an important link between resource efficiency and a city's compactness. Increasing compactness and density does not necessarily mean a decrease in liveability. The 'compact city' model supports resource efficiency and can be highly liveable at the same time (Singapore and Tokyo are good examples of this). Increasing compactness allows for a more efficient integration of transport and power systems. The compact city model means that growth boundaries are implemented so that, instead of continuing to expand city boundaries as populations increase, regulations and incentives are used to encourage the 'infilling' of existing districts. Thus, these districts and their infrastructure become progressively more efficient as they become more densely populated, absorbing future growth.

Tomorrow's urban precincts will have to generate at least half of their power themselves, locally and on site. This will not be possible without zero-energy and plus-energy buildings. These are already state-of-the-art, and the next stage is now emerging, where city districts themselves act as autonomous 'power plants'. Precincts that generate all the power they need on their own through decentralized systems of renewable energy (using solar PV, biomass, micro-wind turbines or geothermal technology) will make city planners' age-old dreams of inexhaustible clean energy come true. The secure provision of renewable energy and green urban transformation is fast becoming a reality for our society and a major planning focus around the globe. However, realizing this dream requires the input of policy makers, power suppliers, researchers, architects and planners, and citizens. The 'low carbon precinct' concept must be developed further into a plus-energy city district that is compact, mixed use and well connected to public transport. A large portion of each district (minimum 30 to 40 per cent of the area) should be dedicated to public green space to maximize open space while allowing for higher densities. Barangaroo in Sydney is just such a new precinct; it will be introduced later in this chapter.

The climate-friendly, energy-optimized and resource-efficient compact city

All around the world, cities face similar challenges and need to take action – even if the contexts differ from Australia to the Asia-Pacific region, from the Americas to Europe and Africa. What a climate-friendly and energy-optimized city should look like is one of today's central questions (Brown 2009). The main locations of Asia's urban population explosion are cities in China, India and the Philippines (think, for instance, of Manila, an underprivileged, crowded city that is urbanizing rapidly). In the Asian region, the main issues are rapid growth and urbanization processes and the related rural-to-urban migration of people. The US, Canada and Australia are exploring strategies to reduce their unsustainable urban sprawl and enormous dependency on automobiles (think, for instance, of a sprawling low-density metropolis like Houston). In Germany and other European countries, the focus is mainly on the energy-efficient conversion of the existing building stock and on optimizing material and energy flows, as the new construction rate is only around 2 per cent (Lehmann 2012). Almost 40 per cent of European cities are now shrinking or stagnating, while a few popular cities enjoy growth and investment. Those growing are the ones with high liveability indicators, which are popular with a skilled workforce in a globalized and increasingly knowledge-based society.

Design concepts for new urban precincts will need to guide the inclusion and reintroduction of biodiversity into the urban built environment to ensure easy access to recreational green space. Preserving biodiversity in the face of urbanization, habitat fragmentation, environmental degradation and climate change is probably one of the greatest challenges of our time. Tomorrow's urban precincts will have to offer new forms of green space for recreation, but also to mitigate the warmer urban climate and risk of heatwave impacts and the urban heat island effect.

How, then, can the transition to the post-industrial, plus-energy city be achieved? Tomorrow's plus-energy urban precincts will be based on the consistent, sound use of energy, land, water, green spaces, materials and mobility. Their long-term goals will be zero emissions, zero waste and the avoidance of energy/water/material wastage. This has already been successfully implemented in Vauban, a green precinct of Freiburg im Breisgau (Germany), in Hammarby Sjöstad in Stockholm and Malmö (Sweden), and in Copenhagen (Denmark). They all include sustainable and walkable precincts with short distances between living and working. In fact, the 'city of short distances' is a sustainable model that is now being emulated all over the world.

Sustainable urban growth: the holistic framework of green urbanism

Green urbanism is a holistic concept of sustainable urban systems and interconnected features, enabling existence and change (growth or shrinkage) without negatively impacting upon planetary life support systems and ecosystems. It is a particular form of urbanism that is concerned with a healthy balance between the city, the peri-urban area and its surrounding (and supporting) hinterland. Green urbanism also underpins practical action in order to shape the urban environment in a sustainable way (see Chapter 22 by Lehmann and Thornton, which discusses green urbanism and its principles in more detail).

The principle-based concept of green urbanism was developed in the late 1990s, promoting more compact and energy-efficient urban development, seeking to transform and re-engineer existing city districts and to regenerate the post-industrial city centre. It was advocated simultaneously by Beatley in the USA in 2004 and by Lehmann in Europe and Australia in 2005. Since then, the concept of green urbanism has been adopted in many parts of the world. For example, several Australian cities have recently formulated metropolitan plans that espouse strategies such as alternative energy generation, using combined-heat-and-power co-generation within urban districts, offsetting construction impact with low carbon structural systems (using fast-regrowing timber), green roofs, green façades and integrating more green space to mitigate the urban heat island effect. These public green spaces can come in all sizes, from intimate small gardens, to urban farming and community gardens, to more formal parks, to meadows and urban forests – all reintroducing biodiversity into the built environment. The principles of green urbanism are based on the triple-zero framework of the following:

- zero fossil fuel energy use;
- zero waste;
- zero carbon emissions.

The following list outlines the 15 principles of green urbanism (Lehmann 2010); see also Chapter 22.

The 15 principles of green urbanism are as follows:

1. Harmony with local climate and context.
2. Renewable energy for zero CO_2 emissions.
3. Zero waste.
4. Water-sensitive urban design.
5. Landscaping, gardens, green roofs and biodiversity.
6. Efficient public transport and public open space.
7. Local materials.
8. Density and retrofitting.
9. Green buildings and districts.
10. Liveability, healthy communities and mixed-use districts.
11. Local food.
12. Protecting cultural heritage.
13. Best practice governance and leadership.
14. Education, research and knowledge sharing.
15. Particular strategies for cities in developing countries.

Precinct-wide decentralized systems transforming cities district by district

Our cities will continue to demand increasing amounts of land and energy (Scheer 2006; Droege 2008; Dhakal 2009), consume large quantities of raw materials and produce increasing amounts of waste. At the same time, however, the economies of scale of cities give us the opportunity to make renewable energy sources more economically viable. How can this be done? Because cities require enormous amounts of energy, our energy and transport systems must be transformed quickly so that they can run as much as possible – but at least 50 per cent – on local renewable energy sources. The energy mix should take into account the costs and availability of technologies. Generated and stored locally, electricity can be transmitted and distributed through an intelligent power network, a so-called smart grid. A critical component in this shift – new battery technology for low-cost mass storage – is improving. Thus, precincts in the zero-emissions city will transform from power consumers into power producers. They will become local, decentralized 'power plants' and will use photovoltaics, solar heat generation and refrigeration, wind energy, biomass, anaerobic digestion of organics, geothermal energy, energy from micro-hydropower plants and other clean technologies. The role of large power suppliers will also have to change from that of monopolists to service providers for many small, decentralized and interconnected producers and consumers.

Precinct-wide decentralized systems will become more common and allow districts to generate their own energy, collect and clean their own water and even produce their own food through urban agriculture, therefore enabling precincts to be more autonomous and self-sufficient. What will be crucial in a precinct of this type is the technological interaction and the interconnectedness between individual buildings (IEA 2013). New infrastructure concepts such as co-generation for district heating or district cooling (or even tri-generation, which produces district-wide chilled water for cooling), the integration of new mobility concepts such as car sharing, car pooling, bike stations and cycling networks, and an increased share of electric mobility, will also be necessary. The concept includes local energy associations in which individual buildings support and balance each other through intelligent power distribution and integration by following exergy principles to recover waste heat, for instance, contained in sewage and wastewater. They will optimize the share of self-generated power for their own consumption so that entire precincts will eventually become independent from the public grid. Electro-mobility will reduce carbon emissions, air pollution and street noise, and will enable us to ventilate buildings naturally again. Thus, for example, balconies can again be built overlooking city streets and not facing away from them.

Even if we succeed in solving the energy problem this way for the time being, modern cities will require massive amounts of additional resources (Moewes 1995; Girardet 1999). This is why the zero-waste concept is also important. It implies putting an end to material wastage in construction and production processes and recovering 100 per cent of resources and nutrients (Lehmann 2013b). Waste will be treated as a valuable resource that must not be incinerated or buried but will be fully recovered and reutilized, where necessary through urban mining of former landfills. Material consumption must be decoupled from economic growth. Thus, zero waste should already be taken into account in the initial development of products, buildings and processes. This requires a manufacturing and architectural design approach that combines prefabricated building components in such a manner that they can be disassembled easily, recycled and reutilized later. In the zero-waste concept, the urban mass balance and an input–output analysis of urban material flows are of particular importance. For example, fast-growing local timbers should be used in order to keep environmental pressure from construction as low as possible. The timber building components can be planned and fitted in such a way that they remain untreated, have a long, useful life, and can be removed and separated again in an ecologically and economically efficient way with few losses for subsequent reutilization, recycling or thermal exploitation (taking into account 'design for disassembly' and 'design for recycling' principles) (Merl 2005).

Urban renewal with a sustainable value system, behaviour change and social innovation

Clean technology alone is not enough. We also need sustainable consumption and a reduction in wasteful behaviour; that is, the transformation of our value systems and consumer behaviour (Deloitte Touche Tohmatsu and World Economic Forum 2011; Rifkin 2011). Unsustainable consumption, especially in the cities of Asia (for example, in the People's Republic of China, where consumption levels are rising particularly fast), is one of the big unsolved challenges that must be addressed urgently in order to avert a supply disaster.

The transformation of the modern city must also integrate its inhabitants and their needs. The creation of a more sustainable society will require both changing the beliefs and values of consumers, as well as progressively reshaping the social and technological infrastructures that deliver and enforce the current systemic over consumption (Lehmann and Crocker 2012). It is crucial that urban dwellers have a say in this transformation through participatory processes. Moreover, the instigators of integrated urban development need to have full support from administrators and policy makers to be able to implement fully the city of tomorrow's ideas, concepts and approaches. Politicians and citizens therefore need to act quickly to create the enabling conditions for the implementation of 'plus-energy precincts' (Lehmann 2013b). Local actors must be engaged and encouraged to participate in the transformation. An important goal is to make cities attractive once again as living spaces for diverse and vibrant population groups, as places of social inclusion and integration that include improved services for their ageing inhabitants. In Australia, Canada and the USA, we have recently seen a growing focus on community values and place making, meaning the creation of public spaces with character. All this will help transform city dwellers from passive consumers into active and engaged citizens.

Science and technology play an important role in the transformation of the city. They can address the big issues of social transformation (by making green technologies affordable to everybody), energy production and the needs of the community without having to place the interests of private investors first. The energy-efficient transformation of cities and the associated new tasks will require upskilling at the technical and political level. Technical expertise will have to be disseminated and skills developed. This includes an intensive exchange of experience as well as an expansion of research in key fields at universities (a point I will return to in the conclusion). Tools, frameworks and indicators for evaluating the ecological performance of cities should continue to be developed and possibilities for improving their performance (via benchmarking and standard setting) should be explored in sustainable urban development research centres with the aid of best practice examples. Universities can take on an important role as think-tanks for this transformation of our cities (Head 2008).

In order to achieve climate protection goals, measures will have to be taken that go beyond energy-efficient building refurbishment and retrofitting (UN ESCAP 2012). In the long term, strategic planning approaches to viable, sustainable urban transformation, low carbon cities and zero-waste cities will gradually lead to zero emissions and maximum material efficiency and waste recovery. The aim is not just to lower energy consumption but also to improve quality of life and well-being. In the context of the zero-carbon city and the important role that energy efficiency plays in the city of tomorrow, I will focus here on ten of the previously introduced 15 principles of green urbanism:

- Promote integrated urban development as a catalyst for climate protection, including electro-mobility, cycling and walking.
- Create decentralized systems using local renewable energy sources for climate-friendly power generation so that active roofs, façades and canopies transform buildings into small power generators.
- Radically use new technologies and create 'plus-energy precincts'.
- Adapt political leadership and urban management principles.
- Secure the transformation to green urbanism by developing a green economy.
- Preserve, reuse and physically integrate cultural heritage and identity by adaptively reusing existing buildings.
- Revitalize, intermix and re-densify existing precincts, and adapt existing buildings to the needs of older people.
- Transform consumer behaviour and value systems (towards long-lasting behaviour change).
- Become a zero-waste city, including waste avoidance and 100 per cent recycling with consistent on-site resource recovery (meaning no waste incineration or landfill disposal, with only the incineration of non-recyclable waste timber).
- Promote knowledge and skill development; build think-tanks where knowledge is valued and shared.

Carbon engineering through cross-laminated timber: the use of massive prefabricated timber panels

Concrete, steel and aluminium enabled the advent of taller buildings, but these materials have significant environmental impacts that cannot be ignored. Alternatively, reconsidering wood in the form of engineered high-performance timber means using modern methods of off-site production and digital prefabrication. One of the most exciting new developments is the prefabrication of buildings using low carbon, engineered timber construction systems, such as cross-laminated timber (CLT). Introducing CLT systems means rethinking design and construction standards, a renewed focus on modular prefabrication and material culture, and a different understanding of residential infill development in general (Lehmann and Fitzgerald 2013). Engineered wood systems from certified sources (for instance, LVL, Glue Lam, Concrete Timber Composite and CLT) use large-format, modular, lightweight,

Figures 9.1a, 9.1b, 9.1c and 9.1d Recent examples of urban infill using prefabricated engineered timber panels as a low carbon construction system. Fig. 9.1a: The large and lightweight panels allow for faster construction. Figs 9.1b to 9.1d show examples of engineered timber buildings in London, Vienna and Zurich, constructed entirely in timber. In London: housing Whitmore Road (architects: Waugh and Thistelton); in Vienna: housing Wagramer Strasse (architects: Schluder); in Zurich: office headquarters (architects: Shigeru Ban).

Source: Courtesy of the architects, supplied by the author.

load-bearing, prefabricated components of structural integrity that emit low levels of carbon in their manufacture and store and sequester CO_2 over their life cycle.

Some of the recent infill developments using solid engineered timber panels in low carbon buildings are shown in Figures 9.1a to 9.1d. The wood for these structures has not been sourced from precious and irreplaceable virgin forests, but from certified, sustainably managed and harvested forests, where more trees are replanted than felled.

Advances in urban timber-building technology allows for new applications through environmentally concerned city planning, where the following three factors are of particular advantage:

1 *Prefabrication*: timber components are being prefabricated off-site in ever-increasing sizes, thus allowing for rapid assembly and construction, and for clean and dry sites in urban areas, with greatly reduced noise emissions compared to conventional building.
2 *Modular flexibility*: wood-based houses can be easily changed and remodelled. If functional requirements change, they can be adapted without much complication, providing a useful extension of the building's life cycle.
3 *Sustainability*: Sourced from sustainably managed forests, wood is a fast regrowing material that stores carbon dioxide and positively influences a building's overall eco-balance, a factor of increasing importance.

Cross-laminated timber is a multi-layered solid timber panel system that may be used as load-bearing wall, floor and roof elements of a building's structure. The panels use polyurethane adhesive to glue sheets (boards) together, which has no risk of toxic emissions at any stage in the product's life cycle. CLT panels can be manufactured in three, five, seven or more layers with typical widths of 3 metres and lengths of up to 20 metres long; the result is massive wood sections that behave like sheer walls. The number of layers and the panel width depend on structural and fire engineering requirements of the elements (wall, floor or roof) of the building for which they are designed. Off-site prefabrication has recently developed due to innovations in digital fabrication, tools and software. Each CLT panel manufactured is cut to the exact dimensions required using state-of-the-art CNC technologies, thus almost completely avoiding waste of material (Frangi *et al.* 2009; Lehmann and Hamilton 2011).

The use of timber in construction is potentially 'carbon negative'; every cubic metre of wood stores more than three-quarters of a ton of carbon (depending on the exact type of timber). Timber also has the advantage that it is easily recyclable and easier to handle on site.

Since timber is a material with the capacity to store carbon in large quantities over long periods, engineered solid wood panel construction offers the opportunity to turn buildings and districts into 'carbon sinks'. Thus, the historically negative environmental impact of urban development and construction can be turned around. All this gives one of the world's oldest construction materials a new and important role to play. Although wood has been used to construct buildings for over 1,000 years, the development and production of large-format solid panels was the first time that an engineered wood product matched the structural and technical requirements of more conventional building materials such as concrete, while being a truly sustainable alternative. The wider applications of cross-laminated timber buildings or hybrid structures, and the potential for mass-customization and construction heights for the innovative application of this low carbon and truly sustainable twenty-first-century building system, are beginning to achieve global recognition.

Cross-laminated timber was first developed by a couple of Austrian wood product manufacturing facilities in the early 1990s (Kaufmann 2011). Large-format CLT panel production commenced around 1998 with a large-scale CLT manufacturing facility in Austria. Today, it is a fully accepted sustainable construction system within the European and Canadian market, and demand has continuously increased. The reasons for this are new legislation (allowing for much higher timber structures), the momentum of the sustainable building movement, enhanced training and marketing, and the publicity received by successfully completed landmark buildings such as the much-publicized Stadthaus and Bridport House in London (Van De Kuilen *et al.* 2011). Awareness of the sustainable credentials of CLT has grown, most notably with Forté, the ten-storey residential timber development in Melbourne.

Vegetation, green roofs and the effect on the urban microclimate

Increased urbanization has led to changes in land cover, which directly influences the urban microclimate through landscape consumption, heat-storing materials and displacement of nature. Many local effects in cities, such as topography, building geometry and dimensions, density, façade and roof materials, the albedo of surfaces, the creation of urban canyons, pollution from transport and the integration of greenery can have great impacts on the temperature profile of any precinct or district (see Chapter 17 by Jusuf, Wong and Chong in Part III).

With global warming, temperatures will continue to increase; a recent report by the Australian Bureau of Meteorology (2013) warned that heatwaves like those that Australia experienced in the summers of 2013 and 2014, with temperatures of 46° Celsius, will become the 'new normal' in summer. Their report, *Australian Heatwaves: Hotter, longer, earlier and more often*, found that Australian heatwaves were becoming more frequent, intense and longer lasting. The key findings were as follows:

- Hot days, hot nights and heatwaves are direct consequences of climate change.
- Heatwaves have increased across Australia.
- Climate change is making many extreme events worse in terms of their impacts on people, property, communities and the environment.
- Record hot days and prolonged heatwaves are expected to increase in the future.

A built-up environment has significant influence on urban air temperature, which has been found to be considerably warmer than its surrounding rural or peri-urban areas. This phenomenon is called the urban heat island (UHI) effect, where urban structures absorb solar heat (radiation) during the daytime and release it back to the environment at night-time (Oke 1981). The UHI effect is present in all cities of all climatic regions and occurs due to increased urbanization where anthropogenic heat and human modifications have altered the characteristics of surfaces and atmosphere (see Figures 9.2a and 9.2b). (The urban microclimate is the subject of the chapters in Part III.)

New design concepts for urban infill buildings will guide the reintroduction of greenery and biodiversity in the urban built environment. However, preserving biodiversity in the face of urbanization, habitat fragmentation, environmental degradation and climate change is probably one of the greatest challenges of our time. The integration of trees, shrubs and flora into green spaces and gardens in the city is particularly important in helping to keep the urban environment cool, because buildings and pavements increase heat absorption, so that heat becomes trapped (causing the urban heat island effect). Prefabricated urban infill using massive timber construction systems as a way to store carbon in buildings is a way for carbon engineering.

A living wall on a façade or a green roof offers an additional insulating layer to a building, helping to provide a buffer against exterior temperatures, maintaining the temperatures inside. Various forms of greenery may exist in urban precincts, such as parklands, gardens, green roofs, vertical greenery, urban farming, nature reserves and planting of extensive vegetation, just to name a few. All of these act as sources of moisture for evapotranspiration (evaporative cooling), where the absorbed solar radiation can be dissipated as latent heat and thus aid in reducing urban temperature (Chen and Wong 2006). Green roofs can greatly reduce roof-top surface temperatures, reduce ambient air temperatures, improve human thermal comfort and create energy savings for buildings. In addition, green roofs provide stormwater management and may also give building occupants the opportunity to enjoy green space where they live or work.

The green roof has a long history in architecture, for instance, in cities and towns around the Mediterranean Sea, where it has been commonplace for centuries to use roof-tops as living and garden spaces (roof gardens), compensating for a lack of space on the ground. In the 1930s, the modernist vision of Le Corbusier and others reintroduced the green roof as a garden substitute on top of the building. In *La ville radieuse* (1935), Le Corbusier discussed the idea of the fifth façade or 'living on the roof', inspired by Mediterranean vernacular architecture and its miniature gardens; in the same year, his design for an apartment on the roof of a building in Paris was one of the first Modern Movement projects featuring living on the roof. In the early 1990s, the green roof reached technical perfection when Germany established generally acknowledged standards for green roof construction, including how to properly detail a safe and successful engineered green roof that does not leak. These standards still define engineered roof-top systems and their layers, with a waterproof membrane, thermal insulation, a root barrier, a moisture retention layer, a drainage layer, filter fabric and a planting medium on top.

Green roofs are more expensive than other roof types. The economic comparison of roof types quickly reveals that white roofs are the most cost-effective solution over a building's life cycle; while black-coloured roofs pose a health risk, white roofs are the most simple way to reduce energy costs, reflecting three times more sunlight back into the atmosphere and therefore absorbing much less sunlight. However, there are plenty of other reasons why green roofs are more attractive.

There is a huge amount of literature on green roofs and an extensive literature supports the fact that vegetation and greenery mitigates the urban heat island (UHI) effect (Oke 2006; Gartland 2008; Ashie 2008; Butera 2008; Dahl 2010; Erell *et al.* 2011). It is well established that plants have a cooling effect and may therefore be used strategically in urban and architectural design. There is also evidence that a combination of vegetation and lighter-coloured material surfaces can keep buildings and roofs cooler using the albedo effect. In addition, as Ong explains, 'plants convert carbon dioxide and are nature's primary strategy in carbon sequestration. Studies have found that greenery [can even] increase the value of urban properties . . . based on visual and social value, [so that we can now] put a price on urban greenery' (quoted in Ong from Killicoat *et al.* 2002; see also Chapter 19, this volume). Green roofs and walls are known to provide urban heat mitigation at low cost and, because they cool buildings through their insulation effect, they reduce heat energy transfer into buildings.

However, the success of green roofs and walls always depends on the design detail and how they are built.

In general, roofs with deeper soils are heavier, but are better able to retain moisture and maintain more stable soil temperatures, thereby offering a more stable ecosystem for plant materials and accommodating a broader range of species. Classical roof gardens, planted with intensive greenery such as larger plant species, are heavy in weight with deep soil profiles (over 150 mm/6 inches) and intensive in maintenance. The contemporary green roof is quite different: it costs a fraction of a roof garden, is lightweight with thin soil profiles and requires minimal maintenance. Usually, these types of green roofs are not accessible and are mainly installed for their environmental performance and aesthetic qualities (Dunnett and Kingsbury 2004; Werthmann 2007).

Green roofs have often disappointed when their performance was measured; for example, they displayed a low biodiversity, as many green roofs do not contribute much to the diversity of plant species or animal habitat (Coutts *et al.* 2013). The most crucial parameters for green roofs are therefore: roof-top surface albedo, substrate depth, vegetation species, density of planting, watering regime/soil moisture and whether the roof is generally an extensive or intensive green roof. Tests have shown that regular irrigation improves the performance of green roofs due to increased evapotranspiration, but this uses more water (which can, of course, be resolved by recycling greywater and harvesting stormwater).

Introducing greenery into cities is seen as the most cost-effective strategy for mitigating the urban heat island effect, because greenery helps to cool the environment through the process of evapotranspiration where large amounts of solar radiation can be converted into latent heat. In Singapore and Seoul, for instance, the integration of vegetation and greenery into architecture has successfully been used for more sustainable urban development, cooling buildings, districts and reducing energy loads (see Figures 9.2c and 9.2d). Furthermore, greenery creates a pleasant urban environment and has a positive impact on users' comfort level by reducing the air temperature. During daytime, trees and planting provide shading, while at night-time, the cooling effect of trees and planting comes from the evapotranspiration process. For instance, in the city of Athens, where the mean heat island intensity is found to exceed 8 to 10° Celsius, the cooling load of buildings in the urban area has been found to be double – and the peak electricity load for cooling may even be triple – the usual amount due to the higher ambient temperatures (Santamouris 2001; WMO 2013). Consequently, such higher temperatures caused by the UHI effect have an impact through an increased demand for cooling energy, while this increased demand for air-conditioning will cost users more in order to maintain their comfort levels, and creates more greenhouse gas emmissions.

A new demonstration project: Barangaroo waterfront precinct, Sydney

In Australia in 2014, despite ideal conditions (plenty of sun, wind power and biomass), renewable energy supplies only around 11 per cent of total electricity demand. The government often lacks the will to implement policies that would increase this and the coal lobby is all too powerful. However, this is now starting to change. In Sydney, in the central harbour district of Barangaroo South, a plus-energy precinct is currently being built according to plans drawn up by British architect Richard Rogers and others. This extension of the existing waterfront and central business district is a mixed-use precinct that will generate a large portion of its power itself using solar and co-generation systems (tri-generation technology);

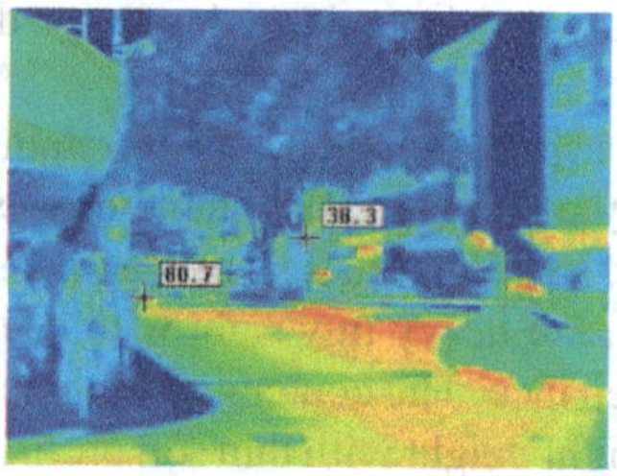

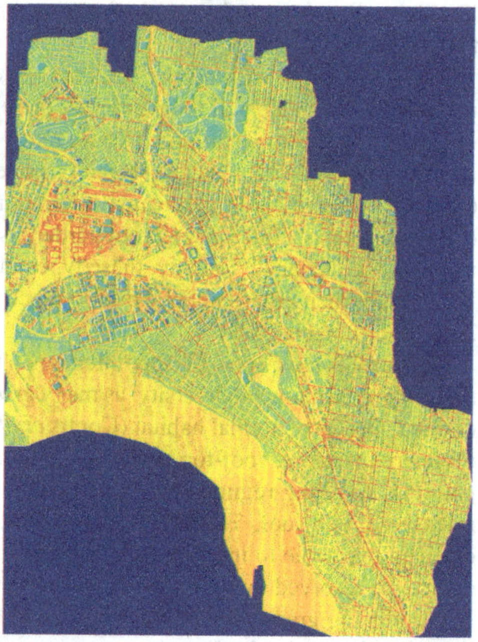

Figures 9.2a and 9.2b: Left: Thermal photography of urban spaces reveals the stored heat (orange- and yellow-coloured), for instance the hot black asphalt of roads, and the cooler (blue-coloured) vegetation. Right: Aerial thermal photography of parts of the City of Melbourne's city centre.

Source: Courtesy City of Melbourne, 2013, supplied by the author.

Figures 9.2c and 9.2d Greenery integrated into façades has been successfully applied in buildings in Singapore and Seoul: façade of an office building in Singapore (left) and the internal green wall in the atrium of Seoul City Hall (right)

Source: Photos by the author.

Figure 9.2e Opportunities for mitigating the UHI effect may be found in the following five areas: integrating vegetation, water bodies, optimized built form, selection of materials and in general urbanization strategies.

Source: Diagram by Lehmann and Hatvani-Kovacs (2012).

it is scheduled to be completed in 2020. Around half of the 22-hectare site is being converted into public space with public parklands and water taxi terminals. This new harbour precinct will have excellent connections to the existing public transport system and will be home to around 1,200 residents and 23,000 office workers and will offer more than 2.5 hectares of new public waterfront space and foreshore promenade (see Figures 9.3a and 9.3b). In the past, Sydney has seen several high-profile urban waterfront redevelopment projects, from Darling Harbour to Circular Quay and Walsh Bay, but Barangaroo is the first one with the ambition to become carbon neutral (a state the developers claim it will achieve after around 36 years of use). The developers expect that 'this dense precinct will create an engaging, active and vibrant waterfront precinct, over half of the new precinct will be assigned for public and recreational space' (Barangaroo Delivery Authority and Lend Lease 2013).

The total approved building footprint of this development is 120,000 sqm, which allows for a total of combined green roofscapes of almost 60,000 sqm. The landscape architects give the well-established reasons for making extensive use of green roofs: longer roof lifespan, greater heat and sound insulation, reduced cooling loads and heating system needs (hence reduced greenhouse gas emissions), and a cutback in stormwater run-off. It is also expected that the green roofs will deliver health benefits, decrease dust in the city and offer visually appealing outdoor roof spaces. In Barangaroo, the lessons learnt from less successful green

Figures 9.3a and 9.3b Visualization of the new Barangaroo South precinct, East Darling Harbour, Sydney, scheduled to open by 2020.

Source: Courtesy of Lend Lease, Sydney, supplied by the author.

roofs have been taken on board to maximize overall benefits in terms of urban heat mitigation and energy transfers into buildings; as the project is currently under construction, we will have to wait to see the outcome.

Urban infill, not suburban sprawl

There is evidence that cities of lower density produce more greenhouse gas emissions per capita, and that higher density allows for a greater variety of services at comparatively low per capita costs (Jenks *et al.* 1996). Urban infill is a useful method to densify existing cities and revitalize city centres.

The urban infill apartment block has many advantages. Apartments share one circulation system, and building blocks between three and ten storeys high improve thermal performance and save on construction materials. The ratio of surface area to volume is much better than one would get in a set of smaller low-rise buildings housing the same volume. In addition, living in an inner-city location close to work and public transport means that the energy use of a city precinct's population can drop by 30 per cent if people move from free-standing suburban houses to inner-city apartments (Lehmann 2010). While urban infill results in higher densities, it can actually increase the walkability of a more compact city and create the possibility of a better quality of life at a lower carbon footprint through more efficient infrastructure and land use.

As our city centres no longer provide affordable living space for families and are less and less places of social inclusion, residents move to suburbs scattered on the outskirts. Suburbs are expanding fast, with developments sprawling out into the landscape, destroying precious agricultural land, making us even more dependent on cars and food imports. Is the alleged energy-efficient rehabilitation of the suburbs a big hoax invented by the insulating materials industry? Would it not be more ecologically sustainable to densify cities intelligently and convert unused flat roofs into gardens and green roofs so that commuters could remain in the city instead of driving out to the suburbs in their fossil fuel-burning cars, a trip on which they burn ten times more energy than they save with their insulating plaster? And all of this just to be in a 'house with a garden', which is often a major self-deception too. After discounting the house and two carports, a shed for bicycles, a lawnmower, a barbecue and a veranda, many properties of 500 sqm offer just 50 sqm of left-over garden space.

In terms of sustainability, detached, energy-efficient buildings outside the city can never be superior to an urban, re-densified precinct. Instead of considering individual plus-energy houses outside their urban context, we must appraise these buildings in relation to the energy needed for their use within their location. Thus, for example, a plus-energy house far outside the city centre is always less efficient because the costs involved in its development to provide infrastructure and the long travel distances for shopping and other daily necessities must be added to the calculation (Hegger *et al.* 2007). Living in a compact inner-city location is always more favourable because it consumes fewer resources, particularly when precincts can generate most of their own power from renewable energy sources.

The economization of construction has been aided and abetted by the de politicization of the construction discourse. There has been little public debate over what kind of spaces we want to meet in, what our cities and houses should offer in the future, and who is currently preventing this from happening. When a debate on architecture does take place, it mostly revolves around spectacular individual projects but is rarely about the city or the urban context. Instead, we should tackle the complex issue of our suburbs and their future. The discussion about cities, suburbs and precincts is entirely insufficient – so it is no surprise that nostalgia and seclusion have become the predominant strategies where housing is concerned.

Conclusion and areas of further research

As urban form locks in long-term consequences, decision making on urban developments requires careful planning and far-reaching strategies. Urban planners, architects, engineers and designers should find ways to preserve biodiversity as cities expand, and subsequently modify natural habitats. Such efforts would most likely focus on preserving as much remnant natural habitat as possible, as opposed to most current land development techniques, which remove most natural vegetation during construction. Ecological connections may be created and reinstated on a smaller scale within urban environments, for example, in the form of urban farming, roof-top gardening and green walls.

Sustainable urbanization will not be possible without zero-energy and even plus-energy precincts. Integrated urban development needs to focus on energy, water, greenery and the urban microclimate, and designers must engage with policy makers in order to drastically reduce our cities' consumption of energy and resources. The effectiveness and benefits of green roofs and green walls have been studied extensively and are well documented. Given its natural cooling effect through the process of evapotranspiration, introducing greenery is one of the most cost-effective strategies for mitigating the urban heat island effect. Green roofs and walls are commonly reported to provide urban heat mitigation at low cost, and to cool buildings through their insulation effect, hence reducing energy transfer into buildings. However, the effectiveness of green roofs and walls depends on the details of their design and how they are built.

A promising opportunity discussed in this chapter is the emergence of low carbon construction systems using massive timber panels for carbon engineering. Since timber is a material with the capacity to store carbon in large quantities over long periods, engineered solid wood panel construction offers the opportunity to turn buildings into 'carbon sinks'. Thus, some of the historically negative environmental impacts of urban development and construction can be turned around (Sathre and Gustavsson 2009).

Despite its many advantages, however, there are still plenty of barriers to CLT's uptake. There is a slow uptake of CLT systems by the construction industry (except in Canada and some parts of Europe; Germany, Austria, Switzerland, Italy and the UK have embraced this new technology) and consumer resistance to denser inner-city living is still strong. It will be crucial to persuade consumers to accept three- to ten-storey infill apartment buildings constructed entirely of timber (Östman and Källsner 2011). Post-occupancy evaluation of new sustainable housing projects made from solid wood panel systems is a promising approach in the analysis of energy consumption and occupant comfort; it will enable researchers to develop and validate the 'ideal' model for infill housing using engineered panel systems based on evidence (e.g. the annual energy use per person). Further research is also needed to analyse whether residents are more likely to accept higher densities when they know it is a comprehensively 'green' building.

The argument for structural innovation, greenhouse gas emission reductions and industry transformation will accelerate the adoption of design and construction techniques for urban infill development that utilizes low carbon construction systems with reduced GHG emissions, global warming potential and construction waste. However, more research is necessary on optimizing such urban systems, including for renewable energies, green roofs and construction systems using timber and off-site manufacturing. The risk averseness of the construction sector and bureaucracy around new technologies frequently puts the brakes on innovation, and little research is conducted in areas in which no or few subsidies beckon, for example, storage technology to store surplus renewable energy more efficiently. As a consequence, technical progress is often hobbled by the ploddingly predictable imagination of civil servants.

Visionary leadership, dynamic urban governance and integrated master planning can guide urban growth in a more environmentally sustainable way. A reform of building codes and standards is necessary, with enabling policies to make resource-efficient buildings and precincts the norm. Clearly, more knowledge is required in many areas to advance planning theory in the development of future precincts and management of contemporary urban change, to support more informed planning policy and the evidence base for new low carbon precincts. Green urbanism principles can assist with these endeavours if they are applied with integrated design principles and a focus on regeneration, reconciling economic, social and environmental objectives.

References

Ashie, Y. 2008, 'Management of urban heat environment', *Urban Environmental Management and Technology*, vol. 1, pp. 215–238.

Australian Bureau of Meteorology (BOM) 2013, *Australian Heatwaves: Hotter, longer, earlier and more often*, BOM, Sydney, NSW.

Barangaroo Delivery Authority and Lend Lease 2013, *Barangaroo boosts floor space in updated master plan*, 6 November, www.thefifthestate.com.au/archives/56640/, viewed 10 November 2013.

Brown, L.R. 2009, *Plan B 4.0: Mobilizing to save civilization*, Norton, New York.

Butera, F. 2008, 'Towards the renewable built environment', in P. Droege (ed.), *Urban Energy Transition: From fossil fuels to renewable power*, Elsevier, Amsterdam, the Netherlands.

Chen, Y. and Wong, N.H. 2006, 'Thermal benefits of city parks', *Energy and Buildings*, vol. 38, pp. 105–120.

Coutts, A., Daly, E., Beringer, J. and Tapper, N. 2013, 'Assessing practical measures to reduce urban heat: green and cool roofs', *Building and Environment*, vol. 70, pp. 266–276.

Dahl, T. 2010, *Climate and Architecture*, Routledge, London.

Deloitte Touche Tohmatsu and World Economic Forum 2011, *The Consumption Dilemma: Leverage points for accelerating sustainable growth*, World Economic Forum, Geneva, www.weforum.org/reports/consumption?dilemmaleverage?points?accelerating?sustainable?growth, viewed 10 January 2012.

Dhakal, S. 2009, 'Urban energy use and carbon emissions from cities in China and policy implications', *Energy Policy*, vol. 37, no. 11, pp. 4208–4219.

Droege, P. (ed.) 2008, *Urban Energy Transition: From fossil fuels to renewable power*, Elsevier, Oxford.

Dunnett, N. and Kingsbury, N. 2004, *Planting Green Roofs and Living Walls*, Timber Press, Portland, OR.

Erell, E., Pearlmutter, D. and Williamson, T. 2011, *Urban Microclimate: Designing the spaces between buildings*, Earthscan, London.

Frangi, A., Fontana, M., Hugi, E. and Joebstl, R. 2009, 'Experimental analysis of cross-laminated timber panels in fire', *Fire Safety Journal*, vol. 44, pp. 1078–1087.

Gartland, L. 2008, *Heat Islands: Understanding and mitigating heat in urban areas*, Washington, DC, Earthscan.

Girardet, H. 1999, *Creating Sustainable Cities*, Totnes, UK, Green Books.

Head, P. 2008, 'Entering the ecological age: the engineer's role', Brunel Lecture 2008, London, www.arup.com/Publications/Entering_the_Ecological_Age.aspx, viewed 10 January 2012.

Hegger, M., Fuchs, M., Stark, T. and Zeumer, M. 2007, *Energie Atlas*, Birkhäuser Verlag/Edition Detail, Basel, Switzwerland.

ICLEI 2007, *Liveable Cities: The benefits of urban environmental planning*, Cities Alliance, UN-EP and ICLEI, Bonn, Germany.

International Energy Agency (IEA) 2013, *World Energy Outlook 2013*, IEA, Paris.

Jenks, M., Burton, E. and Williams, K. 1996, *The Compact City: A sustainable urban form?*, Spon Press, London.

Kaufmann, H. 2011, *Wood Works: Ãkorationale Baukunst – architecture durable*, Springer, Berlin.

Killicoat, P., Puzio, E. and Stringer, R. 2002, 'The economic value of trees in urban areas: estimating the benefits of Adelaide's street trees', in *Proceedings: Treenet Symposium*, Treenet, Adelaide, SA.

Le Corbusier 1935, *La ville radieuse* (*The Radiant City*), L'Architecture d'Aujourd'hui, Paris.

Lehmann, S. 2010, *The Principles of Green Urbanism: Transforming the city for sustainability*, Earthscan, London.

—— 2012, 'Sustainable building design and systems integration: combining energy efficiency with material efficiency', in S. Lehmann and R. Crocker (eds), *Designing for Zero Waste: Consumption, technologies and the built environment*, Routledge, London.

—— 2013a, 'Low-to-no carbon city: lessons from Western urban projects for the rapid transformation of Shanghai', *Habitat International*, vol. 37, pp. 61–69.

—— 2013b, 'Kraftwerk statt Verschwender', in M. Hegger, C. Fafflock, J. Hegger and I. Passig (eds), *Aktiv Haus*, Callwey, Munich.

Lehmann, S. and Crocker, R. (eds) 2012, *Designing for Zero Waste: Consumption, technologies and the built environment*, Routledge, London.

Lehmann, S. and Fitzgerald, G. 2013, 'Wood in the city: sustainable urban infill development using low carbon prefabricated timber construction systems', in R. Crocker and S. Lehmann (eds), *Motivating Change: Sustainable design and behaviour in the built environment*, Routledge, London.

Lehmann, S. and Hamilton, C. 2011, *Sustainable Infill Development Using Low Carbon CLT Prefabrication: Adaptation for the South Australian context*, Zero Waste SA Research Centre for Sustainable Design and Behaviour, University of South Australia, Adelaide, SA.

Merl, A. 2005, *Bau-Ressourcenmanagement in urbanen Räumen, Fallstudie Wien. Nachhaltiger Einsatz von Holz im Rohbau*, Dissertation an der TU Wien [*Building Resources Management in Urban Spaces, Case Study Vienna: Sustainable use of wood in building shells*, doctoral thesis, Vienna Technical University].

Moewes, G. 1995, *Weder Hütten noch Paläste. Architektur und Ökologie in der Arbeitsgesellschaft* [*Neither Huts nor Palaces: Architecture and Ecology in the Working Society*], Birkhäuser, Basel, Switzerland.

Oke, T.R. 1981, 'Canyon geometry and the nocturnal urban heat island: comparison of scale model and field observations', *International Journal of Climatology*, vol. 1, no. 3, pp. 237–254.

—— 2006, 'Towards better scientific communication in urban climate', *Theoretical and Applied Climatology*, vol. 84, no. 1, pp. 179–190, doi: 10.1007/s00704-005-0153-0.

Östman, B. and Källsner, B. 2011, *National Building Regulations in Relation to Multi-Storey Wooden Building in Europe*, SP Trätek and Växjö University, Sweden.

Rifkin, J. 2011, *The Third Industrial Revolution: How lateral power is transforming energy, the economy and the world*, Palgrave Macmillan, New York.

Santamouris, M. 2001, 'The canyon effect', in M. Santamouris (ed.), *Energy and Climate in the Urban Built Environment*, James & James Science Publishers, London.

Sathre, R. and Gustavsson, L. 2009, 'Using wood products to mitigate climate change: external costs and structural change', *Applied Energy*, vol. 86, pp. 251–257.

Scheer, H. 2006, *Energy Autonomy: The economic, social and technological case for renewable energy*, Earthscan, London.

United Nations Economic and Social Commission for Asia and the Pacific (UN ESCAP) 2012, *Low Carbon Green Growth Roadmap for Asia and the Pacific*, UN ESCAP, Bangkok.

Van De Kuilen, J.W.G., Ceccotti, A., Xia, Z.Y. and He, M.J. 2011, 'Very tall wooden buildings with cross-laminated timber', *Procedia Engineering*, vol. 14, pp. 1621–1628.

Werthmann, C. 2007, *Green Roof: A case study*, Princeton Architectural Press, New York.
World Metereological Organisation (WMO) 2013, *The global climate 2001–2010, A Decade of Climate Extremes*, World Metereological Organisation, Geneva, http://library.wmo.int/pmb_ged/wmo_1119_en.pdf, viewed 10 October 2013.

Chapter 10

From green buildings to sustainable urban design

Two case studies

Christoph Ingenhoven, Martin Reuter and Ben Dieckmann

Summary

This chapter presents two recent designs, an office building and a university campus, which could contribute to the creation of low carbon cities and are compatible with the principles of green urbanism.

Introduction

Ingenhoven architects are based in Düsseldorf, Germany and focus on sustainable design (for a discussion of past projects see Bauer *et al.* 2010; Feireiss 2003; Lehmann 2009; Yudelson 2009; Yudelson and Meyer 2013). This chapter presents two projects: an unrealized suburban university campus in Ireland and a completed inner-city high-rise building in Australia. The projects, 'Gateway' at University College in Dublin and 1 Bligh in Sydney, differ greatly in their functionality, urbanity, climate and culture. Nevertheless, they both follow a holistic approach to sustainable design to serve their end-users and improve the context of their location.

In order to monitor and document the sustainability of our buildings, we use the locally available standards. We have achieved the highest possible ratings for the projects we have completed in Japan (Casbee), Singapore (Green Mark), Australia (Green Star) and also under BREEAM and the recently introduced German DGNB system. However, we aspire to go beyond the set requirements. Our aims are: to make the most of the location's natural resources to minimize the use of technology; to create an architecture that is appropriate for the specific climatic conditions; to provide a durable and flexible building that allows a long service life and reversibility; to reduce energy losses and consumption in concept, production and lifetime; to use regrowable, recyclable and recycled material wherever possible; to use natural energy sources; to do good for the city and the people; to reduce transportation by providing mixed-use buildings; and to provide a healthy and joyful environment where the users are happy. These aspirations are consistent with the principles of green urbanism (Lehmann 2010).

University College Dublin: 'Gateway'

The client

The origins of University College Dublin (UCD) date back to the Catholic University of Ireland which was founded in 1854. Today, UCD is Ireland's largest university. The writer

James Joyce is UCD's most famous alumnus and Kevin Roche is a distinguished graduate of its architecture school. The original foundation had several downtown locations. In the 1940s, UCD began to acquire parcels of open land in Dublin's south-eastern suburbs. In the 1960s, the university decided to move from its inner-city location to its present suburban site, which was known as Belfield. This relocation took place over the subsequent 40 years and is now substantially complete.

The project

The role of UCD as Ireland's premier educational establishment is changing in response to a rapidly evolving national and international social and economic framework. Universities are no longer seen simply as places of learning, but rather as focal points for knowledge and industry, providing strategic locations where links between industry and education can develop and thrive. In many ways, campuses have become microcosms of cities, extending the university experience beyond the traditional classroom and opening to the outside community.

Since its 1964 campus master plan, UCD has recognized the need for the campus to grow and develop in response to broader social, economic and physical changes. UCD named this project 'Gateway', as the main construction area is at the entrance of the campus. Gateway was the first project in the planned redevelopment of University College Dublin's Belfield campus, a 132-hectare site located four kilometres south of Dublin city centre. One of the main aims was to harmonize the existing structures and, more importantly, to make a distinct contribution towards enhancing the urban context.

The themes

Approach

Work on a project of this scope may be characterized by two dominant principles: first, concern to reduce the impact of buildings on the environment and people; second, careful attention to urban form and scale allied to concern for landscape, both urban and natural. The best and most beautiful, elegant and sophisticated solutions in nature are created by evolution; they are a temporary best answer to a given problem, which is what we aim to achieve. To this end, we are constantly looking for new, better, lighter, brighter and more efficient solutions.

Inspiring future generations

The main aim of an environmentally friendly, ecologically orientated, flexible, sustainable and aesthetically satisfying university architecture is to inspire future generations to be highly sensitive to and responsible for the life cycles of all living things and our shared global condition.

Open-minded communication

University architecture could be the means to facilitate an interdisciplinary, open-minded form of communication between people, regardless of their nationality, religious affiliation or age. Creating physical and spatial platforms to encourage informal communication and intensive discussions was a basic challenge of this project.

Environmental qualities

There are many environmental qualities involved in building design but, before addressing any details, the main issue is developing a complex that is suitable for both the students and the staff. An unmistakable identity should create a long-lasting site corresponding to the interdependencies of the Belfield campus, based on a sustainable master plan.

Woven into the urban context

Interlinking public pedestrian paths and boulevards throughout the campus with the Belfield agglomeration and creating a futuristic infrastructure network will provide for virtual and actual communication. The new Belfield campus could provide an open communicative centre for the wider community as well as the university's students.

New work spaces

A clear architectural language, an easy and readable building orientation, a congenial environment and optimally functional technical equipment will ensure a high degree of student and employee satisfaction and identification with the buildings and the institution. A new kind of communication would enable people to work in a more interdisciplinary yet specific way, to be motivated and to cultivate direct and informal contact in order to avoid misunderstandings. Building complexes like the UCD Gateway should offer a wide range of flexible spaces for a variety of different uses. This would enable students, lecturers and professors to work more cooperatively while maintaining a specific focus. The buildings of Gateway I will offer a wide range of flexible spaces for a variety of different uses, such as seminar rooms, workshops, offices, laboratories and workshops, as well as meeting areas.

Urban life

The UCD Gateway with its new inner campus, marked by the surrounding beltway, will provide a good quality of urban life by improving interaction. Integrating university-related facilities such as residential accommodation and hotels, refectories or restaurants, shopping and retail uses, medical centres and health spas, as well as theatres, galleries and museums is absolutely vital to the success of the new precinct.

The design

The new Gateway is structured as a system of rings surrounding the existing main educational facilities at the centre of the campus (see Figure 10.1). The inner ring or 'belt walk' is reserved for pedestrian and cycling access, the outer ring for vehicles, thereby keeping the inner zone free of motorized traffic. With its landscaped areas and newly planted trees, this will effectively turn the inner zone into a green lung that will significantly enhance the microclimate of the whole campus and its buildings, and create a high-quality, multifunctional campus.

Figure 10.1 An aerial view of the redesigned UCD campus.

Source: Rendering by Peter Wels, supplied by the authors.

The campus is entered from a national road that passes alongside the campus and separates it from the immediate vicinity. Access to the ring system is via a newly designed central bridge over the national road, to enhance the university's connection with the city. Buildings for both academic and non-academic use are lined in three rows along – but within the boundaries of – the outer ring, forming an entrance to the Gateway as well as a protective wall against the outer ring road. All traffic functions (such as bus stops and taxi ranks) are catered for here in order to keep the inner campus zone free of motorized vehicles. This way, the actual campus can be kept free of vehicles, noise and carbon emissions, which in turn will encourage students, staff and visitors to stay and enjoy its facilities. This turns the campus as a whole into a new social hub welcoming everyone into its heart. Everyone will be able to use and enjoy a safe, easily accessible and healthy campus.

The stepped rows of buildings are covered by a curved roof – the canopy – that groups the individual blocks into a single expressive structure. The multi-use building comprises several occupancies: UCD facilities, UCD exhibition, student residences, a medical centre, an arthouse cinema, the Global Irish Institute, a conference hotel, an apartment hotel, a crèche, and retail, catering and office areas. This canopy's wood construction is the concept's main design element, which is themed across all buildings and incorporates integrated photovoltaic, and solar energy equipment, wind turbines and extensive planting.

The atriums and garden streets between these rows provide pedestrian access to the individual buildings as well as weather-protected recreational space for students, university staff and visitors. A mix of uses in the buildings encourages a bustling urban life culminating at the main Gateway plaza, the main entrance point for pedestrians and cyclists to the inner campus site.

The ecological concept

Mindful of future generations, the project's energy concept is ecologically balanced. One of its main aims is to preserve resources by reducing energy consumption. In addition, all the energy needed for heating and cooling, ventilation and lighting is to be generated on site, making UCD entirely self-sustaining and free of carbon emissions. This will be achieved by using advanced technologies that meet the needs of the campus' users and at the same time address the challenges of the future. All this is done in the light of the local climate's very moderate temperature, but also taking into account the rainy, misty and windy conditions on the topographically exposed site on the hills close to Dublin Bay.

The Gateway's ecological concept is rooted in its ambition to be self-sustaining by generating as much energy as is needed on site, by reducing energy consumption and by preserving natural resources. Reducing the consumption of the environment is one aim; the other is to significantly improve the comfort and well-being of the campus' users. This is achieved by utilizing renewable energy sources, such as solar and wind power as well as geothermal technologies. Intelligent façades provide efficient insulation and allow for natural ventilation and maximum daylight penetration, thereby reducing heating and cooling requirements. Rainwater is collected and reintroduced into the buildings' greywater cycle in order to preserve drinking water.

Several elements were chosen and incorporated for the design's ecological sustainability. They include: optimal thermal insulation for summer and winter; optimal use of daylight while reducing solar heating and providing effective glare protection; optimal thermal comfort via low temperature heating and high temperature cooling systems; low-energy electricity supplies; zero carbon energy generation for heating, cooling, ventilation and lighting; heating and cooling generated through renewable techniques like groundwater used in combination with heat pump and direct cooling; electrical power generated by photovoltaic panels and wind turbines; a centralized building control system; a minimal use of stand-by mode equipment; optimal water-saving and rainwater-using technologies; and 'healthy' (i.e. ecologically and biologically harmless) materials.

A key aspect is the façade and roof design in order to obtain: optimum use of daylight; as much natural ventilation as possible; use of solid building materials (no suspended ceilings, mass); night-time cooling in summer; and the use of buffer zones. An advanced stormwater management system utilizes the natural precipitation cycle of green roof constructions and periodically wet landscaped zones, filtrating particulate matter for unpolluted groundwater resources.

The biggest effort to reach a zero carbon state comes from the emission-free generation of electricity through renewable energy sources. Wind turbines on site or nearby could nearly meet the electricity power requirements. Additional demand would be covered by photovoltaics: 10,000 m^2 are to be installed in the canopy. All of these elements are integrated into the canopy, which also serves as protection against heat, noise and the weather. The canopy also collects rainwater and admits high levels of daylight into the atria and plaza and adjacent spaces. Additional functions installed in the canopy are solar heating, extensive greened areas and wind turbines.

The UCD Gateway is planned to balance economic concerns with ecological considerations in a responsible and future-oriented manner. For all elements of the buildings' load-bearing structure, the design has been specifically developed to be simple and economical in order to facilitate changes at a later stage. The majority of materials will be from renewable sources; for example, wood is used in all façades, to ensure a high-quality build and a long life cycle with minimal maintenance requirements.

The new UCD Gateway was designed to make an important contribution not only to the future development of the Belfield campus but also to the way the university strives to define itself in the twenty-first century. The design harmonizes with the existing structures but, more importantly, it gives the campus an appropriate entrance setting that succeeds in reconnecting the university with its urban context. While the curved roof creates an unmistakable elegance and identity, the university will ultimately be defined by the bold ecological approach of its campus' design, which goes far beyond a mere aesthetic vision. Optimization, minimization, efficiency and finesse, in both technical and ecological terms, were due to create a symbol of excellence. Due to the financial crisis, which had a substantial impact on the Irish economy, construction and the real estate market, the project was put on hold during tendering for investors in 2008.

1 Bligh Street, Sydney

The design for 1 Bligh Street, Sydney is the result of an international competition held in 2006. The design was developed collaboratively by Ingenhoven architects and Architectus, a Sydney-based firm. 1 Bligh is a 139-metre-high office tower located in the middle of Sydney's business district and has gained a six-star Green Star rating.

The view

Located at the centre of the 'valley' created by the Macquarie Street ridge and the tall buildings located in and around George Street, the site is highly visible in its context when viewed from the bridge and harbour. Likewise, these views are a key asset on which the form of the building has drawn. Views along Bligh and O'Connell Streets to the city and an important view over Bent Street to the harbour have also been a direct influence on the position and orientation of the elliptical form. The site is easily viewed from all angles, and as such is deserving of a 'building in the round' approach. Not only does this form maximize exposure to views, it also ensures that, from any angle, the building has a face which interacts positively with the spaces around it.

While the ellipse opens the view to the harbour for occupants, it also allows the placement of a strong corner emphasis at the one point where all the major views converge. This is combined with the main entrance location to provide a strong address for the building in a way that directly responds to its urban context. By shifting the axis of the building's form to align with the grid to the north, views towards the bridge and the harbour are opened up, and the square of Farrer Place is more clearly defined. At the same time, the elliptical shape is able to maximize the area of façade with a direct orientation to the view, thereby improving the value of office space and improving the quality of the working environment. The building's single unified form naturally addresses the harbour but is also more compact than a rectangular building of the same volume due to its 12 per cent shorter contour (see Figure 10.2).

Despite its simple and calm outlook, the design had basically reinterpreted most of the planning guidelines, including setbacks and podium lines. Given the obvious positive results for the streetscape, Sydney's planners accepted the design approach – appreciating not only the fact that the ground plane was designed as being nearly 100 per cent public but also the

Figure 10.2 Street view of 1 Bligh, Sydney.

Source: Photo by Dexus Property Group, supplied by the authors.

Figure 10.3 Atrium of 1 Bligh, Sydney.

Source: Photo by Dexus Property Group, supplied by the authors.

positive contribution made to Farrer Place by providing a responsive counterpart. The remarkable slope of the site along Bent Street is celebrated with a sweeping curve around the O'Connell Street corner. Broad, curving steps provide an ideal place to sit in the sun in winter or in the shade in summer and appreciate the sandstone façades of the Education and Lands Department buildings. The stairs rise to the winter garden platform at the Bligh Street level, which is sheltered by a louvred glass curtain. From here, the full expanse of Farrer Place can be appreciated.

The lower ground floor and ground floor are nearly fully public, the only exemption being the childcare facility and the secured lift lobbies, which are minimized within the building footprint so that people can still pass through the ground floor from Bligh Street to Farrer Place or vice versa (see Figure 10.3). A café on Bligh Street and a café/restaurant at the top of the curved stairs have become well-known and frequently used places for people to meet.

Green building design

As a holistic concept, sustainability is more than a design with energy-efficient plans and plants. The considerations include social, urban planning and energy-related aspects. The first decisions are made when the building footprint is located within the city grid; orientation, compactness and access to public transport are only a few of the important factors. Apparently not all of those can be influenced by every design, but we believe that 1 Bligh is an excellent example of a synergetic balance of building form in both urban planning and energy-related aspects. The building's elliptical shape with its elevated tower volume not

Figure 10.4 Interior of 1 Bligh, Sydney.

Source: Photo by E G Esch Hennef, supplied by the authors.

only created a new landmark in the CBD; the site also allowed for a design that could take advantage of the orientation by locating the cores and the cold air pond (atrium) in the south of the building, and therefore keeping the north elevation free for a maximum use of the views.

High energy efficiency and social quality do not necessary contradict each other. It was important to us that the work environment in 1 Bligh was designed around the principles of flexibility, efficiency, communication, transparency and a high level of indoor environment quality (IEQ). A naturally ventilated glass atrium running the full height of the building enhances connectivity and access to natural light, and features break-out spaces with access to fresh air. The building footprint means that a majority of the occupants are in close proximity to external views and views into the atrium with high levels of natural daylight, comfort and fresh air.

With 42,000 m^2 across 27 typical office floors, averaging 1,600 m^2, the building achieves 92 per cent NLA to FSA (nett lettable area to floor space area). The typical floor plate is designed as an arc, maximizing usable space, bringing the cores closer than in a rectangular building and minimizing distance to amenities. Forty per cent of office space is within five metres of the façade with 1,000 m^2 of contiguous column-free space creating a high level of office layout flexibility and cellular potential. The floor plate provides for a range of tenant types and fit-out densities from 1:10 to 1:20 in any mix of open-plan and office environments.

Vertical communication between different levels (an unusual feature in a high-rise) is further encouraged by the glazed lift cars that ride in the atrium (see Figure 10.4). The transparent atrium with the balcony spaces and lift cars are elements that create a sense of community in the building, simply by allowing the users to see into each other's space.

Environmentally sustainable design

The building incorporates environmentally sustainable design (ESD) principles in design, construction and operation, with a focus on indoor environmental quality and energy efficiency. 1 Bligh has achieved six stars (the highest available rating) from Green Star, Australia's most widely recognized environmental rating tool. It is also targeting a five-star energy efficiency rating from the National Australian Built Environment Rating Scheme (NABERS) for energy in the base building. While the competition brief had originally asked for a five-star rating, it became obvious during the design development that a six-star rating would be possible with minor additional investment. Thus the client asked the team to work towards the higher rating and to prepare tender documents accordingly. To achieve this, the building design has been optimized with regard to energy consumption and sustainability in passive and active terms.

Double-skin façade

Extensive development work was invested in the naturally ventilated double-skin glass façade, which has an automated cavity blind programmed to block all direct sun and minimize daylight glare, while affording excellent daylight levels in the retracted position. The tower operation was optimized by rigorous computer simulation to balance daylight, glare, thermal comfort, views and air-conditioning energy (see Figure 10.5).

In the competition design, we had proposed a double-skin façade for the entire office space and a single-layer louvred glass façade for south of the tower. Here, 'double skin' describes

Figure 10.5 | Bligh, Sydney, section.

Source: Supplied by the authors.

a façade system with a single-layer glass plane on the outside of the building which protects the blinds that sit in front of a double-glazed, thermally active inner glass plane. Although well known in Europe and on low-rise projects around the world, a double-skin façade had not been tried before on a commercial development of this size in Australia. Hence the advantages with regard to energy savings and glare control had been proven during the design phase.

The main reason for the introduction of the cavity façade system was to have the sun screen outside the thermal hull of the building instead of having a solid blind inside the office space. The concept has the great advantage that the solar heat gain in the internal environment is drastically reduced. The overall façade system achieves a shading coefficient (SC) of 0.15. The inner DGU's SC without the shades is 0.32, although the 60 per cent visual light transition is a value that exceeds any used in Sydney to date.

However, the retractable and tilt-able blinds are not the only devices that reduce solar heat gain and glare; the 700 mm construction depth and the cantilevering aluminium fins considerably reduce the solar impact, decreasing the need to lower the blinds. The blind system and the façade have been designed to maintain the views to the maximum extent possible so that, for 90 per cent of the workday, the blinds only need to be tilted to a 20-degree angle, allowing an uninterrupted harbour view for the majority of the time.

In addition to sun and glare protection aspects, the façade has been designed to allow for natural ventilation of the perimeter zones. Sydney's climate (mean temperature and humidity levels) allows for comfortable ventilation without air-conditioning so the architects and engineers had ensured that the detailed layout of the façade (e.g. the shape of the fins within the cavity openings) would avoid overheating and also re-contaminating the air by used air from the floor below getting back into the building on the next level. This had been modelled and tested with computer-aided fluid dynamic simulations. As a result, the positioning of the sun screen inside the cavity was shifted closer to the outer layer of the façade, since the overheating between the inner glass surfaces and blind was identified as the critical factor. The multi-finned airfoil-shaped louvres were also mocked up for wind tunnel and performance tests to ensure that wind would not impact on the blinds and that they would effectively work under all weather conditions. Eventually the operable windows were not implemented, but designed as an option that tenants could purchase as an add-on.

Ventilation and comfort

A full-height, naturally ventilated atrium opens out into public space. Naturally ventilated balconies are located at either side of the atrium. These areas are used as an extension of the office work space or as break-out spaces. During summer and the transition periods, the climate and temperature in the atrium can be kept at comfortable levels by the regulated opening and closing of the glass louvres at the south façade of the atrium. In addition, the conditions can be tempered by relief air from the office levels, and heated in-slab using waste heat from the tri-generation plant.

Waste heat is also used for space heating, and passed through an absorption chiller for cooling, while gas-fired electricity provides for base building demand. The tri-generation operation is supplemented by 500 m^2 of solar thermal evacuated tubes for 'solar cooling', the first system of its kind in a commercial office tower in Australia. A hybrid chilled beam/low-temperature VAV air-conditioning system provides 150 per cent more fresh air, with CO_2 sensors and high air change effectiveness. All finishes were selected to minimize indoor contamination, and copy rooms separately exhausted.

Water efficiency

A sewer mining system treats up to 100 kL/day to a high quality for use in the toilets, urinals and cooling tower, while rainwater is harvested for landscape irrigation. Thus the use of potable water could be reduced drastically. The use of a blackwater treatment plant instead of a greywater system (where used potable water is used for toilet flushing) was actually onc of the key elements that moved the building from five to six stars. In addition, water-efficient appliances are used throughout the building, which is fitted with four-star showerheads, six-star hand-basin taps and four-star toilets.

Recycling material

Other initiatives focused on materials: over 94 per cent of demolition waste was recycled, while the proportion of recycled concrete, steel and PVC minimization exceeded all targets. Although the specification had asked for an extensive use of recycled material, the contractor proposed an even higher use of recycled material, based on their specific concrete mixture.

Figure 10.6 A green high-rise commercial building at 1 Bligh Street, Sydney, in its urban context.

FSC-certified timber was used for the building's finishes. We had also looked into the use of recycled timber for the terrace decks, but the amount of timber available at that point in time was not sufficient; however, recycled timber of lower quality had been used for temporary construction purposes.

The account describes the key elements of green building design; details like bicycle parking spaces, shower facilities and electric car fuelling have been integrated into the building design as well. Other items like thermally insulated façade profiles would have helped to gain another green star point under the innovation category, but that investment would not have been feasible in terms of energy saving; therefore this measure was not undertaken, which leaves room for future developments to incorporate this innovation.

Reflection: beyond green

We see 1 Bligh Street, Sydney as an example of a highly integrated approach to green building that addresses social, urban, energy-related and economic aspects in a synergetic and unforced manner. It is a building that offers the highest occupant qualities, adds value to the city by enriching the public domain, meets the client's goals and does all of this with a significantly reduced carbon footprint.

References

Bauer, M., Mösle, P. and Schwarz, M. 2010, *Green Building: Guidebook for sustainable architecture*, Springer-Verlag, Berlin.

Feireiss, K. (ed.) 2003, *Ingenhoven Overdiek und Partner: Energies*, Birkhäuser, Basel.
Lehmann, S. 2009, 'Beauty in necessity: a conversation with Christoph Ingenhoven', *Architectural Review* [Australia], 19 August, pp. 48–52.
—— 2010, *The Principles of Green Urbanism: Transforming the city for sustainability*, Earthscan, London.
Yudelson, J. 2009, *Green Building Trends: Europe*, Island Press, Washington, DC.
Yudelson, J. and Meyer, U. 2013, *The World's Greenest Buildings*, Routledge, London.

Chapter 11

Urban reclamation and regeneration in Seoul, Republic of Korea

Robert Cervero

Summary

The forces towards rapid-pace suburbanization are universal but have been particularly pronounced in East Asia's mega-cities, including one of the world's largest urban agglomerations, Seoul, Republic of Korea. Seoul's policy-makers created a protective greenbelt around the city and built large-scale, master-planned new towns on virgin land. While these new towns achieved their intended objectives of accommodating new households and moderating housing prices, with time ever-worsening traffic conditions and environmental degradation appeared. Since the turn of the twenty-first century, Seoul has embarked on a bold new experiment in urban transformation, reclaiming space given to the automobile during the 1960s to the 1990s. Seoul has sought a balance between transport infrastructure as mobility provider and public space as urban amenity. Experiences in Seoul suggest that the provision of high-quality public transport services and nicely designed, pedestrian-friendly greenways can diminish the necessity for high-capacity elevated freeway structures. Planners hope that more balanced infill development will make Seoul an attractive city on the global stage that appeals to international finance and businesses as well as tourists, professionals and foreign investors. In Seoul (and elsewhere), freeway deconstruction and land reclamation is tied to the reordering of urban priorities that give preference to planning for people and neighbourhoods, not car-based mobility.

Since the turn of the twenty-first century, Seoul, the capital city of the Republic of Korea, has embarked on a bold new experiment in urban regeneration. This has principally involved reclaiming urban space given to the automobile in the post-Korean war era. Through the leadership of Myung-Bak Lee, former mayor of Seoul and now president of the Republic of Korea, the city has sought to strike a balance between transport infrastructure as a mobility provider and public space as an urban amenity. In large part, this has been motivated by a desire to be globally competitive with the likes of Hong Kong, Shanghai and other cities in East Asia's rapidly growing 'tiger' economies by emphasizing quality of life every bit as much as mobility and large-scale infrastructure development.

Like many modern metropolises in East Asia, over the past several decades Seoul has followed a pattern of American-style sprawl, fuelled by steady economic growth and a concomitant meteoric rise in private automobile ownership. However, population densities in the city of Seoul (currently 10.4 million inhabitants spread over a land area of 605 km^2) have historically been and remain quite high by global standards. The city of Seoul itself, along with the port city of Inchon to the west and surrounding Kyunggi province, formally constitute the Seoul Metropolitan Area (also called the Seoul 'National Capital Area'), with some 23 million inhabitants – the world's second largest urban agglomeration. In 2006, Seoul and Inchon combined had the sixth highest population density worldwide: 16,700 people per km^2. Because of the city-region's historically high densities, the Seoul metropolitan government has over the years aggressively sought to decentralize growth, mainly by building master-planned new towns sited on the region's periphery.

New-town development

In all, 26 new towns have been built in the greater Seoul Metropolitan Area over the past three decades. The five most prominent towns – Bundang, Ilsan, Jungdong, Pyungchon and Sanbon – lie within a ring of 20 to 28 km from Seoul City Hall (the heart of downtown Seoul). Most new towns were built in a modernist Le Corbusier style, as mid- to high-rise 'towers in the park'. A serious housing shortage and rising housing rents (fuelled mostly by land speculation) during the 1980s prompted the Korean government to build new towns quickly and as a mass-produced commodity. In 1987, then newly elected Korean president Tae Woo Roh promised the construction of two million new housing units in the Seoul Metropolitan Area during his tenure. The shortage of central-city land combined with the presence of a protective greenbelt that collars the city of Seoul resulted in most new towns being built on the region's far-flung fringes (Jun and Hur 2001).

Table 11.1 shows that the region's five largest new towns, built between 1989 and 1995, had population densities in the range of 18,600 to 41,000 per km^2, between five and ten times denser than post-war new towns built in England. In all, two million people moved out to these five new towns between 1992 and 1999 at an average of 250,000 people per year (Lee 2003). As shown in Figure 11.2, the impact on the distribution of residents within the region was decisive; while the city of Seoul's population fell slightly during the 1992 to 2007 period, the population in surrounding Kyunggi province jumped by more than 60 per cent.

New-town development achieved its goal of stabilizing housing prices, but at a cost – notably dramatic increases in commuting traffic and the problems associated with auto-oriented development: extreme congestion on radial links to the urban centre, rising tailpipe emissions and fuel consumption, and increased demand for expensive highway infrastructure. Between 1990 and 1996, the average commuting distance of new-town residents increased in the range of 12 per cent to 70 per cent (Jun 2000). Longer trips combined with rising car ownership have inevitably translated into steadily worsening traffic congestion; average speeds during evening hours in the Seoul Metropolitan Area fell from 24 kph in 1998 to 17 kph in 2003. In addition, daily commuting expenses have risen sharply; one study pegged the

Table 11.1 Five major new towns in Seoul Metropolitan Area

Name	*Distance to CBD (Direction)*	*Construction period*	*Project area (sq. km)*	*Population density in 1996 (per sq. km)*
Bundang	25 km (SE)	1989–1995	20	18,621
Ilsan	28 km (NW)	1990–1995	16	22,267
Jungdong	25 km (W)	1994–1995	5	40,599
Pyungchon	20 km (SW)	1989–1995	5	31,911
Sanbon	25 km (SW)	1989–1995	4	41,067

Source: Jun and Hur (2001); population densities estimated by the author.

total out-of-pocket cost (ignoring the value of time and externalities) incurred by new-town residents living outside of Seoul's greenbelt at US$12 per day (Jun and Bae 2000).

'New towns/in town': urban reclamation

By the late 1990s, rumblings were starting to be heard within public policy circles that the region's new towns were a failed experiment, exacerbating traffic congestion and environmental quality. Some feared that such factors, along with the productivity losses from inordinately long commutes, would ultimately be a drag on economic growth, prompting disaffected companies and their workers to relocate elsewhere in the country and possibly even abroad. The idea of re-urbanizing Seoul's central areas through 'new towns/in town' began to surface.

The person who led the charge of reinvesting in the central city and regenerating Seoul city was Myung-Bak Lee. In 2001, Lee ran for mayor of Seoul, largely on a platform of reinvigorating Seoul's central city as means of creating a more sustainable yet productive city. Lee campaigned on the premise that Seoul could achieve a better balance between function and the environment by reordering public priorities so as to emphasize quality of place. Prior to becoming major, Lee was the founding leader of the Hyundai Group (the Republic of Korea's largest builder of public works and infrastructure projects) for three decades. During this time, he earned the nickname 'Bulldozer' Lee, partly due to the company's legacy of constructing massive roadways throughout the country but also because he reputedly once took apart a bulldozer to study its mechanism and figure out why it kept breaking down.

Lee won a decisive victory for the mayoral seat, and upon assuming office in early 2002, moved quickly on his campaign promises. His vision for Seoul's future urban transportation called for not only expanding public transit services but also reducing the ecological footprint of the private car by reclaiming urban space consumed by roads and highways – notably, space used to funnel new-town inhabitants in and out of the central city. Why scar the interior of the city, he reasoned, to funnel suburbanites to well-paid white-collar office jobs in the core? A notable culprit was the network of elevated freeways that converged on central Seoul – facilities that severed long-standing neighbourhoods, formed barriers and visual blight, cast shadows, and sprayed noise, fumes and vibrations on surrounding areas. While freeways provided important mobility benefits, Lee recognized that these have to be weighed against their nuisance effects, particularly in today's amenity-conscious workplace.

The removal of a six-kilometre-long elevated freeway, Cheong Gye Cheon (CGC), in the heart of Seoul and the restoration of an urban stream and pedestrian-friendly greenway was a natural choice to launch Mayor Lee's vision of a more sustainable urban landscape for the city. Change was swift. By February 2003, a plan for the freeway removal was completed and five months later the freeway was completely dismantled. Some two years later, in September 2005, the restored CGC stream and linear greenway was opened to the public following a major public celebration and ribbon cutting by Mayor Lee. The entire cost of the freeway demolition and stream restoration was US$313 million (this will be discussed in detail below).

Although not as large and expensive, equally important in symbolic terms was the mayor's decision to convert a massive 1.3 hectare surface-street intersection to an oval-shaped grass park in front of Seoul's City Hall, the nerve centre of the city. The huge swathe of land devoted to car manoeuvres in front of City Hall, an architectural icon and one of the busiest locations in the city, was an extremely pedestrian-hostile environment. Many residents en route to City Hall had to take a circuitous route in a car-dominated milieu. Today, the former traffic circle is a popular leisure spot for Seoul residents, directly connected to City Hall and frequently used for public celebrations, cultural performances and even student demonstrations. During the winter months, a large outdoor ice rink occupies the oval, drawing thousands of Korean families and teens. Once populated by rows of cars, the entrance to Seoul's cherished City Hall is now populated by people.

Mayor Lee made his policy intentions clear when publicly stating that the transformation of space for cars to space for people represented 'a new way to manage Seoul in the twenty-first century'. His views were partly shaped by what was happening in several Latin American cities at the time. Echoing the sentiments of urban visionaries like Jaime Lerner of Curitiba, Brazil and Enrique Penalosa of Bogotá, Colombia, both of whom staked their political careers on curbing the presence of cars in their own central cities, Mayor Lee defended the roadway removal projects on the grounds that 'we want to make a city where people come first, not cars'. The diminution of roadway capacity represents, in many ways, a recasting of public priorities. In Seoul's case, it marked a shift from building infrastructure that enhances automobility to infrastructure that enhances public amenities and quality of urban living. A longer-term objective was to encourage more and more households to settle in the central city and in redevelopment districts, thus reversing the centrifugal flow of residents to Seoul's outskirts and beyond – in effect, creating a new form of urban development, what Harvey Perloff, former dean of the urban planning programme at UCLA, coined 'new towns/in town' (Perloff 1975).

Coping with reduced road capacity

Of course, the withdrawal of road capacity in an increasingly automobile-dependent society, in and of itself, is unlikely to enhance urban living. Public opinion polls tell us that being stuck in traffic is among the most disliked aspects of urban living, and traffic congestion tops the list of factors associated with a declining quality of life (Downs 2004). To Mayor Lee's credit, he understood that public transportation had to be substantially expanded and upgraded to absorb the traffic (169,000 daily cars in the case of the CGC freeway) displaced by large-scale reductions in roadway capacity. This was partly done through the extension of Subway Line 7 (28 km) and the opening of Line 6 (35 km). At least as important, however, was the 2004 opening of seven new bus-exclusive median-lanes (stretching 84 km, later expanded to 162 km) and 294 km of dedicated curbside bus lanes on Seoul's roads. Known as Bus Rapid Transit, or BRT, this allows buses to have right-of-way over cars; thus they function like and enjoy the speed advantages of trains, at a fraction of a railway's construction costs. In all, 74 lineal km of road lanes were expropriated to accommodate Seoul's BRT network. In addition, many regular bus routes were reconfigured to better feed into the city's extensive subway system, and an integrated fare and transfer system between buses and the metrorail network was introduced (Pucher *et al.* 2005).

Seoul's BRT investments have paid off handsomely. Along BRT corridors, bus operating speeds have increased from an average of 11 to more than 21 kph, and have even increased on some passenger car lanes. BRT buses, moreover, carry more than six times as many passengers per hour as buses operating on regular, mixed-traffic lanes. Buses operating in dedicated lanes have become more reliable because they are less subject to the vagaries of ambient traffic flows; the travel-time variation of Seoul's BRT buses is, on average, five times less than that of buses operating on non-exclusive lanes (Seoul Development Institute 2005). In addition, protective lanes have reduced accidents – down 27 per cent one year after BRT services were introduced. Because of these service enhancements and safety improvements, ridership on BRT buses has increased 60 per cent faster than on non-BRT buses over the past three years.

Transit service expansion was not the only relief valve that Seoul's transportation planners turned to. A real-time traffic information system with message boards and in-vehicle navigation aids was introduced to guide traffic flows and alert motorists to downstream hot spots. Several one-way arterial couplets were also created and curbside parking was substantially curtailed to help expedite traffic flows. More draconian was the introduction of a licence plate scheme that, based on the last number on their plates, required motorists to leave their cars at home once every ten days.

Cheng Gye Cheon freeway-to-greenway conversion

The centre-piece of Mayor Lee's urban reclamation policy has been the Cheng Gye Cheon (CGC) freeway-to-greenway conversion. Cheong Gye Cheon, which means 'clear valley stream', has long been a source of fresh water and the 'heart and soul' of urban life in Seoul, going back to the fourteenth century. During the Chosun dynasty (1392–1910), city dwellers did their laundry in the stream and frequently socialized on its banks. Following the Korean War (1950–1953), the stream's character quickly changed when temporary refugee housing was built along its banks. Untreated waste was dumped directly into the waterway, turning it into a veritable cesspool and eventually prompting city officials to cover the stream with an elevated freeway.

The Cheong Gye Cheon freeway, 50 to 80 metres in width and six kilometres in length, opened in 1971 in the heart of central Seoul. Below the road were the running stream and a sewer trunk-line. The CGC freeway quickly became an important conduit for movement to and within central-city Seoul, gaining all the more importance as new towns began to populate the region's periphery in the 1980s and 1990s. However, time quickly took its toll on the facility. A 1992 study by the Korean Society of Civil Engineers found that more than 20 per cent of the freeway's steel beams were seriously corroded and in need of urgent repair (CGC Restoration Project Headquarters 2004). The Seoul Metropolitan Government immediately began repairing the road's understructure; however, due to concerns about the road's long-term safety and stability, this was seen as a stopgap to either totally reconstructing the freeway or tearing it down altogether.

At about the same time the CGC freeway was being repaired and facing increased scrutiny, calls for urban regeneration were starting to be heard, reflecting growing dissatisfaction with the suburban new-town model of urban planning and growth management. In the midst of the debate about the CGC freeway's future, its outright removal was a perfect platform for incoming Mayor Lee to advance his vision of a greener, more liveable and pedestrian-friendly city.

The rapidity of the freeway-to-greenway conversion was impressive by any standard. Mayor Lee took office in mid-2002 and by February 2003 a master plan for the project was completed. Less than six months later, the entire freeway had been completely removed and in September 2005, 26 months after the freeway had been razed, the linear park and restored stream was open to the public. By comparison, the time between initial planning and projection completion for the US$15 billion 'Big Dig' freeway burial mega-project in Boston was 25 years. A combination of environmental reviews, legal challenges, cost overruns and delays in securing land entitlements delayed the conversion of San Francisco's Embarcadero Freeway to a tree-lined surface boulevard by more than a decade.

The restoration of the CGC stream has proven to be far more than the greening of central-city Seoul and a political commitment to urban regeneration. For many local residents, it has marked a rediscovery of the city's past. A number of long-forgotten treasures were hidden by the freeway, including 22 historical footbridges as well as myriad stone carvings and relics. Today, the CGC stream and greenway is Seoul's second most popular tourist attraction. On weekends and summer evenings, thousands of residents and tourists may be seen strolling along the flowing stream's banks, enjoying a small slice of tranquility in an otherwise dense, bustling city.

The CGC project has not been without controversy. Besides concerns over possible increases in traffic congestion, many small shopkeepers and merchants were against the project out of fear of losing business. Alongside the former elevated freeway was an assembly of small-scale shops and markets selling shoes, apparel, tools, electronic goods and appliances. In 2000, more than 200,000 merchants and 60,000 shops were within two kilometres of the freeway. To some, the freeway-to-greenway conversion threatened to alter the existing trade shed and disrupt the flow of customers and logistics. Informal vendors, moreover, would lose their spots under the freeway, in the past an unwanted place where they, and they alone, could ply their trade rent-free (Hwang *et al.* 2005). Following intensive negotiations, the Seoul Metropolitan Government was able to head off opposition by financially compensating merchants and relocating a number of shops to a newly constructed market centre south of the Han River that was easily accessible by highways and public transport.

Impacts of the CGC greenway

In general, past research shows that freeways increase land values for nearby commercial properties due to the accessibility benefits conferred; however, residences within an impact zone – often 0.5 to 1 km away from the facility – depreciate in value due to the noise, fumes and other nuisances (Boarnet 1997). What about urban amenities, like a greenway? Empirical evidence on their land-price impacts is slimmer. Open space can increase land prices not only by its intrinsic qualities (e.g. greenery, spaciousness) but also by reducing the amount of developable land available. However, the noise and foot traffic generated by nearby popular parks and open areas may be viewed as a nuisance by some residential property owners (Frech and Lafferty 1984). A study in the USA showed that average land prices were no different in well-landscaped districts than in less appealing ones with similar levels of traffic congestion (Polinsky and Shavell 1976). The benefits of open space are most capitalized in residential property values in areas that are denser, have higher household incomes and are closer to central business areas (Anderson and West 2006; Dehring and Dunse 2006). The impact of open space and public amenities on non-residential properties is less clear. Urban economists such as Glaeser and colleagues (Glaeser *et al.* 2001) contend that urban parks, open space and waterfront improvements can help cities attract skilled workers and knowledge-based industries, in addition to stabilizing declining neighborhoods. One study in greater Los Angeles found that public amenities like parks did influence the location patterns of firms (Sivitanidou 1995).

Even less is known about the land-price impacts of converting from one form of public amenity to another, such as replacing an elevated freeway with a linear parkway. A study of

Boston's notorious 'Big Dig' project (wherein an elevated freeway was replaced by a tunnel and a linear park) found that proximity to open space had a positive impact on both residential and commercial property values (Tajima 2003).

According to the Seoul Development Institute (2005), the CGC conversion to greenway resulted in property price gains. From 2003 to 2005, office rents went up by an average of 10 per cent while land prices jumped by nearly 40 per cent within the CGC impact zone (approximately 2 km in each direction from the corridor). The number of businesses and mixed-use buildings within the zone also rose. So did mean condominium prices.

A more recent, in-depth analysis of the land-price effects of the CGC freeway-to-greenway conversion was carried out by Kang and Cervero (2008), using multi-level hedonic price modelling. For office, commercial and other non-residential uses, land-value premiums were found for parcels within 500 metres of the corridor for both the former freeway and present-day urban greenway (see Figure 11.1). However, premiums were notably higher for parcels within the 500-metre walk shed of the urban greenway entrance points than the freeway on-ramps. Figure 11.2 reveals that housing was worth less, all else being equal, within 3 km of the elevated freeway, reflecting a dis-amenity. However, the opposite held once the corridor was transformed to a greenway; homes within 2 km were worth as much as 8 per cent more. Clearly, Seoul's freeway disinvestment/greenway investment scheme conferred net benefits to both residential and non-residential property owners. From the standpoint of land market performance, one could argue that an urban amenity, and specifically an improved urban space, was more highly valued in central Seoul than a significant component of urban

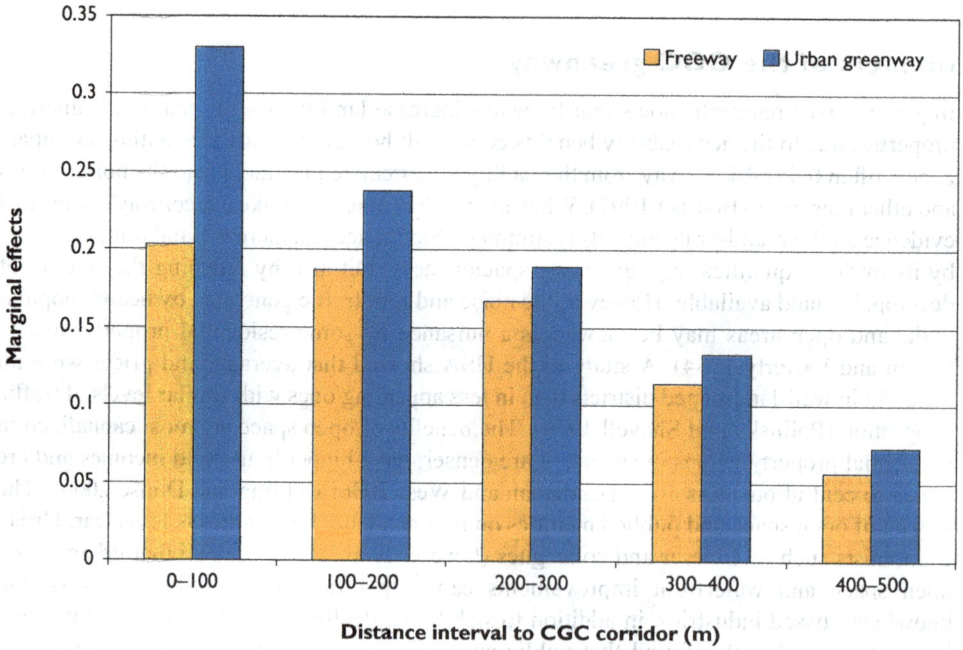

Figure 11.1 Marginal effects of the CGC freeway and urban greenway on non-residential property values by distance intervals, 2001 to 2006.

Source: Kang and Cervero (2008).

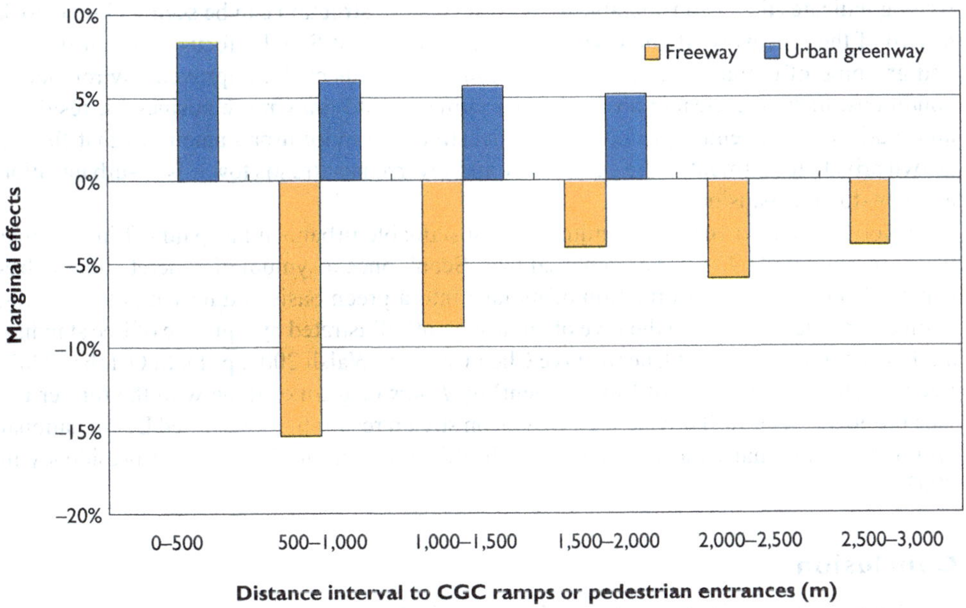

Figure 11.2 Marginal effects of the CGC freeway and urban greenway on residential property values by distance intervals, 2001 to 2006.

Source: Kang and Cervero (2008).

infrastructure, namely a freeway. That is, quality of place has won out over automobility as a desirable urban attribute.

Besides the attractions of greenery and open space, part of the land-value gains conferred by the freeway-to-greenway conversion were likely also due to indirect environmental benefits. For one, air pollution levels along the CGC corridor have fallen since the stream restoration and greenway conversion. Concentrations of fine-grained particulate matter (PM_{10}) along the CGC corridor were 13 per cent higher than Seoul's regional average before the conversion; afterwards, they were 4 per cent below the region's average (Seoul Development Institute 2005). In the case of nitrogen dioxide (NO_2), a precursor to the formation of photochemical smog, concentrations went from 2 per cent above the regional average when the freeway was in operation to 17 per cent below when the greenway was in place.

Many urban centres suffer a heat island effect, with temperatures higher than surrounding suburban and rural areas due to greater surface-area coverage, and Seoul is no exception. The spillover 'cooling' benefits from the CGC transformation are revealed by a heat island study that found ambient temperatures along the central-city stream and greenway were 3.3 and 5.9° Centigrade lower than along a parallel arterial road surface five blocks away (Hwang 2006).

In sum, the evidence to date suggests that Mayor Lee's bold experiment with land reclamation and urban transformation has been an unqualified success. Land prices have risen and traffic conditions have not materially changed – thanks in large part to vastly expanded public transport services that absorbed some of the former car traffic. The Cheng Gye Cheong running stream and pedestrian promenade has become an attractive destination for tourists and residents, and, while statistics are hard to come by, anecdotal evidence at the time of writing indicates that a considerable amount of new construction can be seen within a block or two of the greenway. It could very well be the case that Seoul's focus on beautification and greening of urban environments is creating a virtuous cycle; improved environmental conditions, including cleaner air and cooler climes, could draw more businesses, residents and tourists to the centre, prompting the investment in more urban amenities that further spawn private investment in the central city and reverse past trends towards ex-urbanization and new-town expansion.

Mayor Lee's unwavering commitment to sustainable urbanism has paid off in personal ways. In May 2006, *Times Asia* reported that 'Seoul, once a symbol of concrete jungle, has achieved successful transformation of its face into a green oasis and now it is inculcating upon other Asian cities with the love of environment', illustrated by a picture of Lee standing ankle-deep in the waters of Cheong Gye Cheon stream (Walsh 2006, p. 1). In October 2007, Lee was chosen as a 'Hero of Environment' by *Times* magazine, along with the former US vice president Al Gore. The Cheong Gye Cheon stream restoration catapulted Lee to national prominence, and many local observers credit this with helping him win the presidency in 2007.

Conclusion

The factors that combine to create rapid suburbanization – rising incomes, rapid population growth, the location-liberating influences of information technologies and so on – are universal but have been particularly pronounced in East Asia's mega-cities. Such trends certainly characterize the past three decades of unfolding settlement patterns in one of the world's largest urban agglomeration, Seoul, Republic of Korea. In the tradition of British town planning, policy-makers at both the central and metropolitan level of government embarked upon an ambitious campaign over the last quarter of the twentieth century: creating a protective greenbelt around Seoul and building large-scale, master-planned new towns on virgin lands beyond the greenbelt. While these new towns achieved their intended objectives of accommodating newly formed households and moderating housing price increases, with time the cumulative social costs began to rear their ugly head – notably, ever-worsening traffic conditions and environmental degradation.

As is often the case, bold, new urban visions require bold, new visionaries. In Seoul's case, this was in the person of Myung-Bak Lee, who oddly enough made a career out of building mega-scale infrastructure projects throughout the Republic of Korea as the head of Hyundai Group, but, once he retired and entered public life, 'found religion', as they say. When running for mayor of Seoul in the early 2000s, Lee recognized that the daily tidal pattern of moving large segments of Seoul's workforce in and out of new towns to the core of the city was unsustainable – not only in terms of worsening traffic and air quality, but also 'time pollution', a term coined by British planners that recognizes that time stuck in traffic is time robbed from one's personal life, whether being with one's family, or for leisure or civic engagements (Whitelegg 1993; Putnam 2000).

Mayor Lee's vision involved redefining the role between public infrastructure and community living. Long considered a public asset, land-hungry freeways and roadway interchanges can over time become a public liability. Mayor Lee and his contemporaries in a handful of other global cities have in recent times turned to a different kind of public asset to grow local economies – notably public amenities, urban parks and other civic functions that enhance aesthetics and quality of life. In an increasingly competitive, knowledge-based global marketplace, improved civic spaces and expansion of the arts and cultural entertainment offerings, proponents contend, appeal to highly sought-after professional class workers, Richard Florida's (2004) so-called 'creative class'. The economic dividends, some contend, from civic investments can be considerable.

The longer-term consequences of Seoul's bold experiment with urban land reclamation are yet to be seen. The hope and expectation of many urban planners and political greens is that it will slow the pace of new-town development, reduce the spatial mismatch between development north (primarily employment) and south (primarily housing) of the Han River, and spur the redevelopment of former industrial land. More balanced infill development, planners hope, will make Seoul an attractive city on the global stage that appeals to international finance and businesses as well as tourists, professional class workers and foreign investors.

Experiences not only in Seoul but also in other global cities that have opted to dismantle freeways and expropriate road space – San Francisco, Boston, Paris and New York, among others – suggest that the era of indiscriminate large-scale infrastructure construction and a blind devotion to mobility-based planning is over. The urban footprint of the transport sector – freeways, interchanges, surface roads and parking – is immense, consuming as much as half of all land area in many first-world cities (and in so doing separating activities from each other and thus creating greater demands for motorized transportation as well as roadway infrastructure). Whatever freeways and mega-scale urban infrastructure projects are built in coming years will have to be strategically sited and designed to pass a stricter litmus test of contributing to the larger urban development objectives of the cities and neighbourhoods they serve. In Seoul and elsewhere, freeway deconstruction and land reclamation is tied to the reordering of urban priorities that gives preference to planning for people and neighbourhoods, not mobility. Experiences in Seoul suggest that the provision of high-quality public transport services and well-designed pedestrian-friendly greenways can diminish the necessity for high-capacity elevated freeway structures. In so doing, Seoul and similarly adventurous cities are likely to find themselves on a far more sustainable pathway – not only in terms of environmental quality but economic prosperity and quality of life as well.

Acknowledgement

I would like to thank Chang Deok Kang, a doctoral student in the Department of City and Regional Planning at the University of California, Berkeley, for his assistance in compiling information and data used to carry out this research.

References

Anderson, S.T. and West, S.E. 2006, 'Open space, residential property values, and spatial context', *Regional Science and Urban Economics*, vol. 36, pp. 773–789.

Boarnet, M. 1997, 'Highways and economic productivity: interpreting recent evidence', *Journal of Planning Literature*, vol. 11, no. 4, pp. 476–486.

CGC Restoration Project Headquarters 2004, *CGC Restoration*, Seoul Metropolitan Government, Seoul.

Dehring, C. and Dunse, N. 2006, 'Housing density and the effect of proximity to public open space in Aberdeen, Scotland', *Real Estate Economics*, vol. 34, no. 4, pp. 553–566.

Downs, A. 2004, *Still Stuck in Traffic: Coping with peak-hour traffic congestion*, Brookings Institution Press, Washington, DC.

Florida, R. 2004, *The Rise of the Creative Class*, Basic Books, New York.

Frech III, H.E. and Lafferty, R.N. 1984, 'The effect of the California coastal commission on housing prices', *Journal of Urban Economics*, vol. 16, pp. 105–123.

Glaeser, E.L., Kolko, J. and Saiz, A. 2001, 'Consumer city', *Journal of Economic Geography*, vol. 1, pp. 27–50.

Hwang, K. 2006, *Cheong Gye Cheon Restoration and City Regeneration: Cheong Gye Cheon, urban revitalization and future vision*, Seoul Metropolitan Government, Seoul.

Hwang, K., Byun, M. and Nah, T. 2005, *Project Cheong Gye Cheon*, Nanam Publishing House, Seoul.

Jun, M.J. 2000, 'Commuting pattern of new town residents in the Seoul metropolitan areas', *Journal of Korean Regional Development Association*, vol. 12, no. 2, pp. 157–170.

Jun, M.J. and Bae, C. 2000, 'Estimating commuting costs of Seoul's greenbelt', *International Regional Science Review*, vol. 23, no. 3, pp. 300–315.

Jun, M.J. and Hur, J.W. 2001, 'Commuting cost of leap-frog new town development in Korea', *Cities*, vol. 18, no. 3, pp. 151–158.

Kang, C.D. and Cervero, R. 2008, 'From elevated freeway to linear park: land price impacts of Seoul, Korea's CGC project', working paper, UC Berkeley Center for Urban Future Transport, Berkeley, CA.

Lee, K.S. 2003, 'Seoul's urban growth in the 20th century: from a pre-modern city to a global metropolis', in K.J. Kim (ed.), *Seoul, 20th Century: Growth & change of the last 100 years*, Seoul Development Institute, Seoul.

Perloff, H. 1975, *Modernizing the Central City: New towns, intown . . . and beyond*, Ballinger Publishing Company, Cambridge, MA.

Polinsky, M. and Shavell, S. 1976, 'Amenities and property values in a model of an urban area', *Journal of Public Economics*, vol. 5, pp. 119–129.

Pucher, J., Park, H.Y., Kim, M.H. and Song, J. 2005, 'Public transport reforms in Seoul: innovations motivated by funding crisis', *Journal of Public Transportation*, vol. 8, no. 5, pp. 41–62.

Putnam, S. 2000, *Bowling Alone: The collapse and revival of American community*, Simon & Schuster, New York.

Seoul Development Institute 2005, *Toward Better Public Transport*, Seoul Development Institute, Seoul.

Sivitanidou, R. 1995, 'Urban spatial variation in office-commercial rents: the role of spatial amenities and commercial zoning', *Journal of Urban Economics*, vol. 38, pp. 23–49.

Tajima, K. 2003, 'New estimates of the demand for urban green space: implication for valuing the environmental benefits of Boston's Big Dig project', *Journal of Urban Affairs*, vol. 25, no. 5, pp. 641–655.

Walsh, B. 2006, 'Saving Seoul', *Time Asia*, 8 May, pp. 1–2.

Whitelegg, J. 1993, *Transport for a Sustainable Future: The case for Europe*, Belhaven Press, London.

Chapter 12

From sustainable to low carbon cities

Is India's urban transformation triggering a paradigm shift?

Shipra Narang Suri

Summary

India's urbanization is paradoxical. While the country's urban population is undoubtedly vast, at 377 million, its urbanization pattern can hardly be described as rapid. From 23.3 per cent in 1981, India's urban population grew rather slowly to 31 per cent in 2011, which is lower than, for example, China, Indonesia, Mexico or Brazil. However, informal settlements, informal livelihoods, homelessness, insecurity, various types of pollution and declining green areas are the bane of most Indian cities. When these are combined with concerns about climate change and the regular occurrence of natural disasters, Indian towns and cities seem to be sitting on a ticking time bomb. Over several decades, the mantra of sustainability – the importance of balancing social, economic and environmental concerns – has gradually made it onto the policy agenda in India. Now, cities are becoming ever more active and innovative, perhaps because of the urgent need to address the vast range of problems they face and greater exposure to new ideas, concepts and innovative practices for dealing with them, or due to the imperative of becoming 'world class' while demonstrating a commitment to contemporary global concerns. India's federal structure allows subnational governments a lot of authority in dealing with areas such as urban planning and management, environment and transport, among others, whereas the national government has a rather more prosaic role of formulating broad policy prescriptions in these areas. Quite a few states and cities have therefore elaborated a low carbon agenda over the past few years, and several have initiated projects for its implementation. With the help of a few illustrations, this chapter explores the extent and nature of this paradigm shift, which is simultaneously extremely promising and challenging for India's urban future.

Massive and chaotic: India's urban transformation in the twenty-first century

India's urbanization is best described as massive and chaotic. Although the level of urbanization stands at a mere 31 per cent, which is lower than, for example, China, Indonesia, Mexico or Brazil (HPEC 2011), according to the latest Census figures (2011), the country's urban population is undoubtedly vast, at 377 million. Policies and policy-makers seem unable to deal with the problems of urbanization – a quarter of the total urban population lives in slums, and the lack of infrastructure and basic services, skyrocketing costs of land, informality, homelessness, insecurity, various types of pollution, and declining green areas and public spaces are the bane of most Indian cities. Combined with emerging concerns about climate change and the regular occurrence of natural disasters, Indian towns and cities seem to be sitting on a ticking time bomb (Narang Suri 2011).

The other characteristic of India's urbanization is that the larger cities (metropolises, which have over a million residents, and Class I cities, which have 100,000 or more residents) have witnessed much more rapid growth than smaller towns. In 2001, nearly 70 per cent of the country's urban population resided in the 393 Class I cities. The unequal spatial distribution of urban populations across the country has been a cause for concern for policy-makers at the highest level in successive Five-Year Plans (Kundu 2011). Despite this, urbanization strategies currently being implemented or proposed will potentially maintain this trend. For example, the proposals put forth by a recently constituted High-Powered Expert Committee for Estimating the Investment Requirements for Urban Infrastructure Services (HPEC) are likely to create 87 cities with a population of one million and above by 2030 (HPEC 2011).

These outcomes of India's urban transformation pose enormous challenges for the country's planners and policy-makers. The formal planning system, based on the British town planning system of the early twentieth century, has seen little change since India's independence. Most towns and cities rely on inflexible master plans, which are usually outdated by the time they are completed. Low-density development and continued segregation of uses, rigid development control norms that are flouted at every step and a weak governance system that can neither guide nor enforce completes the sorry picture. This planning system is hardly fit for purpose for the twenty-first century.

While the planning system has remained largely stagnant over decades, some contemporary concerns – the most prominent being sustainable development – have certainly crept onto the policy agenda. After the Earth Summit in Rio in 1992, which committed governments to adopt a path to sustainable development, the Indian government's ninth Five-Year Plan (1997–2002) explicitly recognized the synergy between environment, health and development. The Plan identified as one of its core objectives the need for ensuring environmental sustainability of the development process through social mobilization and participation of people at all levels. Practical efforts, however, focused largely on improved biodiversity and natural resource management, afforestation and sustainable rural development. Little attention was paid to issues of urban poverty or expansion of basic services to the poor, critical elements to achieve sustainable *urban* development.

In 2005, for the first time, a massive programme focusing on urban development and urban renewal was launched by the Government of India, covering 63 cities across the country. The overarching goal of the Jawaharlal Nehru National Urban Renewal Mission (JNNURM) was to encourage reforms and fast-track planned development of the identified cities, with a focus on improving efficiency in the delivery of urban infrastructure and services. Slum upgrading, community participation and making local authorities accountable to citizens were among the Mission's other objectives (Government of India 2005a).

The policy focus on urbanization in India could not be substantially expanded and deepened however, as the false urban–rural dichotomy raised its head once again. With nearly 70 per cent of the country's population still living in rural areas, policy-makers at the national level are often torn between the need to focus on rural livelihoods and access to health, education and other basic services, on the one hand, and the urbanization of poverty and the unchecked growth of slums and informality in cities, on the other. A full-scale overhaul of the urban and regional planning system also seems unlikely in the near future, given India's federal structure and its current fractious political scenario. Given this situation, several Indian cities have taken it upon themselves to introduce innovative measures in urban planning, management and governance, demonstrating vision, creativity and a departure from business as usual (Narang Suri 2011).

As the messy urban transformation in India is countered by increasingly vocal and self-sufficient subnational governments at state and local levels, the question is: Are these efforts simply a continuation of the mantra of sustainability? Or is there a growing acceptance of the climate change imperative at the local level, and a conscious shift towards adopting low carbon policies and practices? Is India's urbanization finally leading to a paradigm shift in its approach towards urban planning, management and governance? These are the questions that this chapter seeks to explore.

Major development challenges faced by Indian cities

Several recent reports have comprehensively outlined the challenges faced by Indian towns and cities, for the first time presenting extensive and up-to-date information on the scale and nature of the challenges (see e.g. HPEC 2011; IDFC 2010; UNDP and Government of India 2009). As they make clear, urban development is a massive and multi-faceted task. The paragraphs below summarize only selected issues that are of direct relevance to the theme of this chapter because the limited space available does not permit a comprehensive discussion of the entire range of development challenges facing Indian cities.

Housing and basic urban services

The challenges of urban housing, especially for low-income populations; infrastructure provision; and the delivery of a variety of urban services, including (among others) water, sanitation, education and health, are long-standing concerns for Indian policy-makers. As per the 2011 to 2012 National Sample Survey reports, there are over 50 million poor people living in the cities and towns of India (Government of India 2013). While this figure is contentious and debated, at the same time it is widely accepted that 'shelter poverty' is much larger than income poverty in urban areas. This is mainly a result of heavily distorted land markets and an exclusionary regulatory system that fails to accommodate the needs of the poor, or to adequately address challenges of slums, informal settlements and pavement dwellings. Adequate, affordable formal housing supply for the urban poor does not seem to be a priority area for policy-makers, even though land values are escalating sharply and the 'market' is clearly excluding the poor. A resettlement policy is urgently required that lays down guidelines to minimize displacements and ensure housing for development project-affected persons based on human rights to adequate shelter. As slums are a state government responsibility under the Indian Constitution, state slum laws also need to be reviewed across the country (Risbud 2009).

In terms of services, too, Indian cities lag behind on almost all counts. It is estimated that water supply is available for an average of 2.9 hours per day across all Indian cities and towns; less than 20 per cent of wastewater is treated; and solid waste management is grossly inadequate. National benchmarks have only recently been developed for the four key service sectors: water supply, sewerage, stormwater drainage and solid waste management (Vaidya 2009).

Environment, ecology and disasters

As mentioned in the introductory section of this chapter, over the past few decades, initiatives focused on the environment have sought to protect biodiversity, increase forest area (or, at least, arrest deforestation), check land degradation, and control various other forms of pollution – most notably that of air and water. The *urban* environment has hardly received the importance that it should, despite several governmental and non-governmental initiatives, not least because its condition is intricately connected with the state of urban infrastructure. Surface water drains that serve as sewage and solid waste collectors are a common sight in many cities. Landfills are unsanitary and poorly managed. Greenbelts are regularly encroached upon, and water bodies are often choked.

Although India's per capita greenhouse gas (GHG) emissions are among the lowest in the world, in absolute terms, this country is the fourth largest emitter globally. Cities are a major contributor. Although unwilling to compromise on its development priorities in order to cut its GHG emissions, the national government has attempted to develop its own strategy to reduce these emissions and combat climate change. These issues will be discussed later in the chapter.

Cities across India are also experiencing the impacts of climate change in the form of more frequent and intense natural disasters, especially heavy rainfall and urban floods. Large parts of the city of Delhi lie in the floodplain of the River Yamuna, and are especially vulnerable during the monsoon season. Kolkata has witnessed frequent and increasingly severe cyclonic storms over the past few years, and is also facing a freshwater crisis, while Chennai is increasingly at risk of being struck by tidal waves. Mumbai is susceptible to rising sea levels, as large sections of the city are built on land reclaimed from the sea, and to heavy monsoon rains, which can cause serious havoc as the natural stormwater drainage systems of the city have been haphazardly built over (Banerjee 2011). This vulnerability was evident during the flash floods of July 2005, when 994 mm of rainfall was recorded over a period of 24 hours in Mumbai. It resulted in waterlogging in several suburbs, overflow of the Mithi River, and flooding of the Western Express Highway. Nearly 500 people were killed in the flash floods and landslides, and about 200 others died due to various deluge-related illnesses. 2,000 dwellings were completely destroyed while over 50,000 were partially damaged. Some 40,000 commercial establishments were also affected. Railway tracks were submerged and services on all railway lines were suspended as a consequence. The airport was also flooded and temporarily closed. The power supply was disconnected in most parts of Mumbai's western suburbs due to the danger of electrocution (Narang Suri 2011).[1]

Urban transport and mobility

For Indian cities, the major challenges in the urban transport sector include road congestion (and the resulting economic losses), local air pollution, noise and road safety/accidents. These are mainly due to inadequate transport infrastructure, an unbalanced modal split heavily favouring private transport and little integration between land use and transport planning (Narang Suri 2011). The absence of an integrated public transportation system across all modes is equally problematic. At the moment, different agencies provide and/or operate different services independently of each other; in Delhi, for instance, the metropolitan railway is operated by the Delhi Metro Rail Corporation, the suburban rail service by Northern Railway, the bus transport service by Delhi Transport Corporation and assorted private operators, and taxi and auto-rickshaw by private (individual or corporatized) operators, without any coordination (Singh 2008).

Past efforts to improve mobility have largely (and misguidedly) been focused on building bigger and better road infrastructure: highways, expressways and grade separators. While this serves to ease congestion initially, the new infrastructure soon becomes just as congested as before, while at the same time pushing non-motorized traffic – pedestrians, bicycles and cycle-rickshaws – off the main roads.

Several cities and towns are experimenting with innovative practices, but once again, as transport is a responsibility of subnational governments, progress varies from one state to another. Still, the national government launched a programme in 2009 'to monitor the progress of cities for public transport, non-motorized transport and pedestrian facilities, usage of ITS, travel speeds, parking spaces, road safety, pollution levels, integration of land-use and transport, and financial sustainability of public transport' (UNEP *et al.* 2010, p. vii).

Urban planning practice in Indian cities

Despite the multitude of challenges facing Indian cities as they expand, urban planning practice has remained largely unchanged in India since independence in 1947. Larger cities rely on master plans, usually with a 20-year perspective, that are developed and implemented by specially constituted development authorities that are outside the purview of the local administration and hence not directly accountable to the local population. The ineffectiveness of planning in Indian cities has become endemic. Ansari (2004) points out some key drawbacks of the conventional 'Master Plan', including the limited attention paid to social and economic development aspects, and to financial resource mobilization for the implementation of the Plan, as well as the long lead-time needed for plan preparation and limited stakeholder involvement in its creation. Economic planning or local economic development strategies are rarely incorporated into the spatial planning exercise, with the result that the plans are unrealistic and impossible to implement (Sridharan 2008).

The implementation of urban plans in Indian cities is further hampered by the fact that water and sewerage systems, power and telecommunication services, roads and public transport, housing and slums are controlled by other 'para-statal' bodies or line departments of central and state governments. Local governments are responsible only for solid waste management, maintenance of public spaces and some basic repair and maintenance of other services such as roads, street lighting and drainage systems.

The disconnect between urban planning and local governance was supposed to be corrected by the adoption of the 74th Constitutional Amendment Act (CAA) in 1992, which proposed that Urban Local Bodies (ULBs) be established and strengthened in order to improve the quality of the urban environment, provide services in a more responsive and effective manner, and enhance participation of local stakeholders in decision-making processes (Savage and Dasgupta 2006). While the Act directly addressed the issue of planning and explicitly assigned the function to local governments, this aspect has continued to be marginal to the development of India's towns, cities, districts and regions.

The launch of the Jawaharlal Nehru National Urban Renewal Mission (JNNURM) towards the end of 2005 was the next important step forward, and it significantly influenced the local governance and urban planning systems in India's large cities. One of the prerequisites for any city to access funds under the JNNURM was the preparation of a City Development Plan (CDP). JNNURM guidelines for CDP preparation suggested that a multi-stage process be adopted for its creation, including an assessment of the current situation, the development of a vision for the future in consultation with stakeholders, strategy formulation and the development of a City Investment Plan (Government of India 2005b).

A revised version of the JNNURM is now in the pipeline, which will apparently target all cities and towns, small or large, and will focus on capacity creation within local governments. To improve the governance of planning, it proposes to strengthen metropolitan and district planning committees, with urban development authorities and unified metropolitan transport authorities as their technical arms. City planning would become a function of elected local governments rather than Urban Development Authorities (HPEC 2011). It remains to be seen if this version of the Urban Renewal Mission is any more effective than the previous one, especially vis-à-vis planning.

Sustainable urban development in policy and practice

This chapter does not attempt to provide a complete review of all policies and practices related to the theme or aim of sustainability in India. The fact that these exist at various levels of government – national, state and local – and in a very wide range of sectors makes the task of providing such an overview impossible. However, in this section, I will describe some of the policies and plans that have integrated the concerns of sustainability and sustainable *urban* development. I also discuss how recent policy is responding to issues of climate change mitigation and adaptation, and disaster risk reduction, and what cities are doing to address these issues as well.

Sustainable urban development as a theme of national plans and policies

As noted above, the Indian government's ninth Five-Year Plan (1997–2002) was the first to explicitly recognize the synergy between environment, health and development, and to identify environmental sustainability as one of its core objectives. However, it was only after the World Summit on Sustainable Development (WSSD) in 2002 that a process of preparing and implementing a national strategy for sustainable development was initiated. The National Environmental Policy (NEP), adopted in 2006, elaborated on the theme of sustainable development, highlighting the link between development, ecological constraints and social justice. The policy asserted that the most viable basis for environmental sustainability is to ensure that people gain better livelihoods from the conservation of natural resources than from environmental degradation.

Interestingly, however, the NEP barely alluded to urban environmental management, except in the context of addressing water pollution and developing a national strategy for urban transport. Sustainable *urban* development did not even receive a passing mention. This could be the result of the low priority accorded to sustainable urbanization by the national government, the poor understanding of urban sustainability issues and concerns, or the silo mentality of the government institutions responsible for it (this policy was prepared by the Ministry of Environment and Forests, probably without much input from the Ministry of Urban Development). In all probability, it is a combination of all three factors which has led to India's 'urban-blind' environmental strategy.

On the other hand, urban sustainability concerns are apparently addressed through other policies, most importantly the National Urban Housing and Habitat Policy of 2007, which 'sought to promote sustainable development of habitat in the country with a view to ensuring equitable supply of land, shelter and services at affordable prices to all sections of society' (Government of India 2011a, p. 11), and the National Urban Transport Policy of 2006, which aimed to 'ensure safe, affordable, quick, comfortable, reliable and sustainable access for the growing number of city residents to jobs, education, recreation and such other needs within cities' (Government of India 2011a, p. 12).

The most explicit recognition of the challenges of sustainable development and climate change in India has come through the adoption of the National Action Plan on Climate Change (NAPCC) in June 2008. This Plan comprises eight missions that focus on key areas: solar energy, energy efficiency, sustainable habitat, water, sustaining the Himalayan ecosystem, Green India, sustainable agriculture and strategic knowledge for climate care. These eight missions are implemented by different ministries at the national level.

Sustainable urban development in local plans and projects

As highlighted in the discussion above, sustainable urban development concerns have only made it on to the policy agenda in the past six or seven years. At the local level, however, there have been several attempts to incorporate sustainability concerns and objectives into city-level plans and regional plans. A few cities have also initiated specific actions or projects that make their cities demonstrably more sustainable. In many cities, the initiative has been taken by civil society organizations or the private sector.

One prominent initiative of the past decade was the Sustainable Chennai project, part of the Sustainable Cities Programme coordinated by the United Nations Human Settlements Program (UN-Habitat) and the United Nations Environment Program (UNEP). Initiated in 1995, the project focused on improving the ability of individuals and organizations to identify, understand and analyse environmental issues and to incorporate their considerations into sectoral development planning activities. It emphasized the need for a more integrated approach to all urban development activities among sectoral agencies, public as well as private. A rapid assessment of development concerns, natural resource utilization, environmental hazards and institutional responsibilities was compiled in an environmental profile, and several cross-sectoral consultative groups were established in order to access information, review past efforts and mobilize broad institutional support for an Environmental Planning and Management (EPM) process. Since 2000, several community-based demonstration projects have been implemented across the city in the areas of solid and liquid waste management, improvement of waterways, reduction of traffic congestion and improvement in air quality (UN-Habitat and UNEP 2001).

More recently, many cities have introduced policies to enhance sustainability, particularly in the transport sector, largely through the provision of public transport; a range of options are being tried, including improved and environment-friendly bus services, the introduction of Bus Rapid Transit (BRT) systems and the development of metropolitan rail systems. The Delhi Metro, for instance, now ferries up to 2.3 million commuters every day on seven lines, whose length adds up to nearly 200 kilometres. At the same time, however, several analysts are beginning to question the disproportionate emphasis (and resources) being allocated to build or expand expensive mass transit systems such as the Metro, often at the expense of improvement of infrastructure for non-motorized traffic and the expansion of BRT systems, which are both significantly less expensive than the Metro and more affordable for the poorer sections of the urban population (Siemiatycki 2006).

Along with its Metro, Delhi has also successfully implemented and enforced a policy on the use of cleaner fuels across the capital. Since 2002, all public service vehicles in the capital use compressed natural gas (CNG) instead of diesel. When introduced a decade ago under directives of the Supreme Court of India, this policy was extremely unpopular, and riddled with several glitches, the most important one being limited availability of CNG and the long queues that snaked for several kilometres outside the handful of stations that supplied the fuel. These issues have now been addressed effectively, and the capital is relatively free of diesel pollution. Between 2000 and 2008, carbon emissions in Delhi plummeted by 72 per cent while sulphur dioxide emissions decreased by 57 per cent, thanks to the Metro and the 3,500 CNG buses, 12,000 taxis, 65,000 auto-rickshaws and 5,000 minibuses running on CNG (Hohne *et al.* 2009).

Although Delhi's BRT experiment famously failed, the city of Ahmedabad in western India has successfully designed and implemented a BRT system that carries 125,000 passengers per day using 70 buses. The Janmarg BRTS, as it is known in Ahmedabad, was developed under the broader vision of 'Accessible Ahmedabad', and inaugurated in 2009. It attracts users from all classes of society, and at the same time extensively deploys modern technology such as automatic vehicle tracking and passenger information systems, the use of smart cards, surveillance and security systems, and area traffic control systems at junctions.

Figure 12.1 Delhi's shift to CNG has played a major role in reducing air pollution and cutting its GHG emissions.

Source: Photograph © Harish Narang, reproduced by permission.

Figure 12.2 The Ahmedabad Janmarg BRTS is a model of low carbon transport for India.

Source: Photograph © CEPT University, Ahmedabad, reproduced by permission.

Dedicated right-of-way for the buses and median bus stations with barrier-free access and at-level boarding enhance accessibility and save time (Narang Suri 2011).

From sustainable to low carbon cities

As noted earlier in this chapter, India is currently the fourth largest greenhouse gas (GHG) emitter in the world, with its annual GHG emissions amounting to 1.727 billion tonnes of CO_2 equivalent. Its per capita emissions are less than half of the world average, however, due to its large population. It is estimated that GHG emissions from India will increase to around 5.7 billion tonnes of CO_2 equivalent by 2030 (ICLEI 2012). An illustrative breakdown of emissions by sector (see Table 12.1) is provided in a 41-city study recently conducted by ICLEI South Asia and the British High Commission in New Delhi.

The ICLEI report cited above also makes a strong case for low carbon urban development in India, highlighting the financing opportunities available to make such a shift. Indeed, Indian local and state governments have been slow to capitalize on the availability of a huge pool of international resources – finance, technology, research and other support – to lead their constituencies into a low carbon future.

Table 12.1 GHG emissions by sector in 41 Indian cities

Sector	*Contribution of emissions from each sector*
Community-level activities	
Residential	20–55%
Commercial	5–20%
Industrial	10–60%
Transport	10–40%
Solid water	3–15%
Corporation-level activities	
Street lighting	40–70%
Water supply	15–25%
Building and facilities	20–40%
Transportation	5–15%

Source: ICLEI (2012).

A major reason for this is that in India, as in many other countries in the world, there is neither a commonly applicable description of a low carbon city or urban infrastructure, nor an established formula for assessing a city's carbon performance. Such a formula would admittedly be rather complex, as it would have to be based on the range of factors which influence GHG emissions in urban settings, such as economic structure, population, urban services, energy mix and land use.

Having said that, the concept of low carbon cities does seem to be gaining currency in both academic and policy circles. There is a common understanding that low carbon development and infrastructure maximizes the utilization of low carbon energy resources and enhances the efficiency of urban service delivery (Padigala and Kraleti 2013). There are also significant improvements in the data available, its breadth and its coverage, as well as its quality and accessibility. Government organizations, civil society and academic institutions have all been empowered by the knowledge available.

The *India Infrastructure Report 2010* focused on infrastructure development in a low carbon economy, pointing out that progressing to a low carbon future will not be possible without addressing the issue of infrastructure development, which is essential for inclusive growth and modernization, but has usually been quite carbon intensive. Noting that the energy sector is responsible for 58 per cent of the country's total greenhouse gas emissions, which are equivalent to 1.9 billion tons of CO_2 (electricity: 38 per cent; transport: 8 per cent; urban and rural residential energy consumption: 7 per cent; and others, such as petroleum refining: 5 per cent), it asserted that a low carbon growth agenda would warrant building infrastructure with a lower carbon footprint. In a welcome departure from the norm, the report considers the urban sector as a whole, and the relationships between urban planning, transport, water and waste are clearly established. Several of the report's contributing authors observe that urban planning policies in India have traditionally encouraged sprawl, and although the virtues of mixed land use, densification (especially in city cores) and the promotion of public transport as well as non-motorized transport are now being recognized, this requires intensive infrastructure development as well as a significant change of mindset (IDFC 2010).

In 2010, the Planning Commission, the highest plan-making body of the Government of India, also turned their attention to low carbon urban development and established an Expert Group on low carbon strategies for inclusive growth. The Group's interim report, submitted in May 2011, summarizes the international discussion on low carbon futures and its implications for India, as well as the present status of GHG emissions from several key sectors. It then goes on to propose low carbon strategies for the power, transport, industry, buildings and forestry sectors (urban challenges and interventions are dealt with in a fragmented manner under the first four themes). It projects a range for GHG emission intensity reduction in 2020 for both 8 and 9 per cent real GDP growth based on two scenarios: the 'Determined Effort Scenario', which calls for more effective implementation of mitigation policies that have either been put into place or are presently contemplated by the Government; and the 'Aggressive Effort Scenario', which aims at a higher level of reduction in emissions through ongoing and new interventions (Government of India 2011b).

While this is promising, it has been noted that there is no explicit link between the work of the Planning Commission and the NAPCC or its missions, mentioned earlier in this chapter (Dubash 2012). Policy coherence at the national level, as well as between the national, provincial and local levels, is a major issue in climate governance and the pursuit of a low carbon agenda in India.

At the same time, however, there is a growing interest in international knowledge transfer and peer-to-peer learning at various levels. One example is the Low Carbon Asia Research Project, funded by the Environment Research and Development Fund of the Japanese government's Ministry of the Environment. According to the project website, its objectives are to establish a Low Carbon Society (LCS) in Asia and 'to develop methodologies to design and evaluate policy options based on scientific knowledge and the multifaceted needs of each country' (NIES 2013). The project supports four areas of research: (1) LCS scenario development at various levels (global, national, subnational and city); (2) institutional architecture for LCS; (3) resources management; and (4) low carbon transport systems. Several studies have already been published, one of which focuses on the city of Bhopal in central India. Undertaken by the School of Planning and Architecture, Bhopal, this path-breaking study develops an LCS scenario for Bhopal for the year 2035. This is clearly a shift from business as usual, as a large amount of data was collected and analysed using complex models, to generate possible future scenarios (see Figure 12.3). These were then translated into expected policy outcomes and an action plan, which is very different from the traditional master planning process (Deshpande *et al.* 2011).

In 2011, another project to develop ways of embedding low carbon techniques into the planning of Indian cities was launched with the support of the British government. Focused on density, layout and the design of infrastructure schemes with the clear aim of reducing CO_2 emissions, the project aims to help in measuring CO_2 savings and demonstrating the relationship between better urban planning and emissions reduction. This will mainly be done through modelling the carbon footprint of development for any city-wide master plan, and testing a wide variety of alternative courses of action. Identifying best practices, what works, what does not and how various techniques can be mainstreamed through integration with the Comprehensive Development Plan process is also expected to be an integral part of the exercise (Atkins 2011).

One of the immediate outcomes of these initiatives is that city administrators are beginning to understand that achieving a low carbon future depends largely on the way in which their cities are (master) planned and managed – with mixed land use and high-density

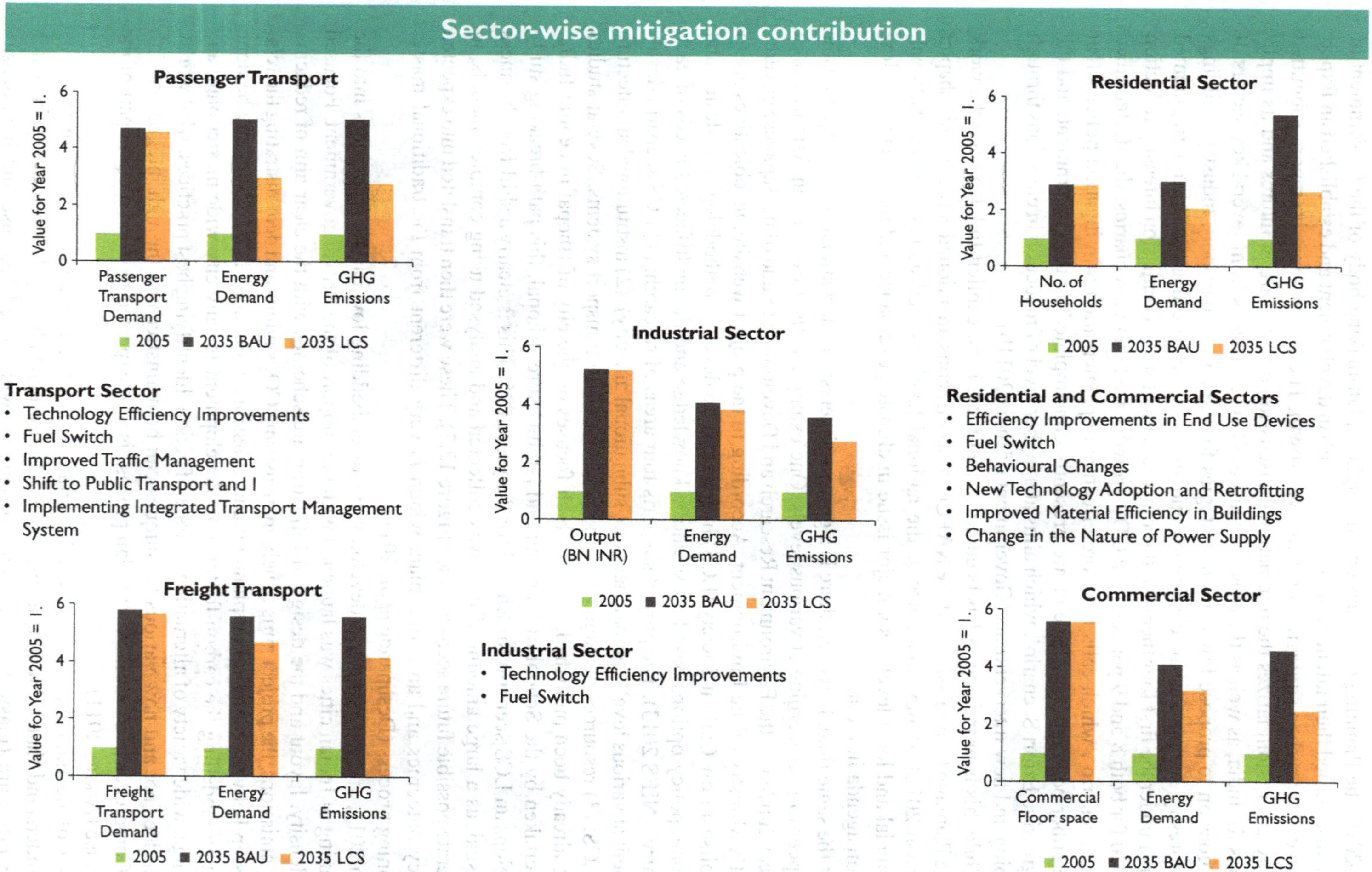

Figure 12.3 Mitigation of emission contributions from various sectors in Business As Usual (BAU) and Low Carbon Society (LCS) scenarios, Bhopal, 2035.

Source: Kapshe (2012), reproduced by permission.

development, accompanied by adequate public transport, as essential starting points. Through improvements in day-to-day service delivery activities, such as effective water management and waste disposal, more efficient street lighting, etc., cities can significantly reduce their emissions without major financial investments. Green buildings, city greening and urban agriculture can also play an important role in cutting urban emissions.

City managers and professional planners are beginning to recognize that a low carbon approach also brings important social and economic benefits, even though these may not be easy to compute or demonstrate. The collection of good-quality data and building GHG inventories and trends, and correlating them with other human development indicators, will be critical in this regard.

A true paradigm shift?

There seems to be little doubt that India is at the cusp of a major transformation in its approach towards sustainable development in general and sustainable urbanization in particular. The national government, as in many other countries, is obliged to collect more information, against a new and diverse set of indicators, and report regularly to international organizations and forums. A significant amount of this information pertains to subnational and local levels, and much of it is available freely online.

Local governments are also participating in peer-learning networks and building local–local partnerships. This has led to greater awareness of the low carbon agenda, of what cities can do by themselves and the areas in which they can demand national and international support (and funding), as compared to, for example, the sustainable development agenda of the 1990s. A range of initiatives, large and small, are helping cities and communities envision a low carbon future. This presents an exciting opportunity for India's urban future.

At the same time, however, it is an opportunity that can easily be undermined by several challenges. Policy coherence is certainly one of these. So far, the agenda of low carbon growth or indeed mitigating the effects of, or adapting to, climate change, has been largely a national one, with various policy interventions envisioned at the national level and their implementation devolved to state levels. Some states have also launched their own initiatives, but these are few and far between. In addition, as shown in the preceding discussion, various policy processes are ongoing concurrently, which points towards a need for coordination within various arms and between various levels of government.

Furthermore, thus far, local governments have had a marginal role, and a feeble voice, in this discourse. This is despite the fact that urban local governments have been performing many functions that are implicitly related to this agenda. These include, for instance, the provision of housing and basic services, slum upgrading and poverty alleviation, and, in many cases, the development and management of infrastructure for transport, water and sanitation. Transforming India's urban local governments into smart, empowered, quality institutions that can not only envision their own future but also implement such a vision in partnership with key stakeholders is the key to a low carbon future.

Linked to the above is the imperative of renovating the colonial-era planning system into something that is more strategic, and at the same time more flexible and action oriented. Planning processes need to incorporate new approaches and ideas for better outcomes. Finally, building awareness, commitment and capacity within national and local governments, civil society, the private sector and academia is equally critical to propel India's cities to a new, low carbon future. Unless these dimensions are adequately addressed, the opportunity for a paradigm shift and lasting change could well be lost for several years to come.

Note

1 As this chapter was being written, heavy monsoon rains lashed India and several rivers broke their banks to flood, or, worse, simply sweep away large sections of riverside settlements. The worst affected were the smaller towns and cities in the north of the country, particularly in the state of Uttarakhand. Early estimates indicate that over 1.6 million people were affected in the disaster, with 2,052 houses wiped out, 147 bridges collapsing and 1,307 roads destroyed in Uttarakhand alone (see www.downtoearth.org.in/themes/DTE/ukparralax/uk_flood.htm).

References

Ansari, J.H. 2004, 'Time for a new approach in India', *Habitat Debate*, vol. 10, no. 4, p. 15.

Atkins 2011, news release, www.atkinsglobal.com/media-centre/news-releases/2011/2011-10-06, viewed 22 August 2013.

Banerjee, I.R. 2011, 'How safe are our cities?', *The Times of India*, Delhi edition, 27 March.

Deshpande, A., Kapshe, M., Mitra, S. and Puntambekar, K. 2011, *Low Carbon Society Scenario Bhopal 2035*, SPA Press, Bhopal.

Dubash, N.K. 2012, *The Politics of Climate Change in India: Narratives of equity and co-benefits*, Working Paper 2012/1, Centre For Policy Research Climate Initiative, Centre for Policy Research, New Delhi.

Government of India 2005a, *Jawaharlal Nehru National Urban Renewal Mission (JNNURM): Overview*, Ministry of Urban Development, Ministry of Housing and Urban Poverty Alleviation, New Delhi.

—— 2005b, *Formulation of a City Development Plan*, Ministry of Urban Development, Ministry of Housing and Urban Poverty Alleviation, New Delhi.

—— 2011a, *Sustainable Development in India: Stocktaking in the run up to Rio+20*, Ministry of Environment and Forests, New Delhi.

—— 2011b, *Interim Report of the Expert Group on Low Carbon Strategies for Inclusive Growth*, Planning Commission, New Delhi.

—— 2013, *Press Note on Poverty Estimates, 2011–12*, Planning Commission, Government of India, New Delhi.

High Powered Expert Committee on Indian Urban Infrastructure and Services (HPEC) 2011, *Report on Indian Urban Infrastructure and Services*, National Institute of Urban Affairs, New Delhi.

Hohne, N., Burck, J., Eisbrenner, K., Vieweg, M., Grießhaber, L. and Ecofys, X. 2009, *Scorecards on Best and Worst Policies for Green New Deal*, WWF and E3G, www.wwf.org.au/news_resources/resource_library/?1369/Scorecards-on-best-and-worst-policies-for-a-green-new-deal, viewed 8 October 2013.

Infrastructure Development Finance Company (IDFC) 2010, *India Infrastructure Report 2010: Infrastructure development in a low carbon economy*, Oxford University Press, New Delhi.

International Council for Local Environmental Initiatives (ICLEI) 2012, *Urban Low Carbon Growth: Financing opportunities for Indian cities*, ICLEI South Asia and the British High Commission, New Delhi.

Kapshe, M. 2012, 'Developing low carbon vision for Indian cities: a case of Bhopal, India', paper presented at *Responding to Climate Change: UK-Indian perspectives*, 20–21 June, Queens University, Belfast.

Kundu, A. 2011, 'Politics and economics of urban growth', *Economic and Political Weekly,* vol. 46, no. 20, pp. 10–12.

Narang Suri, S. 2011, 'Making Indian cities liveable: the challenges of India's urban transformation', in C. Gossop and S. Nan (eds), *Liveable Cities: Urbanising world*, Routledge and ISOCARP, The Hague.

National Institute for Environmental Studies (NIES) 2013, 'Outline', *Low Carbon Societies Research Project*, 1 July 2013, http://2050.nies.go.jp/, viewed 22 August 2013.

Padigala, B. and Kraleti, S. 2013, 'Financing low carbon urban development through clean development mechanism', *Urban India*, vol. 33, no. 1.

Risbud, N. 2009, 'The poor and morphology of cities', in *India Urban Poverty Report 2009*, Oxford University Press, New Delhi.

Savage, D. and Dasgupta, S. 2006, 'Governance framework for delivery of urban services', in *India Infrastructure Report 2006: Urban infrastructure*, Oxford University Press, New Delhi.

Siemiatycki, M. 2006, 'Message in a metro: building urban rail infrastructure and image in Delhi, India', *International Journal of Urban and Regional Research*, vol. 30, no. 2, pp. 277–292.

Singh, S.K. 2008, 'Review of urban transportation in India', *Journal of Public Transportation*, vol. 8, no. 1, pp. 79–97.

Sridharan, N. 2008, 'Globalisation of urban India', *Economic and Political Weekly*, vol. 43, no. 10, pp. 26–31.

UNDP and Government of India 2009, *India Urban Poverty Report 2009*, Oxford University Press, New Delhi.

UNEP, DTIE Paris and UNEP Risoe Centre, Roskilde 2010, *Workshop on Promoting Low Carbon Transport in India*, UNEP, Paris.

UN-Habitat and UNEP 2001, *Sustainable Cities Programme, 1990–2000. A decade of United Nations support for broad-based participatory management of urban development*, UN-Habitat and UNEP, Nairobi.

Vaidya, C. 2009, *Urban Issues, Reforms and Way Forward in India*, Department of Economic Affairs, Ministry of Finance, Government of India, New Delhi.

Part III

Urban microclimates

Mitigating urban heat stress

Introduction to Part III

Part III's nine chapters tackle a key challenge for sustainable and low carbon cities: the urban heat island (UHI) effect. Because cities are covered in heat-absorbing surfaces and materials, they retain more heat, making urban areas warmer than their surroundings, especially at night. As urban microclimates are heating up, creating urban heat stress, this part of the book explains the phenomenon and how we might respond to it.

Although cities occupy less than 3 per cent of the available land area of the planet, urbanization and changes in land cover have significantly modified urban microclimates. Cities' higher density can bring efficiency gains, but there is interplay between the increased risk of the urban heat island effect and these higher densities that needs to be understood. The UHI effect and heat trapped in heat-storing building materials affect energy consumption for cooling, leading to extensive use of air-conditioners, which again increases greenhouse gas emissions.

In Chapter 13, Gerald Mills provides an overview of the UHI, its interactions with global climate change and how low carbon cities could affect it. In Chapter 14, David J. Sailor reviews how best to mitigate the UHI when taking a holistic approach to urban sustainability. In Chapter 15, Matheos Santamouris and Dionysia-Denia Kolokotsa explain the latest developments in cool materials and techniques to deal with a hotter built environment. In Chapter 16, Yukiko Yoshida and Toshiaki Ichinose demonstrate that urban planning, mindful of the need to maintain air flow and natural ventilation, can pre-empt the UHI. In Chapter 17, Steve Kardinal Jusuf, Nyuk Hien Wong and Zhun Min Adrian Chong explore how urban planning and building design can work in tropical urban microclimates to ameliorate the UHI. In Chapter 18, Lutz Katzschner and Sabrina Campe describe mapping urban climates as a way to collate information planners can use to mitigate urban heat, and show the utility of this tool by outlining the results of projects completed in Germany, the Netherlands and Vietnam. In Chapter 19, Boon Lay Ong assesses the Green Plot Ratio concept in relation to urban greenery and biodiversity. In Chapter 20, Nigel Tapper, Andrew Coutts, Margaret Loughnan and Devna Pankhania show how policy developments, technology and water-sensitive urban design can help protect urban populations during heat waves. In Chapter 21, Ehsan Sharifi, Conrad Philipp and Steffen Lehmann report on investigations of the surface-layer UHI in Sydney to determine what combinations of streets, density and greenery increase urban precincts' resilience to heat.

In lieu of a traditional conclusion, in Chapter 22 Steffen Lehmann and Katharine Thornton 'look ahead' by arguing that low carbon principles will help cities in the Asia-Pacific region make the transition towards green urbanism.

Chapter 13

The urban heat island and low carbon cities

Gerald Mills

Summary

Cities occupy less than 3 per cent of the available land area of the planet but modify climate from local to global scales. At a global scale, the contribution of cities to the concentration of greenhouse gases is significant as they are focused areas of human occupation, energy consumption and carbon emissions. At a local scale, cities modify nearly every aspect of climate, including airflow, radiation exchange, air quality and temperature. The urban effect on temperature creates a phenomenon known as the urban heat island (UHI) and has been identified in surface, sub-surface and air temperatures. Each type is measured as a difference between the temperatures inside and outside the urbanized area, where the latter is considered to represent the climate that would exist in the absence of the city. The UHI is a ubiquitous feature of urbanization but its magnitude depends on both the character of the city (its physical form and function) and of the background climate. This chapter surveys work on the nature of the UHI, its temporal and geographic variations, and its impact on human health. It will consider how the UHI could modify as global climate changes (altering the background climate) and consider how plans for low carbon cities, designed to modify the character of the city and mitigate the global impact of cites, could affect its magnitude.

Introduction

Cities occupy a very small proportion of the Earth's land area (less than 3 per cent) but they are home to the majority of the world's population. The proportion of the population that live in cities will continue to increase, even as the population living outside urban areas declines. Most of this growth will occur in the economically developing, tropical areas of the world, where the urban infrastructure (dwellings, offices, roads, etc.) has yet to be built. By 2050, it is projected that the global population will have increased to 9.3 billion and the urban population will be 6.3 billion (UN 2012). Thus, the greater proportion of humanity in the near future will live in places where the local environment has been profoundly modified by the cities we have already built or will build.

Many attributes of cities that affect climate are illustrated in Figures 13.1a and 13.1b, which show two views of Chicago, USA. The city centre is characterized by the absence of vegetation and the dominant presence of hard, impermeable surfaces composed of manufactured materials. Where the buildings are tall and closely spaced, access to the sky is greatly restricted for the building and street surfaces and airflow retarded (although more turbulent) at street level. Outside of this central business district, the landscape changes; buildings are

Figures 13.1a and 13.1b Two views of Chicago (41.9° N, 87.6° W) showing the densely built downtown (upper) and a sparsely built green suburb (lower). Each urban landscape has a distinctly different effect on the temperatures of the surface and overlying air.

Source: Jupiterimages at photos.com, supplied by the author.

much shorter, their plan area is larger and roads are wider. Here, the landscape is relatively smooth (from an aerodynamic point of view). On the other hand, suburban Chicago illustrates an entirely different urban landscape, one that is green, almost rural. Buildings are relatively small and widely spaced but what dominates the scene is the proportion of the landscape that is vegetated; the trees are mature and are taller than most buildings. What these images imply but do not show is the intensity of human activities across the city; the transport system suggests traffic that injects waste heat, gases and materials into the near-surface atmosphere, and each of the buildings requires energy for heating, cooling, lighting, manufacturing that is eventually lost to the outdoors.

In this chapter, urbanization refers to the replacement of natural land cover by manufactured materials (form) and the concentration of human activities (function) in a small area. The result is to cause an urban climate effect at a hierarchy of scales that represents a modification of the 'background' climate; that is, the climate that would exist at a place in the absence of the city (Mills 2007). The magnitude and type of this modification depends on both the background climate and the form and functions of the city. Form refers to the fabric of the city (including the proportion of the urban footprint), which usually consists of materials that form pavements, roads, car-parks and buildings. This fabric is arranged into facets (walls, roof, etc.), each of which has distinct surface properties creating a complex assemblage of micro-scale environments, such as well-vegetated suburban tracts or densely built city centres. Function refers to the flows of energy, water and materials that sustain the city. This metabolism results in stores (urban growth or redevelopment) and wastes (depending on the time scale) that are deposited into water, land or air. The best known of the urban effects on climate is the urban heat island (UHI). This chapter will discuss the urban heat island (UHI) and its management in the context of low carbon cities.

The role of cities in global climate change

The majority of the energy used to support the functions of the city is derived from carbon-based fuels. When ignited, these generate a great number of wastes, including carbon dioxide and a host of other gases (sulphur dioxide (SO_2), nitrogen oxides (NO_x)) and particulate matter (PM). The intense use of, and demand for, energy within urban areas means that they are a major source of global emissions for these materials. Many of these wastes will become concentrated in the near-surface air and, if they exceed thresholds, constitute a major public health risk. Carbon dioxide (CO_2) is not a direct threat to human health in normal circumstances and, until recently, has not been monitored at this scale. However, as a long-lived greenhouse gas (GHG), it accumulates in the lower troposphere and is the main driver of anthropogenic climate change.

The role of cities in driving anthropogenic climate change is still a topic of debate as, until recently, there has been no protocol for gathering information on cities in a consistent manner. As a result, the GHG emission values attributed to cities vary considerably depending on whether, for example, emissions are assigned to the place of production or the place of consumption (Satterthwaite 2008). Moreover, individual city economies give rise to different per capita emissions based on their energy base and reliance on particular industries. Kennedy and colleagues (Kennedy *et al.* 2009) assembled comparable city-level data for ten cities (most located in the mid-latitudes) and estimated per capita equivalent CO_2 emissions of between 4.2 (for Barcelona) to 21.5 tonnes (Denver) based on end-use activities. Overall, it was found that GHG emissions from heating and industrial fuel use were correlated with climate, being higher where heating needs were greater. Moreover, emissions from ground transportation fuels were inversely related to population density. This research complements work on sustainable and low carbon cities, which argues that higher-density, mixed-use cities are inherently more energy efficient (Kenworthy 2006; UN-Habitat 2011).

The projections of climate change are based on emission scenarios and global climate models, which can inform us of potential changes to the regional scale, background and climates within which the urban effect occurs. The projected changes will have substantial impacts on cities both directly (e.g. sea level rise) and indirectly (e.g. altered food supply), particularly as, globally, cities tend to be situated in low-elevation, coastal zones. For our purposes here, the most relevant projections are those on temperature rises that indicate the following (IPCC 2007):

- Air temperature will rise over most land areas, resulting in fewer cold days and more hot days. This is expected to result in: reduced human mortality due to hypothermia; reduced energy demand for heating but increased demand for cooling; worsening air quality in cities; and reduced disruption to transport due to snow and ice.
- Warm spells and heatwaves will increase in frequency resulting in increased risk of heat-related mortality, especially for vulnerable populations, notably the old, sick and very young.

These changes have obvious implications for cities where the urban heat island is likely to add to the expected global warming (e.g. McCarthy *et al.* 2010).

The urban climate effect

To understand the effect that urban areas have on climate, it is important to appreciate the structure of the urban atmosphere and the role of scale. The urban boundary layer (UBL) is formed as regional-scale airflow crosses the upwind edge of the city and begins to respond. The rapid change in the surface roughness as the air encounters buildings results in turbulence that mixes the unique properties of the urban surface into the overlying air. The UBL grows in depth with distance from the edge, growing to a depth of 1–2 km over a large city by the afternoon (Figure 13.2). The properties of the UBL record the presence of the city below and its character; it is generally warmer, more turbid, enriched by a suite of pollutants (including CO_2) and more turbulent. This envelope of air drifts downwind beyond the city as an elevated plume. Its contents become progressively diluted with distance from the urban source; however, in certain meteorological situations, the plume can retain its distinct character hundreds of kilometres downwind. The UBL can be divided into a mixed layer (ML) and a surface layer (SL). The latter consists of the lowest 10 per cent of the UBL; the SL extends to several times the heights of buildings and is closely attached to the complex three-dimensional urban surface, which is multi-faceted. The top of the SL may be regarded as an active 'surface' for the ML above, where the effects of the city are mixed with the wider atmosphere.

The physical urban surface describes the interface between the solid substrate and the atmosphere (Figure 13.2). This surface has many facets, each distinguished by its slope, orientation and material composition. Many of these facets comprise the walls and roofs of buildings that are hollow. Other facets (car parks, streets, etc.) are at ground level and cover a soil substrate. The urban canopy layer (UCL) refers to the layer of atmosphere below roof level and includes both the indoor and outdoor airspaces, which are connected by exchanges

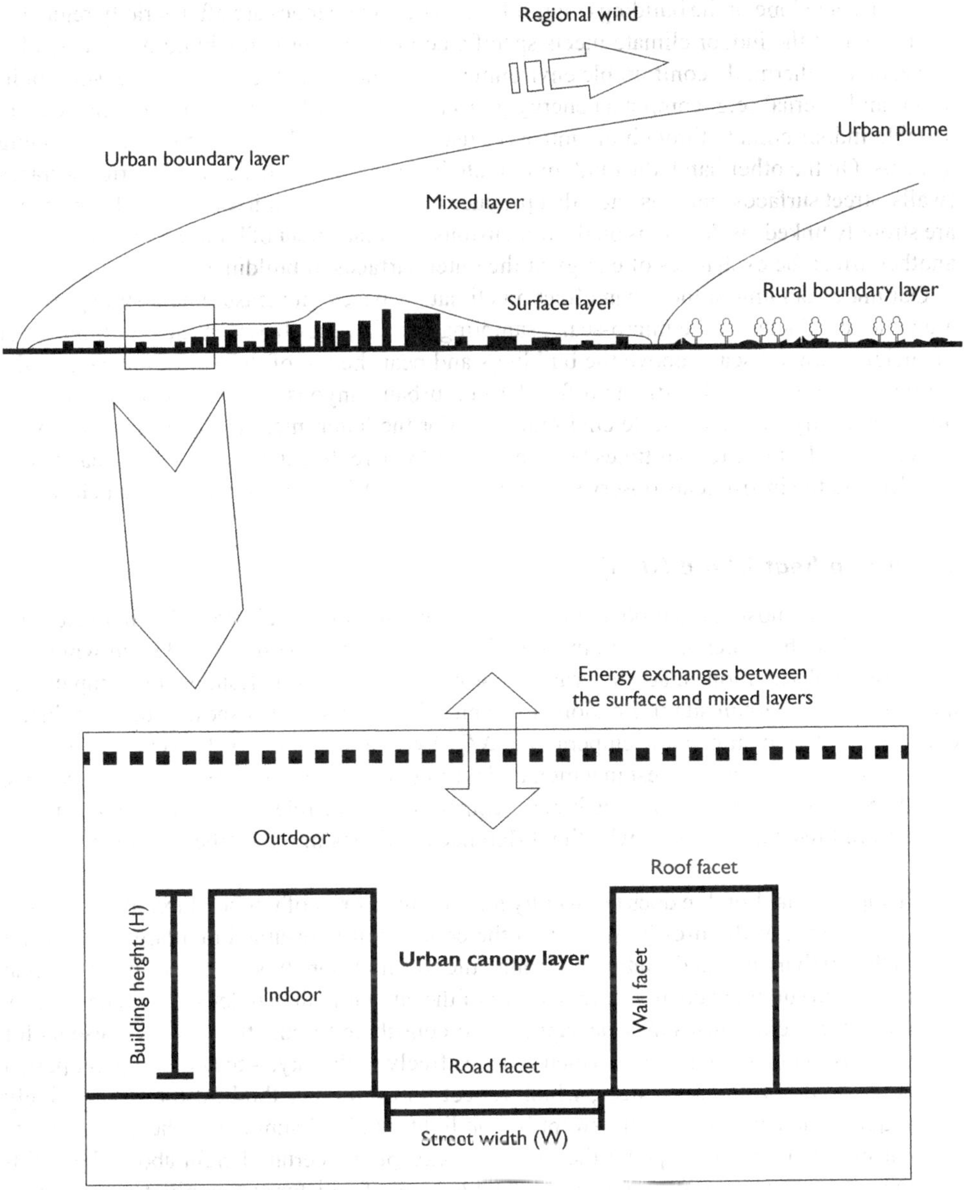

Figure 13.2 The vertical structure of the urban atmosphere. The upper diagram shows the layering of the urban boundary layer into mixed and surface layers above an urban surface. The lower diagram represents part of the urban canopy that consists of facets that are configured to create a street. Within the street, microclimatic variations affect the near-surface air. Above the street, exchanges with the overlying air transfer the effects of the urban canopy layer into the boundary layer (and vice versa).

across the envelope of the building (Figure 13.2). These exchanges are often strictly regulated to ensure that the indoor climate meets specific conditions. For many buildings, the goal is to establish a thermally comfortable environment by controlling the external (e.g. solar radiation) and internal (e.g. computer) energy gains and losses. Many modern buildings ensure a stable indoor climate through the intensive use of energy in heating and air-conditioning systems. On the other hand, the outdoor climate is a result of exchanges at exterior surfaces (walls, street surfaces, gardens, etc.) that produce myriad distinct microclimates. Both spaces are strongly linked, as decisions on the dimensions and placement of buildings relative to one another affect the exchanges of energy at the outer surfaces of buildings.

Our understanding of the urban effect on climate is based on measurement strategies that are either conducted at the micro-scale, sampling the airspaces between the buildings, or at the neighbourhood scale, above the buildings and near the top of the SL. In the case of the former, scientists use city streets (often termed urban canyons) as fundamental structural units that typify the micro-scale environments. For the latter, measurements are made well above the UCL (three to four times building height) where the micro-scale effects have been blended and the instruments observe an integrated signal from the urban surface below.

The urban heat island (UHI)

The UHI is the most studied phenomenon of all urban climatic effects. The first scientific report on this phenomenon occurs in *The Climate of London* (Howard 1833), in which air temperature observations made in a rural (pastoral) setting outside London are compared to those made at an urban site in London city centre. Howard's results showed that the difference in monthly averaged air temperature (ΔT_{u-r}) was about 1°C for the year and that the excess temperature was greatest in winter and during the night but was present in all seasons. He attributes the chief cause for the higher temperature to the release of heat into the atmosphere from fires, but questions why the difference continues into the other seasons:

> It appears that London does not wholly lose its superiority of temperature, by the extinction of most of the fires in Spring: on the contrary, it is resumed in a large proportion in the Sixth month, and continues through the warm season. It is probable, therefore, that the Sun in summer actually warms the air of the city more than it does that of the country around. Several causes may be supposed to contribute to this: the country presents for the most part a plain surface, which radiates freely to the sky, – the city, in great part, a collection of vertical surfaces, which reflect on each other the heat they respectively acquire: the country is freely swept by the light winds of summer, – the city, from its construction, greatly impedes their passage, except at a certain height above the buildings: the country has an almost inexhaustible store of moisture to supply its evaporation – that of the city is very speedily exhausted, even after heavy rain. When we consider that radiation to the sky, the contact of fresh breezes, and evaporation, are the three principal impediments to the daily accumulation of heat at the surface, we shall perceive that a city like London ought to be more heated by the summer sun than the country around it.
>
> (Howard 1833, pp. 9–10)

Howard's analysis is quite sophisticated and identifies anthropogenic heat as the main cause of the UHI, followed by radiation exchanges within city streets, the slowing of wind that inhibits heat loss and the absence of vegetation that depresses evaporation. Much of the research since Howard has followed his methodological path by using fixed stations to represent urban (u) and non-urban (r) and describing the UHI in terms of ΔT_{u-r} (Mills 2008). The use of mobile platforms (e.g. cars and bikes) for temperature observations allowed a more spatial examination of the phenomenon (e.g. Sundborg 1950) and revealed the shape of the UHI; the outer boundary of the UHI corresponded to the edge of the urban footprint and it increased in magnitude toward the city centre (Figure 13.3). Where water (rivers, lakes, etc.) and vegetated areas are present, they appeared as relatively cool places within the UHI.

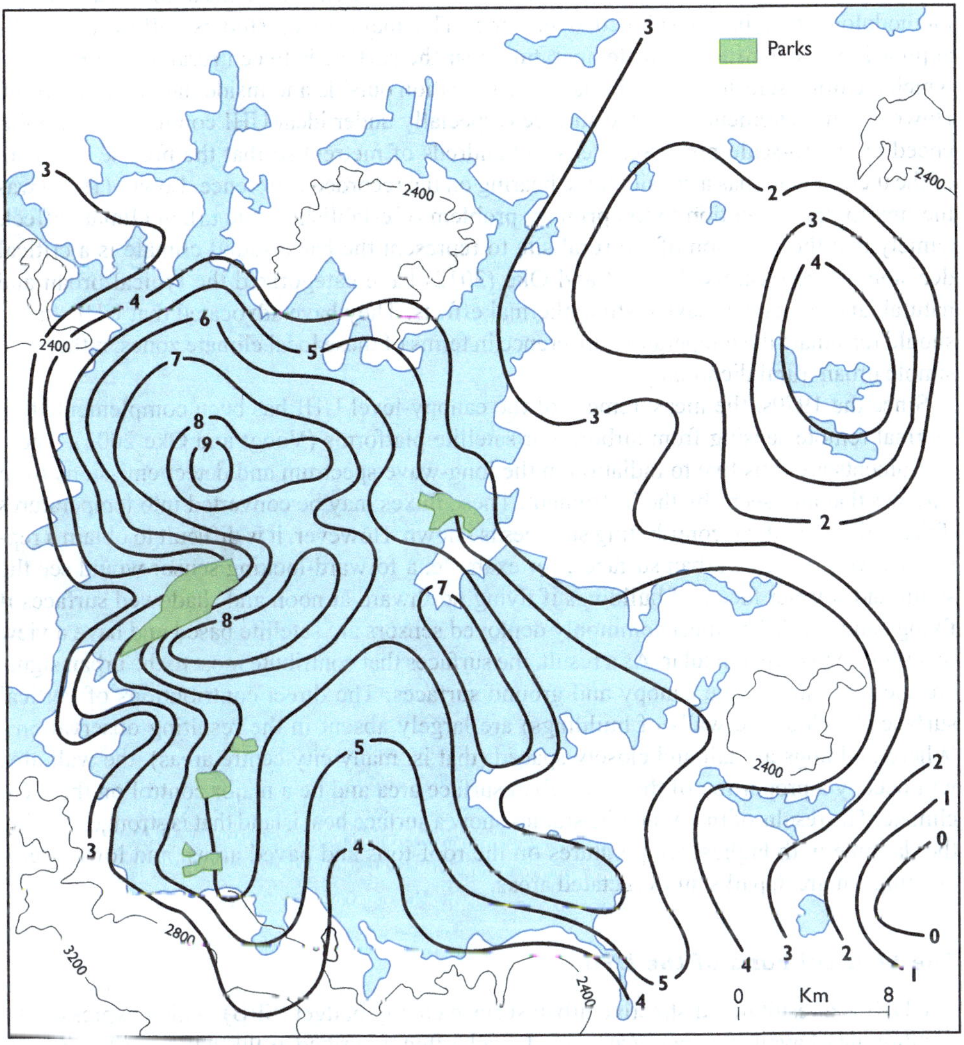

Figure 13.3 Mean minimum temperature for November 1981 in Mexico City. The isotherms are in degrees Celsius and the light grey indicates the area of the city. The dark grey areas are parks.

Source: Redrawn from Jauregui (1986).

Overall, observational studies show that the magnitude of the UHI varies with the weather, being strongest at night under clear skies and calm conditions. When ΔT_{u-r} is plotted over the course of a day, the difference is smallest (and may even be negative) in the period after sunrise when shadowing in city streets postpones the local sunrise. The difference is generally small but positive during the daytime but increases dramatically after sunset. While the rural environment cools exponentially, the urban environment cools linearly so that about four hours after sunset ΔT_{u-r} is at a maximum (Oke 1987). Observations of the canopy-level UHI continue and have expanded the geographical coverage while confirming this basic pattern (Roth 2007).

Despite the longevity of UHI study, there have been few examinations of the observational methodology that has developed over time. The majority of studies follow Howard's approach of comparing the air temperature near the surface between rural and urban sites, which are often selected on the basis of their location outside and inside the urban footprint. However, measurements near the surface (especially under ideal UHI conditions) are influenced by micro-scale processes (tens to hundreds of metres) so that the precise exposure of the thermometer has a considerable bearing on the recorded difference. Lowry (1977) was the first to draw attention to the primary problem of establishing the urban climate effect, namely that the selection of the rural site to represent the background climate is a critical decision. More recently, Stewart and Oke (2012) have categorized the typical urban and natural landscapes that have distinct thermal effects. They have advocated that UHI studies should formulate the temperature difference in terms of these local climate zones, rather than simple urban–rural dichotomy.

Since the 1970s, the measurement of the canopy-level UHI has been complemented by thermal remote sensing from airborne or satellite platforms (Voogt and Oke 2003). These instruments are sensitive to radiation in the long-wave spectrum and detect emissions from surfaces that are 'seen' by the instrument. These fluxes may be converted into temperatures if the emissivity of the contributing surfaces is known. However, it is difficult to obtain a representative sample of urban surfaces; for example, a forward-looking sensor would see the sunlit and warmer facets of buildings if flying northward at noon and shadowed surfaces if flying southward. The most commonly deployed sensors are satellite based and have a view that is close to perpendicular. As a result, the surfaces that contribute most to the urban signal are the roof, street, tree canopy and ground surfaces. The direct contributions of vertical surfaces (such as the walls of buildings) are largely absent in the resulting observations. Where buildings are tall and closely spaced (that is, many city centre areas), the wall area will occupy a proportion of the total urban surface area and be a major control on the UCL climate. The results of these satellite studies show a surface heat island that is strongest during the daytime with highest temperatures on the roof-tops and paved areas, and lowest temperatures in green parks and vegetated areas.

The physical basis of the UHI

The UHI is a result of a distinctive urban surface energy budget (SEB), which expresses the partitioning of available energy at ground level, often described as the active surface, as it is the interface between the atmosphere and the substrate (Oke 1982):

$$Q^* = Q_H + Q_E + Q_G \qquad \text{W m}^{-2} \ (1)$$

where Q* is net radiation, Q_H and Q_E are the turbulent sensible and latent heat exchanges with the atmosphere and Q_G is sensible heat exchange with the substrate by conduction.

Net radiation is the sum of incoming (↓) and outgoing (↑) solar (or short-wave) and terrestrial (or long-wave) radiation:

$$Q^{*}= K\downarrow - K\uparrow + L\downarrow - L\uparrow \tag{2}$$

The short-wave radiation receipt at a surface (K↓) comprises a direct (*S*) and diffuse (*D*) component. The magnitude of *S* depends on the angle that the solar beam strikes a surface and is at a maximum when the beam is perpendicular. For a flat surface with no obstructions, the altitude of the sun above the horizon governs this angle; hence, all other things being equal, maximum solar receipt will occur at noon. The urban surface is multi-faceted, where each facet (such as a wall) has a unique aspect and orientation and may experience local 'sunsets' and 'sunrises' during the day as it moves in and out of shadow. The diffuse radiation receipt arises from two sources: the sky vault and the surrounding built environment. The former consists of solar radiation that is scattered downward and the latter from radiation that is reflected. The contribution of each to *D* depends on the magnitudes of the source and the proportion intercepted by the surface of interest. As a first approximation, the sky view factor (ψ_{sky}) is an accurate measure of the relative roles of each source. For a point on an unobstructed flat surface, the hemispheric dome that represents its view is occupied entirely by sky and ψ_{sky} equals unity. On the other hand, for a street surface separating two buildings at the base of a narrow canyon (such as Chicago's city centre, Figure 13.1a), ψ_{sky} is small and the great proportion of the dome overhead is occupied by building walls.

Reflected solar radiation ($K\uparrow$) is a function of both $K\downarrow$ and surface albedo (α), which is a measure of the reflectivity of the surface to solar radiation. By and large, untreated urban materials have a low albedo; asphalt, for example, has an α value of between 0.05 (fresh) and 0.27 (weathered), compared to 0.26 for short grass or greater than 0.5 for snow. As a result, urban facets are generally efficient solar collectors. Moreover, within the UCL the facets reflect solar radiation to each other in a multiple reflection process, partly compensating for the shade. As a consequence, the albedo of a street system (walls and street) is lower than might be expected from the albedos of the constituent facets. Roofs, at the top of the UCL, are largely unaffected by this process.

The long-wave emission depends on the temperature (*T*) of the source and its emissivity (ε). At a solid surface, $L\uparrow$ is largely a function of its surface temperature (T_a) as the emissivity of most natural and manufactured materials is close to unity (>0.90). On the other hand, the average ε value for the atmopshere is about 0.83 but this varies considerably, being higher when the atmosphere is more humid or where droplets are present. The effect of the UHI is to increase the temperature of the overlying atmosphere, thus increasing $L\downarrow$; however, as the urban surface is also warmed, $L\uparrow$ is also increased. Within the UCL, a substantial proportion of $L\downarrow$ at a facet comes from the surrounding buildings and ψ_{sky} is a good measure of the relative contributions of the atmosphere and the urban environment.

The magnitude of the non-radiative terms is regulated by a gradient of a relevant property and an exchange term. For the substrate heat flux,

$$Q_G = -k\,\frac{\Delta T}{\Delta z} \tag{3}$$

Where the exchange is driven by a temperature difference (ΔT) over a depth of soil (Δz), the heat capacity of the soil and the magnitude of the flux for a given gradient is regulated by molecular conductivity (k). The substrate temperature profile adjusts to the flux at a rate that depends on the magnitude of the flux and the heat capacity (C_S). Urban materials generally have high k and C_S values, which means that they can act as good stores of heat. Asphalt, for example, has typical conductivity 0.75 $Wm^{-1}K^{-1}$ and a heat capacity of 1.94 MJ m^{-3} K^{-1}. Their chief attribute, however, is their constancy; whereas the thermal properties of soils change dramatically with water content, urban materials are usually impervious and dry quickly. For comparative purposes, the equivalent values for clay soil (with 40 per cent porosity) are 0.25 and 1.42 when dry and 1.58 and 3.10 when saturated (Oke 1987). As the UHI is measured as a rural–urban difference, this is especially relevant. The magnitude of the UHI is greatest after a dry period when the rural soils are dry and more thermally responsive to energy exchanges owing to their low values of k and C_S. In other words, the magnitude of UHI is related to the rate of cooling in the rural area, rather than any change in the urban area.

Atmospheric fluxes are dominated by convection (mixing of fluid parcels or eddies) driven by wind (forced) and/or thermal (free) effects. Here we can simply describe them using an analogy to conduction and introducing an eddy diffusivity term (K):

$$Q_H = -\rho C_p K_H \frac{\Delta T}{\Delta z} \tag{4}$$

and

$$Q_E = -L\rho K_V \frac{\Delta q}{\Delta z} \tag{5}$$

These state that the turbulent transfers of sensible and latent heat are functions of the surface–air gradients of temperature (ΔT) and specific humidity (Δq), respectively, and the relevant diffusivity term. The other symbols represent the density (ρ) and the heat capacity (C_p) of air and the latent heat of vaporization (L). The urban effect is to reduce average wind-speed in the UCL and impede convective transfers. The higher surface temperature increases $\Delta T/\Delta z$, but the lack of water at typical urban facets means that $\Delta q/\Delta z$ is small. As a result, available energy is directed to heating the atmosphere. More generally in the urban landscape, the presence of vegetation is an important regulator of Q_E as it can draw water from the substrate and release this at the leaf surface. The ratio of Q_E to Q_H is the Bowen ratio (β), which is an important measure of the capacity of a surface to heat or add water vapour to the overlying air. As cities are generally dry compared to their surrounding landscapes, their β value is lower.

The energy balance terms for each urban facet (Q^*, Q_G, Q_H and Q_E) will exhibit different magnitudes and temporal patterns based on its properties (e.g. slope and orientation, material composition, albedo, emissivity, vegetative cover) and those of both the substrate and adjacent air. These interactions give rise to distinct microclimates in the diverse urban environment that are difficult to generalize. To cope with this, researchers have focused on common urban configurations, such as street canyons, to study the UCL. Air temperature within this layer is the result of exchanges at the wall and street surfaces, energy inputs directly into the atmosphere from vehicle exhausts and transfers across the tops and at the ends of streets.

A more comprehensive assessment of the urban climate effect can be achieved by measuring well above the UCL where atmospheric mixing has blended the effects of the micro-scale processes at the urban surface (Oke 2006); that is, at the approximate level of the dashed line in Figure 13.2). The energy budget of the layer of air below this level may be described by

$$Q^* + Q_F = Q_H + Q_E + \Delta Q_S \qquad (6)$$

The heat storage term (ΔQ_S) accounts for energy changes in the layer extending from the level of measurement to the depth in the substrate affected by the energy fluxes. Here, we have neglected advection as we assume that the landscape is sufficiently homogeneous and extensive. Measurements over different urban landscapes (such as those in Figure 13.1) show that the partitioning of available energy ($Q^* + Q_F$) into sensible and latent heat is substantially dependent on the presence of vegetation. In other words, the effect of vegetation is to moderate the heating of the overlying urban atmosphere by directing available energy into Q_E.

The anthropogenic heat flux (Q_F) is not distinguishable in these measurements, as it is present in all the terms; it includes the accumulated heat added by the metabolism of humans, the waste heat and vapour arising from the heating and cooling of buildings, energy generation, and vehicles. This term is usually estimated from inventories of energy use and traffic counts. At an urban scale, Q_F is estimated at between 30 and 50 Wm^{-2} annually but varies seasonally. As much of this energy flux is generated using fossil fuels, it is not surprising that the temporal fluxes of anthropogenic heat (Q_F) and of CO_2 match closely in most cities.

The UHI is a consequence of changes in the surface energy balance and of the energy budget of the overlying atmosphere. However, it should be apparent that there can be several types of UHIs based on whether we are focused on sub-surface, surface or air temperatures, and how and where we observe the urban effect. For example, satellites observe the temperatures of roof-tops, street surfaces and car parks but not those of walls. They are of limited value for evaluating the surface UHI in dense urban canopies where the wall facets are significant elements in creating the UCL climate. On the other hand, air temperature measurements within the UCL respond to the micro-scale exchanges at surfaces below roof level. This form of UHI has been studied exhaustively; it is strongest at night under clear skies and calm conditions when radiation exchanges dominate the surface energy balance. A useful empirical summary of the evidence from cities around the world is as follows (Oke 1981):

$$\Delta T_{u-r(max)} = 7.45 + 3.97 \ln\left(\frac{H}{W}\right) = 15.27 - 13.88\psi_{sky} \qquad (7)$$

This states that the maximum canopy-level UHI ($\Delta T_{u-r(max)}$) that one could expect under ideal conditions may be estimated from the height-to-width ratio (H/W) in the most densely built-up part of the city. This simple relationship implicitly accounts for the changes in materials and surface cover that accompany urbanization. In other words, where built density is greatest, the streets are often narrowest (compared to building heights) and there is relatively little vegetation.

UHI impacts

The UHI is a ubiquitous feature of urbanization but whether it is a beneficial feature or not depends on the background climate. For example, in cooler climates where the need for wintertime heating dominates, it had been regarded as one of the few favourable climatic outcomes of urbanization. Thus, Chandler wrote of London's climate:

> A few of the changes . . . might be considered favourable. Such are the higher autumn, winter and spring night-time temperatures which reduce heating costs and lengthen the frost-free period, but these advantages must surely be outweighed by increased pollution and decreased sunshine.
>
> (Chandler 1965)

However, this view has modified over the last 40 years for a number of reasons. First, the functions of buildings have changed in many ways. In many cities, the commercial functions of buildings rely on electrical appliances such as computers that generate a considerable amount of energy, yet must operate within a managed temperature range. As a result, internal energy gains can be considerable, especially as these gains are greatest during daytime hours when the external gain is also highest. Second, many buildings attempt to ensure ideal (narrowly defined) 'comfort' conditions for occupants, regardless of outdoor conditions. Third, global climate change is expected to raise the average temperature of the background climate. Consequently, the emphasis has shifted from heating to cooling concerns, even in cool climates. An assessment of London's future climate (published in 2002) predicted that in future:

> warming will be most noticeable during summer nights, so there will be less relief from higher day-time temperatures. . . . Increased temperatures could also reduce comfort of occupants in domestic, commercial and public buildings and lead to business disruption. Much of London's building stock will still be in place in 50 years' time and will have to cope with more very hot summer days.
>
> (LCCP 2002, n.p.)

One of the chief concerns about the UHI is the extent to which it results in thermal discomfort for urban citizens. The energy budget of the human body is similar to that presented earlier:

$$Q^* + Q_M = Q_H + Q_E + \Delta Q_S \tag{8}$$

The new term here is the metabolic heat flux (Q_M), which is generated internally and is always positive; the basal rate is about 70 Wm^{-2} and it increases with activity levels. As warm-blooded creatures, the body seeks to ensure a constant internal temperature of 37°C (so that $\Delta Q_S \approx 0$) by modifying each of the energy budget terms. For example, the biophysical response to increased heat load is to raise skin temperature and increase sweating, thereby enhancing Q_H and Q_E, respectively. Thermal discomfort occurs when the body is under strain as it copes with environmental stress. In very warm circumstances, when the strain is too great, the body will experience hyperthermia.

The urban effect is to enhance this stress by raising the air and surface temperatures that the body is exposed to and reducing convective heat and evaporative losses by slowing airflow. These changes are of particular concern during sustained periods of unusually high temperatures or heatwaves. The definition of a heatwave depends on the average climate to which the population is normally exposed and to which it has become acclimatized. There is evidence that for a number of reasons (including socio-demographic factors) the public health effects of heatwaves are exacerbated in cities (Dousset *et al.* 2011).

Mitigating the UHI

Opportunities to mitigate (or enhance) the UHI are best examined by managing the form (geometry, material composition and surface cover) and function (metabolism) of the city in response to its background climate (Figures 13.4a and 13.4b).

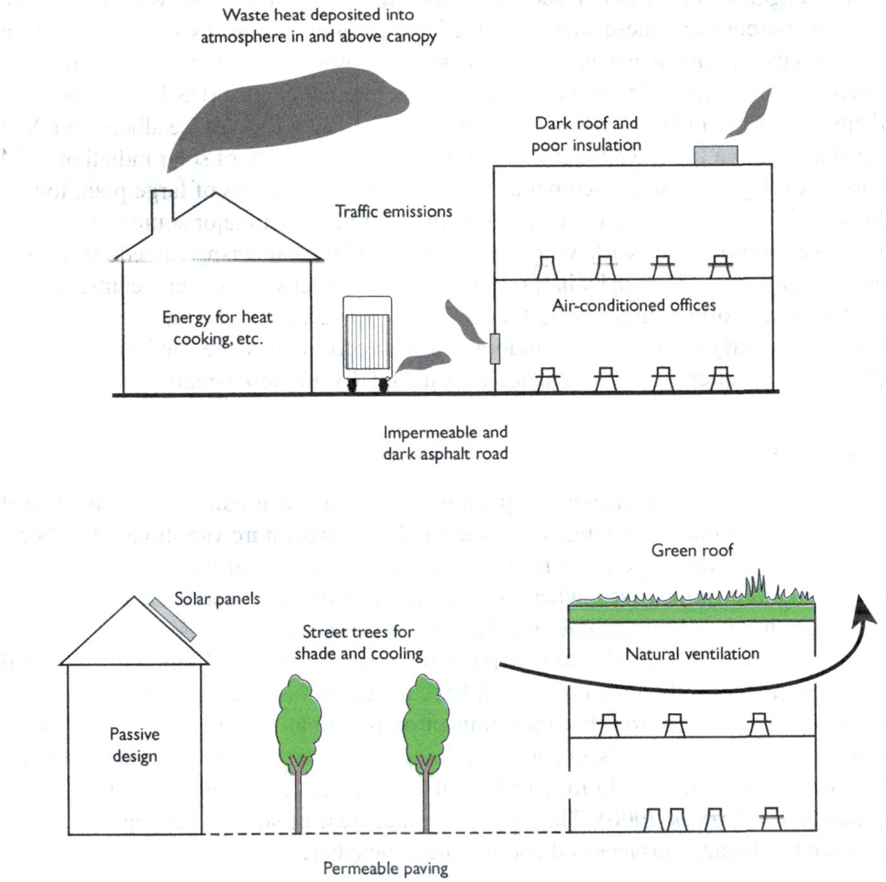

Figures 13.4a and 13.4b The properties of urban areas that can enhance (above) or ameliorate (below) the urban effect on temperature. Narrow streets that restrict night-time cooling, dark surfaces that absorb solar radiation, lack of water and vegetation to evaporate and anthropogenic heating all contribute to the formation of the surface and air heat island. On the other hand, building design that incorporates passive and active energy systems (such as natural ventilation) that limit conventional energy use, vegetation and trees that shade surfaces and evaporate water, and facets that reflect solar radiation and retain water minimizes the heat island.

Geometry

Altering the layout of a city by decreasing built density (and reducing the H/W ratio) will increase ψ_{sky} for the facets with the urban canopy layer and enhance long-wave radiation loss, which is one of the main reasons for the night-time canopy UHI. However, decisions on urban layout are best taken prior to building as, once a building has been finished, comprehensive change is difficult.

Materials

Changing the material composition of urban facets can also modulate surface temperatures. Cities have been built of hard and impermeable materials that were selected for their durability and strength. Precipitation on such surfaces is quickly removed and very little water is stored. As a consequence, these surfaces have little capacity to cool by evaporation. Using porous materials for pavements, car parks and roads allows some water to be retained. This has benefits for hydrological management as well as thermal control (US EPA 2008).

Perhaps the easiest and most effective intervention is to alter the surface albedo (α). Many urban surfaces have a low α so that they absorb the great majority of solar radiation, which is the driver of daytime surface temperatures. The flat roof surfaces of large plan, low-rise buildings are ideally suited to such changes as the roof facet is a major source of heat gain. Replacing a conventional clay tile with a white-coloured tile can change albedo from 0.2 to 0.7, thus reducing the solar gain by half (US EPA 2008). In densely built city centre sites, the total roof area can form a large proportion of the urban surface at the top of the UCL and affect both the overlying air and the indoor space directly below the roof. However, it is unlikely to directly alter the outdoor space within the UCL in these situations.

Surface cover

From a UHI perspective, increasing the green proportion of the urban area boosts Q_E at the expense of Q_H and reduces the range of surface and air temperature variations. In suburban areas of some cities, the large proportion of green area means that the urban temperature effect is weak and in some cases, where gardens are watered during a drought period, may be cooler than the surrounding rural area. In densely built environments, grassy parks will mitigate UHI in their vicinity (behaving like rural areas) but have a limited impact on the surrounding area. Similarly, green roofs will have the greatest impact on the temperature of the indoor space below the roof but their impact on the urban canopy layer will be small, especially where the green roof is atop a tall building. Trees can provide an exceptional range of services (e.g. air quality, hydrology and biodiversity) and can be integrated into existing urban landscapes (Nowak 2006). The judicious placement of street trees can improve the UCL climate by shading surfaces and cooling the atmosphere.

Metabolism

Regulating Q_F will moderate the UHI but it will not offset it. For example, note that the canopy-level UHI is strongest at night in densely built areas of cities where there are often few occupants and little traffic. On the other hand, managing the UHI during warm periods will reduce the need for Q_F from air-conditioning systems. (Re)introducing natural ventilation for buildings is one means of reducing reliance on cooling technologies; however, this must be done carefully in dense urban environments where pollutants from roadways could be drawn into indoor spaces. The main reason to moderate Q_F is to improve air quality, which is an additional environmental stress.

Of course, the best UHI solution is to address all of these issues simultaneously through a strategy that manages the urban climate in the context of its background climate. The best example of this is in the city of Stuttgart, Germany where the climate issues (including air quality, the UHI and climate change) have been integrated into the planning process for more than 50 years (Hebbert and Webb 2012). An important tool in this process is climate mapping of the city and the surrounding area to identify natural airflow pathways that may be employed to ventilate the urban area. As a result, many of the climate problems associated with other cities have been avoided. On the other hand, addressing climate problems after they arise can be difficult. In the case of Hong Kong, which has a warm and humid tropical climate, the outdoor climate at street level has been worsened by the nature of development; very tall buildings and a high built density raise the air temperature but compound the stress by impeding the progression of the afternoon sea breeze, which offers thermal relief by bringing cooler air onshore. The only viable strategy for improving the climate is to take advantage of urban redevelopment to create airflow pathways into the city (Ng 2009).

Conclusions

The urban heat island is just one of the climate changes caused by cities and it may usefully be used as a diagnostic indicator of the degree of modification caused by urbanization as it is correlated with the degree of urban surface cover and building density. In conclusion, let me make some general points regarding strategies that attempt to mitigate the UHI.

First, the background climate is relevant; in some climates the UHI may not be significant or may actually be beneficial. In cold climates, for example, it should reduce the energy used for heating and aid in the melting of snow. However, the warm tropical climates where much of future city growth is to occur, the UHI will add to heat stress. Building cities that limit the UHI in these climates, rather than mitigating after the fact, is best. For many cities in the mid-latitudes, the need for winter heating and summer cooling is nearly equally balanced, although the latter is growing as cooling technology becomes more widely available and expectations for indoor climate change. This growth is likely to continue under global climate change scenarios. Many of these cities are already built and the task is to alter aspects of the urban form and function to mitigate the UHI during the summer season. Ideally, the urban fabric could be designed to respond to the seasons, gathering heat during winter and disposing of it during summer.

Second, any policy designed to mitigate the UHI must take into account the different types of UHI (surface or air, UCL or above) to ensure that strategies address the relevant UHI. Greening roof-tops in dense urban environments will serve many purposes but is unlikely to affect the canopy-layer climate in any significant manner. The best policies for managing the UHI will utilize a range of strategies that include greening the city and modifying materials and surface covers, and will be cognizant of the scales of their effect (street, neighbourhood and city).

Third, good urban design is essential if policies designed to mitigate the UHI are to complement those for low carbon, sustainable cities. For example, a city with a low building density and plentiful green space (typical of suburban tracts in many Western cities) will have a weak UHI. However, if the urban layout separates residences from workplaces, then it is likely that private cars will be the preferred means of transport and the carbon cost will be high. On the other hand, compact, dense cities may be more carbon efficient but if this is realized as closely spaced, tall buildings with little vegetation (Figure 13.1a), then the UHI will be enhanced and more energy will be spent trying to create comfortable indoor conditions.

References

Chandler, T.J. 1965, *The Climate of London*, Hutchinson, London, 1965, www.urban.climate.org, viewed 29 August 2013.

Dousset, B., Gourmelon, F., Laaidi, K., Zeghnoun, A., Giraudet, E., Bretin, P. and Vandentorren, S. 2011, 'Satellite monitoring of summer heat waves in the Paris metropolitan area', *International Journal of Climatology*, vol. 31, no. 2, pp. 313–323.

Hebbert, M. and Webb, B. 2012, 'Towards a liveable urban climate: lessons from Stuttgart', in C. Gossop and S. Nan (eds), *Liveable Cities: Urbanising world: ISOCARP review 07*, Routledge, London, pp. 132–150.

Howard, L. 1833, *The Climate of London*, www.urban.climate.org, viewed 29 August 2013.

Intergovernmental Panel on Climate Change (IPCC) 2007, *Climate Change 2007: Synthesis report. Contribution of working groups I, II and III to the fourth assessment report of the Intergovernmental Panel on Climate Change* [Core Writing Team, R.K. Pachauri and A. Reisinger (eds)]. IPCC, Geneva, Switzerland.

Jauregui, E. 1986, 'The urban climate of Mexico city', in *Urban Climatology and its Applications with Special Regard to Tropical Areas*, WMO 652.

Kennedy, C., Steinberger, J., Gasson, B., Hansen, Y., Hillman, T., Havranek, M., Pataki, D., Phdungsilp, A., Ramaswami, A. and Mendez, G.V. 2009, 'Greenhouse gas emissions from global cities', *Environmental Science and Technology*, vol. 43, pp. 7297–7302.

Kenworthy, J.R. 2006, 'The eco-city: ten key transport and planning dimensions for sustainable city development', *Environment and Urbanisation*, vol. 18, pp. 67–85.

Keogh, S., Mills, G. and Fealy, R. 2012, 'The energy budget of the urban surface: two locations in Dublin', *Irish Geography*, vol. 45, pp. 1–23.

LCCP 2002, *London's Warming: The impacts of climate change on London*, summary report, http://climatelondon.org.uk/publications/londons-warming/.

Lowry, W.P. 1977, 'Empirical estimation of urban effects on climate: a problem analysis', *Journal of Applied Meteorology*, vol. 16, pp. 129–135.

McCarthy, M.P., Best, M.J. and Betts, R.A. 2010, 'Climate change in cities due to global warming and urban effects', *Geophysical Research Letters*, vol. 37, no. 9, pp. 1–5.

Mills, G. 2007, 'Cities as agents of global change', *International Journal of Climatology*, vol. 27, pp. 1849–1857.

—— 2008, 'Luke Howard and *The Climate of London*', *Weather*, vol. 63, pp. 153–157.

Ng, E. 2009, 'Policies and technical guidelines for urban planning of high-density cities: air ventilation assessment (AVA) of Hong Kong', *Building and Environment*, vol. 44, pp. 1478–1488.

Nowak, D.J. 2006, 'Institutionalizing urban forestry as a "biotechnology" to improve environmental quality', *Urban Forestry and Urban Greening*, vol. 5, pp. 93–100.

Oke, T.R. 1981, 'Canyon geometry and the nocturnal urban heat island: comparison of scale model and field observations', *International Journal of Climatology*, vol. 1, pp. 237–254.

—— 1982, 'The energetic basis of the urban heat island', *Quarterly Journal of the Royal Meteorological Society*, vol. 108, pp. 1–24.

—— 1987, *Boundary Layer Climates*, 2nd edition, Methuen, London.

—— 2006, *Initial Guidance to Obtain Representative Meteorological Observations at Urban Sites*, WMO, Instruments and Observing Methods No. 81, WMO/TD-No. 1250.

Roth, M. 2007, 'Review of urban climate research in (sub)tropical regions', *International Journal of Climatology*, vol. 27, pp. 1859–1873.

Satterthwaite, D. 2008, 'Cities' contribution to global warming: notes on the allocation of greenhouse gas emissions', *Environment and Urbanisation*, vol. 20, pp. 539–549.

Stewart, I.D. and Oke, T.R. 2012, 'Local climate zones for urban temperature studies', *Bulletin of the American Meteorological Society*, vol. 93, pp. 1879–1900.

Sundborg, A. 1950, 'Local climatological studies of the temperature conditions in an urban area', *Tellus*, vol. 2, pp. 222–232.

UN-Habitat 2011, *Cities and Climate Change: Global report on human settlements, 2011*, Earthscan/UN-Habitat, London.

United Nations (UN) 2012, *World Urbanisation Prospects, the 2011 Revision – Highlights*, United Nations, Department of Economic and Social Affairs, Population Division, New York.

United States Environment Protections Agency (US EPA) 2008, *Reducing Urban Heat Islands: Compendium of strategies*, www.epa.gov/heatisld/resources/compendium.htm, viewed 12 November 2013.

Voogt, J.A. and Oke, T.R. 2003, 'Thermal remote sensing of urban climates', *Remote Sensing of Environment*, vol. 86, pp. 370–384.

Chapter 14

A holistic view of the effects of urban heat island mitigation

David J. Sailor

Summary

Cities tend to be warmer than their unbuilt surroundings. This effect, commonly referred to as the urban heat island (UHI), has consequences for energy use, air quality and human health. As a result, municipalities are increasingly deploying strategies to reduce the undesirable consequences associated with the UHI. These measures typically focus on reducing summertime urban air temperatures by increasing the solar reflectivity of urban surfaces and facilitating evaporative cooling by reducing impervious surface cover and adding urban vegetation. Other sustainability efforts that reduce energy use in cities also affect the urban climate by reducing waste heat and pollutant emissions. While the focus of mitigation strategies tends to be on summer air temperatures, it is important to recognize that there are other factors affecting the well-being of urban populations. These include the effects of urbanization on humidity and atmospheric mixing and the seasonal variations in the effects of UHI mitigation efforts. Thus, successful sustainability efforts must look beyond summertime air temperatures. This chapter provides a summary of the urban climate system. It discusses common urban climate mitigation strategies as well as their intended and unintended consequences. The chapter concludes with a call for a holistic evaluation of all urban sustainability strategies, recognizing that there are many different and sometimes competing goals related to urban sustainability.

Urban climate and the urban heat island phenomenon

Urbanization fundamentally alters the natural surface of the Earth and the corresponding surface heat, mass and momentum exchange processes. These effects are manifested locally, but they also have consequences throughout the urban climate system. These consequences include alterations in surface temperatures, air temperatures, humidity, mixing depths, pollutant concentrations, wind-flow patterns and precipitation. One of the most notable atmospheric effects of urbanization is the tendency of urban areas to be warmer than their surroundings. This phenomenon, referred to as the urban heat island (UHI), has many adverse consequences, particularly in summer. As a result, many municipalities have begun to pursue efforts to reduce UHI magnitudes. For such mitigation efforts to be successful, however, it is important to (1) understand the factors that contribute to this warming; and (2) obtain a full appreciation for the consequences of mitigation strategies, both intended and unintended. This chapter seeks to address these two challenges.

Overview of the urban energy balance

As illustrated in Figure 14.1, the urban climate system is affected by the complex interactions of surface characteristics and human activities. Urban and natural surfaces differ in a number of important respects. First, urban surfaces – walls, roofs and roads – tend to have lower reflectivity to solar radiation (low albedo) as compared with natural surfaces. For example, asphalt roads and asphalt shingle roofs may have albedos as low as 5 to 10 per cent (Bretz *et al.* 1998). In contrast, natural surfaces often have a surface albedo in the 15 to 30 per cent range (e.g. Strugnell *et al.* 2001). The complex morphology of the buildings in a city also results in an increase in net absorption of solar radiation resulting from multiple reflections within urban canyons. The net effect is that the albedo of the city changes with solar angle, and is also lower than the simple average of the albedos of individual surfaces in the urban environment (Sailor and Fan 2002).

Long-wave (LW) radiation exchange (e.g. thermal radiation) is also an important aspect of the urban energy balance. This thermal radiation is the energy emitted by all surfaces as a function of their temperature and their relative ability to emit thermal radiation (emissivity). The ability of surfaces to radiate to the cold night sky enables them to cool down substantially at night – sometimes even sub-cooling below the local air temperature. In an urban canyon setting, thermal radiation emitted from surfaces is partially intercepted by other surfaces. As a result, thermal radiation tends to remain trapped in urban canyons, keeping them warmer at night, thus contributing to the general observation that the UHI magnitude tends to be largest at night and early in the morning hours (Hidalgo *et al.* 2008; Oke 1982).

Urban areas are also characterized by the presence of a large amount of thermally massive materials (e.g. concrete, asphalt and brick) that store substantial amounts of energy during the day, releasing much of this energy at night through both sensible heat and thermal radiation.

Building windows add an interesting complexity to the surface radiation balance, as they typically transmit 50 to 80 per cent of the incoming solar radiation directly into the building – removing it from the surface energy balance and reintroducing it as waste heat as air-conditioning systems remove it and other thermal gains from conditioned interiors of

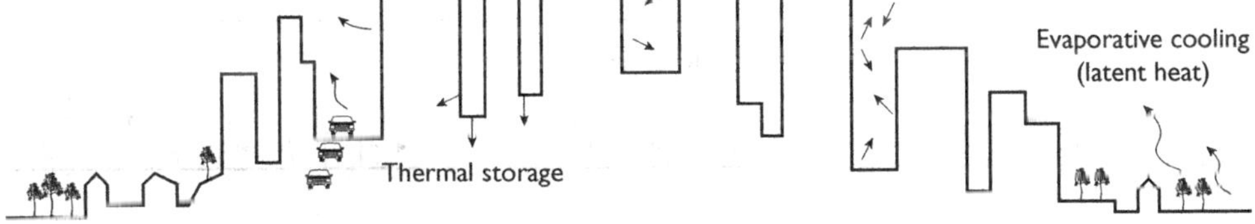

Figure 14.1 Schematic representation of the key drivers of the urban energy balance.

Source: Supplied by the author.

buildings. In addition to this source of waste heat (also referred to as anthropogenic heat), vehicles driven on city streets release large amounts of heat and moisture into the atmosphere.

A final important contributor to the urban energy balance is the significant amount of impervious surface (e.g. concrete and asphalt) combined with the general deficit of vegetative cover. The result is that urban areas have much less capacity for evaporative cooling than do natural surfaces.

These elements of the urban energy balance result in a number of key urban climate attributes. While much attention is focused on the urban heat island (warming) effect, this surface energy balance also contributes to alterations in atmospheric mixing, humidity and pollutant levels, and the development and evolution of convective systems (Grady-Dixon and Mote 2003; Lin *et al.* 2011).

Defining the urban heat island

The concept of the urban heat island has been around for many years. Some of the earliest measurements of the phenomenon date back to Luke Howard's observations of the climate of London (Howard 1833) and Emilien Renou's observations of the climate of Paris (Renou 1855). The concept was only revived as an important research topic more than a century later (Landsberg 1981; Oke 1982, 1987). The generally accepted definition of the urban heat island

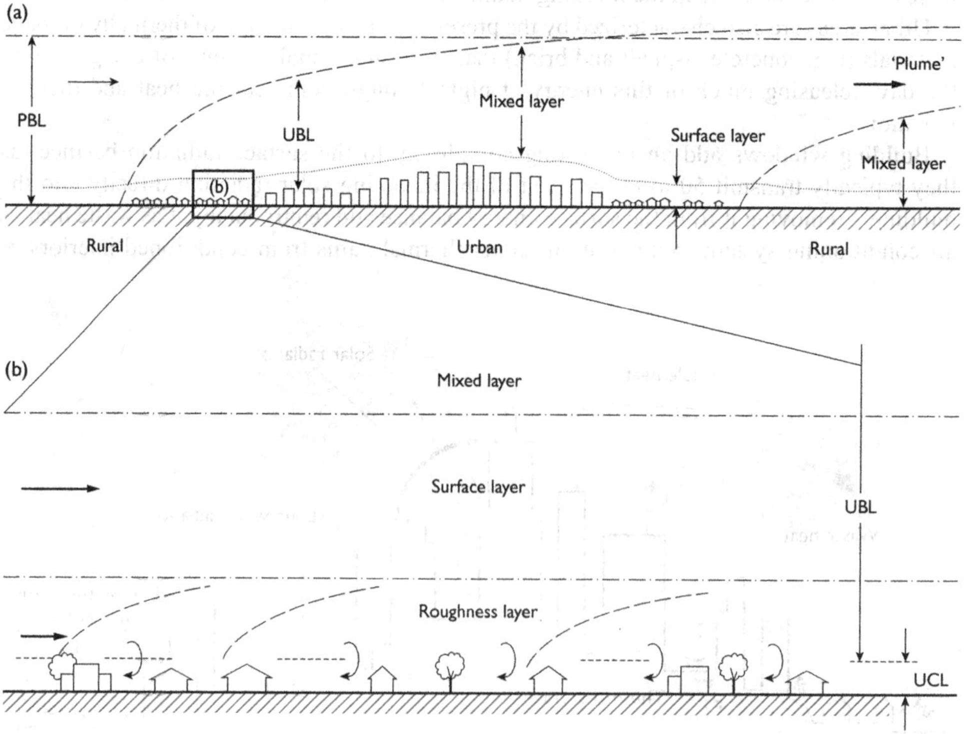

Figure 14.2a and 14.2b Boundary layer structures in the urban climate system (after Oke 1984).
Source: Supplied by the author.

is that it is 'an area of higher temperatures in an urban setting compared to the temperatures of the suburban and rural surroundings' (Geer 1996, p. 272). At first glance, such a simple definition is both appealing and useful. Upon further reflection, however, it immediately becomes apparent that the urban heat island phenomenon is too complex to be captured by such a simple definition. For example, in comparing temperatures in urban and rural settings, one must first decide upon which temperature to use. As pointed out by Oke (1984), there are several important climatic scales and vertical layers within the urban climate system, each of which have distinct drivers and characteristics (see Figures 14.2a and 14.2b). Hence, one could consider defining the urban heat island based on several distinct layers: the surface (skin) temperature; the air temperature within the urban canopy layer (UCL) between buildings; or the air temperature in the roughness or surface layers above the urban canopy. Once the vertical scale of the temperature variable has been selected it is then important to consider the choice of urban and rural locations. Historically, this has often been done in an ad-hoc manner, either taking the locations corresponding to the largest temperature differences, or somewhat arbitrarily defining 'urban' and 'rural' locations. In fact, the general lack of scientific rigour and the inconsistency of locations and temperature variables used in UHI studies has recently been highlighted in a thorough review of the state of the research (Stewart 2011). Finally, it is important to recognize that the urban heat island effect has a distinct seasonal and diurnal variability (Oke 1982, 1987). The heat island in winter and in the early morning hours tends to be the largest. In cooling-dominated climates, however, the heat island effect that is of most interest is the summer daytime heat island. Thus, while it is common to refer to 'the' urban heat island, it is clear that there are, in fact, many possible definitions for the urban heat island, and users of the heat island concept must be careful to provide sufficient clarity when discussing heat island magnitudes.

Goals of UHI mitigation

Despite the wide range of potential definitions for UHI, it is, nevertheless, clear that the elevation of air temperatures in cities leads to undesirable consequences. As a result, city and regional governments have been developing and implementing strategies for cooling cities. These efforts are often referred to as urban heat island mitigation strategies. The goals of UHI mitigation are usually indirect; that is, most policies seek to reduce the adverse consequences of urban climates. These consequences include heat-related morbidity and mortality; poor air quality; and energy use for cooling buildings. Since each of these consequences is highly correlated with urban air temperature, strategies typically focus on reducing average or peak summertime air temperatures. Of course, each of these consequences is also tightly coupled with other atmospheric parameters, including humidity levels and atmospheric mixing. As discussed below, any mitigation strategy that targets summer daytime air temperature is also likely to have impacts on other atmospheric parameters, and to have important implications for urban climates throughout the year.

Mitigation strategies overview

Based on the energy balance introduced in Figure 14.1, it is clear that there are several strategies that hold promise for cooling cities. These strategies are generally based on one of the following categories of approaches: (1) increase the reflectivity of the urban surface to solar radiation; (2) increase the evaporative cooling at the surface; (3) modify the structure and thermal mass of the city; and (4) reduce the amount of artificially generated heat released into the urban environment.

High-albedo surfaces (street level, roof level)

In the simplest representation, urban areas consist of buildings (rooftops and walls), paved surfaces (streets, parking lots and sidewalks), and natural unbuilt areas (soil, rock, grass and other vegetation). The albedo of natural surfaces varies, but it is often in the range of 0.15 to 0.30 for soil and various types of vegetation. The albedo of the man-made surfaces in the urban environment spans a much higher range. For example, new asphalt pavement or dark asphalt roofing shingles will typically have albedo values as low as 0.05 to 0.08. White roof membranes, on the other hand, may have albedo values as high as 0.75 (sometimes higher when new and/or clean). Concrete and various highly reflective asphalt products may have albedo values as high as 0.35. The wide range of albedo values associated with the variety of available building materials creates a unique opportunity for architects, builders and developers to address the UHI. The magnitude of this opportunity, however, depends both on the fraction of the urban surface covered by each type of surface and by the current albedo of existing surfaces. As an example, it has been reported that on an area average basis Sacramento, California is roughly 20 per cent dark roofing and 10 per cent dark pavement (Bretz *et al.* 1998). In general, the albedo of a dark roof can be increased from about 0.05 to 0.65 or higher and the albedo of a dark pavement can be increased from 0.05 to 0.35. Hence, if every dark surface in Sacramento were replaced by a highly reflective surface, the overall albedo of the city could be increased by 0.15. More aggressive estimates (included in Bretz *et al.* 1998) suggest a potential to increase city-wide albedo by up to 0.18. There have been a number of atmospheric modelling studies which suggest that increasing city-wide albedo by this much can result in reductions of peak summer air temperatures by up to a 3° C (Taha 2008a; Taha *et al.* 1999).

It is important to recognize, however, that the albedo of any artificial surface will change over time. Specifically, very dark surfaces (such as asphalt) tend to lighten over time, in part due to wear that reveals more aggregate. Likewise, dark membrane roofs may have initial albedo values of only 0.05, but as they age, through both UV exposure and dust accumulation, their albedo values may increase to 0.10 or higher. Very high-albedo materials have the reverse trend. A high-albedo roof may initially have an albedo of 0.8, but within a few months of soiling will typically have an albedo closer to 0.6 or 0.7. Some of this albedo degradation is permanent, and not recovered with routine cleaning (Bretz and Akbari 1997).

Another important consideration when evaluating prospects of UHI mitigation via high-albedo surfaces is that of implementation. While the potential to increase urban albedo by up to 0.20 may exist, in practice street paving projects take time, funds and long-term coordination. In addition, a significant amount of exposed pavement is in privately owned parking lots and driveways. Thus, it may take many years to replace a significant fraction of low reflectivity pavement with a high-albedo alternative. Similarly, building roofs typically have lifespans of 20 to 25 years. Thus, while high-albedo alternatives to dark roofs are readily available, significant penetration of the roofing market with high-albedo roofing may take a decade.

Vegetation and moisture strategies

While high-albedo strategies target the short-wave radiation aspect of the urban energy balance, vegetation strategies target the evaporative cooling component. Specifically, urbanization tends to replace natural materials with impervious surfaces such as concrete, asphalt and building materials. The result is that precipitation that might normally be absorbed into the natural surface is largely channelled into drainage systems and removed from the urban environment. The general lack of surface moisture and vegetation in cities leads to much lower levels of evaporation and transpiration – processes that would otherwise cool the area through evaporation.

Mitigation strategies to increase evaporative cooling in cities can do so by replacing impervious surfaces, by increasing transpiration from vegetation (e.g. tree-planting programmes), or both (e.g. the introduction of bioswales). While vegetation-focused UHI mitigation has traditionally focused on street-level tree-planting programmes, the growing interest in green (vegetated) roofs and walls is introducing new approaches.

Again, the potential magnitude of UHI mitigation through evaporative cooling depends to a large extent upon the existing urban surface characteristics; the current fraction of the urban surface that is impervious, and the density of the urban vegetation canopy. There have been a number of studies that have explored the effects of urban street trees (Sailor 1998) or green walls or roofs (Alexandri and Jones 2008) on urban air temperatures and cooling demand. Such studies generally find that urban vegetation and moisture strategies have a similar urban cooling potential to albedo strategies.

Urban morphology and thermal mass considerations

The materials used to construct buildings can have an important effect on the urban climate system. Specifically, thermally massive materials such as brick and concrete will store thermal energy during the day and release it into the urban atmosphere at night. Lighter-weight construction materials, such as metal cladding over insulation, has the potential to reduce UHI magnitudes particularly at night, as such materials will cool off more quickly. Similarly, the geometry of the city affects both long- and short-wave radiation exchange and ventilation patterns (Luo *et al.* 2011; Wong *et al.* 2010; Yang *et al.* 2011). Deep urban canyons can essentially trap heat as the surfaces within the canyon are unable to radiate directly to the night sky. Urban areas with a diversity of building heights may facilitate vertical mixing processes, potentially helping to alleviate the urban heat island.

Relatively few studies have investigated the role of urban morphology in affecting the UHI. One notable example is the work of Emmanuel and Fernando (2007), which used a canyon-scale computational model to explore how canyon geometry and surface albedo affect near-surface thermal conditions for both humid and arid climates. Specifically, they found that enhancing urban density was a useful strategy for improving thermal comfort as measured by mean radiant temperature. This study was also unique in that it pointed out the importance of designing UHI mitigation efforts that focus more holistically on human thermal comfort rather than simply on reducing air temperatures. More recent empirical studies have also looked at the role of urban morphology (among other factors) in affecting the magnitude of the UHI (Yang *et al.* 2011).

UHI mitigation through waste heat reduction

A final strategy for UHI mitigation addresses the undesirable release of waste heat (also referred to as anthropogenic heating) in cities. This anthropogenic heat is the by-product of energy consumption in buildings, and by industrial processes and fossil fuel-powered vehicles. Depending on the level of development and the population density of the city, anthropogenic heat emissions can be quite large and some studies suggest that they are responsible for a significant fraction of the heat island effect in large cities (Fan and Sailor 2005; Hamilton *et al.* 2009). Averaged over the entire city, anthropogenic heating is typically small, perhaps in the order of 10 to 50 W/m^2 (Pigeon *et al.* 2007; Sailor *et al.* 2003). However, in particularly dense urban neighbourhoods, the local magnitude of anthropogenic heating may be as high as 1,000 or 1,500 W/m^2 (Heiple and Sailor 2008; Ichinose *et al.* 1999).

There are several plausible methods for UHI mitigation through the reduction of waste heat. The transport sector is responsible for up to one-third of all energy consumed in cities. Improving fuel economy and reducing single-passenger vehicle travel through mass transport and carpooling are strategies that are already being pursued for other sustainability goals. Improving overall thermal efficiency and moving emissions out of the city through the deployment of electrical vehicle fleets is another promising strategy. Some cities have significant industrial activity within their city boundaries. This contributes to waste heat emissions that could potentially be relocated to industrial zones removed from the urban core. Finally, improving energy efficiency in commercial and residential buildings holds considerable promise for further reducing waste heat emissions. Given the possible combinations of these strategies, it would seem reasonable to expect that waste heat emissions in cities could be reduced by 20 to 40 per cent through targeted strategies.

Effects of mitigation strategies

In considering the benefits of UHI mitigation strategies, it is important to take a big picture approach. First of all, most UHI mitigation strategies have effects on many aspects of the urban environment and on sustainable development more generally. While many of these effects are beneficial, some are undesirable. Furthermore, there is a strong tendency to focus on the summertime effects of UHI mitigation. In fact, the UHI effect itself is generally strongest in winter, and strategies that reduce the UHI in summer may have substantial year-round effects that are too important to be ignored.

Building energy use for HVAC

The building sector is responsible for approximately 40 per cent of all energy use in most developed countries. In the US, this is split roughly evenly between residential and commercial buildings (EIA 2010). The fraction of this energy used to meet heating vs. cooling loads varies widely from one city to the next, dictated primarily by local climate. At the national level, however, approximately 20 per cent of end-use energy in the US is for space heating and 18 per cent is for space cooling. It is well established that energy use for either heating or cooling is a strong function of ambient air temperatures (Day and Karayiannis 1999; Lu *et al.* 2009). Thus, UHI mitigation is likely to reduce energy use for cooling in summer, but increase energy use for heating in winter. The relative importance of these counteracting effects depends upon the local climate and the relative size of the heating and cooling demand in any city. In cooling-dominated climates, the reduced air-conditioning energy consumption, combined with the associated reduction in peak electricity demand, may far outweigh the costs associated with increased heating demand in winter. Nevertheless, it is important to consider all aspects of the effects of UHI mitigation on energy use.

Urban air quality

Air pollution, and photochemical smog (tropospheric ozone) in particular, is highly sensitive to ambient air temperature. This sensitivity arises from several factors. First, both biogenic and anthropogenic emissions of precursor pollutants tend to increase with rising ambient temperature. Furthermore, many of the atmospheric reactions that play a role in the formation and destruction of ozone depend on ambient temperature. As a result, independent of the role of sunlight, there is a strong, positive correlation between ozone concentrations and ambient air temperatures. In general, any cooling of the urban heat island would then be expected to reduce concentrations of photochemical smog. There are a few complicating factors, however. First, cooling of the urban atmosphere also results in a suppression of the mixing height, resulting in a smaller volume of air into which surface emissions are mixed – and a corresponding mechanism for increases in precursor concentrations. In addition, it is well known that peak ozone concentrations occur downwind of the peak emissions. In a cooler urban climate, it is therefore reasonable to expect that one effect on pollution may simply be to translate peak ozone concentrations further downwind of the sources – potentially to neighbouring communities (Taha 2008a, 2008b).

Heat-related morbidity and mortality

Heatwaves are associated with increases in heat-related health problems and deaths (e.g. Whitman *et al.* 1997). As a result, there is an interest in the potential role that UHI mitigation may be able to play in alleviating this problem. Heat-related mortality is a strong function of atmospheric air temperatures and humidity. In an extreme heat event, the human body is stressed by daytime peak air temperatures combined with high levels of humidity. During an extreme-heat event that lasts for multiple days, the risk of heat-related mortality increases if the body is not able to cool off at night. Thus, the potential for UHI mitigation to benefit heat-related health depends upon three factors: the daytime cooling; the nocturnal cooling; and any associated changes in humidity. As noted in the above discussion of air quality, UHI mitigation has the effect of also lowering mixing heights and increasing urban canopy-layer concentrations of pollutants and moisture emitted from the surface. In fact, prior studies have illustrated the potential trade-off between the cooling effects of UHI mitigation and the resulting increase in near-surface humidity. For example, Sailor and colleagues (2002) found that while albedo and vegetation mitigation strategies could effectively reduce urban air temperatures, they also had the undesirable effect of reducing urban mixing heights and increasing near-surface humidity levels. The net effect on heat-related health varied depending upon the heatwave under study.

Stormwater management

While the primary goal of UHI mitigation is the reduction of air temperatures, several of the strategies can have significant effects on stormwater. Specifically, any strategy that reduces flow into storm drains either by increasing the fractional coverage of pervious surfaces, or by increasing the evapotranspiration of water from the surface, can result in stormwater management benefits. This issue is particularly important in municipalities that have combined stormwater-sewage systems. During extreme storm events, such systems can become overloaded, resulting in overflows into urban streams or rivers. Aside from reducing the risk of overflows, any decrease in stormwater entering urban water treatment systems also has a direct energy savings benefit associated with the avoided costs of treating that water. Numerous studies have demonstrated the potential stormwater benefits of what is commonly referred to as 'green infrastructure'. In fact, in the case of vegetated roofing, it has been found that even a 10 cm-thick layer of soil on a roof can reduce run-off from that roof by 50 per cent on an annual basis (DeNardo *et al.* 2003; Mentens *et al.* 2005; Van Seters *et al.* 2007).

Public vs. private benefits

Urban heat island mitigation, like efforts to mitigate global climate change, is generally seen as a public benefit. Urban cooling contributions resulting from the actions of one resident or landowner will be small with relatively small private benefits. When mitigation strategies are applied across the city, however, the aggregate of the effect may be a significant public benefit. As a result, it can be challenging to implement UHI mitigation strategies. Fortunately, there are intersections of public and private benefits that can help facilitate the widespread application of UHI mitigation as a combination of individual actions and municipal strategies. For example, installing highly reflective roofing in cooling-dominated climates has both a direct benefit to the building owner and indirect benefits to society. The indirect benefits are those associated with cooling the urban climate, which in turn provides benefits ranging from urban thermal comfort to air quality and city-scale energy consumption. The direct benefit of a roof-top albedo strategy for the building owner is a reduction in roof-top solar heat gain, which can reduce costs for cooling buildings.

Maintenance and cost considerations

A detailed analysis of the economics of mitigation strategies is beyond the scope of this chapter. Nevertheless, an overview of key cost and maintenance issues is necessary to fully appreciate the complexities of urban heat island mitigation efforts. Any implementation strategy will have associated with it a combination of first costs and recurring maintenance costs. A full life-cycle cost–benefit analysis must weigh these costs against public and private benefits, which are often difficult to monetize.

The initial cost of implementing strategies varies widely. Implementing a high-albedo mitigation strategy would be costly if it required the immediate replacement of dark roofs regardless of their age. However, if the strategy targeted roofs at the time of routine replacement (assuming a typical life of 20 years), the incremental first costs of requiring high-albedo roofing may be minimal or zero. By contrast, the initial costs of vegetated roofing may be rather high, particularly if structural retrofits are required to manage the increased roof load. If no structural retrofit is required, a green roof may be installed for an incremental cost of perhaps US$6 per square foot (~US$64/m^2) (Niu *et al.* 2010). Likewise, street-level tree-planting programmes typically have a high first cost, due to labour and materials.

Once implemented, mitigation strategies will accrue benefits at differing rates. While a high-albedo roofing strategy will reach maximum benefit immediately after implementation, this benefit will decrease over time due to ageing and soiling of the roof surface. On the other hand, a vegetation strategy will have relatively little benefit initially until the plantings become established and grow to maturity. For street tree-planting strategies, it may take a decade for the full benefits of the strategy to be realized.

Maintenance costs for mitigation strategies are also quite variable. A high-albedo roof may require very little or no maintenance relative to its dark roof alternative. In fact, due to less thermal cycling, the high-albedo roof may actually require fewer roof repairs, and will generally have a longer life. Vegetated roofs generally require some irrigation during their establishment phase (one to two growing seasons). Depending upon their design and use, vegetated roofs may also require periodic additional maintenance such as weeding, replanting and the application of nutrients. UHI mitigation via street trees is perhaps the most maintenance-intensive strategy. Some level of irrigation may be required during the first several growing seasons and periodic maintenance related to pest and disease management may also be necessary. Some trees may not survive initial transplanting and will require removal and replacement. In addition, as trees mature, there is often a need to maintain their crowns to keep streets and sidewalks clear for cars/trucks and pedestrians, and to keep branches out of power lines. Depending upon the urban infrastructure and tree selection, urban street tree root systems can also contribute to maintenance issues ranging from footpath and street repairs to damaging subsurface natural gas pipelines.

Moving forward: recommendations

The single most important recommendation that this author would offer to planners interested in UHI mitigation is to take a long-term view of any proposed strategy. Consider not only first costs, but also ongoing maintenance costs and longevity. Recognize that any particular strategy will have both direct and indirect effects, and, further, that some of these effects are likely to be unanticipated side-effects – some of which may not be beneficial. Conversely, in many cases there may be co-benefits in addition to the desired outcomes, which, if ignored, may drive policy-makers away from the implementation of a potentially successful strategy.

It is also important to recognize that the performance of strategies will vary widely from one climate to another. A successful strategy in Phoenix, Arizona may not work very well in Dhaka, Bangladesh and vice versa. Finally, it is recommended that policy-makers consider a portfolio of strategies to simultaneously accomplish multiple sustainability goals. Policy preference for one strategy over another (e.g. Toronto's vegetated roofing mandate, or California's Title 24 high-albedo roofing mandate) may not lead to the most beneficial outcomes.

References

Alexandri, E. and Jones, P. 2008, 'Temperature decreases in an urban canyon due to green walls and green roofs in diverse climates', *Building and Environment*, vol. 43, no. 4, pp. 480–493.

Bretz, S.E. and Akbari, H. 1997, 'Long-term performance of high-albedo roof coatings', *Energy and Buildings*, vol. 25, no. 2, pp. 159–167.

Bretz, S., Akbari, H. and Rosenfeld, A. 1998, 'Practical issues for using solar-reflective materials to mitigate urban heat islands', *Atmospheric Environment*, vol. 32, no. 1, pp. 95–101.

Day, A.R. and Karayiannis, T.G. 1999, 'New degree-day model for estimating energy demand in buildings', *Building Services Engineering Research and Technology*, vol. 20, no. 4, pp. 173–178.

DeNardo, J.C., Jarrett, A.R., Manbeck, H.B., Beattie, D.J. and Berghage, R.D. 2003, 'Stormwater detention and retention abilities of green roofs', *World Water and Environmental Resources*

Congress 2003, 23–26 June, American Society of Civil Engineers, Philadelphia, PA, pp. 1639–1645.

Energy Information Administration (EIA) 2010, *Annual Energy Outlook 2010: With Projections to 2035*, Energy Information Administration, U.S. Department of Energy, Washington, DC. DOE/EIA-0384 (2010).

Emmanuel, R. and Fernando, H.J.S. 2007, 'Urban heat islands in humid and arid climates: role of urban form and thermal properties in Colombo, Sri Lanka and Phoenix, USA', *Climate Research*, vol. 34, pp. 241–251.

Fan, H. and Sailor, D.J. 2005, 'Modeling the impacts of anthropogenic heating on the urban climate of Philadelphia: a comparison of implementations in two PBL schemes', *Atmospheric Environment*, vol. 39, no. 1, pp. 73–84.

Geer, I. 1996, *Glossary of Weather and Climate, with Related Oceanic and Hydrologic Terms*, American Meteorological Society, Boston, MA, p. 272.

Grady-Dixon, P. and Mote, T.L. 2003, 'Patterns and causes of Atlanta's urban heat island-initiated precipitation', *Journal of Applied Meteorology*, vol. 42, no. 9, pp. 1273–1284.

Hamilton, I.G., Davies, M., Steadman, P., Stone, A., Ridley, I. and Evans, S. 2009, 'The significance of the anthropogenic heat emissions of London's buildings: a comparison against captured short-wave solar radiation', *Building and Environment*, vol. 44, no. 4, pp. 807–817.

Heiple, S. and Sailor, D.J. 2008, 'Using building energy simulation and geospatial modeling techniques to determine high resolution building sector energy consumption profiles', *Energy and Buildings*, vol. 40, no. 8, pp. 1426–1436.

Hidalgo, J., Pigeon, G. and Masson, V. 2008, 'Urban-breeze circulation during the CAPITOUL experiment: observational data analysis approach', *Meteorology and Atmospheric Physics*, vol. 102, nos 3–4, pp. 223–241.

Howard, L. 1833, *The Climate of London, deduced from meteorological observations, made at different places in the neighbourhood of the metropolis*, vol. 1, Harvey and Darton, London.

Ichinose, T., Shimodozono, K. and Hanaki, K. 1999, 'Impact of anthropogenic heat on urban climate in Tokyo', *Atmospheric Environment*, vol. 33, pp. 3897–3909.

Landsberg, H.E. 1981, *The Urban Climate*, Academic Press, New York.

Lin, C-Y., Chen, W-C., Chang, P-L. and Sheng, Y-F., 2011, 'Impact of the urban heat island effect on precipitation over a complex geographic environment in northern Taiwan', *Journal of Applied Meteorology and Climatology*, vol. 50, no. 2, pp. 339–353.

Lu, N., Taylor, T., Jiang, W., Correia, J., Leung, L.R. and Wong, P.C. 2009, 'The temperature sensitivity of the residential load and commercial building load', *2009 IEEE Power and Energy Society General Meeting (PES '09)*, 26–30 July, IEEE Computer Society, Calgary, AB, Canada.

Luo, Z., Li, Y. and Lu, W-Z. 2011, 'Effects of urban ventilation patterns on the carbon monoxide concentration in a high-rise mega city', *International Journal of Ventilation*, vol. 10, no. 3, pp. 239–250.

Mentens, J., Raes, D. and Hermy, M. 2005, 'Green roofs as a tool for solving the rainwater runoff problem in the urbanized 21st century?', *Landscape and Urban Planning*, vol. 77, pp. 21–26.

Niu, H., Clark, C., Zhou, J. and Adriaens, P. 2010, 'Scaling of economic benefits from green roof implementation in Washington, DC', *Environmental Science and Technology*, vol. 44, no. 11, pp. 4302–4308.

Oke, T.R. 1982, 'The energetic basis of the urban heat island', *Quarterly Journal of the Royal Meteorological Society*, vol. 108, pp. 1–24.

—— 1984, 'Methods in urban climatology', in W. Kirchhofer, A. Ohmura and H. Wanner (eds), *Applied Climatology*, Zurcher Geographische Schriften, Zurich, pp. 19–29.

—— 1987, *Boundary Layer Climates*, 2nd edn, Methuen, London.

Pigeon, G., Legain, D., Durand, P. and Masson, V. 2007, 'Anthropogenic heat release in an old European agglomeration (Toulouse, France)', *International Journal of Climatology*, vol. 27, no. 14, pp. 1969–1981.

Renou, E. 1855, 'Instructions meteorologiques', *Annuaire de la Société Meteorologique de France*, vol. 3, no. 1, pp. 73–160.

Sailor, D.J. 1998, 'Simulations of annual degree day impacts of urban vegetative augmentation', *Atmospheric Environment*, vol. 32, no. 1, pp. 43–52.

Sailor, D.J. and Fan, H. 2002, 'Modeling the diurnal variability of effective albedo for cities', *Atmospheric Environment*, vol. 36, no. 4, pp. 713–725.

Sailor, D.J., Kalkstein, L.S. and Wong, E. 2002, 'The potential of urban heat island mitigation to alleviate heat-related mortality: methodological overview and preliminary modeling results for Philadelphia', *4th Symposium on the Urban Environment*, American Meteorological Society, Norfolk, VA.

Sailor, D.J., Lu, L. and Fan, H. 2003, 'Estimating urban anthropogenic heating profiles and their implications for heat island development', *Fifth International Conference Urban Climate*, Lodz, Poland.

Stewart, I.D. 2011, 'A systematic review and scientific critique of methodology in modern urban heat island literature', *International Journal of Climatology*, vol. 31, no. 2, pp. 200–217.

Strugnell, N.C., Lucht, W. and Schaaf, C. 2001, 'A global albedo data set derived from AVHRR data for use in climate simulations', *Geophysical Research Letters*, vol. 28, no. 1, pp. 191–194.

Taha, H. 2008a, 'Meso-urban meteorological and photochemical modeling of heat island mitigation', *Atmospheric Environment*, vol. 42, no. 38, pp. 8795–8809.

—— 2008b, 'Urban surface modification as a potential ozone air-quality improvement strategy in California: a mesoscale modelling study', *Boundary-Layer Meteorology*, vol. 127, no. 2, pp. 219–239.

Taha, H., Konopacki, S. and Gabersek, S. 1999, 'Impacts of large-scale surface modifications on meteorological conditions and energy use: a 10-region modeling study', *Theoretical and Applied Climatology*, vol. 62, nos 3–4, pp. 175–185.

Van Seters, T., Rocha, L. and MacMillan, G. 2007, 'Evaluation of the runoff quantity and quality performance of an extensive green roof in Toronto Ontario', *Fifth Annual Greening Rooftops for Sustainable Communities Conference, Awards, and Trade Show*.

Whitman, S., Good, G., Donoghue, E.R., Benbow, N., Shou, W. and Mou, S. 1997, 'Mortality in Chicago attributed to the July 1995 heat wave', *American Journal of Public Health*, vol. 87, no. 9, pp. 1515–1518.

Wong, M.S., Nichol, J.E., To, P.H. and Wang, J. 2010, 'A simple method for designation of urban ventilation corridors and its application to urban heat island analysis', *Building and Environment*, vol. 45, no. 8, pp. 1880–1889.

Yang, F., Lau, S.S.Y. and Qian, F. 2011, 'Urban design to lower summertime outdoor temperatures: an empirical study on high-rise housing in Shanghai', *Building and Environment*, vol. 46, no. 3, pp. 769–785.

Chapter 15

Urban microclimates

Mitigating urban heat

Matheos Santamouris and Dionysia-Denia Kolokotsa

Summary

The urban heat island is among the more documented phenomena of climate change. It describes the high ambient temperatures in the dense zones of a city compared to its surroundings. This chapter presents the main developments in the area of urban mitigation technologies and, in particular, the most recent achievements and knowledge in the field of cool roofs and cool pavements, green roofs and urban parks. The chapter highlights the fact that adaptation to and mitigation of climate change related to urban heat, or simply cooling cities, will be a main strategy in the coming years in order to fight the negative effects of energy demand on the environment.

Introduction

Urban settlements are the most important human habitat. Approximately 60 per cent of the world's population lives in cities and towns. The second half of the twentieth century was a period of the most intensive urbanization that our planet has experienced. In fact, urban population has increased from 160 million to about three billion in just 100 years, and it is expected to increase to about five billion by 2025. The migration of people to cities is underway and will continue, particularly in less developed countries, because of the increased opportunities cities offer people and the degradation of rural economies and societies.

The urban heat island (UHI) is among the more documented phenomena of climate change. It measures the ambient temperatures in a city's dense urban zones, compared to its surrounding regions (Santamouris 2001). Studies of cities around the world have measured its intensity, which may reach values of up to 10 K (Santamouris 2007). The UHI effect increases the energy consumption of buildings for cooling purposes, raises the concentration of urban pollutants (particularly of ozone), decreases human comfort and increases the ecological footprint of cities while putting the low-income population of cities at risk (Santamouris 2012).

To counterbalance the impact of the UHI, several mitigation techniques have been designed, developed and tested in real projects with sufficient success to demonstrate their efficacy. Proposed mitigation techniques include the use of reflective and other intelligent materials in buildings and urban structures, the use of environmental heat sinks to dissipate the excess heat into a low-temperature media like the ground or a body of water, and also the use of additional green spaces that may be integrated into the building's structure (e.g. green roofs and façades) or a city's landscape (through urban parks and street trees). UHI mitigation techniques for cities and buildings are depicted in Figure 15.1.

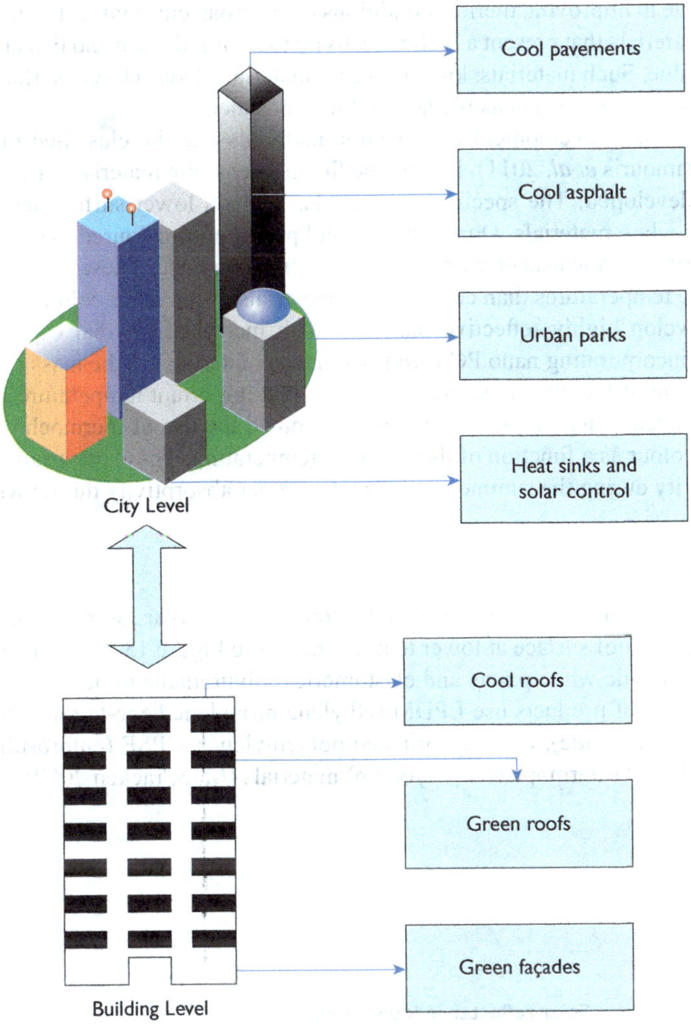

Figure 15.1 Urban heat island mitigation techniques in buildings and at city level.

Urban mitigation technologies have been applied successfully in many urban renovation projects (Fintikakis *et al.* 2011; Gaitani *et al.* 2007, Santamouris *et al.* 2012a, 2012b). This chapter presents the main developments in urban mitigation technologies and, in particular, recent achievements in the field of cool roofs and cool pavements, green roofs and urban parks.

Using smart reflective materials for heat island mitigation

Materials used in urban structures and the envelopes of buildings determine the thermal balance of the urban environment as a whole. Materials absorb and reflect solar radiation, emit infrared radiation and, through convective processes, may heat the urban atmosphere. Important research has recently been carried out to develop and use advanced materials that may contribute to improving thermal conditions in the urban environment. The main aim was to develop materials that present a high reflectivity to solar radiation and that also have a high emissivity value. Such materials, known as cool materials, have a lower surface temperature and thus contribute a lower sensible heat to the atmosphere.

Research on the development of reflective materials may be classified in four specific phases (Santamouris *et al.* 2011). During the first phase, white materials of very high reflectivity were developed. The specific materials have much lower surface temperature than conventional white materials. During the second phase, cool-coloured materials presenting a high reflectivity in the near-infrared spectrum were developed. These materials have much lower surface temperatures than conventional materials of the same colour. The third phase aimed to develop highly reflective materials with increased thermal capacity. This was achieved by incorporating nano PCM (phase-change materials) in the mass of the reflective coating. Testing of the new materials has shown that important temperature surface reductions are obtained. Finally, the fourth phase examined the use of thermochromic materials that change colour as a function of their surface temperature. These materials present a very high reflectivity during the summer period and a proper absorptivity during winter.

Cool roofs

A cool roof is a roofing system that is highly reflective of solar radiation and able to emit heat, keeping the roof surface at lower temperatures (see Figure 15.2). Liquid products used in cool roofs include white paints and elastomeric, polyurethane or acrylic coatings. White single-ply cool roof products use EPDM (ethylene-propylenediene-tetrolymer membrane), PVC (polyvinyl chloride), CPE (chlorinated polyethylene), CPSE (chlorosulfonated polyethylene) and TPO (thermoplastic polyolefin) materials (MacCracken 2009).

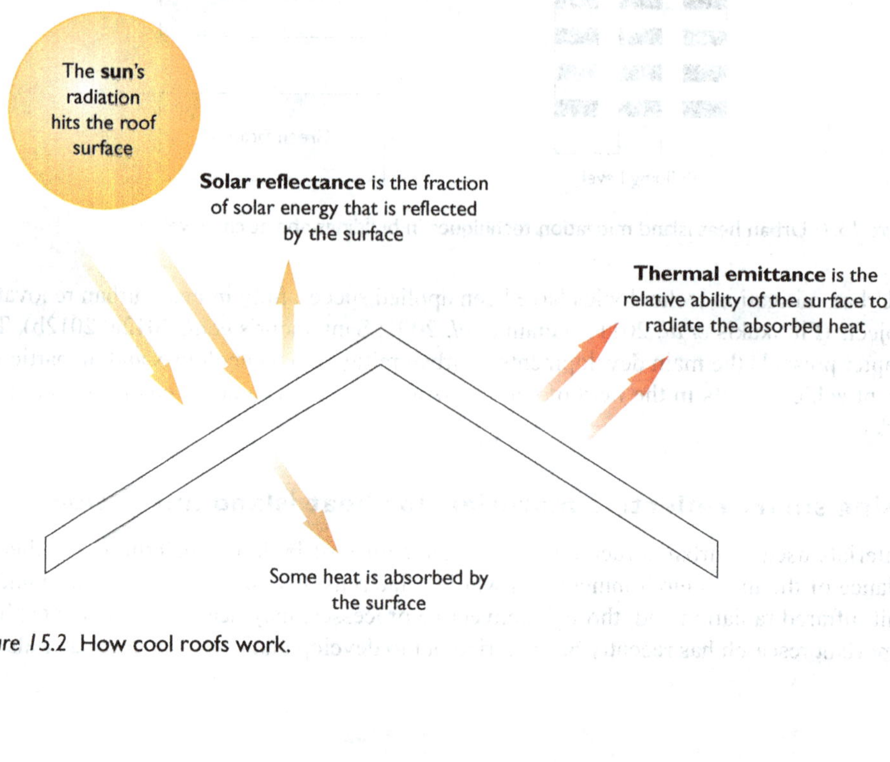

Figure 15.2 How cool roofs work.

Cool roofs substantially decrease the cooling load of buildings while helping to decrease the sensible heat transferred to the atmosphere by buildings, thus mitigating heat islands. A literature review showed that the use of cool roofs in buildings helps to reduce their cooling needs by between 2 and 44 per cent, with an average value close to 20 per cent (Haberl and Cho 2004). As for the mitigation potential of cool roofs, researchers found that the mean decrease of the average ambient temperature for an albedo increase of 0.1 varies between 0.1 K to 0.33 K, with an average value close to 0.2 (Santamouris *et al.* 2012a, 2012b). From a global perspective, Menon and colleagues (Menon *et al.* 2010) have calculated that when the albedo of roofs and pavements increases by 0.25 and 0.15 respectively, the potential carbon dioxide offset is close to 57 Gt.

Cool pavements

The impact of pavements on the development of urban heat islands is significant. Recent studies have shown that paving surfaces play a determinant role on the overall urban thermal balance. Decreasing the pavement's surface temperature substantially contributes to mitigating urban heat islands. This may be achieved by simple shading of paved surfaces to decrease their absorption of the incident solar radiation, or by replacement and rehabilitation of existing pavements with new ones (cool pavements) that have more appropriate thermal properties.

Cool pavements are available for use in cities. Cool paving surfaces use two types of technology; reflective pavements present a high reflectivity to solar radiation combined with a high thermal emissivity, while water-retention pavements use the latent heat of water evaporation to decrease their surface and ambient temperature.

Existing techniques and technologies to increase the albedo of pavements include the following:

- using concrete additives like fly ash and slag cement;
- using conventional cement concrete pavements;
- using light aggregates in asphalt concrete surfaces;
- applying white topping and ultra-thin white topping techniques;
- using colourless and reflective synthetic binders;
- using chip or sand seals with light aggregates, resin-based pavements;
- using roller-compacted concrete pavements;
- using sand/shot blasting and abrading binder surfaces;
- applying colour pigments and seals;
- painting the surfaces with a light colour;
- using micro-surfacing techniques.

Some of the techniques are appropriate for new pavements, while others are for pavement rehabilitation and maintenance (Cambridge Systematics 2005).

Permeable and water-retentive pavements can be vegetated or not. In general, non-vegetated permeable pavements include more voids than the conventional ones to allow water to flow through into the sub-layers and the ground. It is also possible to include water-holding fillers to store water within them. Evaporation of the water reduces the surface temperature of the pavements and contributes to mitigate the urban heat island while also reducing the risk of flooding.

Increasing the green spaces in cities

Planting vegetation in urban areas is one of the main strategies employed to mitigate the urban heat island effect. Vegetation ameliorates the urban climate by moderating the effects of sun, wind and rain. Strategically located plants may reduce the cooling energy consumption between 25 to 80 per cent (Shashua-Bar and Hoffman 2000; Shashua-Bar *et al.* 2010). Usually, green spaces are found in open public areas but they can be in the form of trees located at the eastern or western side of the building's wall, roof-top greenery or vertical greenery. Other advantages of green spaces are improved air quality and lower greenhouse gas emissions (resulting from the reduced energy consumption and from the removal of air pollutants), enhanced stormwater management and water quality, and the improved aesthetic experience of the area's users.

Street trees, green roofs and green façades

Trees alter the environment by moderating the climate, improving air quality, conserving water and protecting wildlife. Trees help cool the environment through shade and evapotranspiration. Trees have impressive shading effects for the built environment. Dense-foliage trees are able to intercept the incoming solar radiation by 70 to 90 per cent (Papadakis *et al.* 2001; Simpson 2002). The shading effect provided by plants on a building's surface lowers its surface temperature and subsequently lowers the consumption of energy for cooling. The technique of planting green roofs on public or private buildings has gained ground during the past decades, as it reduces the energy consumption of the building while improving the microclimate of the wider urban space where the building is situated (Theodosiou 2009; Zinzi and Agnoli 2011). Apart from their impact on building energy consumption, planted roofs contribute to the mitigation of heat islands. Their first advantage is that they increase the total water-permeable surface area in a city, helping to retain water in the soil and allowing larger quantities to be available for evapotranspiration. At the same time, planted roofs present much higher albedo values than dark urban roof surfaces, thus reflecting a greater proportion of the incident solar radiation and not transforming it into heat (Getter and Rowe 2006). According to the United States Environmental Protection Agency (USEPA 2008) the temperature of green roofs can be cooler than the ambient air, whereas conventional roof surfaces can exceed ambient air temperatures by up to 50°C. Wong and colleagues studied the effect of an intensive roof-top garden system and found that surface temperature may reduce by up to 3.1 K and the ambient temperature at 1 m may reduce by up to 1.5 K (Wong *et al.* 2007).

The shift towards roof and external wall planting has increased over recent years, as there have been more installations of roof gardens and vertical greenery in buildings (Wong *et al.* 2010a, 2010b). Roof-top greenery can provide benefits not only to the individual building but also to the surrounding ambient temperature. With an intensive roof-top garden system, the surface temperature may reduce by up to 3.1 K and the ambient temperature at 1 m may reduce by up to 1.5 K. The impact of roof-top greenery is even more pronounced for metal roofs. Without plants, the metal surface can be up to 60–70°C during daytime and lower than 20°C at night. With plants, it ranges only from 24°C to 32°C.

In cities, there is a greater amount of building façade surface than roof surface. For this reason, greening a building façade's surface, known as a vertical greenery system, has great potential in mitigating the UHI effect through evapotranspiration and shading (Takakura *et al.* 2000; Wong *et al.* 2010a, 2010b). Vegetation can dramatically reduce the maximum temperature of a building by shading its walls from the sun, with daily temperature fluctuations being reduced by as much as 50 per cent. In addition, a façade fully covered with greenery is protected from intense solar radiation in the summer and its leaf cover can reflect or absorb between 40 and 80 per cent of the received radiation, depending on the amount and type of greenery (Wong and Jusuf 2008).

Urban parks

It is widely accepted that urban parks are usually cooler than the surrounding urban areas. Lower ambient temperatures in a park help to mitigate the UHI and improve the environmental quality of the park's neighbourhood. In parallel, trees and urban parks filter pollutants, mask noise, stabilize soil and prevent erosion. They also have a significant social impact and contribute to more positive moods for urban residents, improved health conditions and increased property values, and make urban settings more attractive.

Decreased ambient temperatures in parks during the daytime are achieved because trees provide efficient solar protection that decreases surface temperatures and heat transfer to the ambient air and also cools the air temperature through evapotranspiration processes. During the night, the unobstructed sky view decreases surface temperatures and the ambient air. The lack of anthropogenic heat generation in parks also helps to lower temperatures. In several cases, however, parks may have a higher ambient temperature than the surrounding area during the day (Grimmond *et al.* 1996). The reasons for this are: (1) the increased shading of the surrounding urban areas; (2) the decomposition of grass in a park; (3) the possibly increased albedo of green zones; (4) the midday depression of photosynthesis; and (5) the thermal impact of paved surfaces in a park. Skoulika and colleagues (Skoulika *et al.* 2014) report that parks located in dense urban zones in warm Mediterranean climates may be warmer during the daytime than their surroundings for ambient temperatures below 34°C, while for higher temperatures the park tends to be cooler and the park's cooling intensity increases as the ambient temperature rises. This happens for three reasons: (1) higher urban temperatures are always associated with an increased release of anthropogenic heat from air-conditioners in urban areas; (2) in a general way, higher urban temperatures coincide with low wind speeds, and thus the advection in the building layer where the urban stations are installed is reduced; and (3) higher ambient temperatures enhance the evaporation rate in parks. A strong negative correlation has also been found between park-cooling intensity and wind speed, for speeds greater than six metres per second.

The temperature difference developed between the park and the surrounding urban area creates an important atmospheric pressure difference that results in the flow of a breeze from the park to its surroundings that is able to cool the wider zones beyond the borders of the park (Spronken-Smith and Oke 1998). A recent review of the already published studies concluded that, on average, parks are 0.94 K cooler than the reference urban areas during the daytime (Bowler *et al.* 2010).

Many studies have been carried out aiming to identify how far a park's cool climate extends into the surrounding urban area. It has been reported that the magnitude of a park's influence depends on the size of the park and is limited to approximately the width of the park in question (Spronken-Smith 1994). Data from existing studies show that a park's influence extends several hundred metres beyond the park's borders. For example, Jauregui (1990) reported that the influence of a park in Mexico extended up to 2,000 metres from its borders.

However, a more detailed analysis of the existing experimental data shows that, in reality, the cooling impact and the geographical influence of parks are governed by more complex regional and global factors. In fact, they are determined by the characteristics of the surrounding urban area, the locally generated anthropogenic heat, the prevailing climatic conditions and urban density. A very strong correlation between a park's cooling intensity and a city's population density index has been found for both day and night times. The higher the population density index, the higher the park cooling intensity. In parallel, high-rise buildings around a park may block the flow of the cool breeze from the park to its surroundings. In particular, Zoulia and colleagues (Zoulia *et al.* 2009) reported that the impact of a large-scale park in downtown Athens, Greece was very low during the daytime because of the heavy anthropogenic heat generated around it by cars, while similar results are reported by Upmanis and colleagues (Upmanis *et al.* 1998) for a small park in Gothenburg, Sweden that is surrounded by buildings, limiting the air flow in the area.

Using heat sinks to mitigate heat islands

Heat dissipation techniques deal with the disposal of a building's excess heat to a sink (such as the ambient air, water, ground or sky) that has a lower temperature. Effective dissipation of excess heat depends on two main pre-conditions: (1) the availability of a proper environmental heat sink with sufficient temperature difference for the heat transfer; and (2) efficient thermal coupling between the building and the sink. Three main processes of heat dissipation have been thoroughly studied and are well developed:

1 ground cooling based on coupling of buildings with the ground;
2 convective or ventilative cooling based on the use of ambient air;
3 evaporative cooling using water as a heat sink.

Radiative cooling processes using the sky as a heat sink have been proposed; however, they have not yet gained substantial acceptance from the scientific and the technical world.

The ground is characterized by its ability to store incident solar radiation as heat, conserving a constant temperature throughout the year. This way, the underground temperature is lower than the above-ground air temperature during summer and higher than it is during winter. The fact that the ground has a much lower temperature than that of the environment above ground enables it to be used as a natural heat sink during summer by dissipating heat from above-ground constructions to the ground. Taking advantage of the ground temperature is achieved through ground-to-air heat exchangers, tubes that are buried in the earth. The air enters the tubes, is circulated underground and exits the tube with a lower temperature. The designation of this system varies; it may be referred to as 'earth tubes' or as 'ground-coupled air heat exchangers'. Ground cooling for cities has been proposed by various researchers. Fintikakis and colleagues (Fintikakis *et al.* 2011) evaluated the microclimatic modifications that were applied in the historic centre of Tirana, Albania. Earth-to-air heat exchangers were among other mitigation techniques applied. The researchers found that the maximum temperature drop due to all the applied techniques was 3 K, while the maximum contribution of the earth-to-air heat exchangers was found to be close to 0.7 K.

In recent years, the use of evaporative cooling such as wind towers and sprayers for outdoor spaces has increased (Evyatar 2007; Soutullo *et al.* 2011). A study of the effectiveness of evaporative cooling techniques for outdoor spaces was performed at the Universal

Exposition of Seville in 1992 by the University of Seville and CIEMAT (Guerra *et al.* 1995). Evaporative wind towers were designed and built to increase the comfort levels in public spaces. Other studies that analyse the efficiency of evaporative cooling as a passive cooling technique for cities have been carried out by the Centre for Desert Architecture and Urban Planning, Ben-Gurion University of the Negev, Israel (Evyatar 2007). An evaporative wind tower was built in the Negev Sde Boquer and evaluated experimentally. In this study two different cases were analysed, considering the wind tower with and without the evaporative systems (Erell *et al.* 2008; Etzion *et al.* 1997).

Shading is a fundamental and important technique for improving an urban microclimate. Proper shading systems can contribute to a significant reduction of the radiated temperature while simultaneously improving thermal comfort conditions. The use of suitable permeable materials in open spaces can offer a significant reduction of both ambient and radiative temperature by allowing the air to flow and move through them. Urban surfaces can be shaded with various devices and structures (such as tents, pergolas or canopies) or by trees. Shaded urban surfaces receive smaller amounts of solar radiation; thus they have lower temperatures. When shading is applied to a building façade, the result is lower ambient temperatures in open spaces and a reduced rate of heat convection to the building interior. Papadakis and colleagues (Papadakis *et al.* 2001) reported that the radiation incident on building façades is reduced by 70 to 85 per cent when trees are planted close to the building and that temperatures in the area around the trees were significantly lower than those in non-shaded areas.

The presence of a densely built conurbation leads to the establishment of an 'urban boundary layer' above the roof-tops, characterized by significantly lower wind velocities compared to the free flow of air above. The urban heat island phenomenon leads to further augmentation of this problem. Below the building roofs, in the canopy layer, the wind speed is further reduced due to decoupling from the flow above. The local microclimate this forms is characterized by stagnation areas and complex flow fields, and leads to conditions which humans find uncomfortable especially during the warmer months of the year when the wind speed is low (Georgakis and Santamouris 2006; Niachou *et al.* 2005; Yang and Li 2011). Moreover, it has been reported that a temperature increase in urban areas leads to an increase in ozone concentrations in the order of 10 ppb (Stathopoulou *et al.* 2008). For this reason, improvements in urban ventilation will lead to better living conditions. Efforts to design new urban neighbourhoods or redesign existing ones have also been reported, with urban corridors and better street orientation proposed as possible ways to increase urban ventilation (Shaviv *et al.* 2001).

Conclusions

The gathering of a great proportion of the world's population in cities, the need for cheap and readily available housing, extended infrastructure and an alienation from traditional ways of living and working have caused the expansion of the anthropogenic landscape. The consequences of urbanization for the natural landscape are important, the coverage of soil and permeable surfaces with non-permeable materials that absorb heat and the decrease of forests and planted areas dramatically increase surface and air temperatures. The heat island effect is present in both urban and suburban areas, causing city centres to have considerably higher temperatures than their surrounding suburbs. The intensity of the phenomenon (the temperature difference between the centre and the periphery of cities) depends on various factors, including local weather, geomorphology, urban geometry, anthropogenic heat, the quality of the materials used for ground cover and buildings, and the quantity and type of green spaces. Adaptation to and mitigation of climate change related to urban heat, or simply cooling cities, will be a main strategy in the coming years to fight the negative effects of energy demand on the environment.

References

Bowler, D.E., Buyung-Ali, L., Knight, T.M. and Pullin, A.S. 2010, 'Urban greening to cool towns and cities: a systematic review of the empirical evidence', *Landscape and Urban Planning*, vol. 97, no. 3, pp. 147–155.

Cambridge Systematics 2005, *Cool Pavement Report: EPA cool pavements study – task 5,* prepared for Heat Island Reduction Initiative, US Environmental Protection Agency, Cambridge Systematics Inc, Chevy Chase, MD.

Erell, E., Pearlmutter, D. and Etzion, Y. 2008, 'A multi-stage down-draft evaporative cool tower for semi-enclosed spaces: aerodynamic performance', *Solar Energy*, vol. 82, no. 5, pp. 420–429.

Etzion, Y., Pearlmutter, D., Erell, E. and Meir, I.A. 1997, 'Adaptive architecture: integrating low-energy technologies for climate control in the desert', *Automation in Construction*, vol. 6, nos 5–6, pp. 417–425.

Evyatar, E. 2007, 'Evaporative cooling', in M Santamouris (ed.), *Advances in Passive Cooling*, James & James Science Publishers, London.

Fintikakis, N., Gaitani, N., Santamouris, M., Assimakopoulos, M., Assimakopoulos, D.N., Fintikaki, M., Albanis, G., Papadimitriou, K., Chryssochoides, E., Katopodi, K. and Doumas, P. 2011, 'Bioclimatic design of open public spaces in the historic centre of Tirana, Albania', *Sustainable Cities and Society*, vol. 1, no. 1, pp. 54–62.

Gaitani, N., Mihalakakou, G. and Santamouris, M. 2007, 'On the use of bioclimatic architecture principles in order to improve thermal comfort conditions in outdoor spaces', *Building and Environment*, vol. 42, no. 1, pp. 317–324.

Georgakis, C. and Santamouris, M. 2006, 'Experimental investigation of air flow and temperature distribution in deep urban canyons for natural ventilation purposes', *Energy and Buildings*, vol. 38, no. 4, pp. 367–376.

Getter, K.L. and Rowe, D.B. 2006, 'The role of extensive green roofs in sustainable development', *Hortscience*, vol. 41, no. 5, pp. 1276–1285.

Grimmond, C.S.B., Hubble, M.D. and Souch, C. 1996, 'Influence of tree cover on summertime surface energy balance fluxes, San Gabriel Valley, Los Angeles', *Climate Research*, vol. 6, no. 1, p. 45.

Guerra, J., Cejudo, J.M., Molina, J.L., Álvarez, S. and Velázquez, R. 1995, *Control climático en espacios abiertos: evaluación del proyecto Expo '92,* CIEMAT, Madrid.

Haberl, J. and Cho, S. 2004, *Literature Review of Uncertainty of Analysis Methods (Cool Roofs)*, report to the Texas commission on environmental quality, Energy Systems Laboratory, Texas A&M University, TX.

Jauregui, E. 1990, 'Influence of a large urban park on temperature and convective precipitation in a tropical city', *Energy and Buildings*, vol. 16, nos 3–4, pp. 457–463.

MacCracken, M.C. 2009, *Beyond Mitigation: Potential options for counter-balancing the climatic and environmental consequences of the rising concentrations of greenhouse gases,* World Bank Policy Research Working Paper 4938, http://elibrary.worldbank.org/doi/pdf/10.1596/1813-9450-4938, viewed 14 November 2013.

Menon, S., Akbari, H., Mahanama, S., Sednev, I. and Levinson, R. 2010, 'Radiative forcing and temperature response to changes in urban albedos and associated CO_2 offsets', *Environmental Research Letters*, vol. 5, no. 1, http://iopscience.iop.org/1748-9326/5/1/014005/fulltext/, viewed 14 November 2013.

Niachou, K., Hassid, S., Santamouris, M. and Livada, I. 2005, 'Comparative monitoring of natural, hybrid and mechanical ventilation systems in urban canyons', *Energy and Buildings*, vol. 37, no. 5, pp. 503–513.

Papadakis, G., Tsamis, P. and Kyritsis, S. 2001, 'An experimental investigation of the effect of shading with plants for solar control of buildings', *Energy and Buildings*, vol. 33, no. 8, pp. 831–836.

Santamouris, M. (ed.) 2001, *Energy and Climate in the Urban Built Environment*, James & James Science Publishers, London.

—— 2007, 'Heat island research in Europe: the state of the art', *Advances in Building Energy Research*, vol. 1, no. 1, pp. 123–150.

—— 2014, 'Cooling the cities: a review of reflective and green roof mitigation technologies to fight heat island and improve comfort in urban environments', *Solar Energy*, vol. 103, May, pp. 602–703.

Santamouris, M., Synnefa, A. and Karlessi, T. 2011, 'Using advanced cool materials in the urban built environment to mitigate heat islands and improve thermal comfort conditions', *Solar Energy*, vol. 85, no. 12, pp. 3085–3102.

Santamouris, M., Xirafi, F., Gaitani, N., Spanou, A., Saliari, M. and Vassilakopoulou, K. 2012a, 'Improving the microclimate in a dense urban area using experimental and theoretical techniques: the case of Marousi, Athens', *International Journal of Ventilation*, vol. 11, no. 1, pp. 1–16.

Santamouris, M., Gaitani, N., Spanou, A., Saliari, M., Giannopoulou, K., Vasilakopoulou, K. and Kardomateas, T. 2012b, 'Using cool paving materials to improve microclimate of urban areas: design realization and results of the Flisvos project', *Building and Environment*, vol. 53, pp. 128–136.

Shashua-Bar, L. and Hoffman, M.E. 2000, 'Vegetation as a climatic component in the design of an urban street: an empirical model for predicting the cooling effect of urban green areas with trees', *Energy and Buildings*, vol. 31, no. 3, pp. 221–235.

Shashua-Bar, L., Tsiros, I.X. and Hoffman, M.E. 2010, 'A modeling study for evaluating passive cooling scenarios in urban streets with trees. Case study: Athens, Greece', *Building and Environment*, vol. 45, no. 12, pp. 2798–2807.

Shaviv, E., Yezioro, A. and Capeluto, I.G. 2001, 'Thermal mass and night ventilation as passive cooling design strategy', *Renewable Energy*, vol. 24, nos 3–4, pp. 445–452.

Simpson, J.R. 2002, 'Improved estimates of tree-shading effects on residential energy use', *Energy and Buildings*, vol. 34, no. 10, pp. 1067–1076.

Skoulika, F., Santamouris, M., Kolokotsa, D. and Boemi, N. 2014, 'On the thermal characteristics and the mitigation potential of a medium size urban park in Athens, Greece', *Landscape and Urban Planning*, vol. 123, pp. 73–86.

Soutullo, S., Olmedo, R., Sánchez, M.N. and Heras, M.R. 2011, 'Thermal conditioning for urban outdoor spaces through the use of evaporative wind towers', *Building and Environment*, vol. 46, no. 12, pp. 2520–2528.

Spronken-Smith, R.A. 1994, *Energetics and cooling in urban parks*, doctoral thesis, The University of British Columbia, Vancouver.

Spronken-Smith, R.A. and Oke, T.R. 1998, 'The thermal regime of urban parks in two cities with different summer climates', *International Journal of Remote Sensing*, vol. 19, no. 11, pp. 2085–2104.

Stathopoulou, E., Mihalakakou, G., Santamouris, M. and Bagiorgas, H.S. 2008, 'On the impact of temperature on tropospheric ozone concentration levels in urban environments', *Journal of Earth System Science*, vol. 117, no. 3, pp. 227–236.

Takakura, T., Kitade, S. and Goto, E. 2000, 'Cooling effect of greenery cover over a building', *Energy and Buildings*, vol. 31, no. 1, pp. 1–6.

Theodosiou, T. 2009, 'Green roofs in buildings: thermal and environmental behaviour', *Advances in Building Energy Research*, vol. 3, no. 1, pp. 271–288.

United States Environmental Protection Agency (USEPA) 2008, *Reducing Urban Heat Islands: Compendium of Strategies*, www.epa.gov/heatisld/resources/pdf/GreenRoofsCompendium.pdf, viewed 29 October 2013.

Upmanis, H., Eliasson, I. and Lindqvist, S. 1998, 'The influence of green areas on nocturnal temperatures in a high latitude city (Goteborg, Sweden)', *International Journal of Climatology*, vol. 18, no. 6, pp. 681–700.

Wong, N.H and Jusuf, S.K. 2008, 'GIS-based greenery evaluation on campus master plan', *Landscape and Urban Planning*, vol. 84, no. 2, pp. 166–182.

Wong, N.H., Tan, P.Y. and Yu, C. 2007, 'Study of thermal performance of extensive rooftop greenery systems in the tropical climate', *Building and Environment*, vol. 42, no. 1, pp. 25–54.

Wong, N.H.C., Tan, A.Y.K., Chen, Y., Sekar, K., Tan, P.Y., Chan, D., Chiang, K. and Wong, N.C. 2010a, 'Acoustics evaluation of vertical greenery systems for building walls', *Building and Environment*, vol. 45, no. 3, pp. 663–672.

Wong, N.H., Tan, A.Y.K., Tan, P.Y., Sia, A. and Wong, N.C. 2010b, 'Perception studies of vertical greenery systems in Singapore', *Journal of Urban Planning and Development*, vol. 136, no. 4, pp. 330–338.

Yang, L. and Li, Y. 2011, 'Thermal conditions and ventilation in an ideal city model of Hong Kong', *Energy and Buildings*, vol. 43, no. 5, pp. 1139–1148.

Zinzi, M. and Agnoli, S. 2011, 'Cool and green roofs: an energy and comfort comparison between passive cooling and mitigation urban heat island techniques for residential buildings in the Mediterranean region', *Energy and Buildings*, vol. 55, pp. 66–76.

Zoulia, I., Santamouris, M. and Dimoudi, A. 2009, 'Monitoring the effect of urban green areas on the heat island in Athens', *Environmental Monitoring and Assessment*, vol. 156, nos 1–4, pp. 275–292.

Chapter 16

Energy reduction using natural ventilation in city planning

Yukiko Yoshida and Toshiaki Ichinose

Summary

The urbanizing of Asia has led to the need for cooling load to reduce the environmental impact of urbanization. In Japan, the majority of energy consumed in non-residential buildings is for lighting and ventilation. In this chapter, the possible reductions in cooling load are compared to determine whether natural ventilation can be used or not in each building block. Therefore, we use the Urban Climate Simulation System (Ashie and Vu 2004) to calculate the outdoor temperature and wind velocity in part of a building block using a canopy model. By considering the results, we were able to find a way to mitigate urban heat stress via spatial planning.

Introduction

In urban areas, the high population density creates heat islands and causes anthropogenic heat. Figure 16.1 shows that we must develop an Asia-specific strategy instead of adopting European approaches, which focus on reducing heating to achieve low carbon cities. As Asia becomes increasingly urbanized, it requires cooling methods with little environmental impact. In Japan, air-conditioning accounts for about 40 per cent, and lighting and electrical outlets for about 30 per cent of the energy consumption in non-residential buildings (Murakami *et al.* 2006).

Japan has the technological potential to reduce its CO_2 emissions by 70 per cent compared to the 1990 level (NIES 2011), which corresponds to expected demand for energy services by 2050. We must now demonstrate possible ways of meeting these government targets with technical and political measures. Energy and technology systems expert Arnulf Grubler (2009, p. 20) describes the hierarchy of urban energy and CO_2 reduction as follows:

1 spatial division of labour (trade, industry structure);
2 urban form (functional mix, public transport, car ownership);
3 efficiency of energy end use (buildings, appliances, processes);
4 energy systems integration (co-generation, heat cascading);
5 fuel substitution (renewables, nuclear).

Architectural methods that take account of the need for conservation of the global environment generally concentrate on mitigating the heat load of buildings. We therefore evaluated the reduction in energy consumption that may be achieved by improving heating, ventilation

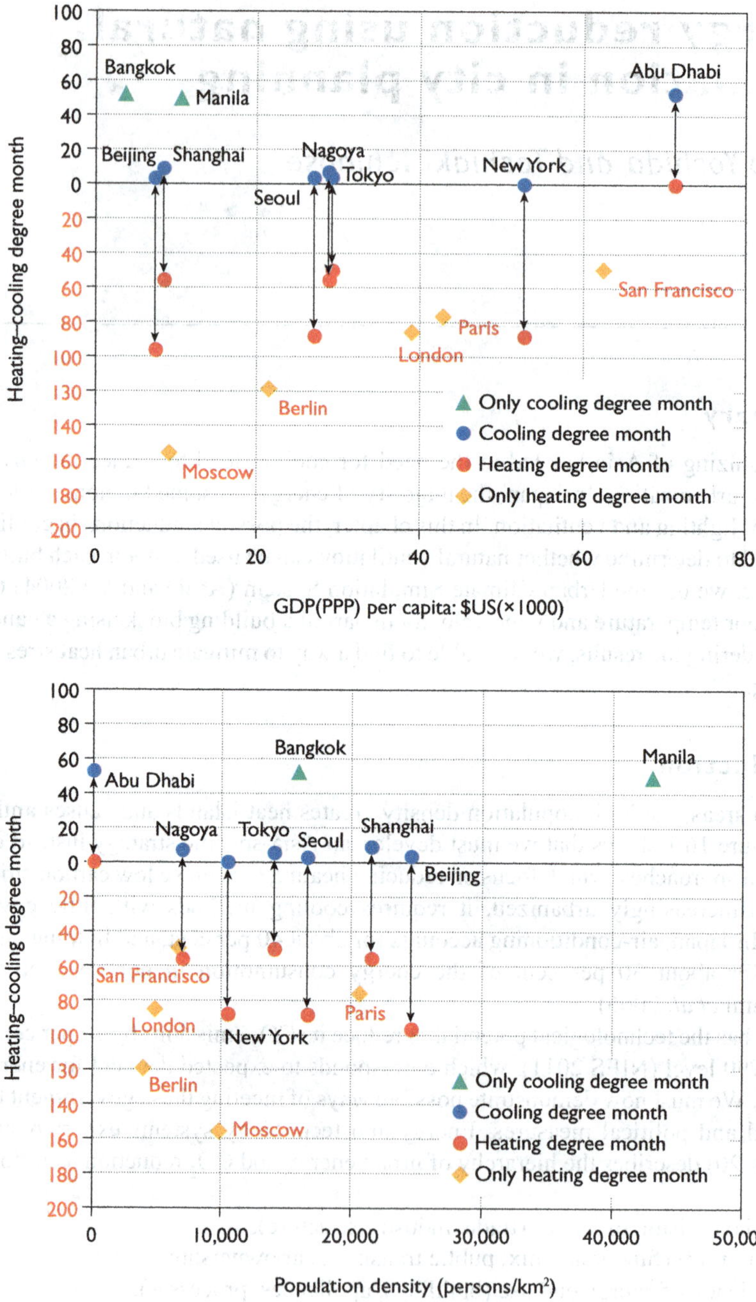

Figure 16.1a and 16.1b Top: GDP (PPP) per capita and heating–cooling degree month. Bottom: Population density and heating–cooling degree month.

and air-conditioning (HVAC) technologies using natural ventilation. To do so, we should ensure that we can use natural ventilation in the outdoor environment (urban microclimate). Therefore, we have developed a building block evaluation method for estimating the level of natural ventilation that is needed to assist with city planning. We have studied city types with a view to realizing measures for effectively reducing the energy consumption of the building sector in Japan. We have also studied ways of evaluating thermal comfort in each type of building block.

How can we mitigate urban heat islands?

We are developing a building block information system, which we call environment and energy around building blocks (EEBB). The logic of this system is based on the optimal control of air-conditioning (Yoshida and Ichinose 2009a, 2009b).

In this system, the energy-saving potential is visualized using Google Maps and Google Earth (see Figure 16.2). Remote sensing is available to the public and statistical data help us to evaluate the simulation's accuracy. Appropriate energy-saving information should be provided for residents and energy managers.

Methods

Urban Climate Simulation System (UCSS)

We used the Urban Climate Simulation System (UCSS), developed by Ashie and Vu (2004) (see Figure 16.3) to try to develop an evaluation technique for a building block environment.

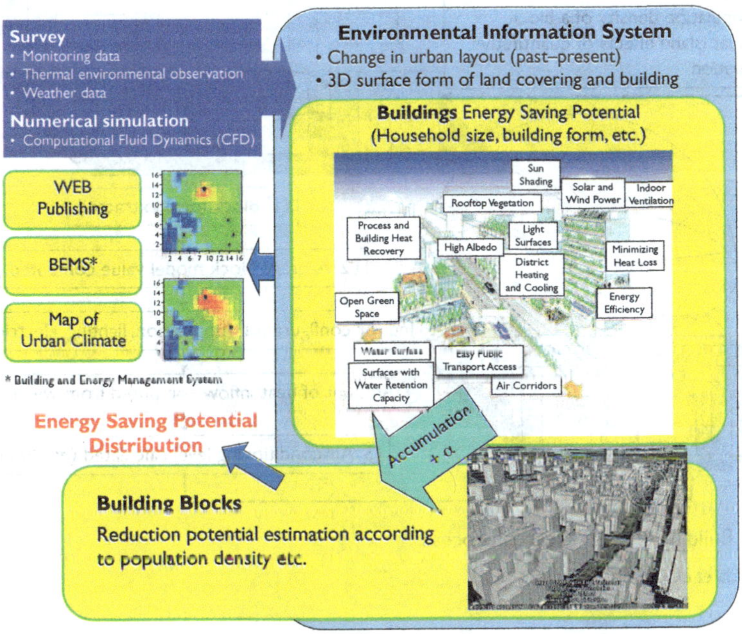

Figure 16.2 Research flow on environment and energy around building blocks (EEBB).

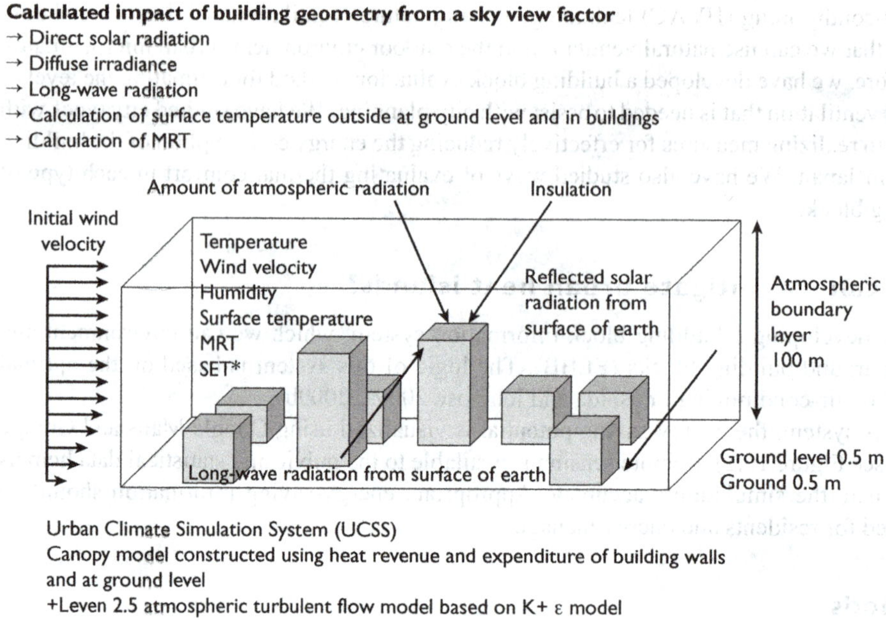

Figure 16.3 The Urban Climate Simulation System (UCSS).

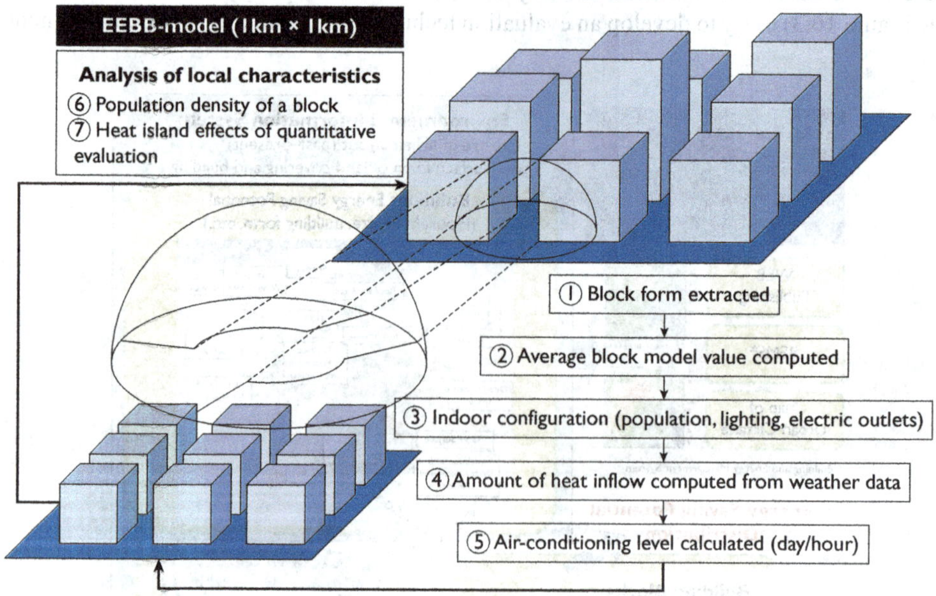

Figure 16.4 Building blocks research process.

Source: Yoshida *et al.* (2009).

Obtaining a large number of results will help to evaluate the natural ventilation level effectively. City planners should be provided with appropriate energy-saving information by using data such as the Automated Meteorological Data Acquisition System (AMeDAS) measurements compiled by the Japanese Meteorological Agency.

We therefore developed a research process (see Figure 16.4), which classifies types of city in terms of the sky view factor (SVF) and weather data. In this study, we calculated the energy consumption by considering specific window types and building block characteristics.

Nagoya case studies

First, we used laser point data provided by the Kokusai Kogyo Corporation to construct a 3D model (Figure 16.5), which uses building blocks (Yoshida 2011; Yoshida *et al.* 2012; Yoshida and Ichinose 2009b) with multi-storey buildings that have an average floor height of 3.2 m. The average building height was 21.5 m. Inside the building blocks, we set a SVF of 20 per cent (Figure 16.6) using the CanopOn2 tool (Takenaka 2009), and the road width was 8 m. When we generate blocks using the canopy model, we have a base model (Figure 16.7), which is 20 m wide by 20 m deep by 20 m high in an area of 1 km^2. A comparison of land use (Table 16.1) shows a good match between a real site and the canopy model.

As Figure 16.8 shows, when the buildings are higher, the anthropogenic heat is greater. Thus, we considered the heat balance with a view to maintaining cities sustainably.

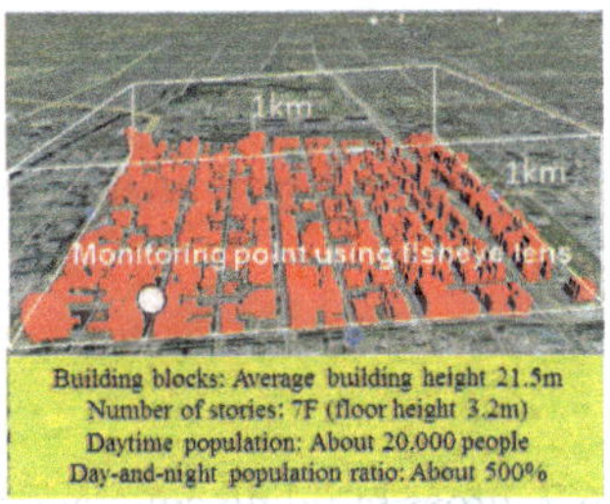

Figure 16.5 Research site in Nagoya on Google Earth.

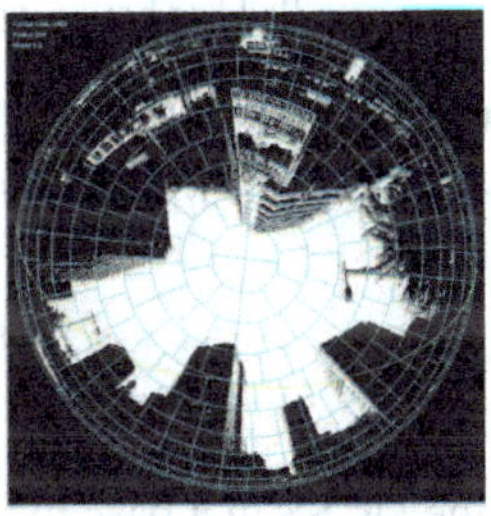

Figure 16.6 Sky view factor of Nagoya from photo obtained with fisheye lens (Nikon CoolPix-P5100 +FC-E8).

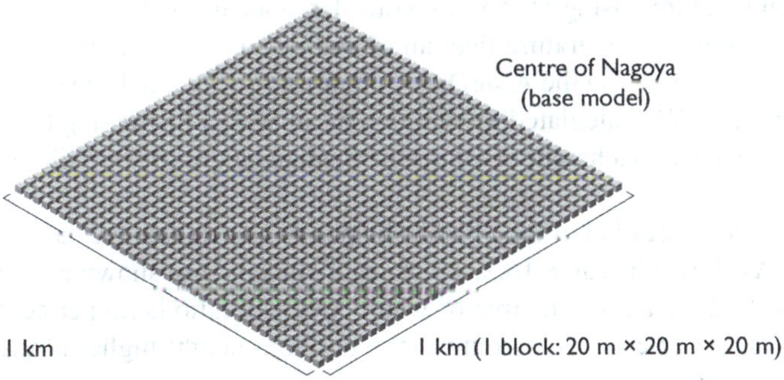

Figure 16.7 Canopy model of Nagoya: building coverage ratio 50% + green ratio 0%.

Table 16.1 Comparison of land use

Comparison of land use	*Building coverage ratio (%)*	*Road coverage ratio (%)*	*Tree coverage ratio (%)*	*Bare land ratio (%)*
Land use on map	55	30	1	14
Canopy model	51	49	–	–

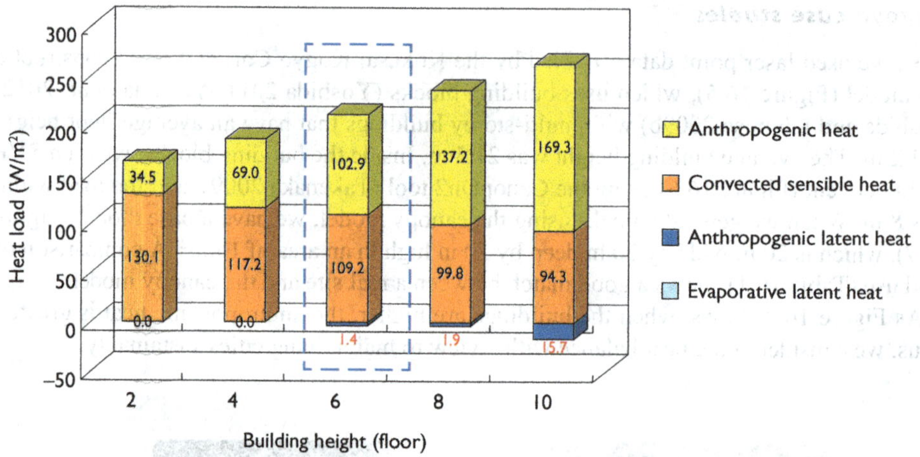

Figure 16.8 Heat load versus building height (Nagoya).

Source: Urban Climate Simulation System (UCSS) developed by Dr Ashie, Japan. MOE developed a simple calculation tool (March 2002).

Results

We calculated the UCSS in a base model (Figure 16.7, gross building coverage ratio 51 per cent and road coverage ratio 49 per cent, current status of Nagoya). Figure 16.9 shows the average temperature of the building blocks and the difference in altitude at an initial wind velocity of 2 m/s and an initial temperature of 30°C. Figure 16.10 follows the same temperature as the AMeDAS monitoring point and the average temperature in the building blocks but with a different initial temperature. Using UCSS, we considered an altitude of 0 m (the surface of the ground) to be a higher temperature than an altitude of 0.5–20 m (above the ground), and there is an equation correlating the AMeDAS monitoring point and the average temperature in the building blocks. We calculated the volume of air with a fluctuating temperature for natural ventilation based on a height of 10 m (the median height of the building blocks) above the road.

Obtaining results for the temperatures in building blocks helped us to evaluate the natural ventilation level effectively. As shown in Table 16.2, we changed the land use shown by the gross building coverage ratio to 25 per cent, the tree or water coverage ratio is 26 per cent and the road coverage ratio is 49 per cent. We arranged the building height higher at the centre (or random) in a base model.

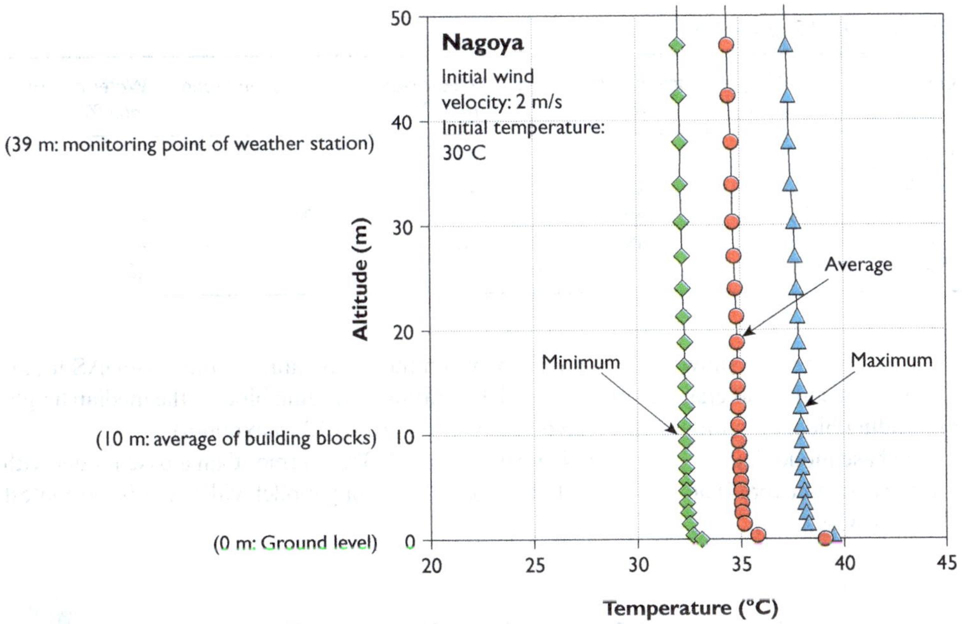

Figure 16.9 Vertical average temperature (UCSS in building blocks).

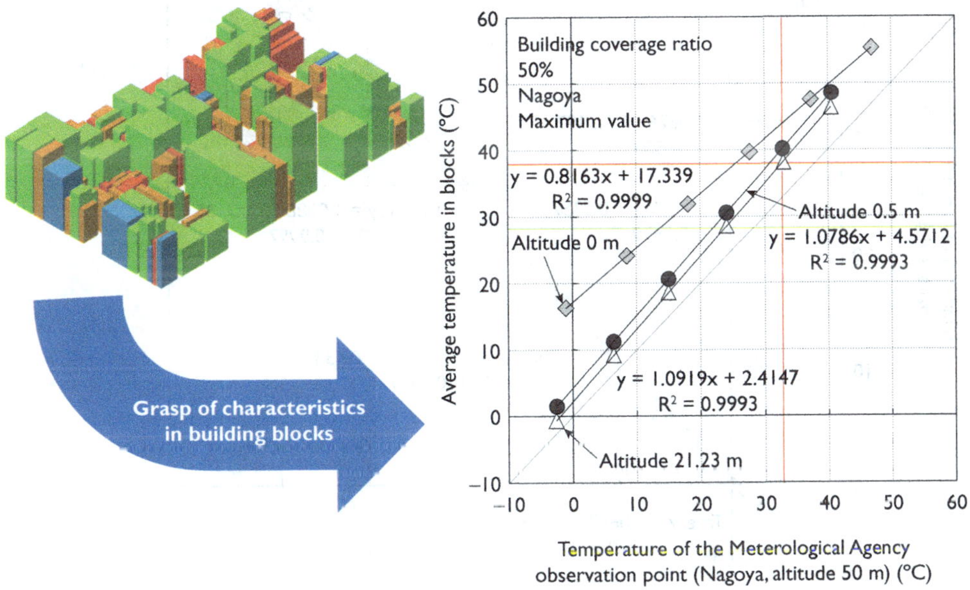

Figure 16.10 Horizontal average temperature (UCSS in building blocks).

Table 16.2 UCSS case study

Case study	*Building coverage ratio %*	*Road coverage ratio %*	*Tree coverage ratio %*	*Bare land ratio %*	*Water coverage ratio %*
Current status	51	49	–	–	–
Random	25	49	–	26	–
Centre	25	49	–	26	–
Tree	25	49	26	–	–
Water	25	49	–	–	26

The UCSS results show the differences between the temperature at the AMeDAS monitoring point and the average temperature in 10.91 m-high building blocks (the median height of building blocks) in Figure 16.11 (average) and Figure 16.12 (maximum).

The base model (current status of Nagoya) is about 0.3°C warmer than a base model with greenery (tree-covered) or with water. In this case, the canopy model with water is the coolest above 20°C.

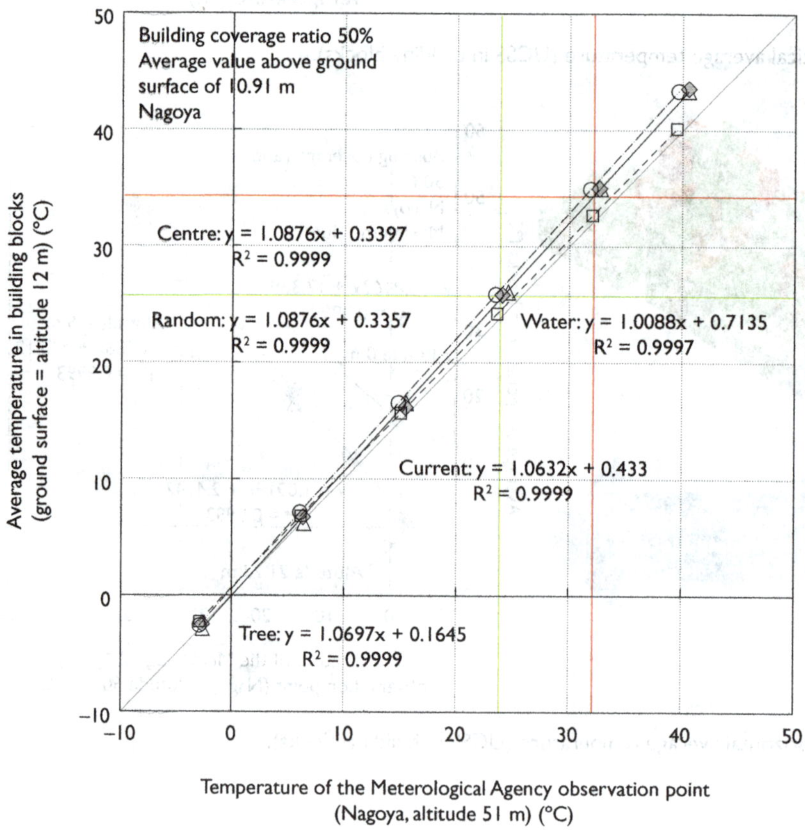

Figure 16.11 Forecast urban climate (horizontal average temperature).

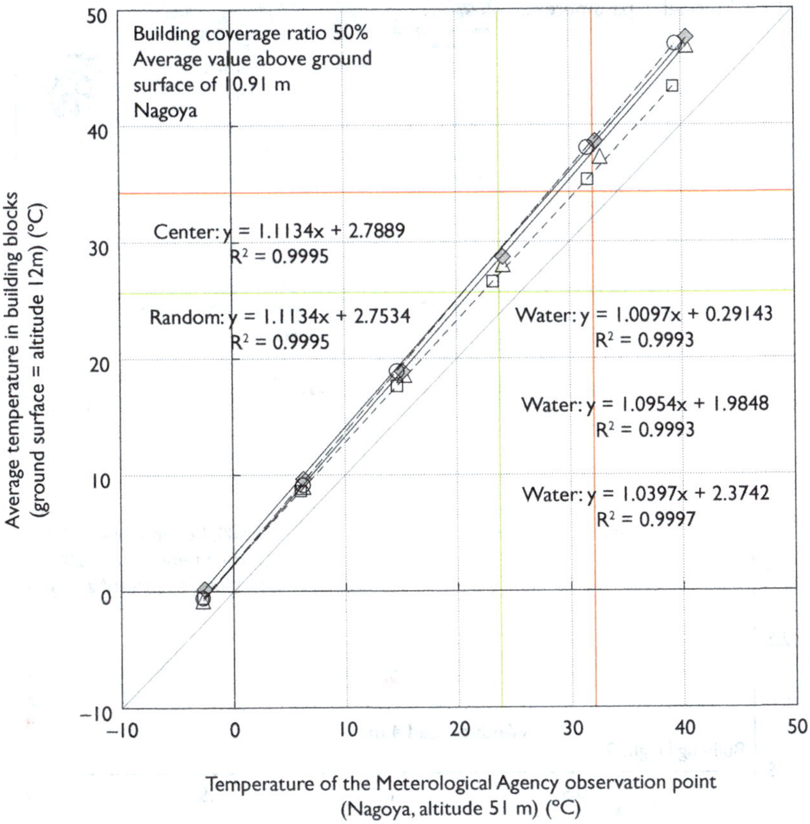

Figure 16.12 Forecast urban climate (horizontal maximum temperature).

The current maximum status (Figure 16.12) is warmer than a base model with water or with greenery (tree-covered). With water, we can prevent a temperature increase more effectively than in any other case. However, the difference between the current status of Nagoya and the case with water is very small. Therefore, we found that even an expensive mitigating measure may have little effect.

Next, we must consider a comfort level that allows people located near a road to avoid heat stress. Using the 3D model (Figure 16.7), which was made by RayMan (Matzarakis *et al.* 2006), with an albedo of 0.3 and a Bowen ratio of 1.3 (see Figure 16.13), we set five values for the Predicted Mean Vote (PMV) index (Fanger 1972) at a cross-street, for five wind velocity values, which were 0, 0.5, 1, 1.5 and 2 m/s (Figure 16.14). When the wind velocity decreased, the PMV value increased, thus increasing the thermal stress.

The situation was compared with the SET* for a general indicator adjusted to the human conditions in Figure 16.15. We found that we need a wind velocity of 1 m/s or more to keep the SET* at 28°C or below. An outside temperature of 28°C is the limit of human comfort, according to ASHRAE Standard 55 (ASHRAE 2010). Therefore, when the outside temperature is higher than 28°C, we need air-conditioning inside buildings and the outside environment requires shade from buildings or trees. The effect of wind is also needed between buildings.

Figure 16.13 RayMan1.2 model of parameters.

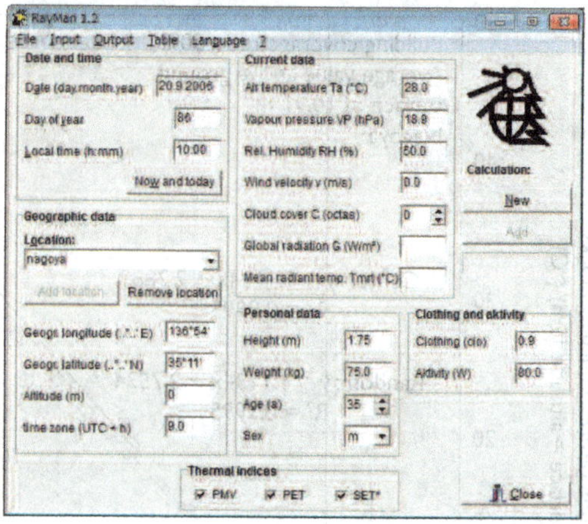

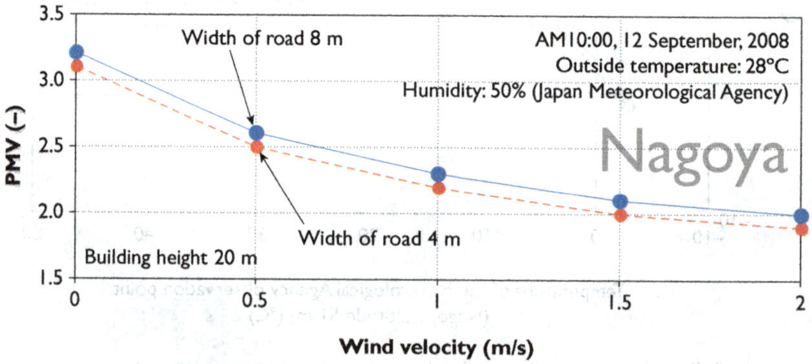

Figure 16.14 PMV and wind velocity in Nagoya.

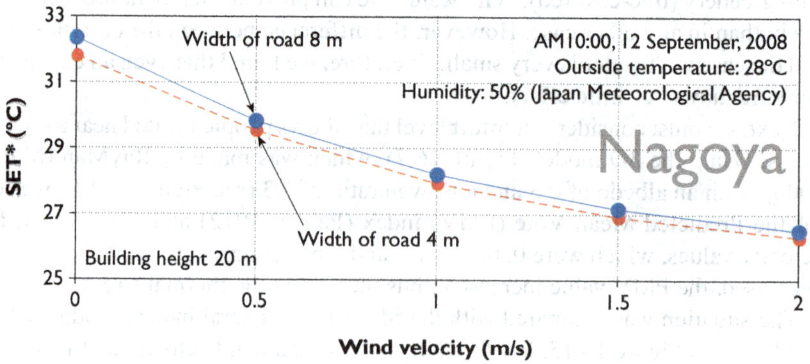

Figure 16.15 SET* and wind velocity in Nagoya.

In terms of building block types, it is easy to set an indicator using the SVF. As shown in Figure 16.16a, when planning the building blocks in Nagoya, we found that SVF values of less than 45 per cent were appropriate in the morning. We also considered the outside environment in the afternoon (see Figure 16.16b). When planning the building blocks in Nagoya, we found that road widths of 6 or 8 m were appropriate. If we are to build low carbon cities, Asia needs to reach a level of comfort with respect to wind values.

We then analysed the air temperature distribution in average blocks using computational fluid dynamics software (CFD2000). We set the parameters as follows: a westerly wind

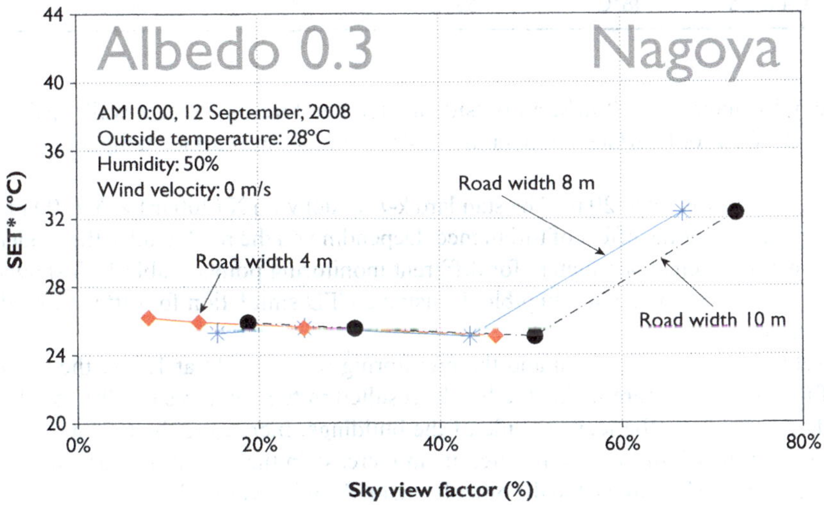

Figure 16.16a SET* and SVF in Nagoya at 10 a.m., 12 September 2008.

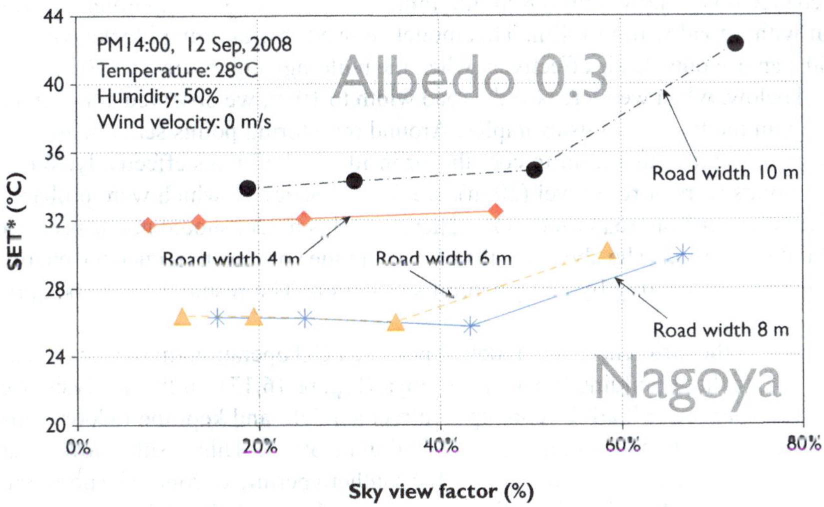

Figure 16.16b SET* and SVF in Nagoya at 2 p.m., 12 September 2008.

Table 16.3 Air temperatures inside building blocks predicted using CFD

Road width	*Building side*	*Monitoring point heights*			
		1.2 m	*5 m*	*10 m*	*20 m*
4 m	Windward	40°C	38°C	38°C	40°C
	Leeward	48°C	50°C	52°C	54°C
8 m	Windward	39°C	38°C	38°C	39°C
	Leeward	52°C	53°C	54°C	55°C
10 m	Windward	40°C	38°C	38°C	38°C
	Leeward	48°C	52°C	54°C	56°C

direction, a wind velocity of 2.0 m/s, an outside air temperature of 35°C, a wall surface temperature of 35°C, a roof surface temperature of 45°C and a road surface temperature of 45°C.

The building height was set at 20 m. The standard k-ε model with X (200 m) $\times$ Y (200 m) $\times$ Z (100 m) accounted for the effect of turbulence. Depending on the road widths, the model provided different turbulence parameters for different monitoring points. Table 16.3 shows predicted air temperatures inside building blocks using a CFD simulation for different road widths and monitoring point heights.

When we set the road width at 4 m and the monitoring point height at 1.2 m, the CFD simulation of the air temperature inside the blocks resulted in temperatures of 40°C on the windward side and 48°C on the leeward side of the buildings. Increasing the height of the monitoring points from 1.2 m to 20 m resulted in an increase in the air temperature of 6°C on the leeward side of the buildings. On the other hand, on the windward side the temperature dropped slightly at monitoring points set 5 and 10 m above the road. Therefore, each building had a different condition as regards the effective use of natural ventilation, which depended on the wind direction.

When we changed the road width to 8 m, the influence of the outside environment was similar to that with a road width of 4 m. This model showed that building blocks with an 8 m road width can use outside air effectively when the building heights are 5 or 10 m.

As described below, when we increased the road width to 10 m, we observed a higher air temperature than in the two previous examples. Around monitoring points set at heights of 5 and 10 m, we can use the outside air to cool the air inside the buildings effectively. When the monitoring points were at roof level (20 m), we found hot zones, which were probably the result of heat transfer from roads and roofs. Therefore, this model should not be used for exploring natural ventilation. The above results provide an idea of how to reduce the energy consumption needed for cooling by using natural ventilation. The results need to be optimized in a future study.

In accordance with the differences in window types, the FCU operation times were calculated for each side of these building blocks in Nagoya (Figure 16.17). In this analysis, we took account of the perimeter load for cooling in direct sunlight and kept the indoor environment comfortable for human beings in each building block. This result shows that good-quality windows have a beneficial effect on the southern perimeter zone. The northern, western and eastern sides of buildings have little effect on cooling with the high-specification windows in the Nagoya building blocks (road width 8 m, building height 20 m).

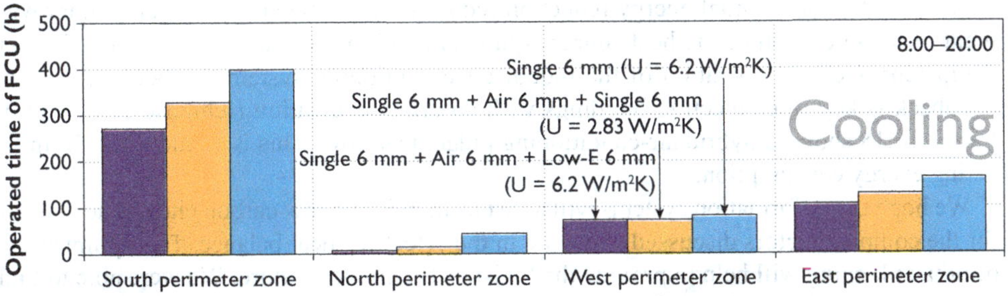

Figure 16.17 FCU operation time for cooling load in each perimeter zone.

Finally, we determined the maximum energy reduction level as regards cooling load when the design conformed with Japanese building law with high-specification windows in Nagoya (road width 8 m, building height 20 m) using natural ventilation. Taking into account available countermeasures and population density, we tried to show how to improve the thermal

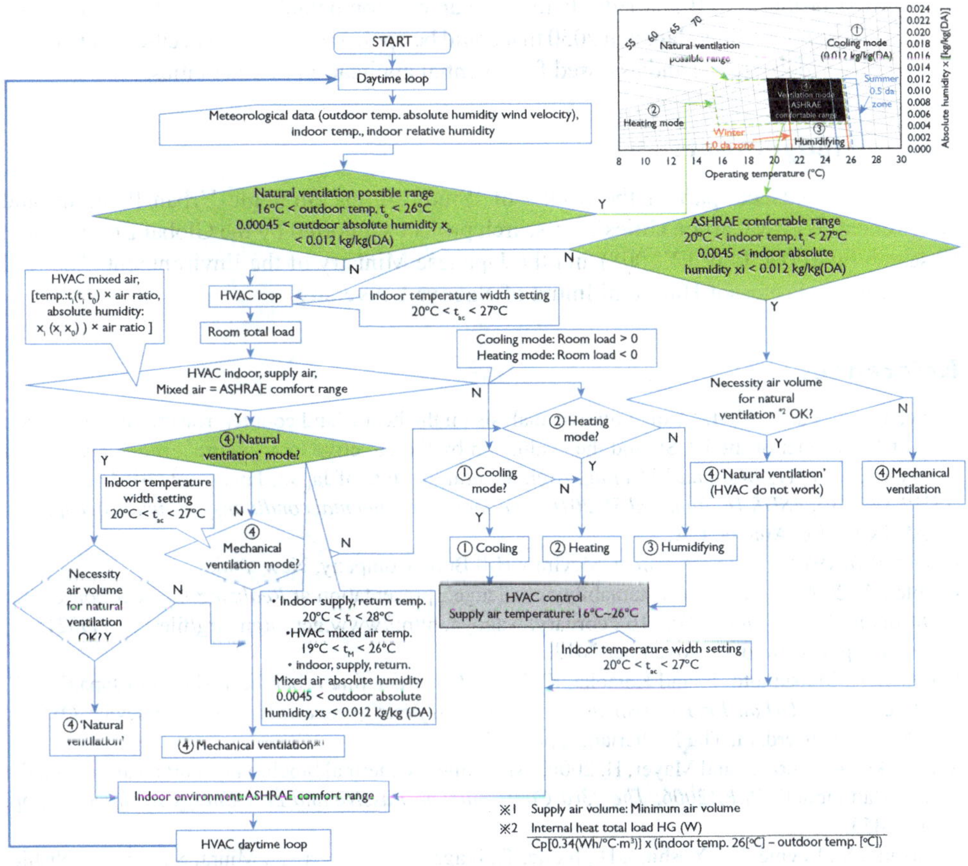

Figure 16.18 Hybrid air-conditioning judgement flow.

environment. A substantial energy reduction requires the deliberate and efficient improvement of the operation of the heat source equipment and the introduction of more efficient equipment once the operation of the indoor air-conditioning system has been improved. Feedback is key in connecting buildings with an urban evaluation (Ichinose *et al.* 2008). Figure 16.18 shows a hybrid air-conditioning judgement flow. This is an indicator for mitigating energy consumption.

We hope that information systems will assist us in building low carbon cities. The reasons for the complex factors discussed above lie in the effects of heat balance. The optimization of urban planning will help to provide the logistics for an ideal future. We were able to find a strategy for mitigating heat islands that ranged from HVAC controls to city planning using natural ventilation. This will, hopefully, lead to the implementation of practical measures designed to combat climate change and heat islands in Asia, which is our primary goal.

Discussion

We believe that anthropogenic heating is an important and essential research topic. Case studies from around the world are needed to understand and deal with the heat island phenomenon. The result of anthropogenic heating is that heat islands and climate change are interacting, and our study contributes to the broader field of urban environmental research. We tried to construct a vision for Nagoya, Japan in 2050 that could be applied in other Asian cities. This vision needs to be widely shared and assessed for potential application in other cities.

Acknowledgements

This chapter presents part of the results of 'Study on the Strategic Urban Planning and Assessment of Low Carbon Cities', a research project supported by the Global Environment Research Fund (Hc-086: FY2008) and the Japanese Ministry of the Environment. The lead investigator is Professor Hidefumi Imura of Nagoya University.

References

Ashie, Y. and Vu, T. 2004, 'A quantitative analysis on the heat island countermeasures by the UCSS; Part 1: Abstract of the UCSS and the evaluation by the effective temperature', *The Summaries of Technical Papers of Annual Meeting*, Architectural Institute of Japan, Japan, pp. 661–662.

ASHRAE 2010, *ASHRAE Standard 55-2010: Thermal environmental conditions for human occupancy*, ASHRAE Inc, Atlanta, GA.

Fanger, P.O. 1972, *Thermal Comfort*, McGraw Hill Book Company, New York.

Grubler, A. 2009, 'The urban sustainability challenge', presentation at *Realizing Low Carbon Cities: Bridging Science and Policy,* 16 February, Nagoya, http://www.gcp-urcm.org/files/A20090216/2_Grubler.pdf, viewed 19 November 2013.

Ichinose, T., Matsumoto, F. and Kataoka, K. 2008, 'Counteracting urban heat islands in Japan', in P. Droege (ed.), *Urban Energy Transition: From fossil fuels to renewable power*, Elsevier, Oxford, UK, and Amsterdam, The Netherlands, pp. 365–380.

Matzarakis, A., Rutz, F. and Mayer, H. 2006, 'Modeling the thermal bioclimate in urban areas with the RayMan model', *PLEA2006: The 23rd Conference on Passive and Low Energy Architecture*, pp. 449–453.

Murakami, S., Levine, M., Yoshino, H., Inoue, T., Ikaga, T., Shimoda, Y., Miura, S., Sera, T., Nishio, M., Sakamota, Y. and Fujisaki, W. 2006, *Energy Consumption Efficiency, Conservation, and*

Greenhouse Gas Mitigation in Japan's Building Sector, Lawrence Berkeley National Laboratory, University of California, Berkeley, CA, http://escholarship.org/uc/item/6gp873s1, viewed 19 November 2013.

National Institute for Environmental Studies (NIES) 2011, *Asian Low Carbon Societies Research Project*, National Institute for Environmental Studies, Ibaraki, Japan, http://2050.nies.go.jp/file/S-6_leaflet_E.pdf, viewed 19 November 2013.

Takenaka, A. 2009, CanopOn2, http://takenaka-akio.org/etc/canopon2/, viewed 19 November 2013.

Yoshida, Y. 2011, 'Toward the realizing for low carbon city in Asia for using natural ventilation', *The 24th World Congress of Architecture*, 25 September–1 October.

Yoshida, Y. and Ichinose, T. 2009a, 'Planning low-carbon society based on building: urban environmental evaluation in Asia', *7th International Science Conference on the Human Dimensions of Global Environmental Change*, IHDP Open Meeting, Bonn.

—— 2009b, 'Research and development of the information system for building-blocks environmental evaluation', *7th International Conference on Urban Climate ICUC-7*, Yokohama, Japan, 29 June–3 July.

Yoshida, Y., Shimizu, H. and Murayama, A. 2012, 'Development of the evaluation for building-blocks environment in Nagoya', *6th Japanese-German Meeting on Urban Climatology*, Hiroshima Institute of Technology, Japan, 22–23 September.

Chapter 17

The impact of increasing urban air temperatures on urban planning and building energy consumption in tropical climates

Steve Kardinal Jusuf, Nyuk Hien Wong and Zhun Min Adrian Chong

Summary

A built-up environment has significant influence on urban air temperature, which is considerably different from the temperature observed in rural areas (a phenomenon known as the urban heat island effect). Urban structures absorb solar heat in the day and release it back to the environment at night. As a result of the city warming, higher ambient temperatures may be observed in the days and nights. Many local aspects of cities, such as topography, building geometry and dimensions, building materials, building density, urban canyons, traffic and greenery, may have a significant impact on urban temperatures. Given the fact of climate change, studying the changes in urban temperature is of great importance. Meanwhile, buildings' increasing energy consumption due to rising urban air temperatures is of great concern, since the world is facing another round of energy crises. The first part of this chapter discusses the influence of urban morphology on urban microclimatic conditions, and outlines strategies to ameliorate the increasing urban air temperature through urban planning. The chapter then discusses the impact of urban air temperature on building design, especially multi-storey commercial and residential buildings, and proposes some mitigation measures to achieve an energy-efficient building.

The urban heat island and climate change

The climate is the weather averaged in some locations over a long period of time, but it is not uniform in different places, such as cities and rural areas, even when they are close to each other. A built-up environment has significant influence on urban air temperature, which is considerably different from the climate observed in proximate rural areas, and this is known as the urban heat island effect. This phenomenon is present in all cities of all climatic regions and occurs due to urbanization where human modifications alter the characteristics of the Earth's surface and atmosphere. Urban structures absorb solar heat in the day and release it back to the environment at night. As a result of a city's structures absorbing solar heat, it warms up, and higher ambient temperatures (than in the surrounding suburban or rural areas) may be observed there both day and night. Many aspects of cities, such as their topography, building geometry and dimensions, building materials, building density, urban canyons, traffic and greenery, have considerable impact on their temperature profile.

In recent years, the urban heat island (UHI) issue has been examined by various researchers reporting the UHI intensity in their respective cities (Giridharan *et al.* 2007; Liangmei *et al.* 2008; Padmanabhamurty 1990–1991; Sani 1990–1991; Wong and Chen 2009; Grimmon

2011). UHI intensity can easily be 10°C. In the city of Athens, where the mean heat island intensity has been found to exceed 10°C, the cooling load of buildings in the urban area was found to be double and the peak electricity load for cooling may be three times higher because of higher ambient temperatures (Santamouris 2001).

The higher temperatures caused by the UHI effect also have an impact on the increasing demand for cooling energy in cities. Increased demand for energy costs users additional money for air-conditioning in order to maintain comfort levels. As a result, more electrical energy production is needed, which triggers the release of more greenhouse gases due to the combustion of fossil fuel. A feedback loop occurs as greenhouse gases are likely contributing to the global warming that is causing climate change, raising the urban air temperature and making people switch on their air-conditioning.

In the near future, the potential risk of the UHI may be enhanced due to climate change. Given a range of greenhouse gas emissions and climate sensitivities to change in radiative forcing, the globally averaged near-surface air temperature is projected to increase by 1.4°C to 5.8°C between 1990 and 2100 (IPCC 2000). Climate change has the potential to alter the intensity and increase the spatial extent of UHI in cities. Since UHI intensification is also associated with more pollution, in the future higher temperatures and changes in cloud cover could lead to higher rates of smog formation, and lower wind speeds may keep pollutants concentrated over urban areas, with negative consequences for human, animal and plant health.

Recent UHI research has been related to climate change issues. Fujibe (2011) studied urban warming in 12 Japanese cities that have a population of more than one million inhabitants. The study analysed data from weather stations that were established in the early years of the twentieth century in cities that have undergone rapid urbanization. The study found that the urban heat island is influenced by the background air temperature, which is increasing due to climate change. A warming trend of 0.3 to 0.4°C per decade was observed, even for cities with a low population density (fewer than 100 people per square kilometre). Micro-scale factors, such as trees and buildings, are also found to affect urban air temperatures.

Emmanuel and Kruger (2012) argued that, given the relatively small land size of cities (only 2.8 per cent of the Earth's surface), cities themselves do not directly cause global climate change. Rather, cities indirectly propagate global climate change through their appetites for energy and raw materials. The further increase in urbanization leads to land cover changes in cities, which directly influence the microclimate and regional climate, and perhaps change the global climate, for example, the El-Nino Southern Oscillation (ENSO) (Vörösmarty *et al.* 2000). However, due to the lack of detailed data on land cover and urban morphology, global climate models (GCMs) do not include urban areas' representation, which leads to underestimation of temperature projections for urban areas (Kershaw *et al.* 2010).

Therefore, climate models with high temporal and spatial resolutions that incorporate the effects of urbanization and account for urban modification of microclimatic conditions are needed. Coupled with data, this could provide a quantitative description of how the intensity of the urban heat island effect may interact with climate projections for the twenty-first century with respect to temperature, precipitation, wind speed and cloud cover.

Mitigation measures that incorporate knowledge of the interaction between climate dynamics, including climate change, and the built environment could provide solutions to air quality and public health problems, thereby improving regional quality of life in the future. UHI mitigation strategies are becoming essential for the future of urban development. It is increasingly important to study urban microclimatic environments and to apply these findings to improve people's comfort and to reduce the harmful impact of climate change.

Methods to mitigate urban temperature profiles and energy consumption

To mitigate the negative effects of the UHI, two different sets of strategies are commonly used: at the urban level and at the building level.

Urban design approaches

A number of studies relate the urban air temperature conditions to the built environment. At a city's micro-scale, buildings and vegetation influence the incident solar radiation received by urban surfaces. Different urban morphology conditions create different urban microclimate conditions. Researchers quantified the urban morphology with several parameters, such as sky view factor (SVF), building height, greenery density and so on, to understand their correlations and subsequently to predict the microclimate condition.

Urban morphology

Bottyan and Unger (2003) developed a statistical model to determine the relationship between UHI intensity and urban surface parameters such as sky view factor (S), building height (H), built-up ratio (B) and water surface ratio (W). A multiple regression analysis was performed and a strong linear relationship was found to exist between the mean maximum UHI intensity and the urban surface parameters, with a multiple r^2 of 80.1 per cent and 84.5 per cent in the heating and non-heating seasons respectively. Application of the model showed that considerable similarities may be seen between the predicted and measured UHI intensity. This served as a basis for its application to other cities and potential development by adding more meteorological parameters.

Giridharan and colleagues (Giridharan *et al.* 2007) revealed that seasonal changes had more influence on Hong Kong's UHI than changes due to geographical characteristics. Their results showed that, among the seasonal models, the peak summer clear sky daytime and nocturnal models displayed the clearest indicators of the impacts of urban design variables with R^2 values of 0.7 and 0.8 respectively. Critical variables that can mitigate both daytime and nocturnal UHI are sky view factor, surface albedo, altitude and vegetation above 1 m in height. These elements were suggested for consideration in the concept stage of urban design.

In Singapore, Jusuf (2009) developed a Screening Tool for Estate Environment Evaluation (STEVE), an empirical model that correlates maximum temperature (T_{max}), average temperature (T_{avg}) and minimum temperature (T_{min}) of a point of interest with a radius of 50 metres in an estate with three urban morphology parameters (building, pavement and greenery), as shown in Equation 1.

$$T_{min}\ (^{\circ}C) = 4.061 + 0.839\ T_{min-r}\ (^{\circ}C) + 0.004\ PAVE\ (\%) - 0.193\ GnPR - 0.029\ HBDG + 1.339E\text{-}06\ WALL\ (m^2)$$

$$T_{avg}\ (^{\circ}C) = 2.347 + 0.904\ T_{avg-r}\ (^{\circ}C) + 5.786E\text{-}05\ SOLAR_{total}\ (W/m^2) + 0.007\ PAVE\ (\%) - 0.06\ GnPR - 0.015\ HBDG + 1.311E\text{-}05\ WALL\ (m^2) + 0.633\ SVF$$

$$T_{max}\,(^{o}C) = 7.542 + 0.684\,T_{max-r}\,(^{o}C) + 0.003\,SOLAR_{max}\,(W/m^2) + 0.005\,PAVE\,(\%) - 0.016\,HBDG + 6.777E\text{-}06\,WALL\,(m^2) + 1.467\,SVF + 1.466\,ALB \qquad (1)$$

The parameters in Equation 1 may be categorized into two groups of predictors. The first, *climate predictors*, are the daily minimum (T_{min-r}), average (T_{avg-r}) and maximum (T_{max-r}) temperatures at the meteorological station. For the SOLAR predictor, the average of total daily solar radiation ($SOLAR_{total}$) was used in T_{avg} models, while the average of the day's solar radiation maximum ($SOLAR_{max}$) was used in the T_{max} model. The second group of predictors are *urban morphology* elements: the percentage of pavement area (PAVE), average height to building area ratio (HBDG), total wall surface area (WALL), Green Plot Ratio (GnPR), sky view factor (SVF) and average surface albedo (ALB). Equation 1 reflects the interactions between these parameters in different time frames. T_{max}, T_{avg} and T_{min} represent the day's peak temperature (i.e. during daytime), its average temperature and lowest temperature (i.e. during night-time).

Extensive use of concrete and other heat-absorbing surfaces (PAVE) increases the urban air temperature by decreasing the surface moisture available for evapotranspiration. Furthermore, more solar radiation is absorbed and reradiated into heat because dry surfaces have low albedo (ALB) value (Santamouris 2001). In terms of building bulk, the higher the thermal mass of the environment (HBDG), the lower the temperature, since it reduces the heat released to the surrounding environment. Meanwhile, large wall surface area (WALL) leads to a higher air temperature, since the wall reflects short-wave and long-wave solar radiation to the environment.

Many researchers have identified the relation between the urban heat island and canyon geometry, especially the SVF. Low SVF leads to high nocturnal air temperature (Oke 1981; Oke *et al.* 1991; Chapman *et al.* 2001). Based on several months of continuous temperature measurement in Singapore, a study (Wong and Jusuf 2009) concluded that, during daytime, there is a strong correlation between the SVF and air temperature. A higher SVF leads to a higher air temperature, which means there is no obstruction for solar radiation to heat up the environment. During night-time, specifically at around 05.00 to 07.00 hours (GMT+8), a very weak correlation between the SVF and air temperature was found.

Greenery is an important factor in Singapore's urban development. It creates a pleasant and more liveable urban environment. Greenery, calculated as Green Plot Ratio (GnPR), has a positive impact in reducing air temperature. During daytime, trees provide shading, while at night-time trees' evapotranspiration process creates a cooling effect. The relationship between SVF and GnPR on trees is estimated as: every increase of GnPR reduces the SVF by 0.2 (Jusuf 2009).

The complex interactions between various aspects of urban morphology are shown in a sensitivity analysis study of an ideal type of urban canyon in order to simplify the variation of building, pavement and greenery distributions (see Figure 17.1a). The *greenery density* (GnPR) may affect the SVF; the *building height* affects the SVF, WALL and HBDG values; and *canyon width* (see Figure 17.1b) affects the SVF, PAVE and HBDG values.

The effect of the SVF reduction due to greenery varies the behaviour of the predicted T_{min}, T_{avg} and T_{max} (see Figure 17.2). The increase of GnPR mainly governs the reduction of minimum air temperature, shown as a straight line. Meanwhile, in the T_{avg} model, both GnPR and SVF influence the average temperature. When there is no greenery on site (GnPR = 0), the sky openness (SVF) is determined mainly by the canyon geometry. Greenery provides

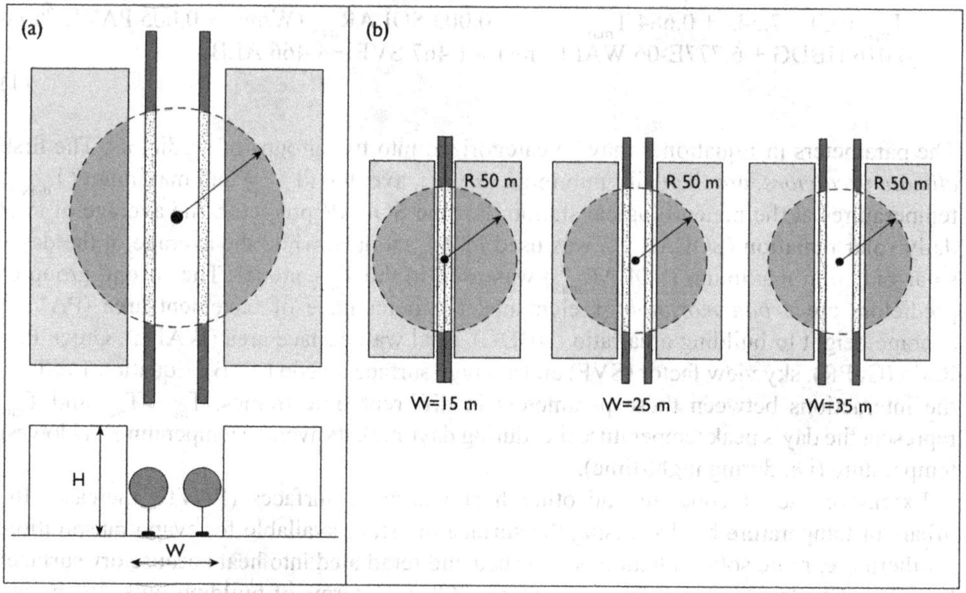

Figures 17.1a and 17.1b Building layout model for sensitivity analysis (left); building layout model 3 on change of canyon width (right).

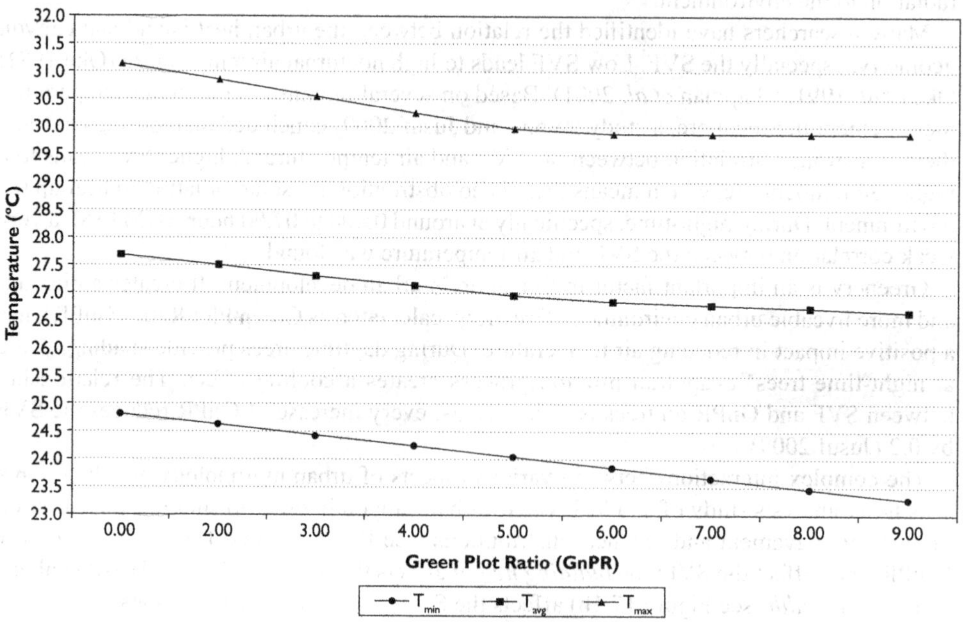

Figure 17.2 Predicted T_{min}, T_{avg} and T_{max} for various values of GnPR.

shading to the environment and reduces the SVF; thus, it reduces the air temperature. The reduction of average temperature is more prominent when GnPR is increased by between 0 and 4 due to the combination factors of GnPR and SVF. Once urban greenery completely covers up openness to the sky, only the greenery density reduces the average air temperature, shown as a broken sloping line. On the other hand, greenery density has an indirect impact on the maximum air temperature. An increase in greenery density affects openness to the sky, which reduces the maximum air temperature through its shading effect. Hence, as seen in Figure 17.2, when trees have provided complete shading to the canyon at GnPR of 4, an increase in greenery density will not reduce the maximum air temperature any further.

The influence of building height on the predicted average air temperature is clear (see Figure 17.3). Looking at the 'V' profile in the graph where GnPR = 0, the average air temperature reduces when the building level is between 2 and 16. It shows that an increase in building height reduces openness to the sky (SVF), i.e. by providing shade. Once the canyon is completely shaded by the surrounding buildings, further increase in building height (which further increases the WALL areas and the built density of the environment) starts having an impact on the average air temperature increase. When the greenery density is increased, the turning point of the graph (from reduced into increased average air temperature) shifts towards lower building heights, showing the relationship of GnPR in reducing SVF value.

Figure 17.4 shows the predicted air temperature due to an increase in canyon width. In the T_{min} model, the increase in canyon width increases the pavement area (PAVE), which also increases the minimum air temperature, shown as an increasing trend when the canyon width is changed progressively from 15 m to 45 m. Meanwhile, in the T_{avg} and T_{max} models, the increase in average and maximum air temperatures is governed by the increase in PAVE and

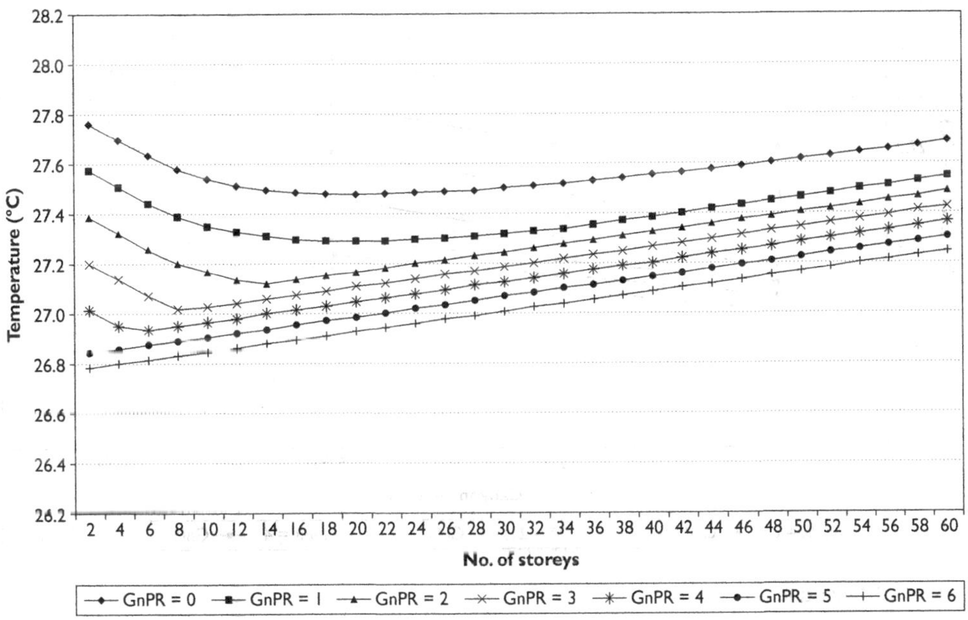

Figure 17.3 Predicted T_{avg} for various values of number of storeys and GnPR.

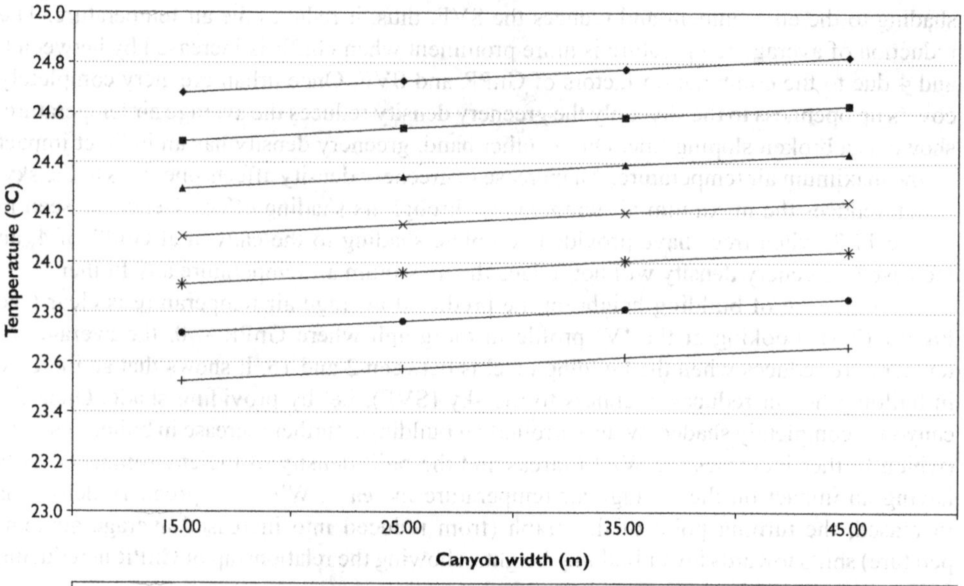

(a)

Figures 17.4a, 17.4b and 17.4c Predicted T_{min} (a), T_{avg} (b) and T_{max} (c) for various values of number of storeys and GnPR

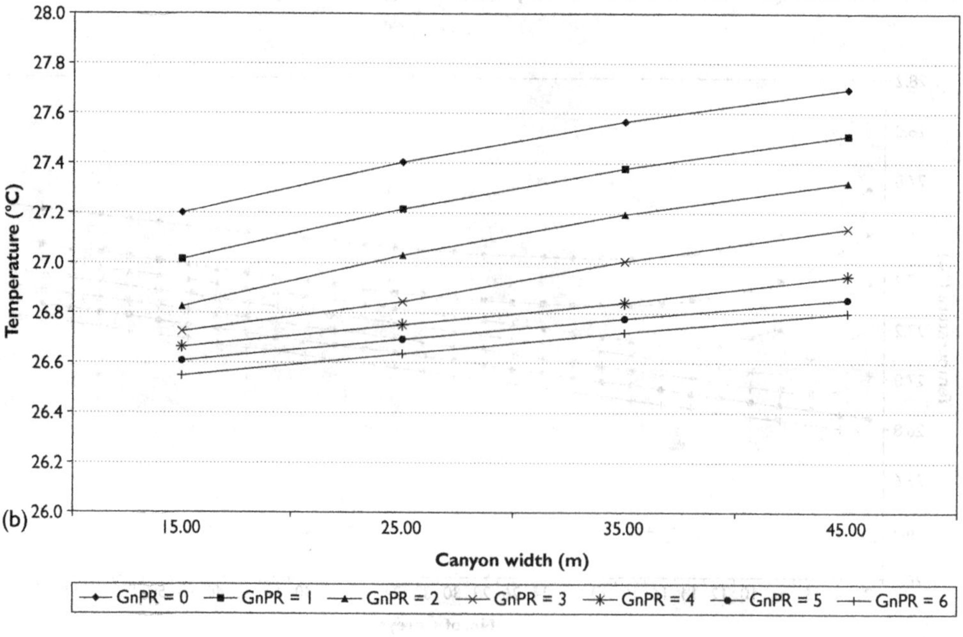

(b)

Figure 17.4 continued

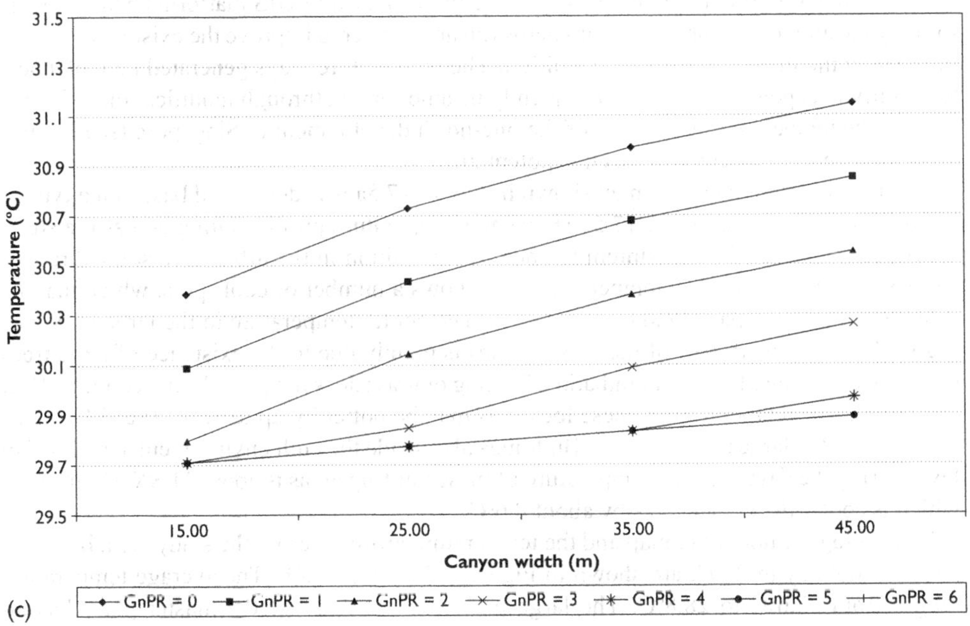

(c)

SVF values (see Figures 17.4a and 17.4b). At the same canyon width (e.g. at 15 m), in the T_{avg} model, the reduction of air temperature is governed by both GnPR and its impact on the reduction of SVF. When the greenery completely shades the canyon (SVF = 0), only the greenery density (GnPR) reduces the air temperature. On the other hand, at the same canyon width in the T_{max} model, once the SVF is 0, further increase in greenery density (GnPR) will not reduce maximum air temperature any further. Figure 17.4c shows the GnPR value yields at GnPR = 5.

This study has shown the relationships between some major elements of urban morphology. It is important for urban planners to find the optimum distribution of greenery, building and pavement. Increasing the building height does not always have a negative impact (i.e. it does not necessarily increase air temperature). To a certain extent, it provides shading to the environment. On the other hand, widening the canyon width may have an adverse impact during daytime, since it increases openness to the sky, which increases the incoming solar radiation. In this situation, greenery plays an important role in reducing the air temperature, through its evaporative cooling and also its shading. In these sensitivity analyses, the greenery density was calculated and limited only to the fixed greenery area and the density of trees. However, in the planning process, within the same radius of 50 m, the greenery area can be increased, for example, by using vertical greenery and roof-top greening, since the Green Plot Ratio (GnPR) concept is three-dimensional.

The degree of impact on air temperature due to different elements of urban morphology can be up to 1.2°C. With carefully designed building configurations, the building's cooling energy consumption is saved between 5 and 10 per cent. Although 5 per cent may seem a small energy saving, a note should be taken that this saving is only from one building. This energy saving can be compounded at the macro level by 5 per cent through a proper master plan design. When the building design aspect is also addressed, a higher energy saving potential can be expected (Wong *et al.* 2011).

Air temperature prediction models (i.e. Equation 1) used in GIS platforms help planners to study the air temperature conditions across an area, either to improve the existing condition or to study the impact of a future condition. The temperature maps generated help planners to identify hot spots, which can subsequently be ameliorated through modifications of urban forms. The temperature map study of the one-north development in Singapore (see Figures 17.5, 17.6 and 17.7) demonstrates this potential.

The maximum temperature map shown in Figure 17.5a was developed based on a typical hot day in 2007, while Figure 17.5b shows the temperature profile cutting across the study area (Section A–A). The maximum temperature condition in the urban area studied ranges from 31.1°C to 32.8°C. The temperature map shows a number of cool spots, which may be categorized as large parks and built-up areas. The cooler temperature in the large park (see Figure 17.5a, northern part of the master plan) is mainly due to the existence of large trees that provide shading from solar radiation. Turfing or grass does not provide any cooling effect (Point 25). The temperature in these locations may be hotter by up to 1.3°C (i.e. 32.4°C) as compared to the large park (31.1°C). Buildings also shade the built environment at pedestrian level during the daytime. The temperature of these built-up areas is about 31.8°C (Point A4), which is cooler than open areas by about 0.6°C.

The average temperature map and the temperature profiles across the study area, based on a typical hot day in 2007, are shown in Figures 17.6a and 17.6b. The average temperature ranges from 27.6°C to 28.8°C. The large park (northern part of the master plan) has an average temperature of around 27.6°C, while the built-up areas (Point A4) have an average temperature of around 28.1°C. As compared to the open areas (Point 25), the park is cooler by 0.8°C. The buildings in the built-up area release the heat that has been absorbed during the daytime, as shown in the minimum temperature map (Figure 17.7a), which reflects the night-time temperature condition. The temperature is around 25.9°C (Point A4). Greenery inside the built-up area (Point 24) helps to ameliorate the temperature condition, by 0.6°C on average, through its evapotranspiration. The following section discusses the impact of greenery on the built environment in more detail.

Urban parks

Introducing greenery in cities may be the most effective strategy for mitigating the heat island effect. The presence of greenery in urban areas helps to cool the environment through the process of evapotranspiration, which converts a large amount of solar radiation into latent heat. Various forms of greenery may exist in urban areas, including green roofs, vertical greenery, nature reserves, urban parks and extensive planting of vegetation. Such greenery provides sources of moisture for evapotranspiration, allowing the absorbed solar radiation to be dissipated as latent heat and thus aid in reducing urban temperature (Chen and Wong 2006). The use of vertical greenery can significantly reduce the incoming solar radiation through a building's concrete façade if the shading coefficients of the plant species are low, leading to better thermal insulation (Wong *et al.* 2009).

In the tropical city of Singapore, Wong and Chen (2005) conducted a macro-level study to explore the cooling impact of greenery on the UHI. They concluded that the presence of large areas of greenery in a city mitigates the UHI effect. A maximum temperature difference of 4.01°C was observed between areas of dense greenery and in the central business district (CBD). At macro levels, the cooling effects of greenery could be observed from the mobile surveys where routes near large green areas experienced lower temperatures than industrial,

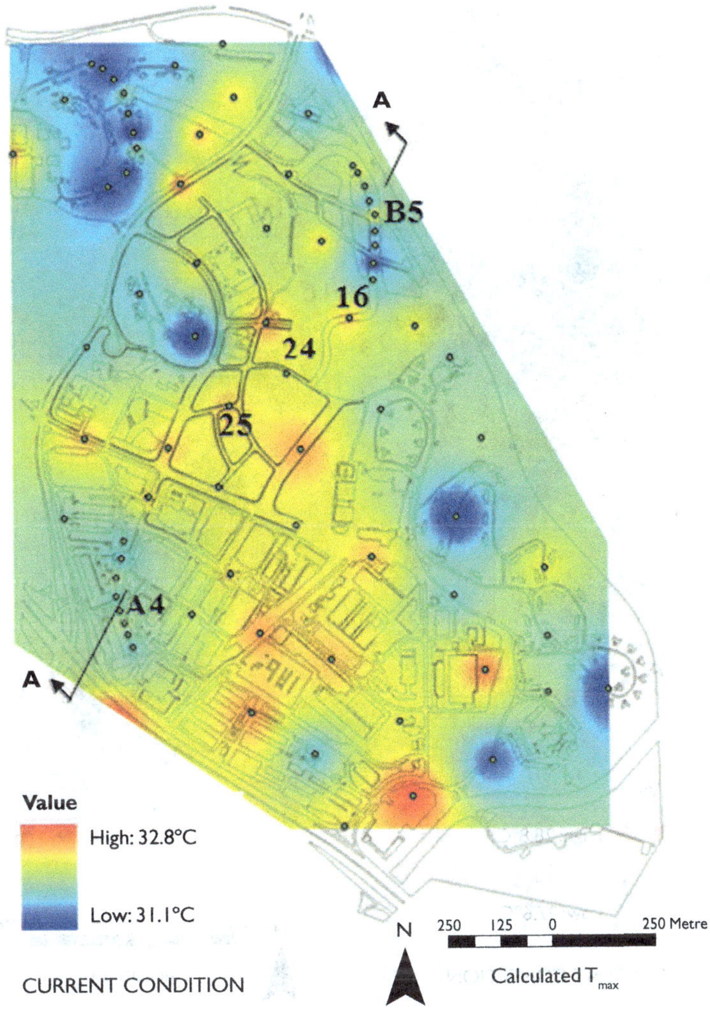

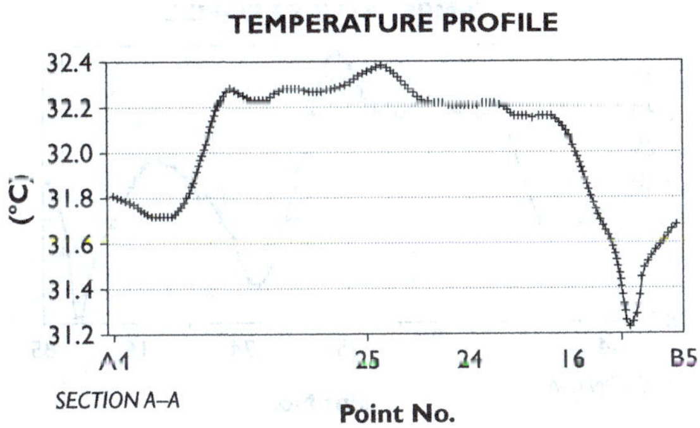

Figures 17.5a and 17.5b Maximum temperature map of one-north, Singapore (top) and section A–A temperature profile (bottom).

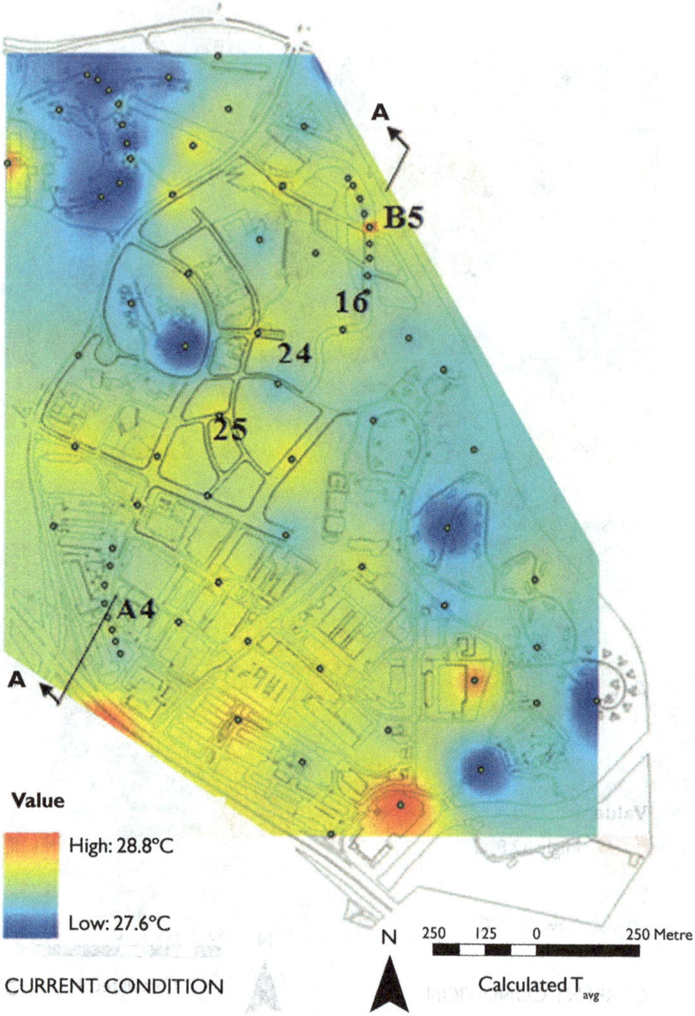

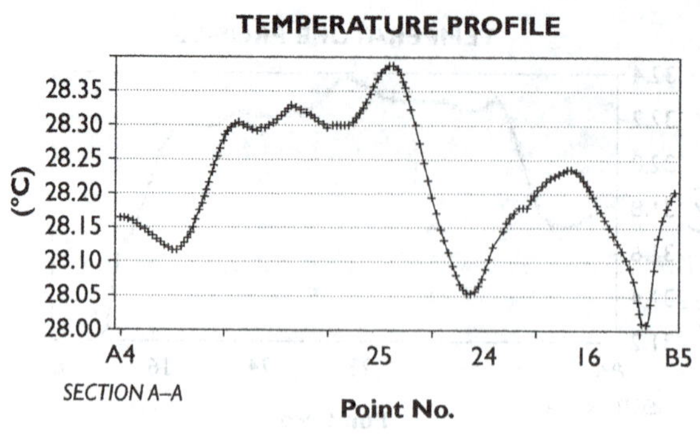

Figures 17.6a and 17.6b Average temperature map of one-north, Singapore (top) and section A–A temperature profile (bottom).

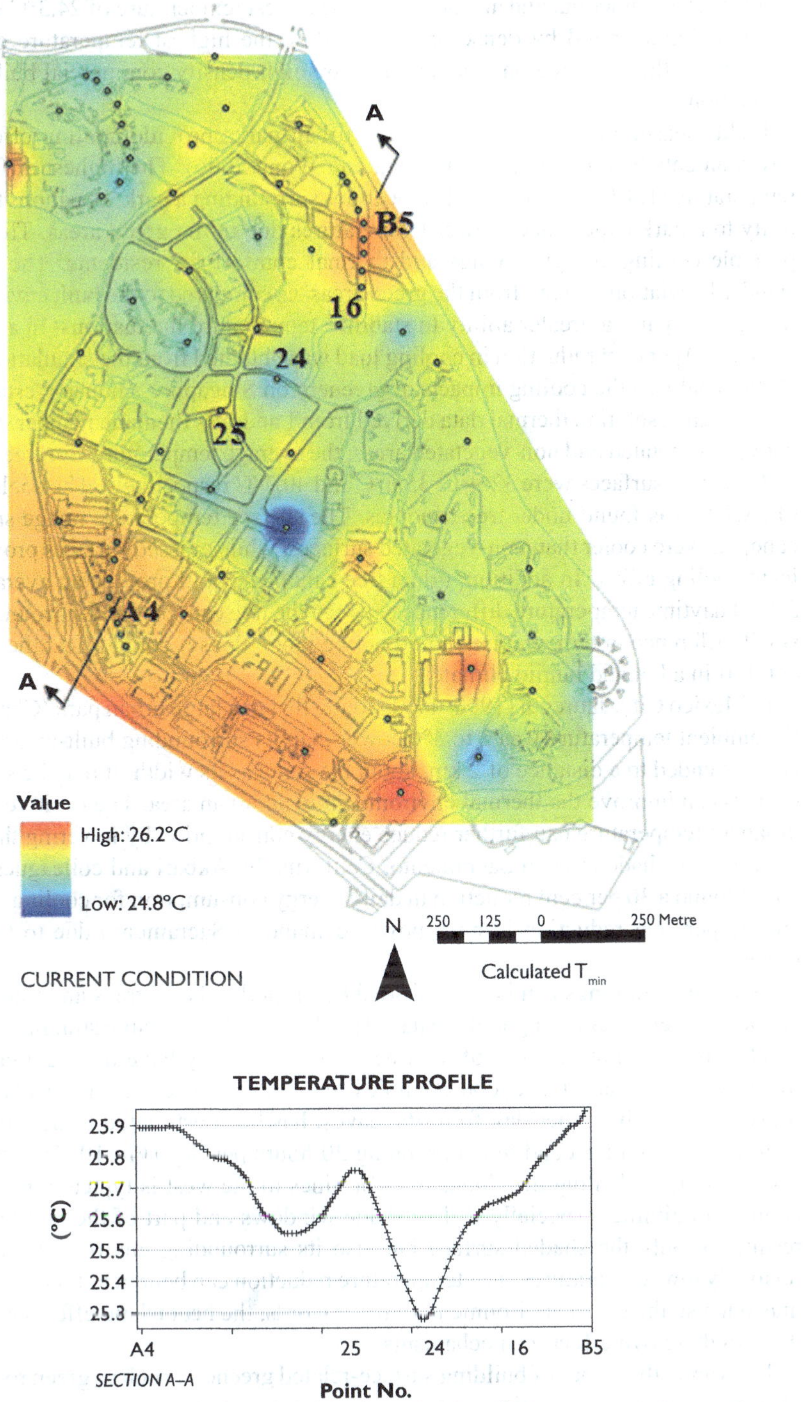

Figures 17.7a and 17.7b Minimum temperature map of one-north, Singapore (top) and section A–A temperature profile (bottom).

residential, commercial and airport areas. The lowest temperature of 24.3°C was observed in an area characterized by dense greenery, while the highest temperature of 28.31°C was recorded in the CBD region, characterized by high-density commercial buildings and little vegetation.

Field measurements have proved that urban parks provide cooling effects for nearby environments both day and night (Chen and Wong 2006). The highest difference in mean temperature (1.3°C) was recorded at locations surrounding a park. Locations that are in proximity to a park experienced lower temperatures due to the green areas. This could lead to possible cooling energy savings and thermal comfort for residents. The small range of standard deviation derived from the green areas, as compared to the built environment, shows that greenery has a greater ability to stabilize temperature fluctuations. In addition, a maximum of 10 per cent reduction in cooling load was obtained from the simulation study. Nichol (1996) analysed the cooling impacts of greenery on Singapore's housing estates through the use of high-resolution thermal data derived from Landsat's thematic mapper. In a comparison between vegetated and non-vegetated areas, the average temperatures for tree canopies, grass and concrete surfaces were 32.9°C, 35.6°C and 40.7°C respectively. The coolest temperature of 28.7°C was found under tree canopies. The surface temperature image showed that tree canopies were cooler than non-vegetated surfaces, while grassed surfaces provided no significant cooling effect. In addition, urban tree canopies could provide an average of 1.5°C to 2°C of daytime temperature difference between the tree canopies and surrounding areas. This small difference in temperature should not be ignored, as it could affect the level of human comfort in a hot and humid climate.

In Mexico City, Jauregui (1990–1991) found that in a large urban park (Chapultepec Park), the ambient temperature was 2 to 3°C lower than its surrounding built-up area and its influence extended to a distance of 2 km, about the same as its width. It may be said that a group of trees can improve the thermal environment in an urban area. The positive impacts of low urban air temperature can further reduce energy consumption by lowering the cooling load. Research on shade trees in Sacramento, California by Akbari and colleagues (Akbari *et al.* 1997) found a 30 per cent reduction in daily energy consumption for cooling and between 27 and 42 per cent reduction in peak power demand in Sacramento due to the city's urban greenery.

Individual buildings can be well shaded by properly placed trees and thus achieve lower surface temperatures throughout the day. The shading effect is quite straightforward and very much depends on the density of the tree. People normally have no quantitative sense of a tree's evaporative ability. A good example is that the cooling effect of an isolated mature tree, transpiring 450 litres per day from its leaves, has been estimated to be equivalent to five average six-room air-conditioners running 20 hours per day (Pitt 1979). Researchers have also found that planting deciduous trees or vines to the west is typically most effective for cooling a building, especially if they shade windows and part of the building's roof. As a result, not only the shaded surface but also its surrounding environment can experience relatively low temperatures. The temperature reduction can benefit not only individual buildings but also the urban environment. In other words, the heat island effect can be moderated through these two effective mechanisms.

Trees and other types of building surface-related greenery, such as green roofs and vertical plantings, can reduce heating and cooling energy use and are associated with lower air pollution and fewer greenhouse gas emissions. They also remove air pollutants, sequester and store carbon, help lower the risk of heat-related illnesses and deaths, improve stormwater control and water quality, reduce noise levels, create habitats, improve aesthetic qualities and increase property values.

Urban design to encourage ventilation through street canyons

Urban design that creates urban canyons, especially deep ones, might encourage ventilation as canyons can act as traps which will increase the loss of both short-wave and long-wave radiation from buildings. Heat trapped inside urban canyons can be dissipated from source areas by turbulent transfer. However, the heat loss from buildings within such streets will be reduced when there is unwanted obstruction of wind flow by urban forms. In this case, ventilation in urban open space is necessary to help reduce the severity of the UHI effect. The major urban design elements that can modify wind conditions are: the overall density of the urban area; the size and height of individual buildings; the presence of high-rise buildings, and the orientation and width of streets (Givoni 1998).

In Tsim Sha Tsui, a densely populated area of Hong Kong, Memon and Leung (2010) investigated the impact of wind speed on urban heat island intensity (UHII) using meteorological parameters collected over a 15-year period (1990 to 2005). Strong negative correlations were obtained in the regression analyses that were performed to understand the influence of wind speed (WS) and cloud cover (CC). The correlation between UHII and WS (see Equation 2) shows a strong negative correlation, where an increase in wind speed of 1 m/s decreased UHII by 1.9°C. Meanwhile, the correlation between UHII and both WS and CC also shows a strong negative correlation, where an increase in wind speed of 1 m/s and 0 oktas of cloud cover decreased UHII by 1.5°C (see Equation 3).

$$UHII = -1.9WS + 6.7; \qquad R^2 > 0.8 \tag{2}$$

$$UHII = 0.6 - 1.5WS + 0.9CC; \qquad R^2 > 0.8 \tag{3}$$

Several urban design strategies may be used to promote good ventilation within the urban setting (Wong and Chen 2009; Ng 2009). The first strategy is to create and maintain air paths and breezeways – especially at ground level. This is to ensure that the prevailing wind travelling along breezeways and major roads can penetrate deep into the district by linking open spaces and maintaining low-rise structures along the prevailing wind direction. Proper building orientation with adequate gaps between buildings also enhances natural ventilation. The main streets and breezeways should be aligned in parallel or up to 30° to the prevailing wind direction, in order to maximize the penetration of prevailing winds through the estate. The building axis should be parallel to the prevailing wind to avoid obstruction (see Figure 17.8). Increasing permeability is another important strategy that can improve the ventilation for pedestrians, for example, provision of void decks at ground level and a garden at mid-level. 'Void deck' is a term used in Singapore for a space in public housing apartments located at the ground level that is left empty for community activity (see Figure 17.9). Varying the height of buildings also significantly improves the wind penetration of buildings in urban settings (see Figure 17.10).

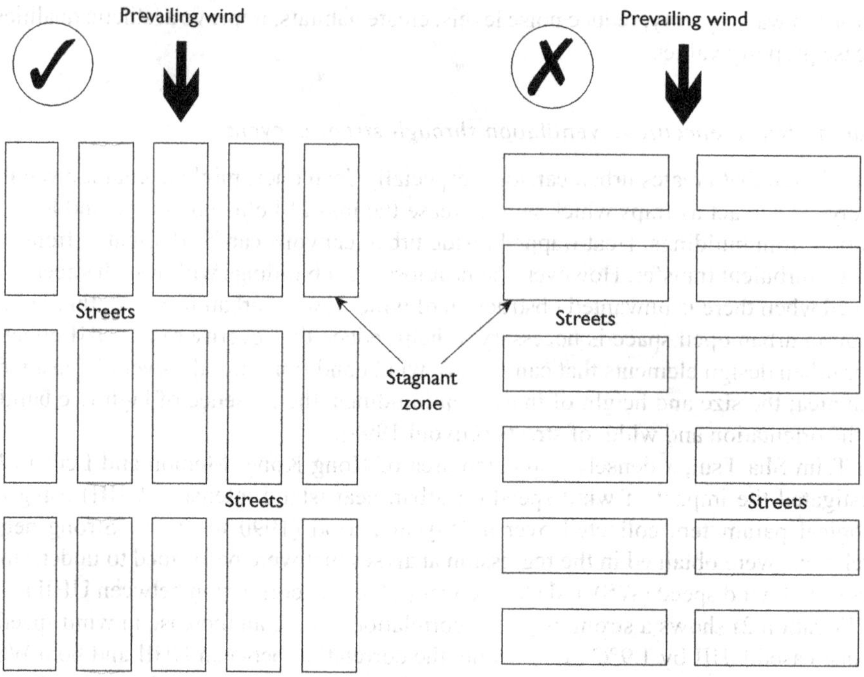

Figure 17.8 Orientation of street grids.

Source: Ng (2009).

Figure 17.9 A public housing apartment in Singapore with an empty space at ground level, known as the void deck.

Building design approaches

The second level of UHI mitigation strategy is through building design. One of the most important strategies is to limit the amount of heat gain through building envelopes. Envelope Thermal Transfer Value (ETTV), along with its predecessor Overall Thermal Transfer Value (OTTV), is used in Singapore as an indicator for buildings' thermal performance. ETTV has been set at 50 W/m^2 by Singapore's Building Construction Authority (BCA). The effects and use of Envelope Thermal Transfer Value (ETTV) to improve a building's energy performance have been studied using DOE-2 (energy simulation software) (Chua and Chou 2010). Chua and Chou have demonstrated that the cooling energy a building requires will drop by 2.5 per cent if the ETTV is reduced from 50 W/m^2 to 45 W/m^2.

ETTV considers the three basic components of building heat gain through external walls and

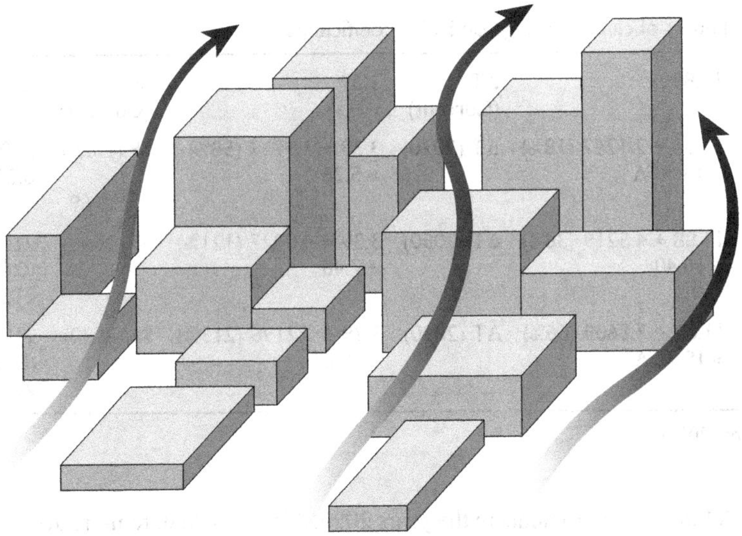

Figure 17.10 Example of varying building heights.

Source: Ng (2009).

windows: (1) heat conduction through opaque walls; (2) heat conduction through fenestration; and (3) solar heat gain through fenestration. These three components are then averaged over the building envelope area to obtain the ETTV, which is given by Equation 4 below.

$$ETTV = TD_{eq(wall)}(1-WWR)U_w + \Delta T(WWR)U_f + SF(WWR)(SC) \quad (4)$$

Whereby:

- ETTV : Envelope Thermal Transfer Value (W/m^2)
- $TD_{eq(wall)}$: heat conduction through opaque walls
- WWR : window-to-wall ratio (fenestration area/gross area of exterior wall)
- U_w : thermal transmittance of opaque wall (W/m^2K)
- ΔT : heat conduction through fenestration
- U_f : thermal transmittance of fenestration (W/m^2K)
- SF : solar heat gain through fenestration
- SC : shading coefficient of fenestration

The coefficients $TD_{eq\,(wall)}$, ΔT and SF vary according to the weather of the locality of interest and have been determined, using computer simulations with Singapore weather files, to be of the values 11.88, 3.39 and 210.92 respectively (Goh 2001).

In view of the increasing air temperature due to climate change, it is imperative that the current regulations on building heat gain should be adjusted in order to keep buildings energy efficient. Based on parametric study, Chong (2012) studied the impact of climate change on

Table 17.1 Impact of climate change on ETTV coefficients

$TD_{eq\ (wall)}$ (Current)	11.88 Δ	ΔT (Current)	3.39	SF (Current)	210.92
$TD_{eq\ (wall)}$ (2020)	11.88 + 2.1267 (18%) ≈ 14.00 Δ	ΔT (2020)	3.39 + 1.9737 (58%) ≈ 5.36	SF (2020) ≈ 210.69	210.92 - 0.2293(0.001%)
$TD_{eq\ (wall)}$ (2050)	11.88 + 4.5219 (38%) ≈ 16.40	ΔT (2050)	3.39 + 4.0927 (121%) ≈ 7.48	SF (2050)	210.92 + 1.8069 (0.008%) ≈ 212.73
$TD_{eq\ (wall)}$ (2080)	11.88 + 7.8609 (66%) ≈ 19.74 Δ	ΔT (2080)	3.39 + 7.2196 (213%) ≈ 10.61	SF (2080)	210.92 + 0.6645 (0.003%) ≈ 211.58

Source: Chong (2012).

$TD_{eq\ (wall)}$, ΔT and SF coefficients in the years 2020, 2050 and 2080 to more accurately predict the total heat gain through the building envelopes. Table 17.1 summarizes the impact of an A2 climate change scenario on each of the three weather-dependent coefficients. From Table 17.1, it may be observed that climate change would have the greatest impact on the weather-dependent component of conduction through fenestration (ΔT), with its coefficient increasing by 58 per cent, 121 per cent and 213 per cent in the years 2020, 2050 and 2080 respectively. Hence, it is recommended that building designers use glazing with a lower U-value to mitigate the effect of this increase. Alternatively, the WWR may also be reduced. This, however, may increase the amount of heat gain through walls, which, from Table 17.1, appears to be substantial in absolute value as well. On the contrary, climate change does not seem to have a significant impact on the solar heat gain into buildings as is evident from the low change in SF values, which ranges from 0.001 per cent to 0.008 per cent. This is reasonable considering that climate change scenarios focus on anthropogenic heat gain, which has a larger impact on the ambient air temperature and humidity levels than solar radiation. Hence, mitigation measures should be aimed at reducing conductive heat gain through the walls and windows of building envelopes.

Assuming a reference office building design with typical office building envelope design variables of $U_w = 2.16\ W/m^2K$, $U_f = 2.97\ W/m^2K$, SC = 0.37 and WWR = 0.40, the calculated ETTV for the reference building using Equation 4 would be 50 W/m^2, which is within the limit set by BCA's code on envelope thermal performance. Table 17.2 illustrates the increase in ETTV in the years 2020, 2050 and 2080, and portrays a significant increase in ETTV with

Table 17.2 Increase in ETTV for reference building

	Reference building	*% increase (base = 50W/m²)*
ETTV (Current)	50.00 W/m²	0.0
ETTV (2020)	55.69 W/m²	11.4
ETTV (2050)	61.62 W/m²	23.2
ETTV (2080)	69.50 W/m²	39.0

Source: Chong (2012).

climate change. Hence, using the ETTV (current) formula (Equation 4) to predict heat gain through building envelopes would not be an adequate representation in the future, due to changes in climatic conditions. Depending on the lifespan of the building, building designers may choose the coefficients that are applicable in the years 2020, 2050 or 2080. Substituting the applicable coefficients presented in Table 17.1 into Equation 4, building designers would then be able to design against these changes in climatic conditions, therefore mitigating against these possible increases in heat gain through building envelopes and reducing their impact on cooling energy consumption, which can be significant in hot and humid Singapore.

The integration of vegetation with building façades and roof-tops is a great way to improve a building's thermal envelope. Gardens can be planted on the roof-tops of existing buildings to reduce the RTTV, and studies (e.g. Wong *et al.* 2003) have shown that the effects are quite significant, reducing building energy consumption by up to 15 per cent, 17 to 79 per cent in space cooling load and reducing RTTV of 17 to 81 per cent, as shown in Figure 17.11 and Figure 17.12 respectively. Shade (provided by plants) from solar radiation, coupled with the additional insulation from the layer of moist soil, can have a positive impact on a building's RTTV and improve its thermal performance (Eumorfopoulo and Aravantinos 1998).

Using building energy simulations TAS, the use of vertical greenery was found to be able to reduce the ETTV of a full-glass façade building by up to 40.68 per cent (Wong *et al.* 2009). The effects of vertical greenery on buildings' cooling energy demand were also studied using TAS. As shown in Figure 17.13, three different types of buildings (1A, 2A, 3A) were compared with fully covered with vegetation versions (1B, 2B, 3B). The results in Table 17.3 show that vertical greenery can reduce building cooling energy by up to 74.29 per cent for Scenario 1, where there are no windows. The reason behind the energy reduction shown in the previous figures is due to the drop in the amount of solar radiation rays received by the building façade, resulting in the reduction of Mean Radiant Temperature (T_{mrt}). As T_{mrt} drops

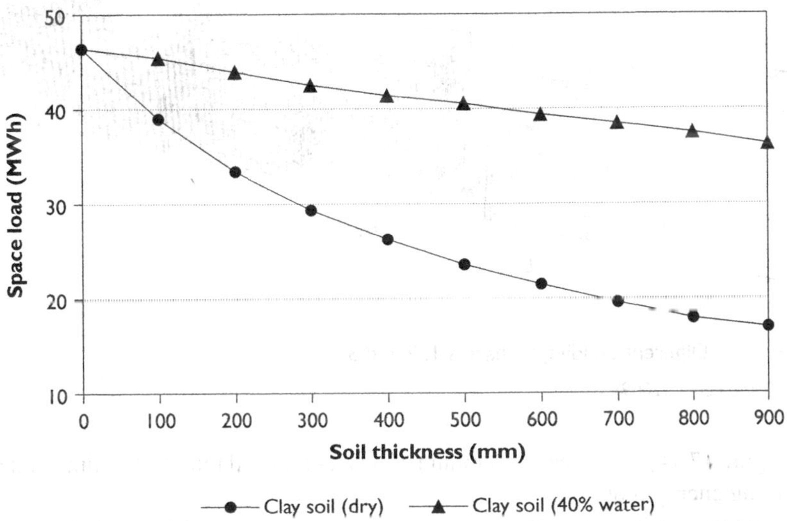

Figure 17.11 Comparison of space load component (total building load) for different soil thickness on roof of a five-storey commercial building.

Source: Wong *et al.* (2003).

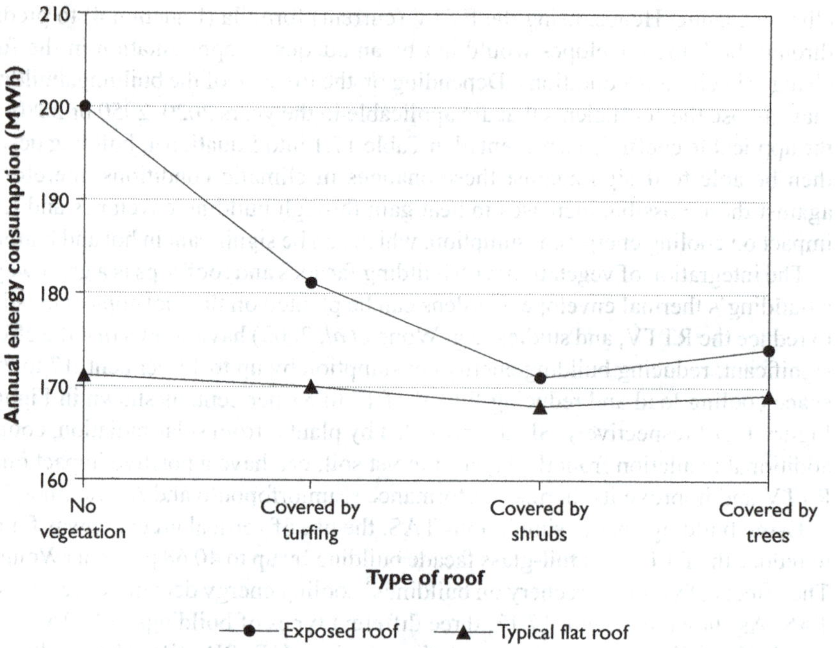

Figure 17.12 Comparison of annual energy consumption for different types of roofs for a five-storey commercial building.

Source: Wong *et al.* (2003).

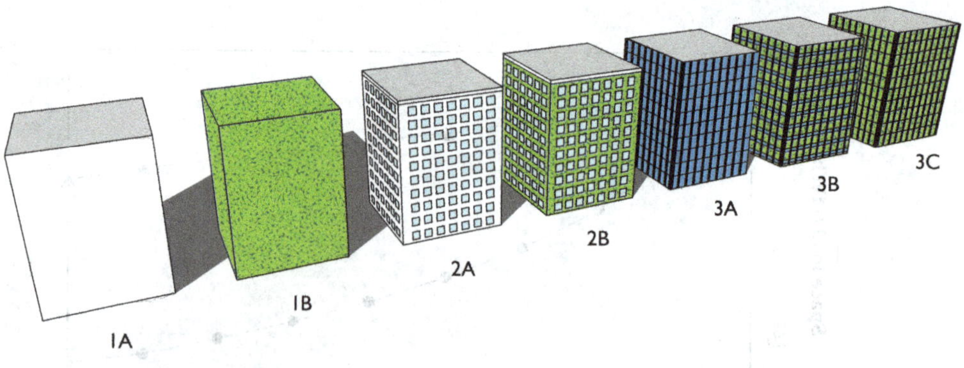

Figure 17.13 Different building scenarios 1, 2 and 3.

Source: Wong *et al.* (2009).

(see Figure 17.14), it reduces the amount of heat conducted into the building and hence a drop in cooling energy required.

The excessive use of fenestration is a recent architectural trend that gives buildings a stylish and modern look but results in a high window-to-wall ratio (WWR). This excessive use of glass can have both positive and negative impacts on the building's energy consumption. The positive impact is the replacement of artificial lighting by natural light; however,

Table 17.3 TAS simulation results

Scenario	*MRT (°C)*		*Energy reduction (%)*
	Minimum	*Maximum*	
1A	30.28	34.39	–
1B	23.20	24.01	74.29
2A	27.71	36.57	–
2B	27.82	35.30	10.35
3A	27.63	49.94	–
3B	27.73	45.81	12.45
3C(i)	27.48	45.03	17.93
3C(ii)	–	–	31.75

Source: Wong et al. (2009).

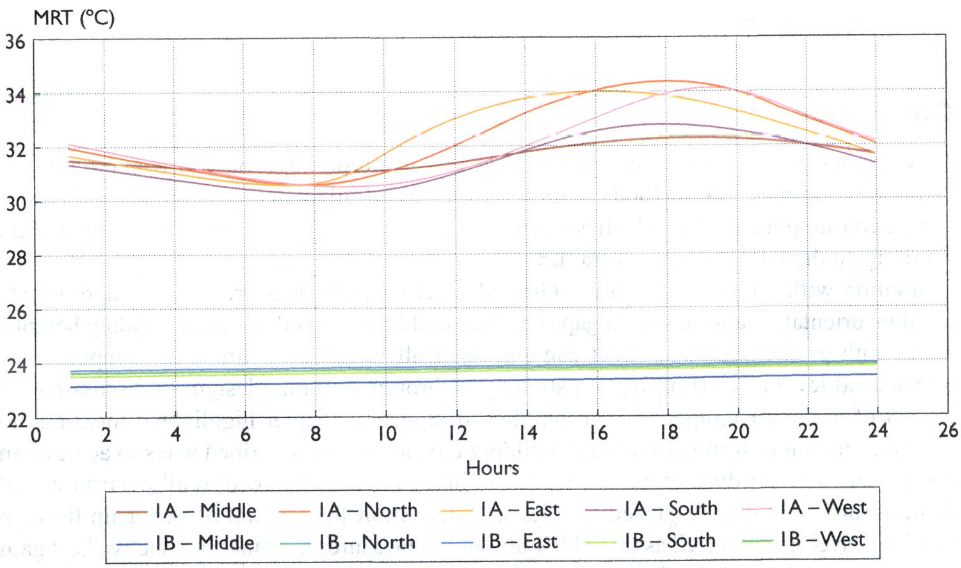

Figure 17.14 Mean Radiant temperature (Tmrt) for Scenarios 1A and 1B of TAS simulation.

Source: Wong et al. (2009).

this can also increase the amount of solar radiation into the building. The use of double-glazed façades can fulfil both aesthetic requirements as well as conserve energy by reducing the amount of internal cooling required to counteract the effect of the solar radiation. In a study of the effects of double-glazed façades on energy consumption, thermal comfort and condensation (Wong *et al.* 2005), results obtained using simulations showed that a double-glazed building's use of energy for cooling can be drastically reduced (by nearly 120 MWh/year: see Figure 17.15) compared to one with a single-glazed façade.

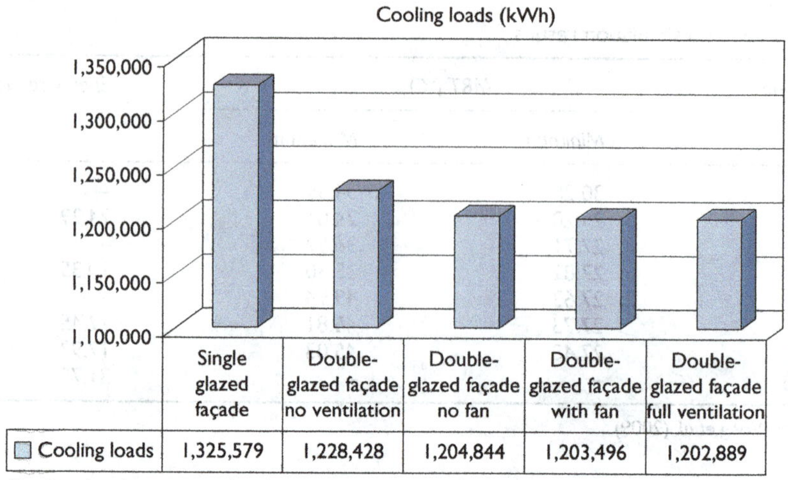

Figure 17.15 Cooling load summary for five scenarios.

Source: Wong *et al.* (2005).

Conclusion

This chapter has examined the influence of urban morphology on urban microclimatic conditions and reviewed methods to mitigate increasing urban air temperatures and reduce energy consumption in tropical cities through urban design and planning and building design. To mitigate the UHI impact, several urban design strategies may be used to promote good ventilation within the urban setting. Ground-level air paths and breezeways, appropriate building orientation and sufficient gaps between buildings as well as varied building heights, along with parks, trees and street canyons, will all help reduce urban air temperatures. The second level of UHI mitigation strategy is through building design. In discussing the impact of urban air temperature on building design, the chapter highlighted measures to limit the amount of heat gain through building envelopes and described ways to achieve an energy-efficient building. It recommended buildings being designed to allow airflow and demonstrated that double-glazed façades will help limit the amount of heat gain through building envelopes. As well as these UHI mitigation measures to reduce conductive heat gain through the walls and windows of building envelopes, and in view of the increasing air temperature due to climate change, it concludes that the current regulations on building heat gain should be adjusted in order to keep buildings energy efficient.

References

Akbari, H., Bretz, S., Kurn, D.M. and Hanford, J. 1997, 'Peak power and cooling energy savings of shade trees', *Energy and Buildings*, vol. 25, no. 2, pp. 139–148.

Bottyan, Z. and Unger, J. 2003, 'A multiple linear statistical model for estimating the mean maximum urban heat island', *Theoretical and Applied Climatology*, vol. 75, pp. 233–243.

Chapman, L., Thornes, J.E. and Bradley, A.V. 2001, 'Rapid determination of canyon geometry parameters for use in surface radiation budgets', *Theoretical and Applied Climatology*, vol. 69, pp. 81–89.

Chen, Y. and Wong, N.H. 2006, 'Thermal benefits of city parks', *Energy and Buildings*, vol. 38, no. 2, pp. 105–120.

Chong, Z.M.A. 2012, *Study on the Impact of Climate Change on Envelope Thermal Transfer Value (ETTV) Regulation of Commercial Building in Singapore*, Master's thesis, National University of Singapore, Singapore.

Chua, K.J. and Chou, S.K. 2010, 'An ETTV-based approach to improving the energy performance of commercial buildings', *Energy and Buildings*, vol. 42, no. 4, pp. 491–499.

Emmanuel, R. and Kruger, E. 2012, 'Urban heat island and its impact on climate change resilience in a shrinking city: the case of Glasgow, UK', *Building and Environment,* vol. 53, pp. 137–149.

Eumorfopoulo, E. and Aravantinos, D. 1998, 'The contribution of a planted roof to the thermal protection of buildings in Greece', *Energy and Buildings,* vol. 27, no. 2, pp. 29–36.

Fujibe, F. 2011, 'Urban warming in Japanese cities and its relation to climate change monitoring', *International Journal of Climatology*, vol. 31, pp. 162–173.

Giridharan, R., Lau, S.S.Y., Ganesan, S. and Givoni, B. 2007, 'Urban design factors influencing heat island intensity in high rise high density environments of Hong Kong', *Building and Environment,* vol. 42, pp. 3669–3684.

Givoni, B. 1998, *Climate Considerations in Building and Urban Design*, Van Nostrand Reinhold, New York.

Goh, T. 2001, 'Enhancing the energy standards' in *BCA Seminar on Energy Efficiency in Building Design*, Building and Construction Authority, Singapore.

Grimmon, C.S.B. 2011, 'London's urban climate: historical and contemporary perspectives', paper presented at *City Weathers: Meteorology and Urban Design 1950–2010,* 23 June, Manchester, www.sed.manchester.ac.uk/architecture/research/csud/workshop/programme/, viewed 1 July 2011.

Inter-Governmental Panel on Climate Change (IPCC) 2000, *Special Report on Emissions Scenarios*, www.ipcc.ch/pdf/special-reports/spm/sres-en.pdf, viewed 24 January 2013.

Jauregui, E. 1990–1991, 'Influence of a large urban park on temperature and convective precipitation in a tropical city', *Energy and Buildings*, vols 15–16, nos 3–4, pp. 457–463.

Jusuf, S.K. 2009, *Development of Estate Level Climate Related Impact Assessment Framework and Air Temperature Prediction within Urban Climatic Mapping Method in Singapore*, Doctoral thesis, National University of Singapore, Singapore.

Kershaw, T., Sanderson, M., Coley, D. and Eames, M. 2010, 'Estimation of the urban heat island for UK climate change projections', *Building Services Engineering Research Technology*, vol. 31, pp. 251–263.

Liangmei, H., Jianlong, L., Dehua, Z. and Jiyu, Z. 2008, 'A fieldwork study on the diurnal changes of urban microclimates in four types of ground cover and urban heat island of Nanjing, China', *Building and Environment*, vol. 43, no. 1, pp. 7–17.

Memon, R.A. and Leung, D.Y. 2010, 'Impacts of environmental factors on urban heating', *Journal of Environmental Sciences,* vol. 22, no. 12, pp. 1903–1909.

Ng, E. 2009, 'Policies and technical guidelines for urban planning of high-density cities: air ventilation assessment (AVA) of Hong Kong', *Building and Environment*, vol. 44, pp. 1478–1488.

Nichol, J.E. 1996, 'High-resolution surface temperature patterns related to urban morphology in a tropical city: a satellite-based study', *Journal of Applied Meteorology*, vol. 35, pp. 135–146.

Oke, T.R. 1981, 'Canyon geometry and the nocturnal urban heat island: comparison of scale model and field observations', *International Journal of Climatology*, vol. 1, nos 1–4, pp. 237–254.

Oke, T.R., Johnson, G.T., Steyn, D.G. and Watson, I.D. 1991, 'Simulation of surface urban heat islands under "ideal" conditions at night. Part 2: Diagnosis of causation', *Boundary-Layer Meteorology*, vol. 56, pp. 339–358.

Padmanabhamurty, B. 1990–1991, 'Microclimates in tropical urban complexes', *Energy and Buildings*. vol. 15, nos 1–2, pp. 83–92.

Pitt, D. 1979, 'Trees in the city', in I.C. Laurie (ed.), *Nature in Cities: The natural environment in the design and development of urban green space*, John Wiley & Sons, New York.

Sani, S. 1990–1991, 'Urban climatology in Malaysia: an overview', *Energy and Buildings*, vol. 15, nos 3–4, pp. 105–117.

Santamouris, M. 2001, 'The canyon effect', in M. Santamouris (ed.), *Energy and Climate in the Urban Built Environment*, James & James Science Publishers, London.

Vörösmarty, C.J., Green, P., Salisbury, J. and Lammers, R.B. 2000, 'Global water resources: vulnerability from climate change and population growth', *Science*, vol. 289, pp. 284–288.

Wong, N.H. and Chen, Y. 2005, 'Study of green areas and urban heat island in a tropical city', *Habitat International*, vol. 29, pp. 547–558.

—— 2009, *Tropical Urban Heat Islands: Climate, buildings and greenery*, Taylor & Francis, New York.

Wong, N.H. and Jusuf, S.K. 2009, 'Air temperature distribution and the influence of sky view factor in a green Singapore estate', *Journal of Urban Planning and Development*, vol. 136, no. 3, pp. 261–272.

Wong, N.H., Cheong, D.W.C., Yan, H., Soh, J., Ong, C.L. and Sia, A. 2003, 'The effects of rooftop gardens on energy consumption of a commercial building of Singapore', *Energy and Buildings*, vol. 35, no. 4, pp. 353–364.

Wong, N.H., Wang, L., Chandra, A.N., Pandey, A.R. and Xiaolin, W. 2005, 'Effects of double glazed facade on energy consumption, thermal comfort and condensation for a typical office building in Singapore', *Energy and Buildings*, vol. 37, no. 6, pp. 563–572.

Wong, N.H., Tan, Y.K., Tan, P.Y. and Wong, N.C. 2009, 'Energy simulation of vertical greenery systems', *Energy and Buildings*, vol. 41, pp. 1401–8140.

Wong, N.H., Jusuf, S.K., Syafii, N.I., Chen, Y., Hajadi, N., Sathyanarayanan, H. and Manickavasagam, Y.V. 2011, 'Evaluation of the impact of the surrounding urban morphology on building energy consumption', *Solar Energy*, vol. 85, pp. 57–71.

Chapter 18

Urban heat islands

Case studies from Frankfurt am Main, Arnhem and Ho Chi Minh City

Lutz Katzschner and Sabrina Campe

Summary

The urban heat island (UHI) and heat load effect are environmental issues that have been documented in many cities. In studies seeking environmental solutions to mitigate the UHI effect, the use of microclimatic conditions has been found to be effective. In this context, thermal comfort is expressed through thermal indices, which can be measured or calculated. A number of research projects (conducted worldwide) have studied the effectiveness of urban geometries and greenery in ameliorating the UHI effect and reducing heat load in open spaces. Their findings show the thermal benefits of greenery in reducing the ambient air temperature and heat load at both meso and micro levels. Urban climate mapping shows how microclimatic conditions can be mitigated independently of urban climate through ventilation and radiation processes. Urban climate analysis shows that heat stress greatly increases in densely built-up areas. Planning measures to reduce this include the improvement of ventilation, the reduction of incoming short-wave radiation and the reduction of long-wave radiation from façades. However, these measures need to be shown in a spatial distribution, so that concrete action may be taken to improve air flow.

Urban climate

Cities in all climates contain urban heat islands due to their density, building structures that store heat, higher barriers against wind and diminished vegetation. This considerably increases thermal and air pollution. Looking at temperature averages, one can observe urban heat islands in city centres, but looking at neighbourhoods reveals that there is even more UHI effect at the micro-scale level. This makes detailed urban heat island studies even more important. The additional impact of climate change in a rapidly developing city needs to be addressed at all scales of urban planning. The dominant contributing factors that can be managed by planning are radiation and ventilation. The urban climate may be described in a thermal and dynamic aspect. Ventilation is needed for cooling, to reduce heat stress radiation.

As well as the regional and meso-scale climate, micro-scale influences may be seen in streets, open spaces and so on. This is where most planning actions take place, so detailed analysis needs to be done at this scale. Climate change will increase heat in cities and lead to even more difficult thermal conditions. To analyse the urban climate, one has to create urban climatic analysis maps to show hotspots and air paths. The urban climatic maps may be used as a tool in urban planning. To this end, the urban climatic conditions were mapped at the meso scale to document urban heat load, mostly expressed with a thermal index. Urban typologies' climate characteristics may only be seen from this level. A spatial analysis of heat load in the German city of Kassel may be seen in Figure 18.1.

Thermal comfort

Heat stress conditions in cities may be expressed using thermal indices, which are based on the heat balance of a human body. These indices allow a clear identification of all influencing factors, which can be changed by open space and architectural planning. In this chapter, the widely known physiological equivalent temperature (PET) is used as a thermal index.

Recent studies of climate change predict a temperature increase in Central Europe. This will be intensified by the urban heat island effect with the consequence that the quality of human life in these countries' urban settlements will be affected more frequently, over longer periods, and to a greater extent in the future. Given these concerns, human comfort is coming to the fore. As cities throughout Central Europe have, hitherto, been more concerned about accumulating heat to keep cities warm during cold winters, there is now the potential for heat stress due to their traditional architecture and urban design. The aim of thermal studies and strategies to mitigate the impacts of climate extremes on the health and well-being of people in cities takes existing urban structures into account, to develop planning strategies to ensure human thermal comfort during extreme heatwaves (Katzschner *et al.* 2007). The research design includes various methods that are carried out by different work groups. This chapter

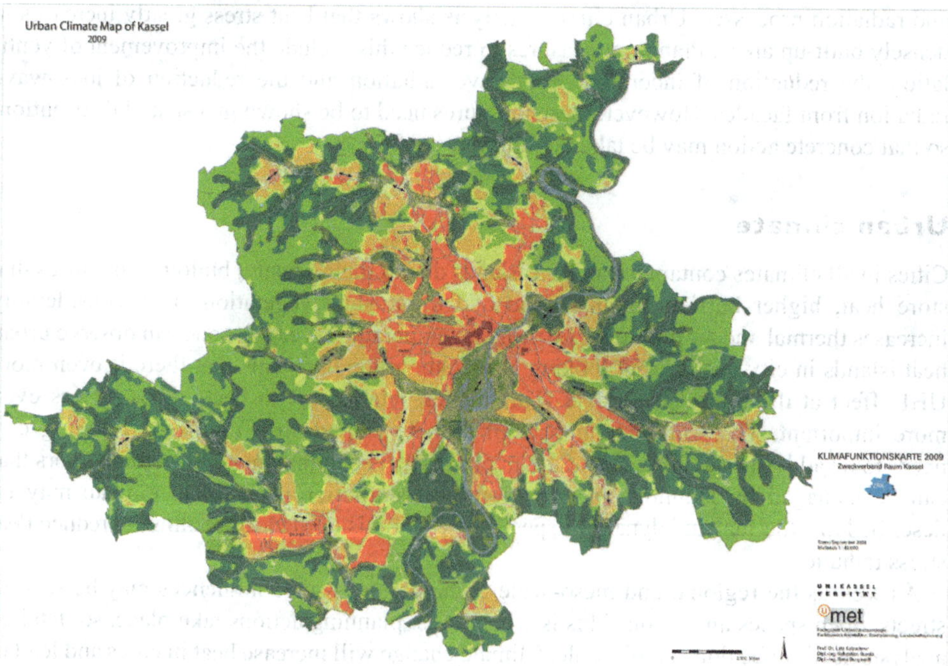

Figure 18.1 Spatial distribution of urban heat load in the city of Kassel, Germany; see Figure 18.2 for legend.

Category	Climatope	Description	Evaluation/Planning Zones
1	Fresh and cool air production zones	Open areas with significant climatic activity, cool and fresh air production; climatically active open sites in direct relation to housing area; very high nocturnal heat degradation	*Areas to preserve carefully:* High sensitivity with respect to intervention which changes land use. Do not allow the increase of surface roughness (e.g. no further constructions or buildings). Keep open of cold/fresh air stream; Minimize the existing barrier on the air streams. The air movement connections must be fully analysed and understood including the source of the air stream channels, which may be far away from the concerning area.
2	Cool air production zone	Open areas with less significant climatic activity; cool and fresh air with effects to neighbourhoods; areas without any emissions; high nocturnal heat degredation	*Areas to preserve:* The increase of surface roughness (e.g. further constructions or buildings) can only be allowed if they respect slope winds and thermal-induced circulation patterns; furthermore, redevelopments should only be allowed in exceptional cases, which is supported by detailed investigation and analysis on climatic function aspect.
3	Mixed and transitional climate zone	Strong daily variation through income radiation, but good cooling effect; Areas with high percentage of vegetation; Low and discontinous emissions; Buffer zones between different climatope; Moderate/good nocturnal heat degradation	*Areas with possible development:* Important linkage areas, foresee the orientation and density of buildings, surface roughness should not be increased due to reduction in ventinlation with effect on neighborhoods.
4	Heat accumulation potential zone	Some heat storage but mainly buffered through greeneries and wind; Dominated construction areas with lots of vegetation in the open spaces; Low nocturnal heat degradation	*Development allowed:* No appreciable sensitivity in terms of climate with respect to intensification of use and building agglomeration. Generally redevelopment is possible if they take care about ventilation and if the ratio between built up area versus green area is maintained/respected.
5	Heat accumulation zone	Heat storage remarkable but still some wind effects and cooling potentials; Density development with little vegetation in open spaces; Very low nocturnal heat degradation	*Areas for improvement and plan actions are necessary:* Risk of future heat stress with some ventilation. So generally the areas should be maintained or improved, and not worsen. Development can only be allowed with compensation for climate effects. The existing air circulation should be analysed before any proposed change so that the urban climate is respected.
6	Overheating zone	Heat storage high; Low cooling potentials and low ventilation; Heavily compressed and sealed inner city areas; No/very low nocturnal heat degradation	*Areas for improvement and plan actions are necessary:* In need of renewal from the point of view of urban climate. Greenery for facades and surfaces are needed. Increasing of existing heat stress, due to the accumulated problems on thermal load in the high dense built-up area, the climatic condition of this zone should be improved. Development in this zone is allowed only if enough compensation is done. Improving air exchange is one major recommendation together with shadow providing design.

Figure 18.2 Heat load classification in Kassel, with corresponding PET values.

presents results from KLIMES, a project that investigated microclimatic conditions experimentally and by interviewing people about their perception of thermal conditions in specific locations.

The investigation begins with an analysis of the urban climate. Figure 18.1 shows the urban climate analysis to locate the urban heat island (UHI) and, to give further information for any microclimatic studies, it also shows the location of increased air temperature in a city. There is strong public interest in open urban space; it is acknowledged that good spaces can contribute to the quality of life within cities, while less good ones compound isolation and social exclusion. The quality of open space relates to the thermal, physical and social environment. In discussions of a new urbanity, the provision of good quality open space is recognized as a means to revitalize cities. In order to increase people's use of outdoor space and thus revitalize cities, environmental conditions and thermal comfort levels, which affect people's use of these spaces, have to be considered and improved.

As well as the above-mentioned project KLIMES, this chapter describes a European and an Asian urban climate research project, which aimed to produce urban climatic maps to assist urban design and planning. The calibration (Figure 18.2) of the urban climatic maps is based on the thermal index PET, which was evaluated using information about people's thermal sensitivity (obtained in interviews) with parallel meteorological measurements.

The PET (Höppe 1999), which expresses the thermal heat load conditions, was calculated based on interviews and measurements carried out at the same time. The results should

Increasing number of hot days in urban areas:	
Calculation based of urban climate map in the city of Kassel, Germany with PET >30°C	
Annual heat stress in Kassel today	46 days
Scenario 2050	
Increase at rural	3 days
Increase city parks	4 days
Increase residential areas	6 days
Increase in dense structures	11 days

Figure 18.3 Increasing number of hot days (PET > 30°C) in Kassel, to 2050.

provide information to help planners integrate climatic considerations into their spatial planning and determine which measures are reasonable. The relevant parameters for the PET valuation index are air temperature, mean radiation temperature and wind. This bio-meteorological parameter, taking thermo-physiological correlations into account, describes people's thermal comfort and is therefore a physical parameter for measuring people's general well-being (which depends on the thermal complex).

The PET is used because it describes the thermal load based on human heat balance. The index reveals how people around the world are able to cope with different temperatures. The comfort level is around a PET value of 24°C. A balanced sense of well-being predominates if the body absorbs as much heat as it is able to release. In Figure 18.3, the number of hot days in German cities is shown. While in Frankfurt 46 hot days (with temperatures above 25°C) were observed before people start complaining, in Hong Kong people see this as normal and it is widely accepted. In the latter city, there was an average of 255 days of air temperatures above 25°C.

Air temperature is not an adequate indicator for understanding thermal comfort. That is why thermal indices are used instead. The PET is used to calibrate urban climate classification in different climate areas and the urban climatic maps.

To evaluate the extent of the urban heat island, the frequency of hot days is counted. To account for the anticipated increase in hot days in the future, global warming scenarios may be used. Figure 18.3 shows the increase in hot days during the period from 1950 to 2050 of a city (Kassel) located in the centre of Germany and the spatial distribution of the increase of hot days. Most of the areas in the city will be affected.

Urban climate and planning

Urban climatic analysis maps may be used to locate the urban heat load (urban heat island) and to give further information for any microclimatic studies as well as the location of the increased heat load in a city. To mitigate heat load and to develop actions, planning levels and meteorological levels have to be designated and respected in investigation methods. For any city, the first step has to be to study and document the meso-scale urban climate. This information may then be used for planning open space and buildings (Table 18.1).

The urban climatic analysis map is a climatic information and evaluation tool that helps planners to understand climatic-environmental conditions and variations. The urban climatic analysis map resolves scientific climatic knowledge into guidelines and planning recommendations, and may be used to guide planning actions and decision making. The concept of an urban climatic map was originally generated by German researchers in the 1970s, who named it the 'synthetic climate function map' (Baumüller *et al.* 1992). At that time it collated meteorological data, climate information, land-use data and terrain information, facilitated a comprehensive climate analysis and evaluation, and presented the results in various maps (Ren *et al.* 2010).

The urban climatic analysis map system consists of two major components: the urban climatic analysis map and the urban climatic planning recommendation map (these two may also be combined in one). The urban climatic analysis map collates meteorological data, planning, land use, topographic and vegetation information; their interrelationships and effects on wind and thermal environment are analysed and evaluated spatially and provide strategic urban planning recommendations, according to the climatic understanding and evaluation acquired from the map (Ren *et al.* 2010, pp. 2215–2216).

The development of the German urban climatic map in the 1970s was the result of intense public debate about and support for environmental protection, which culminated in political willingness to plan for the future of the natural environment in a responsible and sensitive way. According to the German Federal Building Code, urban development should be sustainable and beneficial to an environment, fit for people, and protective of the natural environment and resources. In line with this ambition, planners, meteorologists and scientists have worked together to develop urban climatic maps. Gradually they have succeeded in adding more data to the analysis in order to synergize climatic, topographical and urban parameters with the objective of providing guidelines for planning processes. Since the 1980s, many European cities (in Austria, Sweden, Norway, Switzerland, Hungary and Portugal) have mapped their urban climates. According to literature reviews of urban climatic map studies (Ren *et al.* 2009, 2010), cities in more than 15 countries now have their own urban climatic analysis maps.

Since Germany is a leader in the field, the meteorology of the urban climatic analysis map used in Europe, Asia and South America is mostly adapted from the earlier German model

Table 18.1 Planning and urban climate scales

Administration level	*Planning level*	*Urban climate issue*	*Climatic scale*
City 1:25,000	Urban development; master plan	Heat island effects, ventilation and air paths	Meso-scale
Neighbourhood 1:5,000	Urban structures	Thermal comfort, air pollution	Meso-scale
Block of houses 1:2,000	Open space design	Thermal comfort	Micro-scale
Single building 1:500	Building design	Radiation and ventilation effects	Micro-scale

and follows the 1997 German guidelines VDI 3787 for drawing up their climate and air pollution maps. In the original German version, planning advice could be commissioned by experts and planners aiming not only to improve the urban climatic situation (e.g. mitigating the urban heat island and improving thermal comfort), but also to decrease air pollution and enhance air quality. It was based on available statistical analysis and interpolation techniques.

If using the above-mentioned concept, one should not forget the different levels in planning and climate, as this influences the method and depth of investigations. Table 18.1 shows that urban climate analyses require a special resolution, depending on the planning aims and level of administration. There are three scales of planning to keep in mind. First, *city master plans* characterize areas for urban development. Therefore, climatic analyses have to indicate where air paths or urban heat islands are located. Strategies for ventilation and the mitigation of heat have to be considered as a whole and cannot be separated. Second, *zoning plans* or *neighbourhood plans* or *blocks* need more detailed specifications. The main planning aspects here are the microclimatic conditions that can be influenced by radiation processes from surfaces and walls combined with ventilation near the ground. Third, *building sites* require analysis of the building density, street canyons, greenery and open city spaces on a micro-scale level. Many of these elements can have an isolated impact.

Frankfurt am Main

In the case of Frankfurt, an urban climatic map was developed in order to obtain information for urban development strategies. A major objective was to determine the distribution of new building areas in connection with the existing greenbelt. For this, the urban climatic map was linked to the greenbelt concept in order to get more fresh air into the city centre. The linkage aspect of moving fresh air from the greenbelt to the city also combines recreation and mobility issues, which relate to the quality of open space, as noted above.

Microclimatic analyses and corresponding simulations of these areas provide information and guidance about which measures are the best for reducing heat load and what planning options arise from the city's microclimatic condition. The City of Frankfurt is characterized by excessive overheating due to its dense urbanization and an extensive amount of sealed surfaces. This cannot be entirely compensated for by the circulation of air. The existing wind circulation systems in the Frankfurt area, especially the nocturnal wind flow from the riverbanks, as well as local downhill air movements, are of great importance to the quality of the thermal and aerial ventilation components of the city climate. It is assumed that, due to global warming, thermal load in the affected areas will continue to rise. The impact of climate change will be more extreme in a densely built-up city than in rural areas (Kuttler 2009). The number of hot days with peak temperatures higher than 30° C will increase. As a result, local air exchange and the integration of green areas are particularly important, since they can improve the city's climatic potential even in areas with insufficient thermal comfort.

The climate classifications (six categories) updated in 2009 for Frankfurt am Main's urban climatic map are oriented towards thermal aspects (thermal index) and the ventilation quality (roughness) of the urban areas. In built-up areas, there are various structural environmental factors which, depending on their extent, lead to more or less strong overheating tendencies as well as to reduced ventilation and a reduction or delay of nocturnal cooling. Such factors include surface sealing, building materials, building structures, and volumes and green areas. However, aspects such as the topography, location and size of the relevant land area and (where applicable) the existing compensatory influence of adjacent mitigation sites do play a role. Depending on weather conditions and topography, these aspects (in combination with emissions) may result in high levels of air pollution that can also have a negative impact on adjacent areas. Barriers caused by building structures affecting flowing air masses are particularly important in valleys as air induction corridors' functionality can be tremendously affected if access is blocked.

In order to carry out a microclimate analysis, it is essential to know in which category the investigated areas are situated. In this case, they were found primarily in the overheating areas of Categories 5 and 6 of the urban climatic map. Category 5 describes an overheated area with restricted air exchange. Heat-stress days will greatly increase in these areas in the future. Category 6 includes overheated areas with highly restricted air exchange. Heat-stress days will increase tremendously in these areas in the future. For the analysis of affected heat load

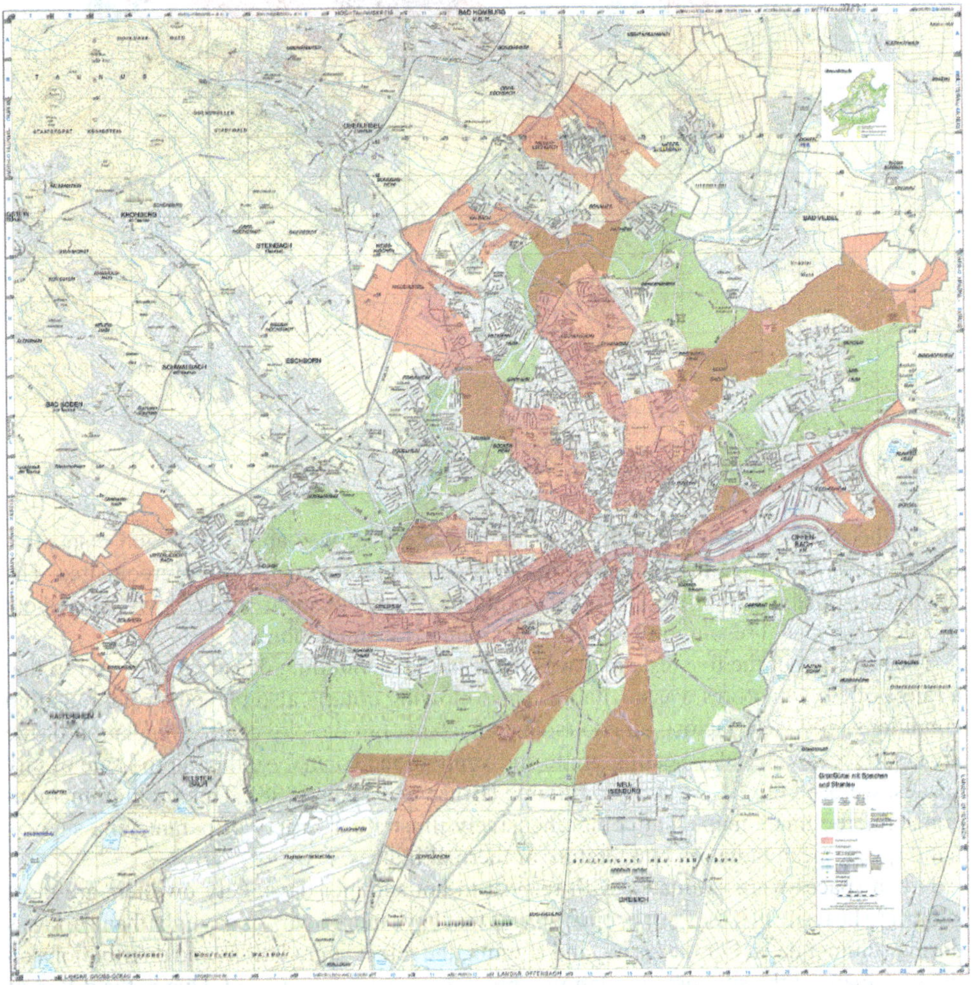

Figure 18.4a Frankfurt's greenbelt concept.

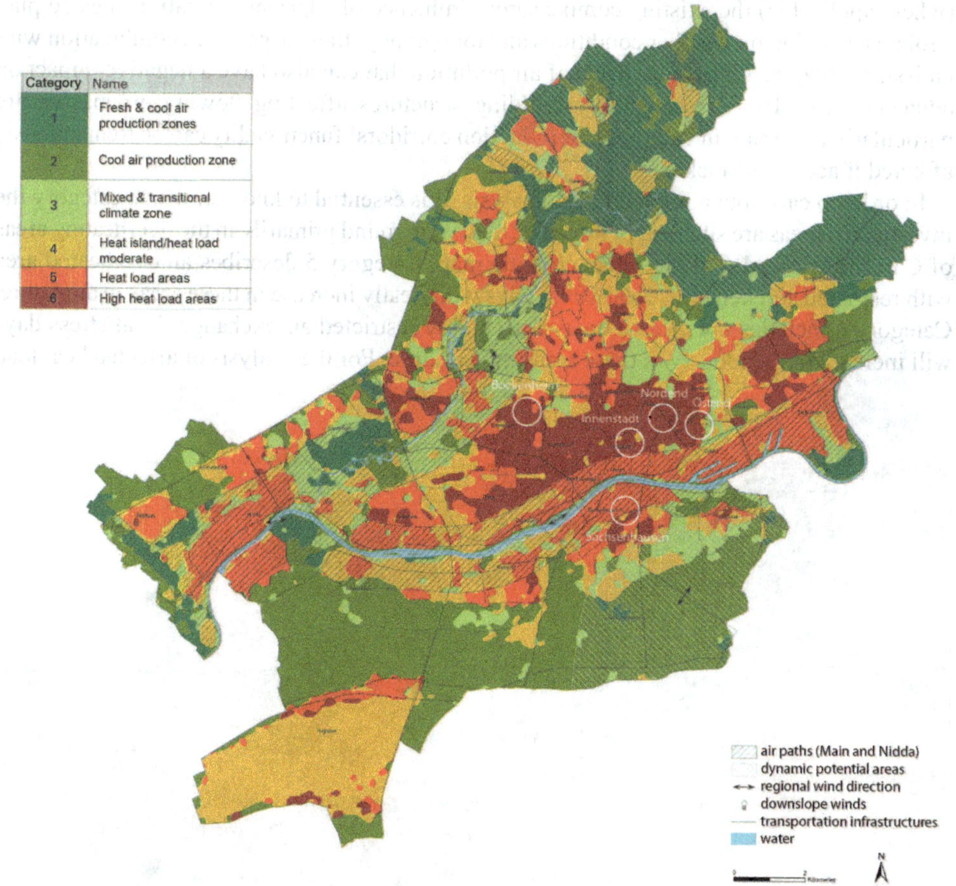

Figure 18.4b Frankfurt's urban climate map with test areas (white circles).

areas, the meso-scale information from the climate map was applied to each respective location in a down-scaling process in order to assess its microclimatic prerequisites. The micro-scale analysis of urban open spaces (plazas, streets, green areas, etc.) was carried out in a detailed spatial resolution by means of a numerical simulation.

The selection of the five test areas was made based on building density and period so that a range of city structures was available. The following criteria also played a role: radiation conditions (sky view factor), albedo (solar reflectance) in the city centre, vegetation of the area, surfaces (materials), sealing grade and ventilation. To analyse the micro-scale of city spaces, we used the ENVI-met programme (Bruse and Fleer 1998). This programme generates microclimate maps that comprehensively illustrate a variety of parameters, such as wind speed, mean radiation temperature or thermal indices.

Urban spaces were evaluated with respect to hot summer days. City dwellers are most subject to heat stress if they are in open spaces that have more hours of daylight than darkness during summer days. For the spring and autumn, the relevant gradations and distributions on the PET maps of the summer simulations apply, though at a lower level. Due to the different solar radiation angle, the shaded areas are greater so that heat load areas become relatively smaller and the remaining gradations, as a consequence, become proportionately larger. Therefore, spaces that are considered stressed in the summer may be experienced as comfort zones during other seasons. These areas are defined by means of graphical presentations of the results.

Based on the PET and wind maps, a specially developed rating scale filtered positive and negative hotspots – depending on the area's vegetation and degree of sealing, height and width ratios, and orientation of city canyons as well as ventilation grades – for each area of Frankfurt tested. According to the simulation results, these hotspots were graded into *very good, good, neutral, bad* and *very bad* (Figure 18.5).

The basis for the subsequent selection of measurements were the PET values averaged throughout the day (8 a.m. to 8 p.m.), as illustrated on corresponding maps, as this period is relevant for planning – the length of a person's stay is, after all, not limited to any one time of the day. A compromise had to be found that offered the best solution for any given time of the day during which heat load would affect open spaces. It is, however, impossible to believe that a universal solution can be found for all locations (in street spaces or on city squares). The areas most frequented by pedestrians should, however, be designed in such a way that they are thermally comfortable. With the help of the evaluation analysis, the potential areas in the respective research area should serve as an example for hotspots with deficits in order to mitigate the impact on stressed areas. Urban thermal comfort conditions can be improved by reconstructing or changing a plaza or open space.

In summary, it may be said that, based on this study, a further densification of the inner city of Frankfurt am Main (especially in the overheating areas) is not feasible from an urban climate point of view. However, the city structure should remain compact to maintain sub-urban cold air influx areas. Effective mitigation measures against heat load in dense urban areas need space to maintain ventilation channels and areas that produce cold air. The measures suggested in the project presented, however, focus on changing existing structures as little as possible, and mitigating the impact on endangered areas with a high heat-stress potential as well as using existing potentials as effectively as possible. With that in mind, space-saving measures such as changing the façades or roof surface and local protection measures against solar radiation were mentioned.

In principle, the proportion of green space has to be increased in order to make use of water evaporation and to lower surface and air temperatures. It is necessary to plant shade trees to promote evaporation, particularly on the north side of east–west-oriented street canyons and open squares. Due to anticipated and increasingly high summer temperatures and lower precipitation levels, plants tolerant of dry climates should be planted (Kuttler 2011). A most useful way to improve the comfort level of city dwellers in urban areas during hot summer periods is to densely line the streets with tall deciduous trees, as this has proved to be an effective means of shading surfaces. Industrial building materials absorb more energy than organic materials. Therefore, the use of plants can mitigate heating tendencies. The network of green areas plays an important role which – depending on the composition and size of vegetation – reduces the volume of solar radiation, promotes evaporative cooling and has a varying thermal range. In a dense city such as Frankfurt am Main, it is crucial to maintain a high quality of existing open spaces to reduce local heat stress for all urban residents. Most of the greenbelts in inner cities are, however, too small to provide a quantitatively measurable amount of air circulation to reach adjacent built-up areas. Nevertheless, even small patches of vegetation are essential as relatively cool outdoor areas for people to go to as a refuge during hot summer days.

Urban trees not only reduce solar radiation but also slow down wind movements that are considered uncomfortable. Coniferous trees provide additional protection against wind in winter, which can be of great importance during this season. On the other hand, the wind supply reduces the heat load during hot summer periods. Barriers should be removed accordingly in particular areas. In this context, the fact that roadside trees can restrict air exchange, as well as the expansion and thinning of air pollutants emitted near the ground, should be taken into account. Since a majority of trees found in Central European cities are deciduous, the maximum output of these pollutants takes place during the winter months when the flow obstruction due to treetops is relatively weak. Urban trees also collect dust, especially during summer (Kuttler 2010).

Surface materials influence radiation and also the direct heat supply. When selecting façade materials and colours, a choice must be made between the thermal comfort in the building and the thermal comfort outside. The quality of building surfaces and ground spaces has a substantial impact on local air temperature quality. The sun's reflection (albedo) thus plays an important role. The more solar radiation is reflected away from surfaces, the less energy from those surfaces is transformed into heat and released into the atmosphere. During planning, it must be considered that, while highly reflective material or colours lower buildings' energy needs, they may also lead to a decrease in the thermal comfort of outdoor areas since the reflected radiation is absorbed by pedestrians, which may be perceived as uncomfortable. The greening of roofs improves the microclimate of the roof area and protects the building from solar radiation because a major part of the received energy is used to evaporate the water stored. This has, however, only a minor effect on thermal levels close to the ground.

The ventilation of a compact urban space has to be improved in several areas. Therefore, according to the data on the urban climatic map, existing air channels must be maintained. This also applies when new buildings are constructed, so planners should consider storey heights, building orientation and building density. Additional heat factors should be prevented and measures taken in order to counteract overheating tendencies such as avoiding further structural densification, building redevelopment, unsealing of surfaces, increasing vegetation and installing artificial shading elements (if trees cannot be planted in certain areas). It also has to be noted that particular local methods cannot automatically be applied to the entire urban space or to other cities. Ventilation, orientation, width and street development as well as surface materials have to be examined to be able to derive suitable measures.

Table 18.2 **Evaluation of legend used in Figure 18.1. The first column shows the assessment, the second the PET, and the third lists windspeed.**

Assessment		***PET***	***Windspeed***
+++	*Very good*	Neutral	Comfortable ventilation
++	*Good*		
O	*OK*	Warm	Slight ventilation
– –	*Poor*		
– – –	*Very poor*	Very hot	Windless

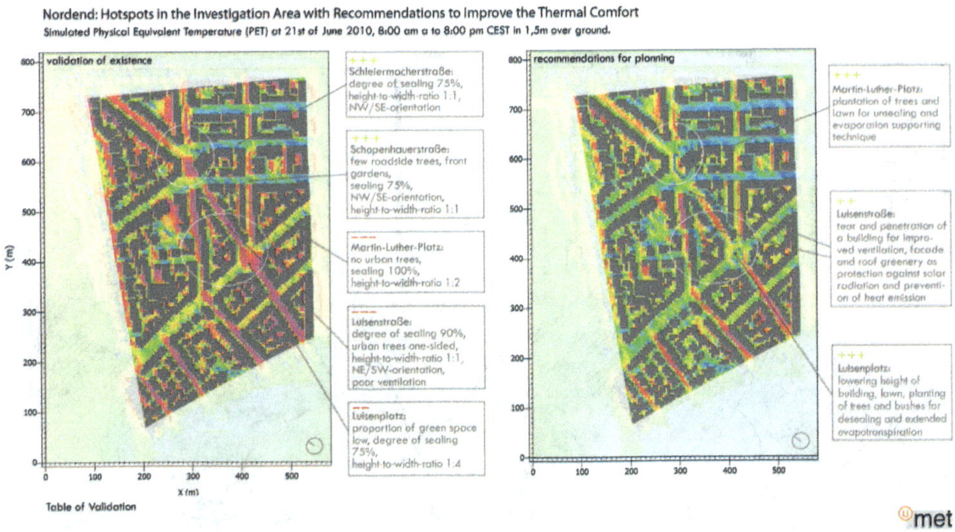

Figures 18.5a and 18.5b Results from an ENVI-met simulation under existing conditions (left) and with climatic oriented actions (right).

Before making planning proposals, an evaluation needs to be done to judge positive or negative effects on open space.

Arnhem

The urban climatic map of Arnhem synergizes all information and analyses collected to present the climatic situation and its evolution. The layers translate to meteorological characteristics like heat storage (thermal load), cooling effects, surface conditions and surface roughness, and dynamic parameters (wind pattern). Different categories were derived and may be described for planning purposes. The evaluation follows the thermal load index PET. For people in The Netherlands, a temperature of 22°C is most suitable. Below and above that threshold, people could feel uncomfortable.

The map's legend has two categories. One comes from the calculation, and the other is wind information with land-use information in addition to that. Green colours mean cold air production with downstream drainage or fresh air with or without relief energy, and the light green and orange areas are mixed zones that can be warm during the day but are able to cool down at night, while the red colours indicate areas with greater risk of heat accumulation. Thus, altogether there are six classifications from strong heat load to mixed climates and cold production areas, including the sub-layers to explain land-use differences. Here one can even differentiate with one classification following the city structure and building density. The wind information as well as the sub-layers are added into the map as an information layer, but they are already factored into the six classification results. They show the direction as well as whether they are part of an air path or a slope circulation system.

The urban climatic map of Arnhem shows three different sections of a heat island: the city centre, residential areas and parts of the built-up outer surroundings. The sub-layers express some variability within a classification due to city structures. Further on in the table as well

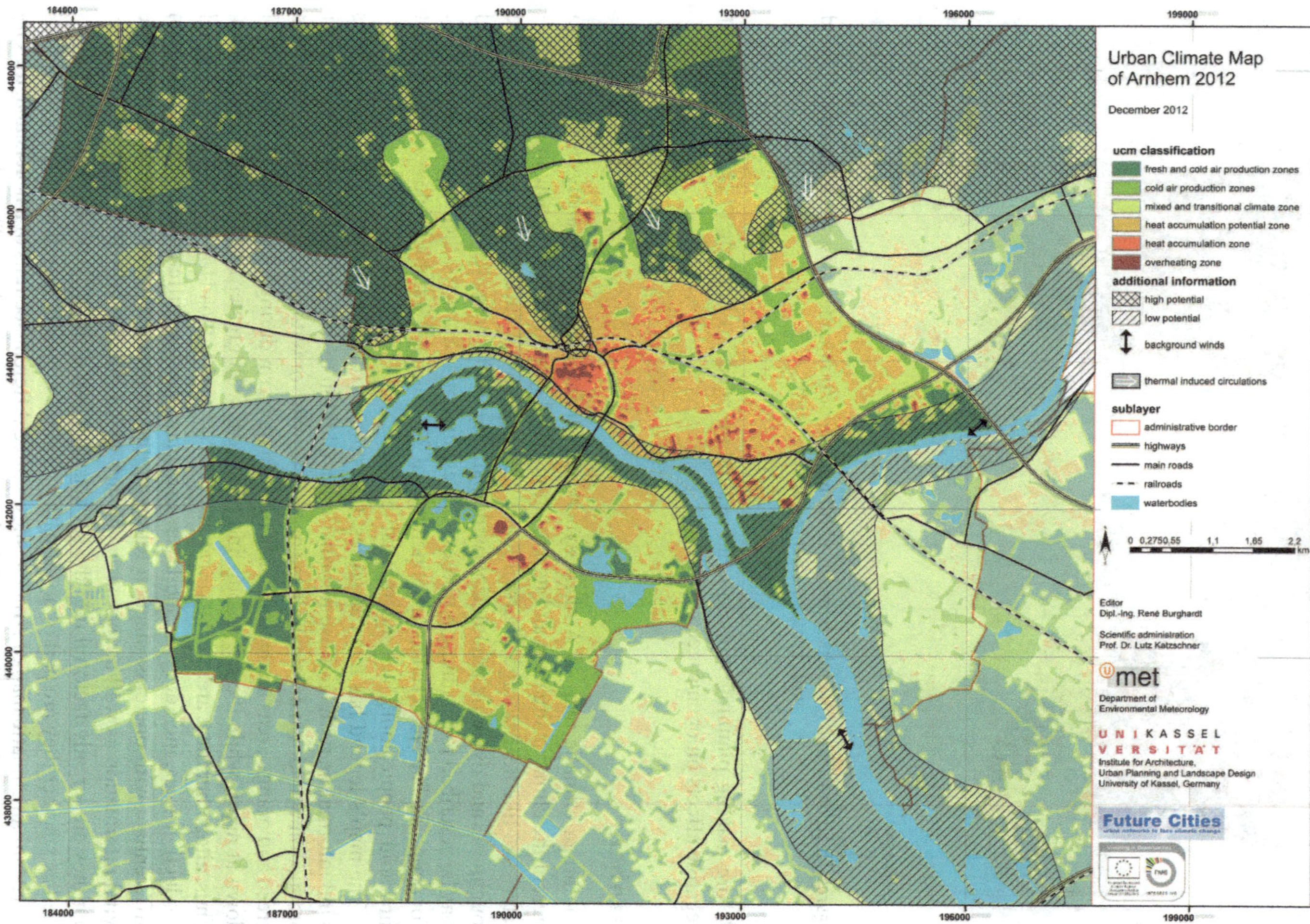

Figure 18.6 Urban climatic map of the city of Arnhem.

as in the map, the sub-layers are additional information added onto the calculation. With this, one can explain the inhomogeneity of one classification but not in detailed spatial accuracy.

Industrial areas have heat load but are somewhat buffered by sufficient ventilation, especially those affected by the main air path with background wind.

The effect of air paths as well as thermally induced circulation in the northern part of the city is remarkable. A 70 m height difference is enough to create downhill movements. The wind (upstream and downstream) and background winds are important factors in connection to the valley. For the map, the ventilation classification is separated into low potential, which needs background winds, and high potential, which is driven by relief and heat balance differences of surfaces.

Conclusions from the urban climatic map of Arnhem

To understand and use the map for urban planning, some evaluations need to be done. The conclusions are written in a table following the map's legend (Figure 18.6). These are general recommendations for the level of the urban climatic map's zones. Cool and fresh air areas have to be preserved or even enhanced, while the urban heat island, especially Zones 5 and 6, should be improved. The first two are sensitive to land-use change, while, in the last two, land-use change should lead to an improvement of heat stress.

Detailed local recommendations follow, but cannot be dealt with here at the urban scale. For Categories 1, 4 and 6, the following description and recommendations may be derived from the map:

- Category 1 description: cold air production; dynamic relevance also for cool air masses, downhill movements.
- Category 1 evaluation: areas to carefully preserve; high sensitivity with respect to intervention that changes land use. Do not allow increase of surface roughness (e.g. no further construction or buildings). Keep open for cold and fresh air streams; minimize the existing barriers to air streams. The air movement connections must be fully analysed and understood, including the source of the air stream channels, which may be far away from the area concerned.
- Category 4 description: moderate urban heat:

  (a) dense row houses of four to five storeys;
  (b) less dense row houses of three to four storeys;
  (c) less dense row houses of up to ten storeys;
  (d) low-rise buildings of up to three storeys.

- Category 4 evaluation: development allowed; no appreciable sensitivity in terms of climate with respect to intensification of use and building agglomeration. Generally, redevelopment is possible if care is taken about ventilation and if the ratio of built-up area to green area is maintained/respected.
- Category 6 description: maximum urban heat island, including:

  (a) inner city;
  (b) industry with tall buildings;
  (c) a shopping mall in the southern part of the city.

- Category 6 evaluation: areas for improvement and planning actions are necessary; in need of renewal from the point of view of urban climate. Greenery is needed for façades and surfaces.
- Increase of existing heat stress, due to the accumulated problems of thermal load in the high-density built-up area; the climatic condition of this zone should be improved. Development in this zone is allowed only if enough compensation is provided. Improving air exchange is one major recommendation together with design to provide shade.

For all categories, additional information for building orientation may be taken from wind arrows.

Ho Chi Minh City

The urban climatic map (UC-Map) for Ho Chi Minh City (HCMC) was created using the methodology outlined above. Additional information is given through information from areas with similar climatic characteristics, which could be derived from regional climate studies. This allows generalized recommendations on evaluating types of urban structure. The HCMC UC-Map includes planning orientation in the legend (Figure 18.7). Based on the information obtained from the urban climatic analyses, climatic zones and air paths could be developed and recommendations derived. Zones D and E are the most dense areas and have the greatest heat load. E is particularly influenced by afternoon ventilation. Thermally induced wind connects areas V and E (mainly during daytime). The ventilation aspect is covered by a special ventilation zone, V. The monsoon winds, which blow in north-easterly and south-westerly directions, have to be respected and taken into account. As there will be considerable development in the near future, the area needs further climatological wind evaluation before any buildings are designed. B receives lower-speed southerly winds, so barriers should not be allowed in that area. For any new development, wind permeability is strongly recommended and the orientation should be considered. Like A, C is open to the north-east and has good ventilation, so the dominant wind direction from the south-west is supported.

The main objective is to link urban climate analysis with the master plan. The urban climatic map shows that there are good and sufficiently ventilated areas in zone V along the channel with low roughness that immediately experience air flow problems inside the building sites. Corners and inhomogeneous height give enough turbulence for some ventilation. Future simulations can provide more recommendations for a building design to maintain ventilation with a sufficient air flow.

How to use the urban climatic map with the table of recommendations:

- First-step urban climatic map: the urban climate is classified into six categories as mapped from analysis.
- Second-step climatic map derives districts A to E and V (letter in legend) that have the same climatic characteristics. The measures underlie the regional climate and are later separated into the six categories.
- Third-step sub-layer for districts with six categories: measures and recommendations.

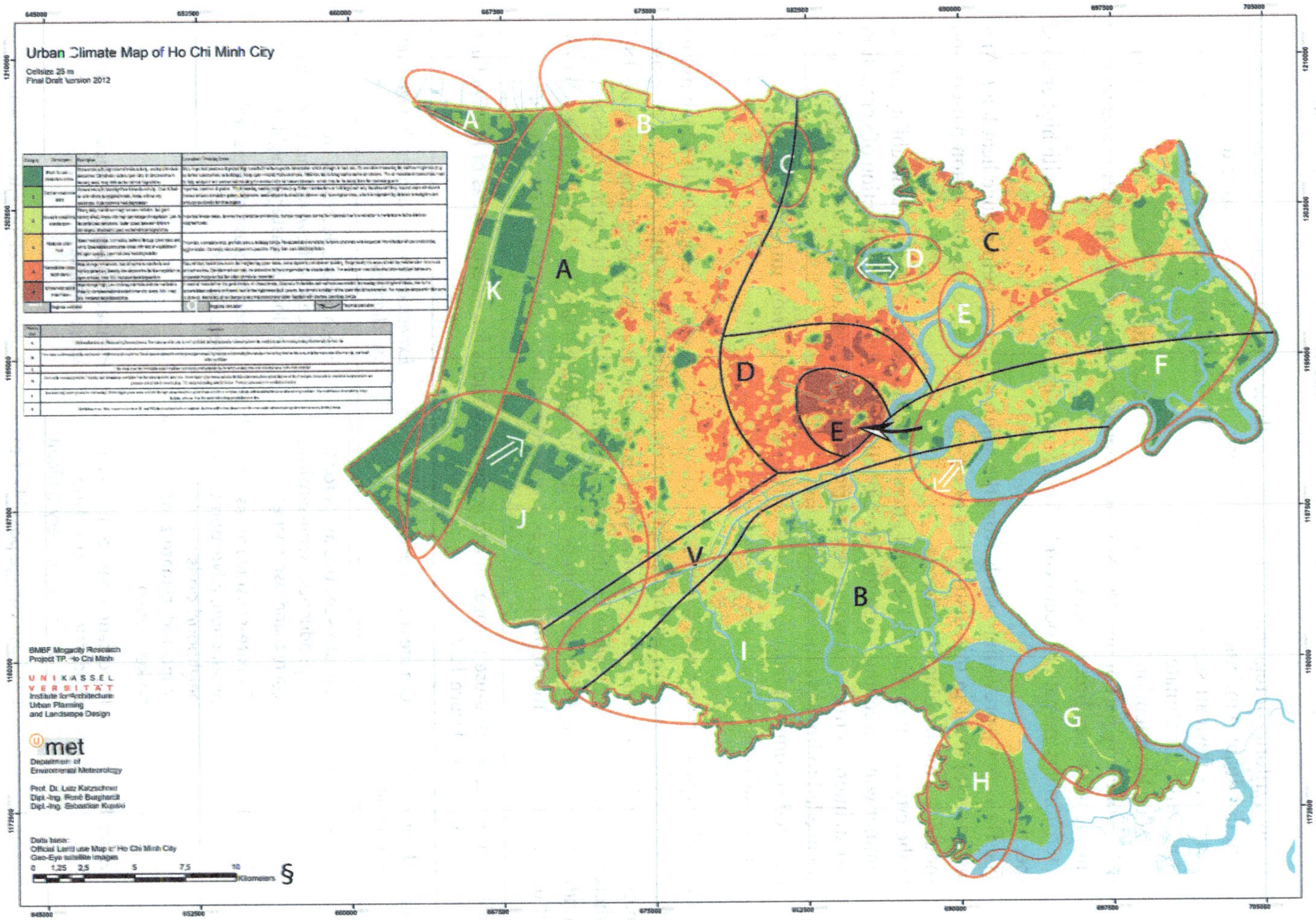

Figure 18.7 Urban climate map of Ho Chi Minh City with climatic regions; conclusions and recommendations from the map are in the table; circles with white letters define planned land-use changes.

Some general recommendations may be derived from the HCMC UC-Map, which are orientated after the separation in the black-marked climatic districts A to E (see map):

- At this territorial or regional level, one can see three main channelling effects (D) coming from dominant wind and land-use conditions. These areas are sensitive to new developments as fresh air supply is generated here. Thus land-use change with buildings will be negative for climate.
- Considerable attention has to be paid to the entrance situation of these areas, especially in the southerly part of the city. New development may prevent ventilation and take the linkage to the sea breeze away.
- A territorial wind pattern occurs in D. Many areas here are undisturbed with higher wind speeds.
- The B district's prevailing winds from the south and north help reduce heat load. Any new development in this district has to recognize the specific ventilation pattern and should not totally block winds.
- In the city climates A and D special attention has to be given to the permeability for the summer wind directions; therefore the city climate is divided into direction of ventilation and an interior climate in which street orientation is important. Corners and open spaces are needed to improve the situation here.
- Detailed recommendations take the territorial classification into account and overlap with the urban climatic analysis map.

Now another question arises: What would happen if areas were developed with new building sites? The HCMC planning authorities gave a map (marked with white letters) of proposed changes, which was evaluated. The land-use changes in areas A–K (see map in Figure 18.7 where they are indicated by white capital letters) will lead to more heat load and to some extent create a barrier effect. Table 18.3 was derived from climate analysis, taking into account the influence of the above-mentioned changes proposed by the government on the city's climate.

Table 18.3 Proposed land-use changes in and recommendations for different areas of Ho Chi Minh City from Figure 18.7

Area	*Urban climate classification*	*Climate recommendations*	*Expected climate changes through land-use changes*
A	Zone 1 and 2	Climate-sensitive areas for fresh air production, semi-sensitive against land-use changes	Fresh air production is stopped, the reduced air flow will increase urban heat load in easterly neighbourhoods
B	Zone 1, 2 and 3	Climate-sensitive areas for fresh air production, advection towards neighbourhoods should be respected, climate zone 1 should be kept open	Planned land-use changes will not affect climate
C	Zone 1	Important areas for fresh air production and circulation, land-use changes very difficult	Planned land-use changes will not affect climate
D	Zone 1	Land-use change possible if east–west orientation of buildings allows ventilation	Building development here will increase heat load through heat storage

Area	*Urban climate classification*	*Climate recommendations*	*Expected climate changes through land-use changes*
E	Zone 1 and 2	No high sensitivity against circulation, moderate land-use change possible	No climate changes
F	Zone 1 and 2; scattered 3	High sensitivity with respect to existing circulation pattern; in the western part thermal sensations; land-use changes should not block north-easterly winds, roughness length should not exceed 0.5 m (openness layer one and two up to 50%)	The high-sensitivity area is changed to mean heat load areas, the fresh air movement is disturbed
G	Zone 1 and 2	High climate sensitivity, land-use changes only allowed with openness of 50% coverage	No climate changes seen
H	Zone 1 and 2	High climate sensitivity, land-use changes only allowed with openness of 50% coverage	Changes here will affect urban heat island downtown, reduction of air movement, the effect of urban climate here can be seen inside the area and outside the area
I	Zone 1 and 2; scattered 3	Very high climate sensitivity of special importance for neighbourhoods, land-use changes have to maintain fresh air production areas and the advections (transportation) towards city centre, development around zone 3 allowed but zone 1 and 2 should be kept open	Changes here will affect urban heat island downtown, reduction of air movement (especially during low wind speed situations), the effect of urban climate here can be seen inside the area and outside the area
J	Zone 1 and 2	From climate point of view a most important area, which should kept free of buildings, in combination with areas K and J; southerly winds have to be prevented and should not be blocked, orientation of buildings is most important	Through maintaining rice paddies, effects are small, south-westerly winds are reduced and will lead to increased urban heat load downtown
K	Zone 1 and 2	From climate point of view most important areas for cool air production; land-use changes should respect orientation of buildings and openness one in the first roughness layer of 25% coverage	Planned land-use changes have low climate effects

Final remarks

It is obvious that conditions of thermal comfort simultaneously benefit well-being, recreation and energy saving. Urban climate analysis and urban climatic maps are important in designing and planning outdoor spaces for well-being and also using energy potentials for renewable sources like ventilation and radiation. They should be part of the basic information gathered for any urban development plan.

The GIS calculation with land-use data and building data has proved to be sufficient to explain heat load conditions. Together with existing measurement data and some phonological data, a good agreement may be discerned. This is a sufficient basis on which to develop more recommendations for cities. Following the analysis, a strategy may be devised to protect cold air areas against further development and to allow winds to enter cities following the main air paths seen on the map.

While well-being or thermal perception relies not only on the microclimate but also on psychological issues (Knez and Thorsson 2006), the conditions of the urban climate help to make a city attractive. To use the urban climatic map effectively, recommendations for each category must be made. The criteria are derived from thermal conditions and the importance of each urban climate classification layer, either to preserve or enhance the climatic conditions. Based on the scientific understanding of each layer, the characteristics for concern are classified. The classification values are basically a numerical assignment (positively and negatively) of the concern's likely physical effects on the urban climate.

References

Baumüller, J., Hoffmann, U., Nagel, T. and Reuter, U. 1992, *Klimauntersuchung des Nachbarschaftsverbandes Stuttgart: Klimaatlas*, Stuttgart, Germany.

Bruse, M. and Fleer, H. 1998, 'Simulating surface–plant–air interactions inside urban environments with a three dimensional numerical model', *Environmental Modeling and Software*, vol. 13, pp. 373–384.

Höppe, P. 1999, 'The physiological equivalent temperature: a universal index for the biometeorological assessment of the thermal environment', *International Journal of Biometeorology*, vol. 43, pp. 71–75.

Katzschner, L., Bruse, M., Drey, C. and Mayer, H. 2007, 'Untersuchung des thermischen Komforts zur Abpufferung von Hitze durch städtebauliche Konzepte', *Berichte des Meteorologischen Instituts der Universität Freiburg*, no. 16, pp. 37–42.

Knez, I. and Thorsson, S. 2006, 'Influences of culture and environmental attitude on thermal, emotional and perceptual evaluations of a public square', *International Journal of Biometeorology*, vol. 50, pp. 258–268.

Kuttler, W. 2009, 'Zum Klima im urbanen Raum', *Deutscher Wetterdienst (Hrsg.): Klimastatusbericht 2008*, S. 6-12. Offenbach/M.

—— 2010, 'Urbanes Klima, Teil 2', *Gefahrenstoffe – Reinhaltung der Luft, Umweltmeteorologie*, vol. 70, no. 9, pp. 378–382.

—— 2011, 'Klimawandel im urbanen Bereich, Teil 2. Maßnahmen' ('Climate change in urban areas, Part 2, Measures'), *Environmental Sciences Europe*, vol. 21, no. 23, pp. 1–15, doi: 10.1186/2190-4715-23-21.

Ren, C., Ng, E. and Katzschner, L. 2009, 'Review of urban climatic map studies around the world', paper presented at the Seventh International Conference on Urban Climate, IAUC, Yokohama, Japan.

—— 2010, 'Urban climatic map studies: a review', *International Journal of Climatology*, vol. 31, pp. 2213–2233, doi: 10.1002/joc.2237.

Verein Deutscher Ingenieure VDI-Guideline 3787 1997, *Part 1, Environmental Meteorology: Climate and Air Pollution Maps for Cities and Regions*, VDI, Beuth Verlag, Berlin.

Chapter 19

The Green Plot Ratio and the role of greenery in low carbon living

Boon Lay Ong

Summary

The Green Plot Ratio (GnPR) was first introduced as a concept in 2002. The idea behind the Green Plot Ratio is to provide a metric to measure the amount of greenery on a particular site. Knowing this value, it should be possible to – at least theoretically – estimate the amount of carbon sequestration, water retention and other environmental benefits of plants. This chapter first presents some benchmarks in the evolution of the GnPR since its initial introduction. It also identifies some of the key concerns in the implementation of the concept as a regulatory tool. Given the variability of Leaf Area Index (LAI) values, what are the criteria for determining the best LAI values for GnPR calculations? How can we protect trees and other preferred plant species within the GnPR structure? What areas of sustainability are not covered by GnPR (or badly covered) and how can we address these areas? Are there alternative models to GnPR we can explore that can give us better outcomes? How can GnPR respond to issues of biodiversity? This chapter concludes with some suggestions for the future.

Introduction

The American Society of Landscape Architects calls greenery 'the missing link in green design' (ASLA 2009). This is odd as, intuitively, the environmental value of greenery is easily perceived. We all know that plants convert carbon dioxide and are nature's primary strategy in carbon sequestration. All our food and most of our water and other resources, like medicines, building materials and clothing, are derived from plants. The value of plants in their rainfall interception, soil erosion control and various ecological services is common knowledge and common sense. Studies have found that greenery increases the value of urban properties from 6 to 28 per cent[1] (Morales 1980; Luttik 2000). It is even possible, based purely on visual and social value, to put a price on greenery (Killicoat *et al.* 2002). Paradoxically, in terms of research findings, the benefits of greenery over technological solutions remain in question.

It is true that for any single benefit plants give, we can probably do better with a technological solution. For example, Rizwan and colleagues' literature review (Rizwan *et al.* 2008) found that 'roof spray cooling' can afford a 13 to 17°C (degrees celsius) reduction compared to a maximum of 3°C reduction due to the combination of vegetation and lighter colours, and less than 2°C reduction with just planting and vegetation. It may even be argued that the aesthetic satisfaction and social desirability of a painting or even a large television set is greater than the view from a window. The crux of the argument for greenery, however, lies not in the individual benefits, nor even in its entire range of benefits, but in the perspective we have on our relationship with the environment.

Ecosystem services

It is a difficult argument to put across. The benefits of greenery are extensive, but, viewed in terms of individual plants or a single landscape, seemingly insignificant. A comparison between the use of air-conditioning and greenery to achieve thermal comfort is instructive. Certainly, air-conditioning will give us better control over the thermal environment, allowing us to specify a common comfort temperature globally (within ±1 °C), regardless of climatic zone and season. However, the cost of that capability is a disproportionate consumption of energy, recirculated air, ill-health and contribution to climate change. The use of plants for the same end results in a moderated thermal comfort temperature (one that varies with the outside conditions), but results in the provision of re-oxygenated and freshened air,[2] visual pleasure and comfort and improvements to our well-being. The right plants even allow us to harvest food and medicine. The technological solution requires us to simply turn a switch and abdicate everything else to a support infrastructure that includes a power supply and factories, maintenance technicians and engineers. The greenery solution requires more work from us and, in terms of temperature reduction alone, produces a comparatively less measurable result. However, it is a well-contained relationship – besides some external sourcing of seeds, the rest is essentially between us and the forces of nature. The support infrastructure is naturally created rather than artificially constructed.

Climate change challenges us to reassess our understanding of how the world works. The expected change in global temperature is an increase of between 3 to 5°C over the next 50 years or so. The rise in global temperature since 1880 has been less than 1°C. Yet such a small temperature change manifests itself in dramatic ways. In a similar way, the use of greenery may seem in each instance a small change. Yet the cumulative effect is substantial and significant. It was long understood that vegetation distribution is determined by climate. The reverse effect of plants on climate was thought to be negligible or even non-existent.[3] However, it was recently discovered that greenery has a positive effect on the amount of rainfall in a particular location and, to that extent, affects climate as well (Scheffer *et al.* 2005). It is another instance of the 'butterfly effect', where the impact of a small change, considered negligible to conventional science, turns out to be the deciding factor in nature's intricate symphony.

To properly appreciate the value of greenery in the human environment, we have to set aside the linear comparison of benefit against benefit and instead see ourselves as part of the natural ecosystem. Experts in the field have taken to describing the value of greenery in terms of ecosystem services (Lovell and Johnston 2009). These services are divided into four categories: *provisioning*, *regulating*, *cultural* and *supporting*. Described in this way, ecosystem services are not restricted to the context of natural ecosystems. They are indicative of the built environment as an ecological entity. These four benefits of greenery are clearly connected to human society. Lovell and Johnson (2009, p. 45) recommend that greenery design should have as its goal the achievement of 'ecological, cultural and production functions within a given space'.

However difficult, there is a need to try to quantify these benefits. This difficulty permeates sustainable design standards. One of the problems with World Green Building Council

(WGBC) standards like BREEAM, LEED and Green Star[4] is that the accummulated credits and rating do not translate directly to a reduction of measurable impacts like energy savings, water treatment and waste recycling. There is even concern that sustainable standards credits do not in fact translate into actual environmental performance improvements. There are moves in these standards towards performance-based ratings, and rightly so. The benefits of greenery, while real enough, are even more difficult to quantify in the same manner. It is especially hard to do within the urban context, as urban greenery often ends up requiring additional energy and costs that counteract the benefits the greenery brings.

A significant paper in this area comes from Gill and colleagues (Gill *et al.* 2007), which discusses the role of green infrastructure for climate change. The description they provide for climate change impacts highlights the complex and ecological nature of the problem:

> Urban areas have distinctive biophysical features in comparison with surrounding rural areas (Bridgman *et al.* 1995). For example, energy exchanges are modified to create an urban heat island, where air temperatures may be several degrees warmer than in the countryside (Wilby 2003; Graves *et al.* 2001). The magnitude of the urban heat island effect varies in time and space as a result of meteorological, locational and urban characteristics (Oke 1987). Hydrological processes are also altered such that there is an increase in the rate and volume of surface runoff of rainwater (Mansell 2003; Whitford *et al.* 2001).
>
> (Gill *et al.* 2007, pp. 115–116)

Even the social and economic benefits of greenery are not attributable to the number of trees or anything quite so easily quantifiable, but to the more subjective perception of an amenable and natural environment (Luttik 2000; Morales 1980). It should be noted that this subjective perception is related to a sense of ecological aesthetics rather than a specific preference for greenery (Kellert *et al.* 2011; Helliwell 2008).

The Green Plot Ratio

It is in this context that the Green Plot Ratio (Ong 2002) is significant. Despite the acknowledged difficulty in quantifying and perhaps even comprehensively determining the benefits of greenery, it is nevertheless important that we try to quantify these benefits where we can and to the extent that we can. Many of the ecological benefits of plants ensue from their living processes. The intensity of these living processes is linked to greenery density, often expressed in terms of Leaf Area Index (LAI). LAI is defined as the total one-sided leaf area per unit of ground surface area. It is calculated by dividing the total one-sided leaf area on a tree by the area of the ground covered by its canopy. It is a notoriously indeterminate number, as the density of leaves differs from plant to plant within a particular species, from time to time during the plant's growth, and even from spot to spot throughout the plant's canopy during any particular measurement.

To make it even more complicated, LAI is not the only variable that determines the plant's responses to the environment (Jones 1992). Many other factors come into play, including the rest of the tree and even the surrounding environment. For example, the amount of water evaporated by a plant into the atmosphere comes from both transpiration and evaporation (from the soil and bark). The effect of a plant on windspeed is not just dependent on leaf density but also on the orientation, structure and shape of the plant. Even when the effect of the leaf canopy alone is studied, leaf shape, size and texture are also important, as is its composition (e.g. density of stomata). Nevertheless, LAI is important, because it

> measures the amount of leaf material in an ecosystem, which imposes important controls on photosynthesis, respiration, rain interception, and other processes that link vegetation to climate. Consequently, LAI appears as a key variable in many models describing vegetation–atmosphere interactions, particularly with respect to the carbon and water cycles.
>
> (FAO n.d.)

Leaf Area Index and leaf area density

Leaf area density is another and related measure widely used in forestry studies (Lalic and Mihailovic 2004; Hosoi and Omasa 2009). Defined simply as the amount of leaf area per unit volume, leaf area density may be considered a more refined measure of LAI. These two measures are related by the following equation (Lalic and Mihailovic 2004):

$$L = ,0\text{-}h\text{-}a,z.dz. \tag{1}$$

where L = Leaf Area Index
$a(z)$ = leaf area density of successive layers
z = layer
h = height of canopy

The LAI is thus the integrated leaf area density of the canopy. For the purpose of urban policy, LAI is the simpler measure as it provides a single integrated value of greenery density.

Leaf Area Index

As noted earlier, the significance of employing LAI is that it is a common denominator in many of the ecological effects of plants. The key to understanding this lies in first understanding the role of LAI in the interception of sunlight through a plant canopy.

Solar interception

The amount of sunlight intercepted by a canopy is mathematically represented by Beer's Law. Beer's Law models light transmission through the canopy in terms of an exponential function linked to *L*, the Leaf Area Index, and *k*, the extinction coefficient of the canopy. The value of *k* depends on several factors, including leaf transparency and the angle of the leaf to the sunlight, or radiant flux. The product *kL* is applied to in Beer's Law to calculate the rate of energy flow through the foliage. This exponential relationship may be regarded as a general equation that describes the absorption of a flux (energy, e.g. light, wind, rain) through a permeable medium (canopy).

Beer's Law is here expressed as follows:

$$I_c = I_0 e^{-kL} \tag{2}$$

where I = flux density, W m^{-2} and subscripts
0 = measured at surface, and
c = canopy
k = extinction coefficient, dimensionless
L = Leaf Area Index

Table 19.1 shows the value of k for different leaf angle distributions. Of the various models, the ellipsoidal model has been found to be most accurate in modelling different types of tree canopies, where x may be adjusted to match different canopy shapes (Campbell 1986).

The percentage of solar radiation transmitted through a canopy is represented in Figure 19.1.

Table 19.1 The extinction coefficient, k, expressed as a function of beam elevation, b, for different leaf angle distributions for radiation flux over a horizontal surface area

Leaf angle distribution	*Extinction coefficient (k)*
Horizontal	$k = 1$
Vertical	$k = (2 \cot\beta) / \pi$
Spherical	$k = 1/(2 \sin b)$
Ellipsoidal	$k = (x^2 + \cot^2 b)^{1/2}/(Ax)$ where x is the ratio of the horizontal to the vertical axis of the ellipsoid and $A \approx (x + 1.774\,(x + 1.182)^{-0.733})/X$
Diaheliotropic	$k = 1/\sin\beta$

Source: Jones (1992, p. 38).

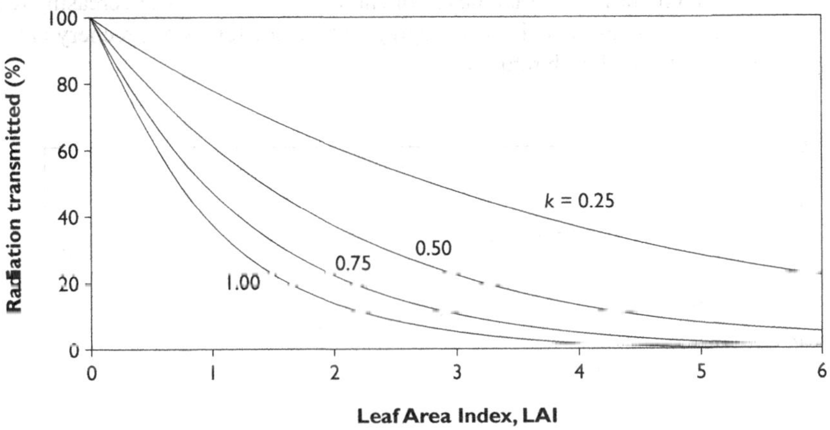

Figure 19.1 Variation in percentage of radiation transmitted for varying values of k and L. k varies between about 0.3 for vertical foliage and 1.0 for canopies of horizontal leaves (Atwell *et al.* 2010). The percentage of radiation transmitted is calculated using $R = e^{-kL} \times 100$.

Temperature

The cooling effect of plants is linked with LAI and Beer's Law. Beer's Law not only describes the amount of sunlight that penetrates through the canopy, it also enables the calculation of the cooling effect of plants through transpiration. The amount of transpiration at the leaf level is complex and dependent on several factors, but may be reduced in terms of environmental factors to the water potential of the air. Most of the heat energy used by the plant is absorbed through transpiration as it involves the latent heat of evaporation of water which is around six times the energy required to heat the same amount of water from 0°C to boiling point. A much smaller amount of energy is absorbed by the plant mass so that, in general, the resultant temperature on the leaf surface is around air temperature.

Windspeeds

The variation of windspeed through a canopy is also linked to LAI. This is described theoretically through the equation:

$$u_0 = u_d \exp [-k'_w L^{1/3}] \tag{3}$$

where u = wind velocity through the canopy and subscripts
0 = measured at ground level, and
d = depth of the canopy
k' = wind extinction coefficient, dimensionless
L = Leaf Area Index

The variation of wind velocity as a function of k' and LAI is shown in Figure 19.2. Again, the value of k' is dependent on several factors, including the size and shape of the leaf as well as the orientation of the leaves with respect to wind direction. Although the data shown here are theoretically derived, they show that the exponential function means a decreasing relative benefit with increasing values. This is encouraging in that low levels of greenery can bring about substantial benefits and high returns.

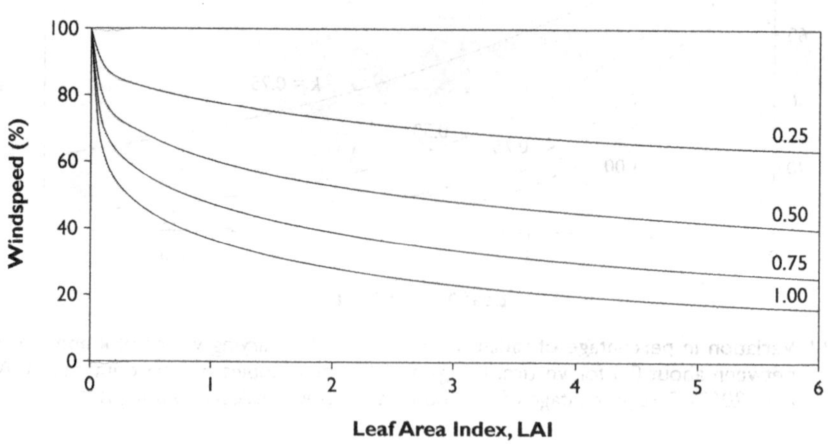

Figure 19.2 Variation in wind velocity as a function of LAI and k'.

One of the applications of this relationship is the effect of vertical greenery on wind behaviour on high-rise buildings.

Rainwater interception

LAI is also linked to rainwater interception (Herwitz 1985). Surprisingly, Herwitz found that bark accounted for more than 50 per cent of total interception storage capacity under still air conditions and more than 80 per cent under turbulent conditions. The latter finding is expected, as wind will both prevent water from collecting in the leaves and shake off the water that is collected.

Other ecosystem effects

It may be seen from the above that there is a general and simple relationship between LAI and many of its interactions with the environment – light, wind and rainwater interception. In many instances, this relationship is exponential, like Beer's Law, while in others it is linear. It is not unreasonable to expect a similar relationship with air pollution uptake, for example, and other processes of which LAI is indicative. Soil stabilization, for example, may be expected to be related to the size of the root system, which is again linked to the canopy and LAI. Even crop production is predicted using LAI (Sinclair *et al.* 1984).

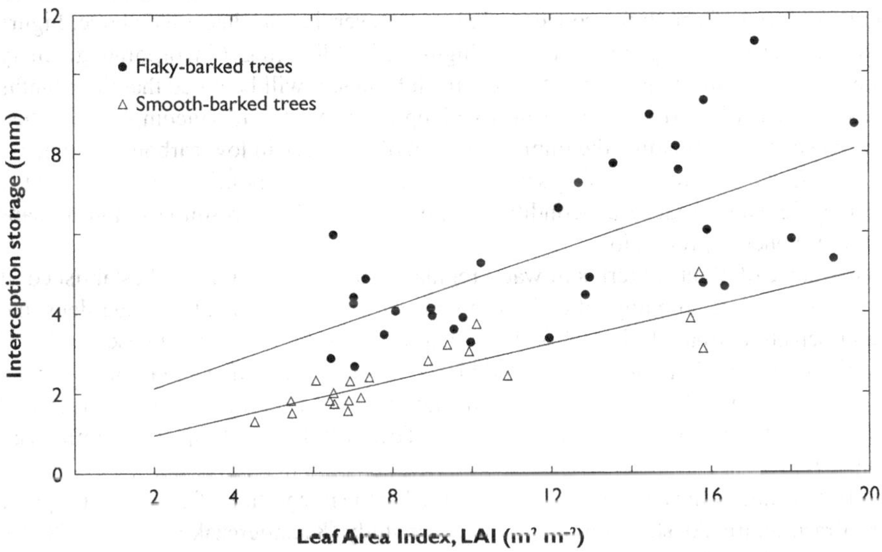

Figure 19.3 Empirical relationships estimated between interception storage and Leaf Area Index for selected trees. The two lines shown are regressions equations expressed as follows (Herwitz 1985).

$y = 1.449 + 0.333x$ for flaky-barked trees ($r = 0.65$, $n = 31$)
$y = 0.462 + 0.228x$ for smooth-barked trees ($r = 0.86$, $n = 20$)

An ecological metric for greenery

Other ways of quantifying the effect of greenery on the environment exist, notably those using iTree (www.itreetools.org/) and urban morphology types (Gill *et al.* 2007; Helliwell 2008; Killicoat *et al.* 2002). None, however, has the wide-ranging potential of LAI (and hence GnPR) as an indicative measure of the environmental benefits of greenery. In developing standards and guidelines for greenery, we need to negotiate between comprehensive accuracy and robust simplicity. There are several mitigating factors to be considered in the calculation of the environmental impacts of greenery, and professionals in the field will be required to take these factors into account. However, the regulation of greenery in cities can benefit from the simplicity and validity of GnPR.

Designing with greenery for low carbon living

Low carbon living implies an overriding concern with the carbon footprint and the production of CO_2 in an already overburdened atmosphere. In a very direct way, it targets the identified culprit of climate change. Low carbon living goes beyond the earlier concern with energy and to some extent negates it with the promise of clean energy. It casts the net wider and may be regarded as a move towards a more ecological approach. At its best, it connects with the various ecological cycles of earth (e.g. carbon, nitrogen, phosphorus and water).

Successful low carbon living thus requires a more comprehensive look at what constitutes our carbon footprint and how to reduce it. One way to make the link between low carbon living and greenery is through the concept of ecosystem services as discussed above. Figure 19.4 shows how such a design proposal for a high-rise building might incorporate greenery as a strategy for low carbon living and the benefits it brings. It will be noted that the benefits listed are somewhat arbitrarily allocated and overlapping. The list is also incomplete. Further, to achieve the expected benefits, the implementation of greenery in low carbon living needs to be integrated with existing building services and additional economic returns – using the plants to recycle water, reduce air-conditioning, provide food and resources, and perhaps even as fuel for energy production.

The importance of plants in terms of water regulation and recycling is well established. It is part of best practices in many cities in terms of permeable pavements, raingardens and other water-sensitive urban design (WSUD) strategies. A useful database of case studies of various cities on the application of green and blue infrastructure as an adaptation to climate change has been prepared by the University of Manchester for the Interreg IVC Green and Blue Space Adaptation for Urban Areas and Eco Towns (GRaBS) Project (Kazmierczak and Carter 2010).

Figure 19.5 identifies the key components involved in the integration of greenery in a piece of water infrastructure. As shown in the Manchester study (Kazmierczak and Carter 2010), it is useful to combine the two – greenery and water – into one green and blue strategy. The key to such a strategy is to separate water into different categories – potable, clean, grey and sewage. Different treatments are required in each case and, in some instances, a combination of engineering and greenery solutions may be expected. As noted in several of the papers quoted, the appropriate solution in each instance has to be tailored to suit the circumstances and location. While nature is robust and adaptive to changes in the environment, there is also death and decay to consider – an outcome that is less problematic to the environment than to the human condition.

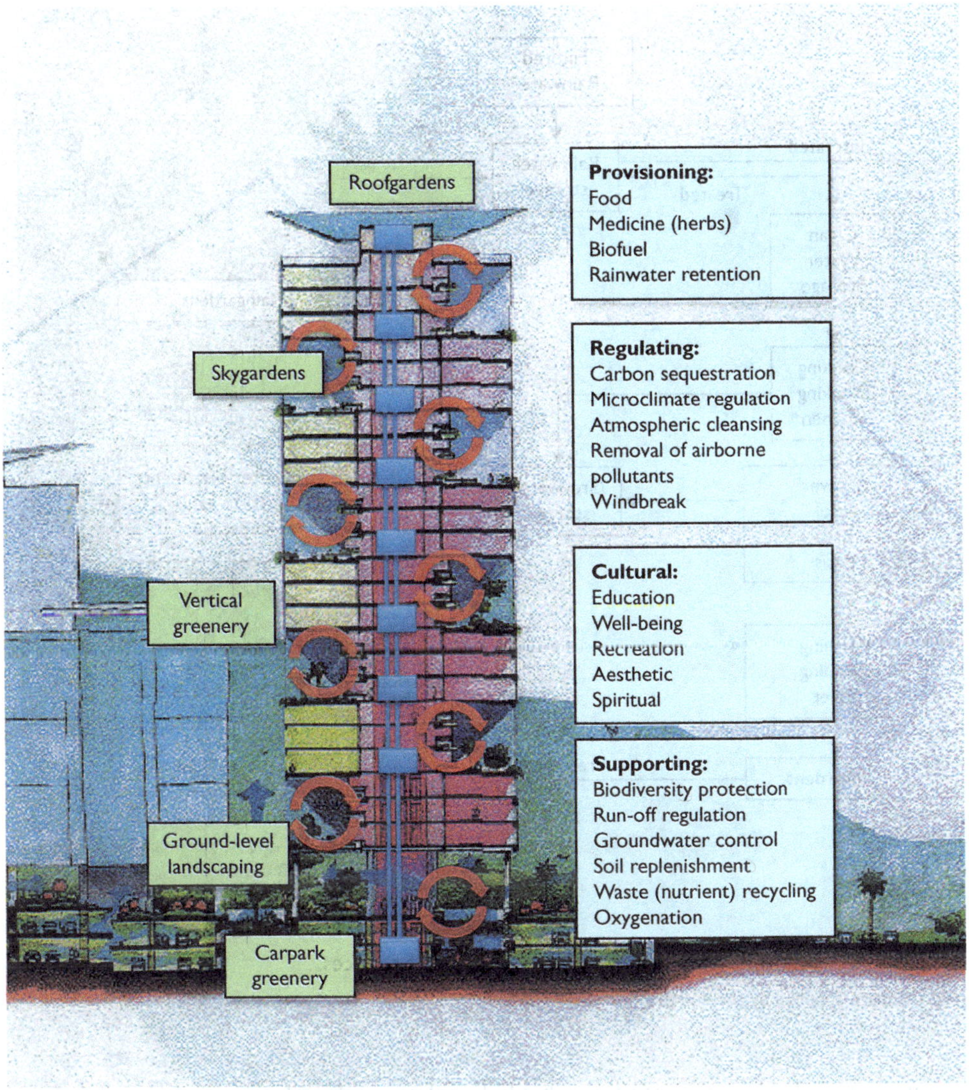

Figure 19.4 The ecological high-rise design: the application of greenery to low carbon living requires the integration of building services and green plants.

Source: Image supplied by the author.

The value of greenery in the other ecosystem services is perhaps less well evolved in terms of our knowledge and understanding. Traditionally, these benefits have been the domain of different fields: horticulture, agriculture, botany, ecology, engineering and gardening. While there is cross-learning and a common body of knowledge, each field has developed its own preferred methodology. For example, while from a purely scientific standpoint carbon sequestration should be calculated from equations involving photosynthesis reactions, sunlight and LAI, forestry has evolved a more robust way to calculate it using trunk dimensions

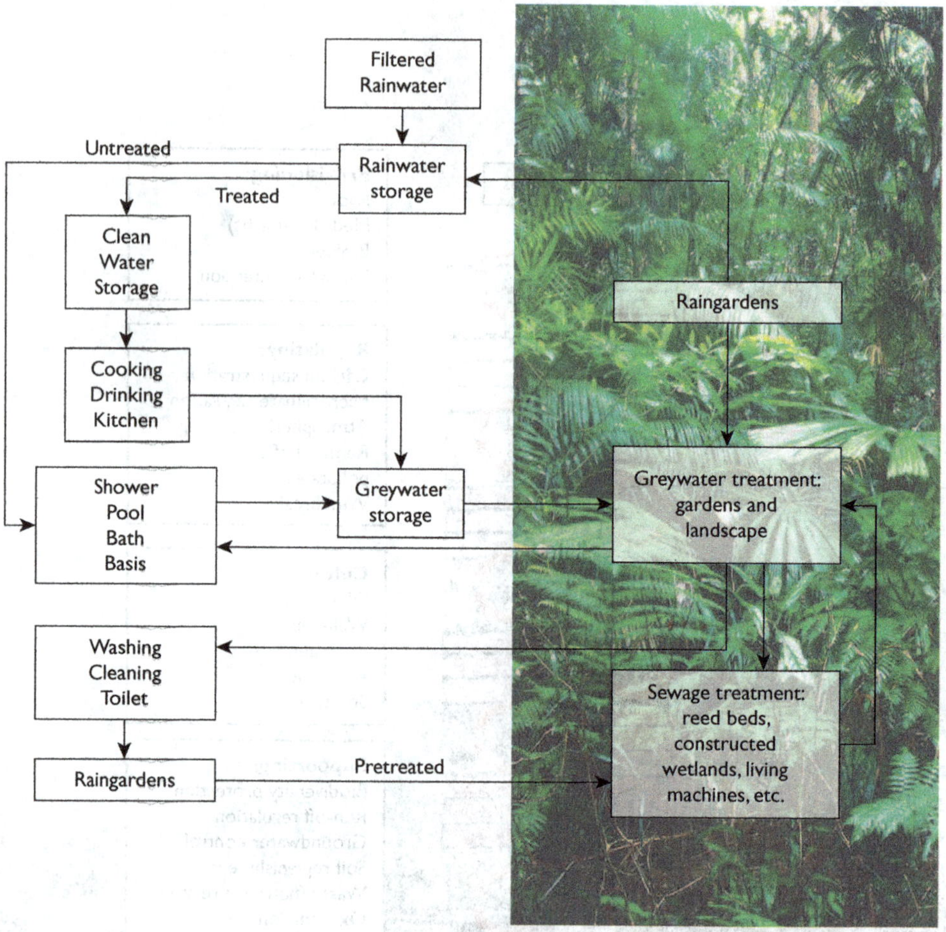

Figure 19.5 Integration of greenery into water and sewage services.

Source: Image supplied by the author.

(Nowak *et al.* 2013). As a result, much additional research and investigation is needed to establish the relationship between LAI and desired ecosystem benefits.

The role of GnPR

While the benefits accrued from the implementation of greenery have been identified, the measurable aspects of it remain to be determined – how large a garden is needed, how large a pond, what species, how many species, how many of each species, where to plant what and so on. At the smaller scale, LAI will be helpful in calculating these parameters. Taken at a larger scale, GnPR will be more useful, as it is a composite LAI of all the greenery on site.

While general formulae relating LAI to various environmental functions have been provided here, they are not yet empirically consolidated and their usefulness as predictive models remains to be tested and established. Despite the amount of work already done, the value of greenery for low carbon living is still in its infancy and experts struggle with the difficulties of providing adequate predictive tools for use in a system as dynamic as an ecosystem. The other difficulty is to develop a common and useful tool that is able to bridge the different disciplines and make use of the knowledge developed therein.

The greatest benefit of GnPR is as a regulatory tool (Ong *et al.* 2012). As such, it needs to be supplemented with greenery guidelines, perhaps along the lines of the ASLA's SITES™. It remains to be worked out how this relationship can be forged. Will general guidelines along the lines of SITES™ be enough for the successful implementation of greenery for low carbon living? Or will a quantitative measure like GnPR be needed as a parameter that traverses different services? The material presented here will hopefully contribute to answering these questions.

Acknowledgements

The formulae presented here were originally developed as part of a Ph.D. study at the University of Cambridge in the 1990s. The study was sponsored by the Overseas Graduate Scholarship from the National University of Singapore and supervised by Dean Hawkes, University of Cambridge, and David Aikman, Horticultural Research International, Wellesbourne. The work we did was greatly inspired by Hamlyn Jones, who was Head for Sites and Director of Crop Science Research at the Horticulture Research International during that period. His key text, *Plants and Microclimate* (Jones 1992), was a primary reference text for the study. Additional work on GnPR came through the support of Jurong Town Corporation, National Parks of Singapore and the Building Control Authority of Singapore. The project shown in Figure 19.4 was a competition entry submitted in collaboration with ArchUrban Architects and Planners in Singapore and a colleague, Ah Fong Foo. I am thankful for the support I have received from these parties and the individuals involved, too many to mention by name here.

Notes

1 The improvement in economic value is dependent on several factors, more usefully lumped together as environmental factors. The best added value comes from access (both visual and physical) to a lake.
2 Bearing in mind the age of our planet, 'fresh' air is simply air that is recycled and rejuvenated by plants.
3 I fondly recall being thrown out of the office of a botanist in the 1990s when I told him I wanted to study the effect of plants on climate.
4 BREEAM, LEED and Green Star are GBC ratings for the UK, US and Australia respectively.

References

American Society of Landscape Architects (ASLA) 2009, *About the Sustainable Sites Initiative*, www.asla.org/sites.aspx, viewed 1 May 2013.

Campbell, G.S. 1986, 'Extinction coefficients for radiation in plant canopies calculated using an ellipsoidal inclination angle distribution', *Agricultural and Forest Meteorology*, vol. 36, pp. 317–321.

Food and Agriculture Organization of the United Nations (FAO) n.d., *Leaf Area Index*, www.fao.org/gtos/doc/ECVs/T11/T11-LAI-report-v02.doc, viewed 1 May 2013.

Gill, S., Handley, J., Ennos, A. and Pauleit, S. 2007, 'Adapting cities for climate change: the role of the green infrastructure', *Built Environment*, vol. 33, pp. 115–133.

Helliwell, R. 2008, 'Amenity valuation of trees and woodlands', *Arboricultural Journal*, vol. 31, pp. 161–168.

Herwitz, S.R. 1985, 'Interception storage capacities of tropical rainforest canopy trees', *Journal of Hydrology*, vol. 77, pp. 237–252.

Hosoi, F. and Omasa, K. 2009, 'Estimating vertical leaf area density profile of tree canopies using three-dimensional portable LiDAR imaging', in *Proceedings of the ISPRS Workshop Laser-scanning*, pp. 152–157.

Jones, H.G. 1992, *Plants and Microclimate: A quantitative approach to environmental plant physiology*, Cambridge University Press, Cambridge.

Kazmierczak, A. and Carter, J. 2010, 'Adaptation to climate change using green and blue infrastructure: a database of case studies', *Interreg IVC Green and Blue Space Adaptation for Urban Areas and Eco Towns (GRaBS) project*, University of Manchester, Manchester.

Kellert, S.R., Heerwagen, J. and Mador, M. 2011, *Biophilic Design: The theory, science and practice of bringing buildings to life*, John Wiley & Sons, Hoboken, NJ.

Killicoat, P., Puzio, E. and Stringer, R. 2002, 'The economic value of trees in urban areas: estimating the benefits of Adelaide's street trees', *Proceedings Treenet Symposium*, 106.

Lalic, B. and Mihailovic, D.T. 2004, 'An empirical relation describing leaf-area density inside the forest for environmental modeling', *Journal of Applied Meteorology*, 43, pp. 641–645.

Lovell, S.T. and Johnston, D.M. 2009, 'Designing landscapes for performance based on emerging principles in landscape ecology', *Ecology and Society*, vol. 14, pp. 1–24.

Luttik, J. 2000, 'The value of trees, water and open space as reflected by house prices in the Netherlands', *Landscape and Urban Planning*, vol. 48, pp. 161–167.

Morales, D.J. 1980, 'The contribution of trees to residential property value', *Journal of Arboriculture*, vol. 6, pp. 305–308.

Nowak, D.J., Greenfield, E.J., Hoehn, R.E. and Lapoint, E. 2013, 'Carbon storage and sequestration by trees in urban and community areas of the United States', *Environmental Pollution*, 178, pp. 229–236.

Ong, B.L. 2002, 'Green Plot Ratio: an ecological measure for architecture and urban planning', *Journal of Landscape and Urban Planning*, vol. 965, pp. 1–15.

Ong, B.L., Ho, A. and Ho, D.K.H. 2012, 'Green Plot Ratio: past, present & future', *iNTA2012 – Tropics 2050*, Singapore.

Rizwan, A.M., Dennis, L.Y.C. and Liu, C. 2008, 'A review on the generation, determination and mitigation of urban heat island', *Journal of Environmental Sciences*, vol. 20, pp. 120–128.

Scheffer, M., Holmgren, M., Brovkin, V. and Claussen, M. 2005, 'Synergy between small- and large-scale feedbacks of vegetation on the water cycle', *Global Change Biology*, vol. 11, pp. 1003–1012.

Sinclair, T.R., Tanner, C. and Bennett, J. 1984. 'Water-use efficiency in crop production', *BioScience*, pp. 36–40.

Chapter 20

Urban populations' vulnerability to climate extremes

Mitigating urban heat through technology and water-sensitive urban design

Nigel Tapper, Andrew Coutts, Margaret Loughnan and Devna Pankhania

Summary

For the first time in global history, more people worldwide are making their homes in the urban rather than the rural realm. As a result of physical processes that are clearly understood scientifically, cities are considerably warmer than their rural surrounds. This is highly significant, since that urban warmth is superimposed on any background global climate warming, with some important implications, including for human health, since people become ill and die under conditions of excessive heat. In addition, a move towards urban consolidation that is evident worldwide, driven by the need to reduce encroachment on rural land and the need to increase transport, energy and carbon efficiency, threatens to further increase urban temperatures. This chapter presents recent research and policy development on these matters and argues strongly that a holistic approach needs to be taken to matters involving building and urban design to develop more resilient and sustainable cities for the future. Those involved in urban development and design must incorporate a range of approaches known to reduce urban temperatures at the same time that consolidation and further urbanization proceeds. However, on a positive note, actively pursuing heat reduction strategies in urban environments, where most of the global population now live, and where temperatures are already significantly elevated, may provide us with some important 'head room' to manage and adapt to the increased temperatures associated with future climate change.

Introduction

For the first time in global history, more people worldwide are making their homes in the urban rather than the rural realm. This is highly significant because cities are considerably warmer environments than their rural surrounds and that urban warmth is superimposed on any background global climate warming. This has important implications, including for human health, since people become ill and/or die under conditions of excessive heat. In addition, urban consolidation processes that are evident worldwide and that are driven by the need to reduce encroachment on rural land and to increase transport, energy and carbon efficiency threaten to further increase urban temperatures.

The critical nexus of urbanization and climate change and the urgent need to develop appropriate urban adaptation responses has been on the international agenda for some time. It was recognized by the United Nations Human Settlements Programme (UN-Habitat) in its *Global Report on Human Settlements* (2011). The Intergovernmental Panel on Climate Change (IPCC)'s Working Groups II and III made the same point when they held an expert meeting in preparation for the fifth Assessment Report (IPCC 2011). Since the last assessment, and with the knowledge that some climate change is now inevitable no matter what happens with future emissions, the focus of the IPCC has now clearly shifted to include adaptation as well as mitigation responses.

This chapter considers matters of human heat exposure, vulnerability and adaptive capacity in cities. Using mainly Australian cities as examples, it presents recent research and thinking about mitigation and management of the effects of urban heat through water-sensitive urban design (WSUD). We argue strongly that a holistic approach needs to be taken to matters involving building and urban design to develop more resilient and sustainable cities for the future. Those involved in urban development and design must incorporate a range of approaches known to reduce urban temperatures (irrigated green infrastructure, smart building materials, etc.) at the same time that consolidation and further urbanization proceeds.

Heat and population health in cities

The European extreme heat event of August 2003 claimed tens of thousands of lives and is arguably the most catastrophic weather event since European climate records began. Similarly an extreme heat event in Southeast Australia in January 2009 claimed 374 lives in three days in the Australian state of Victoria. Many Australians are still unaware of the impacts of extreme heat on human health: heatwaves are silent killers and as such their impacts are often underreported. As climate change progresses, heat exposure is likely to cause additional heat-related illness and death, especially for the most vulnerable groups in the community, such as older people, young children, people with chronic disease and those living in cities. Understanding the relationship between extreme heat and health is complex. However, three factors known to strongly influence risk are exposure, vulnerability and adaptive capacity. Figure 20.1 provides a conceptual framework of heat-related health risk in urban areas.

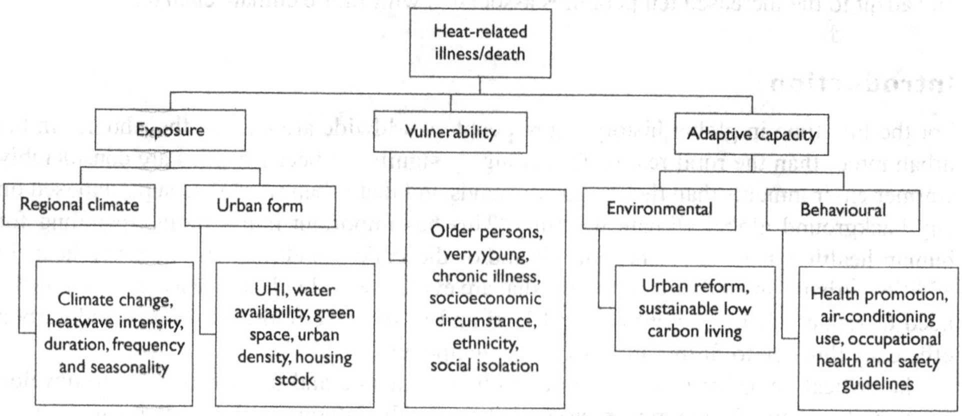

Figure 20.1 Schematic diagram showing the relationship of exposure, vulnerability and adaptive capacity to heat-related human health outcomes.

Source: Supplied by the authors.

Heat exposure in urban areas

As shown in Figure 20.1, the exposure of humans to heat in urban environments is in part related to regional climate and its day-to-day, seasonal and inter-annual variations. Australian cities, especially the southern cities, are noted for their occasional extreme summer conditions, for example, the extreme heat events of Adelaide and Melbourne in January 2009 and Sydney in 2013 when the temperature reached the upper 40s (degrees Celsius) in each of these large cities. This is essentially a consequence of their locations on the margin of an extremely hot continent in summer. While controls on these extreme heat events are largely synoptic (Sturman and Tapper 2006) and broadly unrelated to human influence, there are concerns that atmospheric circulations at a range of scales, and in particular the intensity of extreme heat events, may alter under climate change (discussed below).

As the Australian population has urbanized over the past century, Australian average surface temperatures have also risen by about 1°C from 1910 to 2009 as a consequence of climate change, with the past decade being the warmest on record (CSIRO 2011). By comparison, average global temperatures have warmed during the same period by a more modest 0.7°C. Significantly, as a consequence of continuing climate change, average temperatures are predicted to increase around the world. In Australia, average temperatures are projected to rise by between 1.0 and 5.0°C by 2070 when compared with temperatures of the period 1980 to 1999 (CSIRO and BOM 2012), and extreme heat events are likely to become longer and more severe (Alexander and Arblaster 2009). It is therefore becoming essential to prepare for extreme heat events and to protect communities from excessive heat exposure.

It is critical to note here that urban warmth (the so-called urban heat island or UHI) is of similar magnitude to predicted greenhouse gas-induced climate change during the twenty-first century and, most worrying for the well-being of the majority of humans living in cities, that urban warmth is superimposed on the climate warming that is already underway (Sturman and Tapper 2006; Coutts *et al.* 2010). Fortunately for urban dwellers, the climate processes associated with urban warming are well understood scientifically (e.g. Coutts *et al.* 2007, 2008) and can be mitigated. Urban areas tend to maximize energy inputs from the sun/environment and energy from artificial sources, and minimize energy used in evaporating moisture (there are lots of waterproof surfaces in cities). As a result of these energy shifts, there tends to be much more energy available for direct heating of the urban atmosphere and for daytime storage in urban materials, producing a much warmer urban environment, especially at night. Recent observational work using a network of micrometeorological towers across different urban landscapes in Melbourne, Australia confirms these shifts, showing that storage in building materials increases by ~250 per cent, evaporation reduces by ~50 per cent, and heating of the atmosphere increases by ~30% simultaneously as one moves from a fully rural to a fully urban environment (Coutts *et al.* 2007). As a consequence, a nocturnal UHI of ~4.0°C is typical for Melbourne and is an excess of warmth that is carried over a considerable part of the city.

A further issue for urban environments is a justifiable concern around the globe about sprawling cities that consume productive rural landscapes, hinder the provision of services in a dispersed urban landscape and reduce transport/energy/carbon efficiency, to name a few of many issues. This has led to a widespread drive, especially in developed countries such as Australia, to develop policies to encourage consolidation of our cities. However, work undertaken by Coutts and colleagues (2008) has shown that such consolidation within a fixed urban boundary could exacerbate the magnitude and extent of the UHI through promoting the climate processes that lead to the development of urban warmth. Urban consolidation is inevitable and is required for sound sustainability principles. It will therefore be imperative under climate change to offset the excess warmth of urban areas as much as possible – including that produced by consolidation – by actively promoting building, urban design and planning approaches that can cool the urban landscape. These approaches are discussed in more detail towards the end of this chapter.

Population heat vulnerability

Post-event analysis of extreme heat events has shown that thresholds or tipping points above which mortality and morbidity increases exist across all climate zones, with higher thresholds in hotter climates. For Melbourne, Australia, a city with a dry and occasionally very hot summer climate, Nicholls and colleagues (2008) were able to identify a temperature threshold above which mortality increased for older persons (65+ years). In this instance an average 24-hour (9 a.m. to 9 a.m.) temperature of 30°C (e.g. a 40°C maximum day temperature followed by a 20°C minimum night temperature) resulted in a 15 to 17 per cent increase in the number of deaths in the elderly (see Figure 20.2). In the context of this chapter it is extremely important to note that, because of the abrupt threshold effect shown, *a relatively small reduction in temperature at or close to the threshold can save many lives.* An overnight minimum of 24°C or above resulted in a 19 to 21 per cent increase in deaths.

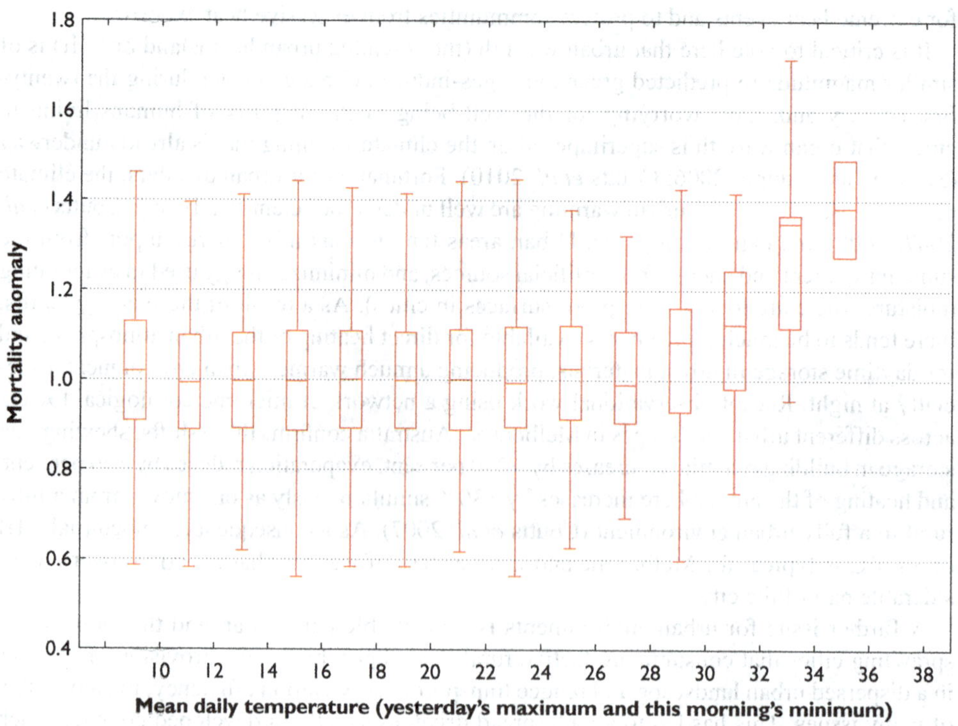

Figure 20.2 Mortality anomalies for Melbourne, Australia, associated with binned daily mean temperatures, 1979 to 2001 (from Nicholls *et al.* 2008)

Source: Supplied by the authors.

Table 20.1 Maximum temperature $T_{(max)}$ thresholds producing excess mortality for all Australian cities, plus modelled days exceeding that temperature for 1990–2010 (current climate), 2020–2040 and 2060–2080, using a high emissions (A2) scenario

Australian city	*$T_{(max)}$ threshold (°C) for excess mortality*	*Modelled number of days exceeding $T_{(max)}$ threshold 1990–2010*	*Modelled number of days projected to exceed $T_{(max)}$ threshold 2020–2040*	*Modelled number of days projected to exceed $T_{(max)}$ threshold 2060–2080*
Adelaide	42	0.81	2.33	2.90
Brisbane	36	5.33	9.84	17.78
Canberra	33	10.60	24.94	32.47
Hobart	35	0.67	1.77	2.91
Melbourne	39	2.62	5.67	8.23
Perth	44	2.29	2.81	6.38
Sydney	38	6.09	8.72	11.34

Source: Adapted from Loughnan *et al.* (2013).

Loughnan and colleagues (2013) recently calculated the threshold temperatures for heat-related mortality and morbidity across all Australian capital cities. Temperature thresholds for daily maximum temperature range from 44°C in Perth to 33°C in Canberra (see Table 20.1). Overall, the thresholds are higher in the southern mainland cities. Calculating thresholds based on daily maximum, minimum or mean temperature may not be suitable for all regions, and use of a heat stress index such as Apparent Temperature (AT) is needed in subtropical and tropical regions. This calculation includes daily maximum temperature and daily maximum dew point temperature and represents a measure of oppressive weather. Overall, there were clear indications that heat-related mortality and morbidity increased with temperature in all Australian capital cities.

In the absence of rapid population acclimatization, climate change will exacerbate heat-related mortality and morbidity. In Australian capital cities, estimations of likely numbers of heat-related deaths are projected to reach up to 15,000 per year by 2100 (AMA 2005), highlighting the public health significance of this issue. In order to prepare communities and infrastructure to cope with future heat events, government and non-government agencies need information about likely changes in extreme heat days. Regionally downscaled climate model output was used by Loughnan and colleagues (2013) to show climate change projections for extreme heat days through until 2080 (see Table 20.1). Most Australian cities (except for Sydney) show approximately three times the number of days with dangerously high maximum temperatures by 2080 compared with the current climate. This is of major concern for emergency management authorities as well as for those with responsibility for designing and planning our cities.

Vulnerability to heat exposure in urban areas is largely place-specific and related to urban form, socio-demographic characteristics of the resident population and adaptive capacity. As the global population becomes older and more urbanized, the adverse impacts of climate change on human health will increase. To understand the influence of 'place' on health during episodes of extreme heat, composite indices of vulnerability to extreme heat events have been developed. These include variables such as age profiles, health status, socioeconomic circumstance, ethnicity and attributes of the built environment. Loughnan and colleagues (2013) have developed vulnerability indices for all Australian capital cities. The indices and associated maps indicate that populations most vulnerable to heat live in areas with the greatest urban heat exposure, and that areas with predicted population increases (including a greater proportion of older people) often coincide with known areas of high vulnerability. Maps of current extreme heat vulnerability in Australian cities, along with predicted changes in vulnerability, may be found on the National Climate Change Adaptation Research Facility website (NCCARF 2013). Maps of population vulnerability in urban areas provide a template to guide future adaptation and mitigation policies. They provide a blueprint for urban planners and landscape architects to reform cities to make them hospitable and productive places that are resilient in the face of climate extremes.

Fostering adaptive capacity to heat

Building resilient urban communities, with the ability to cope with extremes, may be achieved by enhancing behavioural and technological adaptive capacities. 'Adaptive capacity' refers to the processes, actions or outcomes of individuals, communities and institutions moving to cope with and manage the human health impacts of extreme heat. This is especially vital when inherent vulnerability cannot be altered (Smit and Wandal 2006). When exposure to high ambient air temperatures exceeds subjective thresholds of comfort, individuals will actively adapt to their surroundings to remain cool (Loughnan *et al.* 2013). Physiological responses at the individual level (i.e. sweating, vasodilation) occur involuntarily; conversely behavioural adaptations are actively undertaken to reduce thermal stress (i.e. removing clothing, drinking cold liquids) (Kwok and Rajkovich 2010; Loughnan *et al.* 2013). Further opportunities can be created through technological adaptations at the building level which allow occupants greater control over thermal conditions by various design strategies. Table 20.2 characterizes the three key types of adaptation.

State governments and local authorities have also recognized that community resilience to extreme heat events is largely dependent on adaptive capacity. For example, the *Heatwave Plan for Victoria* is designed to provide a coordinated and integrated response of actions and systems to extreme heat to protect the health of those most vulnerable to health impacts (Department of Health 2011). This plan reinforces the role of local government through the municipal public health planning process, through public health alerts when forecasted temperatures are predicted to reach or exceed heat health temperature thresholds. Health messages will be released to the broader community through media releases and interviews. These will actively inform and educate people of the risks to human health posed by excessive heat and promote actions to minimize these risks such as to 'look after yourself and keep

Table 20.2 Examples of three major types of human adaptive capacity to urban heat

Physiological (involuntary)	*Behavioural*	*Technological*
Sweating	Removing clothing	Operable windows
Vasodilation	Decreasing metabolism by reducing exertion	Automated fans
Open body position	Moving to cooler space, particularly to sleep	Shutters/screens
	Ingesting cold liquids, eating less	Solar shading
	Timing behaviour, including work, to avoid heat	Reflective glass
	Taking siestas	Heating/cooling systems

Source: Adapted from Kwok and Rajkovich (2010) and Loughnan *et al.* (2013).

in touch with others, drink plenty of water, keep cool and stay out of the sun' (Department of Health 2011, p. 12).

Arguably, one of the most effective means to protect human health from excessive heat exposure is to increase time in consistently cool, air-conditioned environments such as office buildings, shopping malls and libraries. Studies from the USA indicate that greater accessibility to in-home air-conditioning or evaporative cooling systems has resulted in a significant decrease in heat-related mortality and morbidity over the past two decades. Rogot and colleagues (1992) examined heat-related mortality rates and central or room air-conditioner usage individually across 40 states in the USA. Overall, they found that in hot weather events in Florida, mortality rates were reduced by 42 per cent for households with 'central air-conditioning' compared to those with no air-conditioning. Distinct health benefits were evident in 26 out of 40 states.

Despite these health benefits, individuals in centrally controlled, high-volume air-conditioned buildings may acclimatize to cool, steady-state conditions that limit adaptive opportunities as well as reduce overall resilience to excessive temperatures. A study of adaptation comparing indoor environments (both air-conditioned and naturally ventilated spaces) of building types in Korea found that those acclimatized to air-conditioned environments had lower tolerance for warm indoor environments (Kim *et al.* 2007). In addition, cooler, steady-state environments create more stringent comfort criteria, where occupants are less tolerant of variability and perceive temperatures to be greater than in reality (Nichol and Humphreys 2002).

As well as promoting physiological and psychological dependence for cooler climates through centrally controlled air-conditioning systems, increased technology can also exacerbate the UHI and peak energy demands (Hsieh *et al.* 2011). Greater anthropogenic waste heat released by air-conditioners to alleviate the UHI will likely reinforce the positive feedback between elevated temperatures and cooling energy demands, exacerbating uncomfortable microclimates and increasing dependence on technological adaptation. Munck and colleagues (2013) simulated the anticipated potential impacts of air-conditioning scenarios on air temperature in Paris, France using a meso-scale model coupled with an urban energy balance model. For the most extreme case, where air-conditioning use was likely to double across the city, there would likely be a 2°C increase in city-wide temperature generated by waste heat. This feedback further complicates challenges in meeting urban energy demands, while also raising global greenhouse emissions (GHG) from power generation and use.

Enhancing behavioural and technological adaptive opportunities is most effective when coupled with improved building and urban design and planning to achieve cooler, more comfortable and sustainable environments. Low carbon approaches to building design are becoming vital as climate change presents communities with an increase in extreme heat events associated with higher CO_2 emissions from fossil fuel-based energy. Energy efficient buildings can be achieved through structural and technological modifications that reduce external and internal heat gains. Chwieduk (2003) proposes four key principles: 'minimising heat gains, ensuring equipment is switched off, attenuating peaks of thermal mass and effective ventilation' (see Table 20.3). Implementation of these measures not only minimizes cooling loads for buildings, but also creates adaptive opportunities allowing occupants greater control over their thermal conditions. When individuals are given greater control over their surroundings, it modifies their perception and tolerance of higher temperatures (Brager and de Dear 1998; Nichol and Humphreys 2002). Transitioning from steady-state, air-conditioned environments to more passively cooled environments with greater temperature

Table 20.3 Strategies to promote adaptive capacity for passive and low energy buildings

Approach	*Strategies*
Site analysis	Orientation for solar access
Envelope optimization	Insulation
Daylighting and shading	Top lighting, side lighting, light shelves, overhangs and vertical fins
Passive cooling	Cross, stack and night ventilation of thermal mass, evaporative cooling, courtyards, earth contact, earth tubes and transitional spaces
Low energy 'active' systems	Fans, heat recovery ventilators, evaporative chillers

Source: Adapted from Kwok and Rajkovich (2010)

variability through low carbon building design will promote more resilient and sustainable communities in the face of climate change.

Mitigating heat exposure through WSUD

As described earlier, vulnerability to heat exposure is place-specific, but so is heat exposure itself, and in urban areas it is influenced by urban form and local land surface characteristics. While adapting to heat exposure through technological and behavioural responses can help reduce the risk of heat-related illness and mortality, this does not address the core issue of limiting extreme heat. Heatwave plans are a response to extreme heat, but do not deal with background heat itself – part of which is attributed to the design of the urban environment.

Mitigation efforts for climate change and for increases in heatwave intensity, frequency and duration are topics of local, national and international debate, with approaches such as emissions trading schemes, carbon taxes or reforestation programmes variously adopted. However, the urban form and its contribution to the climate of cities receive much less attention. A number of approaches have been identified that target UHI mitigation. Increasing vegetation cover and urban albedo are two approaches that are regularly promoted. Other approaches include controlling street design (widths and orientation) and building heights, increasing surface emissivity, and reducing the thermal conductivity and heat capacity of urban materials. These approaches aim to intentionally modify surface radiation and energy exchanges with the atmosphere to reduce heating. Mitigation also targets the production of waste heat in urban areas by reducing vehicle use and undertaking many of the building efficiency improvements discussed above.

It should be noted here that the UHI itself refers to a city-scale phenomenon describing the temperature contrast between urban and surrounding rural areas. It is often quantified as the maximum difference in air temperature between the most built-up urban sites (e.g. the CBD) and rural locations (UHI intensity $= T_{urban} - T_{rural}$). This maximum temperature difference tends to occur at night or pre-dawn as rural areas cool rapidly, while urban areas cool more slowly due to heat storage and trapping, and reduced long-wave radiation losses between tall buildings. These urban–rural temperature contrasts are strongest under clear and calm conditions. Because it is a comparative measure, UHI intensity depends as much on the conditions of rural landscapes as it does on urban landscapes. At times, urban cool islands may exist where the city is actually cooler than surrounding rural areas. This may occur, for example, when water is available for irrigating urban landscape during drought but where the rural landscape remains dry. This shows that the practical use of the UHI as a concept and measure is somewhat limited. What we are truly interested in is the spatial and temporal variability in heat exposure across entire urban regions driven by land-use and land-cover characteristics, whether a UHI is present or not. Therefore, minimizing high urban temperatures (heat exposure) should be a key performance objective (rather than mitigating the UHI) and should target all contributing factors to high urban heat exposure (from urban form and extreme heat events). Further, research has shown that variability in air temperatures at the local (neighbourhood) scale can be as large as city-wide differences (Broadbent *et al.* 2013; Erell and Williamson 2007).

A key driver of high heat exposure in urban areas is the modification of the water balance. The replacement of natural vegetated surfaces with hard, impervious urban surfaces restricts infiltration and increases rainfall run-off. To avoid flooding, this vast amount of stormwater is rapidly exported from the urban environment through drainage and waterways. Water that would naturally have infiltrated into soils and subsequently evaporated or progressed to streams through subsurface flows is now efficiently removed from the system. This leads to a water deficit – one that is traditionally met through imported potable water. In fact, stormwater run-off from urban areas can be as high as total water consumption in some cities. In Melbourne, the total urban landscape stormwater run-off in 2010 was 463 GL, compared to the total Melbourne regional consumption of 412 GL (LVMAC 2011). In addition to the export of stormwater, periods of drought leave urban landscapes even dryer, and if potable water restrictions are in place, this deficit cannot be addressed. This leaves urban areas warm and dry, supporting atmospheric heating (Coutts *et al.* 2007) and the development of areas of high heat exposure.

As discussed above, increasing vegetation cover is one of the key approaches for mitigating heat exposure. However, the effect of a warming and drying landscape can have dramatic effects on urban vegetation health, limiting its capacity to cool. Insufficient soil moisture at the root zone due to limited water availability may lead to leaf stomata closure and, if severe, leaf wilt and canopy loss. This limits their potential to provide cooling through transpiration and shading, although a sensible approach to lack of water, promoting drought-tolerant plants and (xeric) gardens, reinforces the creation of desert-like landscapes. The combination of urban development, low water availability, compromised vegetation health and climate change supports high heat exposure and increases the risk of heat-related illness and/or death. Re-vegetation of urban areas must be supported by sufficient water to fully realize the benefits of vegetation (Coutts *et al.* 2010).

Water-sensitive urban design presents an opportunity to address the urban water deficit. WSUD aims to retain water in the urban landscape and restore a more natural water balance, increasing infiltration and evapotranspiration, while reducing stormwater run-off. Forms of WSUD include bio retention systems (i.e. rain-gardens), constructed wetlands and lakes, green roofs and walls, and infiltration systems (i.e. porous pavements). These systems are designed to change the path and reduce the rate of stormwater as it travels to downstream waterways, while some also support pollutant removal. Harvested stormwater may be used as a source of water given its abundance, particularly for landscape irrigation, to further promote infiltration and evapotranspiration. Restoring a more natural water cycle can also support tree health by increasing soil moisture. This concept is illustrated in Figure 20.3, where implementation of WSUD features and stormwater harvesting at the micro and local scale delivers local-scale cooling. A key feature of this approach is the distributed nature of micro-scale implementation as cooling effects are localized.

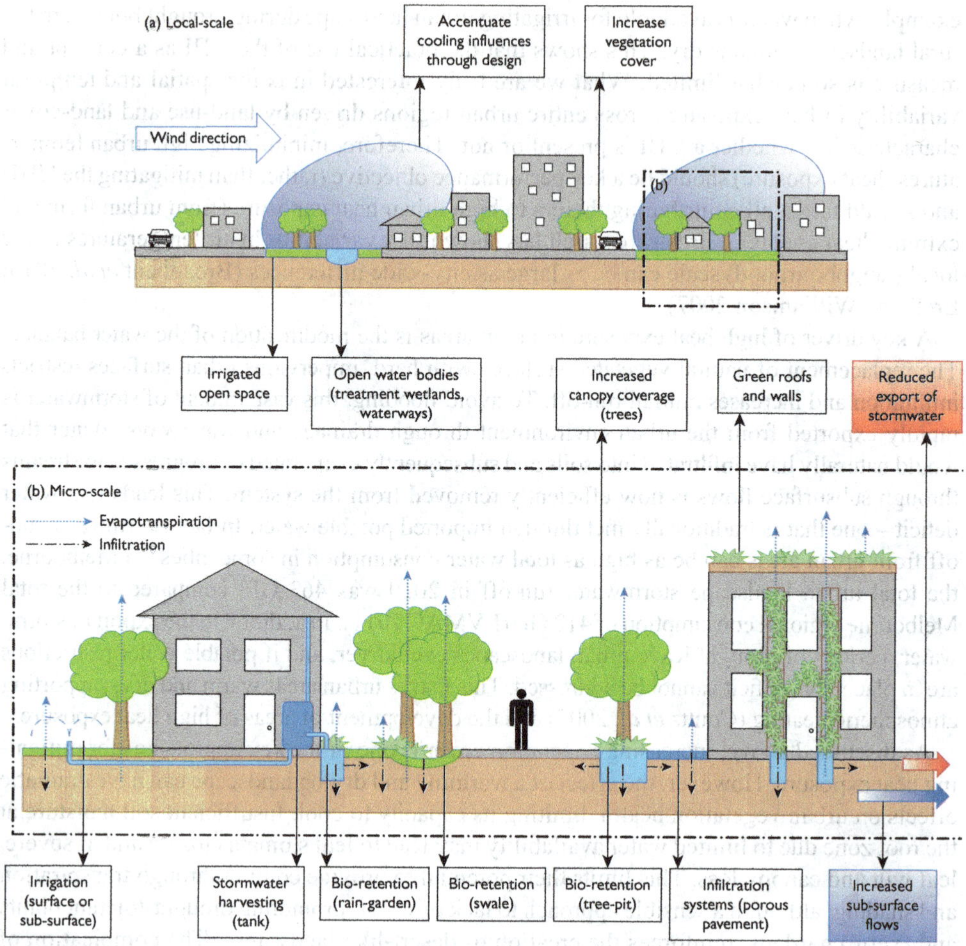

Figure 20.3 Schematic representation of widespread implementation of stormwater harvesting and water-sensitive urban design elements at the microscale for the restoration of a more natural water balance and increased vegetation cover

Source: Coutts et *al.* (2012), used with permission and supplied by the authors.

The concept effectively involves turning our cities into oases and follows influences observed in and around irrigated parklands and the park cool island (PCI) effect (where parks are cooler than their surrounding urban landscapes). Spronken-Smith and colleagues (2000) observed evapotranspiration rates that were three times higher than the surrounding residential neighbourhood. These high rates of evapotranspiration can support cooler air masses that propagate downwind (Honjo and Takakura 1990). It has been suggested that this downwind cooling effect extends up to one park width (Spronken-Smith and Oke 1998). Even at the household scale, an irrigated lawn can support high rates of evapotranspiration (Suckling 1980). Distributing regular, well-watered green spaces (and vegetation) throughout the landscape at the micro scale can encourage neighbourhood-scale evapotranspiration and cooling at the local scale (Figure 20.3). In a comparison of local-scale evapotranspiration rates between a suburban neighbourhood well supplied with water and a rural landscape in Vancouver, British Columbia, Oke and McCaughey (1983) observed that evapotranspiration rates were 40 per cent higher over the suburban neighbourhood during the day.

Strategic implementation of WSUD

Simply increasing vegetation cover and irrigation, and integrating WSUD into the landscape arbitrarily, may not provide significant benefits for cooling, and given limited budgets, implementation of WSUD should be strategically placed to maximize cooling potential. Implementation should be fit-for-place and targeted to the most vulnerable locations (Loughnan *et al.* 2013), working with both the natural and man-made features of the urban landscape, as the amount of cooling depends on both the type of WSUD implemented and the place in which it is situated. As discussed above, trees are especially effective in cooling urban environments (particularly if supported by water) as they provide both transpiration and shading benefits. Air temperatures have been shown to be lower under individual trees and clusters of trees during the day (Bowler *et al.* 2010), and research suggests than even a small fraction of irrigated vegetation in the landscape can have cooling effects (Gober *et al.* 2010). However, placement is critical, because trees are most effective in wide-open streets where solar radiation is high, while planting trees where buildings already provide shade will be less effective. Nevertheless, strategic placement of trees around buildings can also improve energy efficiency within the building.

The extent of cooling depends on wind speed and direction of the prevailing wind, but also the layout and design of surrounding buildings (Chen and Wong 2006), along with the amount of cooling *relative* to the surrounding environment. Buildings can act as blocking features, preventing downwind cooling. Murawaka and colleagues (Murawaka *et al.* 1991) found that the downwind cooling effect of a river in Japan was evident when the building density was low and street widths were wide, allowing increased propagation of cool air. This highlights the importance of working with, and designing, built environments to maximize cooling of WSUD. Considerable benefits may also be achieved by ensuring existing green spaces and vegetation are well supplied with water through irrigation or retrofitting of WSUD to existing vegetated landscapes.

Green roofs are particularly useful in reducing energy transfer into buildings and hence improving their energy efficiency, but their performance also depends on their strategic placement. Green roofs provide greatest benefits on poorly insulated buildings (Jaffal *et al.* 2012). In addition, green roofs are most effective in terms of improving outdoor thermal environments on lower-rise buildings because the higher the building, the less the cooling effect there is at street level where people reside (Santamouris 2014). In addition, as a building becomes taller, the surface area of the roof relative to the wall area is reduced, so the effectiveness for internal energy improvements diminishes. Green walls are likely to provide greater street-level benefits than green roofs (Alexandri and Jones 2008).

WSUD design characteristics

Currently, most WSUD features are designed for stormwater run-off and quality benefits or for supplementing potable water consumption, and any microclimatic benefits are coincidental. Ongoing research is highlighting the importance of design characteristics on the performance of WSUD. If enhanced urban cooling is identified as a key performance objective, this should be considered in system design. For instance, green roofs are often touted as providing UHI mitigation benefits through evapotranspiration. However, many green roofs are designed with shallow soil substrates and drought-tolerant vegetation, and thus evapotranspiration can be low and atmospheric heating from the roof high (Coutts *et al.* 2013; Takebayashi and Moriyama 2007). Irrigating green roofs should be considered if reducing heat exposure is a performance objective.

As another example, bio-retention systems support rapid infiltration, but there is a risk that the soils will also drain too quickly. Bio-retention systems are often designed with surface water detention ponds to enhance water capture. Trees may be incorporated into bio-retention systems to further support pollutant removal and provide amenity. Large fluctuations between wet and dry soil conditions could have negative effects on vegetation health over the long term if selected species are not tolerant of, or accustomed to, these kinds of conditions. If improved microclimate is a performance objective, the design of bio-retention systems or other WSUD features needs to be optimized to support tree health. Soil moisture conditions need to be maintained between the wilting point and the saturation point where water is available for plant root access (Jim and Peng 2012). Supplementary irrigation with harvested stormwater could enhance the performance of these WSUD systems. Clearly, more research is required on designing and testing WSUD to maximize microclimatic benefits, and to measure their performance against other cooling strategies.

Concluding comments

This chapter has shown that human health outcomes related to extreme heat events are a result of the interplay of exposure to heat and the vulnerability and adaptability of the populations to extreme heat. We have also discussed the role that WSUD can play in reducing heat exposure in urban environments and have shown that even small reductions in temperature can reduce human morbidity and/or mortality. However, we note that WSUD is just one mechanism for reducing heat exposure in urban areas, and will be less effective in hot, moist environments than in hot, dry environments. Ultimately, a range of approaches that would also include high albedo surfaces, smart building materials and improved building energy efficiency will be required to reduce high heat exposure in the outdoor urban environment, the exact mix of approaches depending on specific circumstances.

However, on a positive note, actively pursuing heat reduction strategies in urban environments, where most of the global population now live, and where temperatures are already significantly elevated, will offset the impacts of urban consolidation and may provide us with some important ‘head room’ to manage and adapt to the increased temperatures associated with future climate change.

References

Alexander, L.V. and Arblaster, J.M. 2008, ‘Assessing trends in observed and modelled climate extremes over Australia in relation to future projections’, *International Journal of Climatology*, vol. 29, no. 3, pp. 417–435.

Alexandri, E. and Jones, P. 2008, ‘Temperature decreases in an urban canyon due to green walls and green roofs in diverse climates’, *Building and Environment*, vol. 43, no. 4, pp. 480–493.

Australian Medical Association (AMA) 2005, *New Report Shows Climate Change Threatens Human Health*, 22 September, http://ama.com.au/node/2120, viewed 12 November 2013.

Bowler, D.E., Buyung-Ali, L., Knight, T.M. and Pullin, A.S. 2010, 'Urban greening to cool towns and cities: A systematic review of the empirical evidence', *Landscape and Urban Planning*, vol. 97, no. 3, pp. 147–155.

Brager, G.S. and de Dear, R.J. 1998, 'Thermal adaptation in the built environment: a literature review', *Energy and Buildings*, vol. 27, pp. 83–96.

Broadbent, A.M., Coutts, A.M., Demuzere, M., Beringer, J. and Tapper, N. 2013, 'An urban microclimate research campaign in a water sensitive urban design suburb, Mawson Lakes, Australia', *Urban Climate* (submitted).

Chen, Y. and Wong, N.H. 2006, 'Thermal benefits of city parks', *Energy and Buildings*, vol. 38, pp. 105–120.

Chwieduk, D. 2003, 'Towards sustainable-energy buildings', *Applied Energy*, vol. 76, pp. 211–217.

Commonwealth Scientific & Industrial Research Organisation (CSIRO) 2011, *Climate Change: Science and solutions for Australia,* CSIRO Publishing, Collingwood, Victoria.

Commonwealth Scientific & Industrial Research Organisation (CSIRO) and Australian Bureau of Meteorology (BOM) 2012, *State of the climate 2012*, www.csiro.au/en/Outcomes/Climate/Understanding/State-of-the-Climate-2012.aspx, viewed 3 September 2013.

Coutts, A., Beringer, J. and Tapper, N. 2007, 'Impact of increasing urban density on local climate: spatial and temporal variations in the surface energy balance in Melbourne, Australia', *Journal of Applied Meteorology*, vol. 47, no. 4, pp. 477–493.

—— 2008, 'Investigating the climatic impact of urban planning strategies through the use of regional climate modelling: a case study for Melbourne, Australia', *International Journal of Climatology*, vol. 28, no. 14, pp. 1943–1957.

—— 2010, 'Changing urban climate and CO_2 emissions: implications for the development of policies for sustainable cities', *Urban Policy and Research*, vol. 28, no. 1, pp. 34–47.

Coutts, A., Daly, E., Beringer, J. and Tapper, N. 2013, 'Assessing practical measures to reduce urban heat: green and cool roofs', *Building and Environment* (submitted).

Department of Health 2011, *Heatwave Plan for Victoria: Protecting health and reducing harm from heat waves*, Health Protection Branch, Victorian Government, Department of Health, Melbourne, Victoria.

Erell, E. and Williamson, T. 2007, 'Intra-urban differences in canopy layer air temperature at a mid-latitude city', *International Journal of Climatology*, vol. 27, pp. 1243–1255.

Gober, P., Brazel, A., Quay, R., Myint, S., Grossman-Clarke, S., Miller, A. and Rossi, S. 2010, 'Using watered landscapes to manipulate urban heat island effects: how much water will it take to cool Phoenix?', *Journal of the American Planning Association*, vol. 76, pp. 109–121.

Honjo, T. and Takakura, T. 1990, 'Simulation of thermal effects of urban green areas on their surrounding areas', *Energy and Buildings*, vol. 15, pp. 443–446.

Hsieh, C., Aramaki, T. and Hanaki, K. 2011, 'Managing heat rejected from air-conditioning systems to save energy and improve the microclimates of residential buildings', *Computers, Environment and Urban Systems*, vol. 35, pp. 358–367.

Intergovernmental Panel on Climate Change (IPCC) 2011, *Expert Meeting of WGII and WGIII on Human Settlement, Water, Energy and Transport Infrastructure: Mitigation and Adaptation Strategies*, www.ipcc-wg3.de/meetings/expert-meetings-and-workshops/em-infrastructure, viewed 10 June 2013.

Jaffal, I., Ouldboukhitine, S. and Belarbi, R. 2012, 'A comprehensive study of the impact of green roofs on building energy performance', *Renewable Energy*, vol. 43, pp. 157–164.

Jim, C.Y. and Peng, L.L.H. 2012, 'Substrate moisture effect on water balance and thermal regime of a tropical extensive green roof', *Ecological Engineering*, vol. 47, pp. 9–23.

Kim, H., Chun, C. and Kwok, A. 2007, 'Thermal adaptation to air-conditioned spaces', *Proceedings of the Conference on Sustainable Buildings Asia SB07*, 27–29 June, Seoul, Korea.

Kwok, A.G. and Rajkovich, N.B. 2010, 'Addressing climate change in comfort standards', *Building and Environment*, vol. 45, pp. 18–22.

Living Victoria Ministerial Advisory Council (LVMAC) 2011, *Living Melbourne, Living Victoria Roadmap*, Victorian Government, Department of Sustainability and Environment, Melbourne.

Loughnan, M., Tapper, N. and Phan, T. 2014, 'Identifying vulnerable populations in subtropical Brisbane, Australia: a guide for heatwave preparedness and health promotion', *ISRN Epidemiology*, vol. 2014, Article ID 821759, doi:10.1155/2014/821759.

Loughnan, M., Tapper, N., Phan, T., Lynch, K. and McInnes, J. 2013, *Spatial Vulnerability to Extreme Heat Events in Australian Capital Cities*, National Climate Change Adaptation Research Facility, Gold Coast, Queensland.

Munck, C.D., Pigeon, G., Masson, V., Meunier, F., Bousquet, P., Tremeac, B., Merchat, M., Poeuff, P. and Marchadier, C. 2013, 'How much can air conditioning increase air temperatures for a city like Paris, France?', *International Journal of Climatology*, vol. 33, pp. 210–227.

Murakawa, S., Sekine, T., Narita, K.I. and Nishina, D. 1991, 'Study of the effects of a river on the thermal environment in an urban area', *Energy and Buildings*, vol. 16, nos 3–4, pp. 993–1001.

National Climate Change Adaptation Research Facility (NCCARF) 2013, www.nccarf.edu.au/publications/spatial-vulnerability-urban-extreme-heat-events, viewed 3 September 2013.

Nichol, J.F. and Humphreys, M.A. 2002, 'Adaptive thermal comfort and sustainable thermal standards for buildings', *Energy and Buildings*, vol. 34, pp. 563–572.

Nicholls, N., Skinner, C., Loughnan, M. and Tapper, N. 2008, 'A simple heat alert system for Melbourne, Australia', *International Journal of Biometeorology*, vol. 52, no. 5, pp. 375–384.

Oke, T.R. and McCaughey, J.H. 1983, 'Suburban–rural energy balance comparisons for Vancouver, BC: an extreme case', *Boundary-Layer Meteorology*, vol. 26, pp. 337–354.

Rogot, E., Sorlie, P.D. and Backlund, E. 1992, 'Air-conditioning and mortality in hot weather', *American Journal of Epidemiology*, vol. 136, no. 1, pp. 106–116.

Santamouris, M. 2014, 'Cooling the cities: a review of reflective and green roof mitigation technologies to fight heat island and improve comfort in urban environments', *Solar Energy*, vol. 103, May, pp. 602–703.

Smit, B. and Wandal, J. 2006, 'Adaptation, adaptive capacity and vulnerability', *Global Environmental Change*, vol. 16, pp. 282–292.

Spronken-Smith, R.A. and Oke, T.R. 1998, 'The thermal regime of urban parks in two cities with different summer climates', *International Journal of Remote Sensing*, vol. 19, pp. 2085–2104.

Sturman, A. and Tapper, N. 2006, *The Weather and Climate of Australia and New Zealand*, 2nd edn, Oxford University Press, Oxford.

Suckling, P.W. 1980, 'The energy balance microclimate of a suburban lawn', *Journal of Applied Meteorology*, vol. 19, pp. 606–608.

Takebayashi, H. and Moriyama, M. 2007, 'Surface heat budget on green roof and high reflection roof for mitigation of urban heat island', *Building and Environment*, vol. 42, no. 8, pp. 2971–2979.

United Nations Human Settlements Programme (UN-Habitat) 2011, *Cities and Climate Change: Global report on human settlements 2011*, Earthscan, London.

Chapter 21

Multi-scale analysis of the surface-layer urban heat island effect in five higher-density precincts of central Sydney

Ehsan Sharifi, Conrad Philipp and Steffen Lehmann

Summary

The urban heat island (UHI) effect is invariably present in cities, mainly due to increased urbanization. It can result in higher urban densities being significantly hotter (frequently more than 4°C, even up to 10°C) than their peri-urban surroundings. Urban structure and land cover are key contributors to the surface-layer urban heat island (sUHI) effect at city and district scale. This research aims to explore which urban configurations can make urban precincts and their microclimates more resilient to the dangerous sUHI effect. In the context of the city of Sydney, Australia, the research aims to explore the most heat-resilient urban features for neighbourhoods, at precinct scale. The investigation examines five high-density precincts in central Sydney. The analysis of these precincts is based on remote-sensing thermal data of two independent sources: a nocturnal remote-sensing thermal image of central Sydney on 6 February 2009 and diurnal Landsat 7– ETM+ data from 2008 to 2009. Comparing the surface temperature of streetscapes and buildings' roof-top feature layers indicates that open spaces are the urban elements most sensitive to the sUHI effect. Therefore, the correlations between street network intensity, open public space ratio, and urban greenery plot ratio and sUHI effect are being analysed in the five high-density precincts selected. Results indicate that a higher open-space ratio and street network intensity have a significant correlation to a higher sUHI effect at precinct scale. However, higher urban greenery plot ratios could mitigate the sUHI effect in these precincts. In addition, annual variation of land-use features of streetscape, building (roof-top), open space and urban greenery are analysed based on diurnal Landsat 7– ETM+ data for 2008 to 2009. Results indicate that an increase in urban greenery is the most effective strategy for land surface that is more resilient to the sUHI, while open public space is up to 15 per cent less heat resilient. The research outcomes support the importance of increasing urban greenery, particularly in open public spaces, to achieve cooler cities.

Introduction

Cities are anticipated to accommodate up to 70 per cent of the global population by 2050 (UNDESA 2012). Unlike the current urbanization rate of 50 per cent, almost all the expected global population growth will be accommodated in cities. Such rapid urbanization means higher densities in existing cities and many more new urban areas will be needed to accommodate up to two billion new urban dwellers. However, rapid urban development in fast-growing cities tends to overlook the environmental and social aspects of urban life

(Girardet 2008; Lehmann 2010; Register 2002). A considerable amount of natural landscape is being transformed into building mass and hard surfaces, creating environmental threats for existing and future cities.

With huge demands for natural resources (i.e. energy, food, water and materials), cities are contributing up to 80 per cent of greenhouse gas (GhG) emissions, resulting in global warming (UNECE 2011; UN-Habitat 2011). Climate change projections indicate a likely increase of 2 to 5°C in Australian surface temperature by 2050 (CSIRO 2007; OECD 2010). Such an increase in temperature will have a severe impact on natural ecosystems and human life in cities, including public health and quality of public space (Guest *et al.* 1999; Stone 2012).

Cities also suffer from the effect of an additional form of heat, known as the urban heat island (UHI) effect. This human-made heat is trapped in the built environment's thermal mass and can result in higher densities being significantly hotter than their peri-urban surroundings. The urban–rural temperature difference frequently reaches 4°C and can peak at more than 10°C (Gartland 2008; Oke 2006; Wong and Yu 2008). Such additional heat can seriously impact upon citizens' health and the quality of public life in cities.

Heat islands are uneven in their spatial distribution and can vary between industrial, commercial and residential areas. While cities' higher density can bring efficiency gains, there is interplay between the increased risk of the urban heat island effect and higher densities, which needs to be well understood. Because cities are often covered in heat-absorbing surfaces and materials, such as concrete and bitumen, they absorb and store heat (e.g. through solar radiation), making urban areas warmer than the surrounding hinterland and rural areas, especially at night-time.

Background to the research project

The extensive recent literature on the UHI effect indicates that the artificial increase of temperature in cities is happening because of changes in radiative energy and water budget in the built environment (Erell *et al.* 2011; Gartland 2008; Oke 2006; Santamouris and Geros 2006). This artificial temperature increase affects urban microclimates in different layers of the atmosphere, including the surface layer (buildings and land surfaces), the canopy layer (below the canopy of trees or at human scale) and the boundary layer (up to 1,500 metres above the ground surface). These three layers of urban microclimates are tangled in complex climatic systems, while local air circulation in the built environment can moderate the UHI effect by mixing the air in each layer with other adjacent layers (Erell *et al.* 2011). Oke (2006) argues that the UHI effect has four major contributing factors (see Figure 21.1):

1. Urban geometry, which alters heat exchange balance in the built environment by affecting shadow and wind patterns. It affects the exposure of materials to sunlight and the consequent heat storage in thermal mass. This complex heat radiation exchange between building mass and adjacent atmosphere can also change the intensity and patterns of airflow in urban canyons.
2. Urban cover and surface materials, which affect the heat absorption and reflection time-rate in the built environment. Thermodynamic specification, colour, texture and density of materials and their exposure to sunlight can alter the heat flux in outdoor space in complex procedures.
3. Urban landscape, which affects water and heat exchange rate in the built environment. Photosynthesis and evaporation processes in urban greenery contribute to decrease the

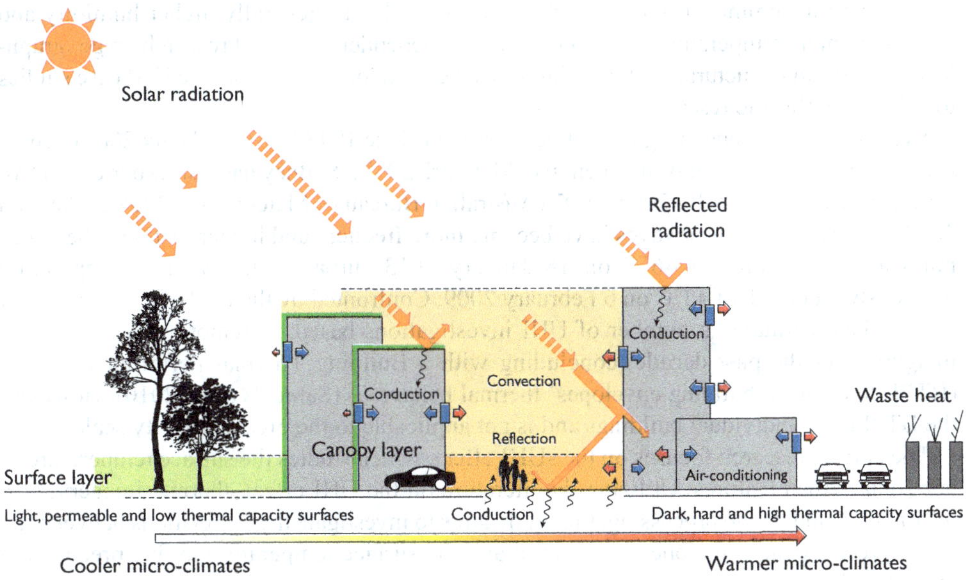

Figure 21.1 Urban structure, landscape, land cover and metabolism contribute to the urban heat island effect in cities.

ambient temperature. Urban greenery typology, distribution and intensity also affect lower atmospheric air turbulence.

4 Urban metabolism and anthropogenic (human-made) waste heat in cities, which is mainly related to mass energy consumption for indoor air-conditioning and motorized transportation.

Existing approaches to the UHI effect are more likely to focus on large-scale monitoring and mitigation strategies or micro-scale material science. More research on the key contributors to the surface-layer UHI (sUHI) effect at precinct scale can provide useful links between UHI investigations at city and material scales.

The temperature of some Australian cities, such as Sydney, Melbourne and Adelaide, is already up to 4°C warmer than surrounding areas. The current investigation discusses ongoing research on Sydney, which is an example of a city facing an increasing UHI effect due to its post-nineteenth-century urban development. Due to the city's subtropical climate and the UHI effect, public spaces in the city are already warmer in summer than humans' thermal comfort, pushing citizens into air-conditioned buildings and creating an ever-increasing rise in outdoor temperatures. Such artificial urban heat stress increases the mortality rate, especially of the elderly (Hu *et al.* 2007). The aim is to investigate the most effective sUHI mitigation strategies at the precinct scale in Sydney.

Materials and methods

Although major UHI contributors may be present in a wide range of regional climates, the effectiveness of urban features for the UHI effect is highly contextual (Oke 2006; Wong and Yu 2008). For example, the UHI effect's behaviour in the canopy layer of a subtropical city like Sydney in summer differs from drier climates, due to generally higher humidity and lower day–night temperature variations. The high dependence of UHI research on geographical, climatic and structural contexts highlights the need for climate-specific UHI case studies to achieve applicable research outcomes.

Sydney has experienced significant development since 1945 (Toon and Falk 2003), which is continuing into the twenty-first century (McGuirk 2003). Sydney has also experienced five severe heatwaves: in 1939, 2004, 2007 (Australian Bureau of Meteorology 2008), 2009 and 2012. In recent years, heatwaves have become more frequent and longer-lasting. The maximum air temperature of 46°C on 18 January 2013 surpasses the highest temperature previously recorded, of 43°C on 6 February 2009. Confronted by the UHI effect, the City of Sydney has facilitated a number of UHI investigations based on remote-sensing thermal imagery over the past decade, concluding with a Building Thermal Performance Index (BTPI) to evaluate building envelopes' thermal behaviour (Samuels *et al.* 2010). However, the BTPI is for individual buildings and is not applicable to the precinct or city scale.

The current research focuses on the sUHI effect, which studies the surface temperature of horizontal urban features. Utilizing the literature on the UHI effect, thermal imagery, GIS information and image processing, this study aims to investigate the correlations between the urban greenery ratio, the open-space ratio and the surface temperature in five precincts in central Sydney.

On 5 February 2009, the temperature reached 31°C at 6 p.m. with a relative humidity of 33 per cent. During the night, wind speed was less than 5 m/s, which was insufficient to cool down the city by the next morning. Consequently, on 6 February 2009, the air temperature reached a record of 43°C at 7 p.m. with a relative humidity of 10 per cent and wind speed of less than 5 m/s. This heat stress continued in Sydney on 7 February with a maximum temperature of 39°C at 6 p.m. and a relative humidity of 12 per cent. Due to higher humidity in lower temperatures, the real feeling of the (apparent) temperature did not fall below 30°C on 6 and 7 February. According to Thom's Discomfort Index (Moran *et al.* 1998; Thom 1959) and the Human Heat Index (ASHRAE 2004), the microclimatic condition in Sydney during the target days was partly in 'heavy discomfort' and mostly in 'emergency discomfort' zones, which can cause heatstroke, especially for elderly and disadvantaged people (Kovats and Hajat 2008).

Aerial thermal imagery of central Sydney was conducted on 6 February 2009 by Digital Mapping Australia for the City of Sydney, available with the resolution of 8 metres. The resulting remote-sensing maps indicate different surface temperatures in central Sydney. Building and population densities, open space and urban greenery primary data are based on GIS information provided by the City of Sydney. Spatial dimension, plot ratio and distribution of open space and urban greenery are extracted from a Google Earth image dated 4 February 2009 (to match the data to the thermal imagery of 6 February 2009).

Thermal imagery of central Sydney on 6 February 2009 maps different surface temperatures of the built environment. It also provides the average surface temperatures of ten precincts (urban districts with identifiable characters), which shape different temperature zones inside central Sydney. From these precincts, five higher-density precincts have been selected for the current research. Sydney Harbour, Haymarket, Harris Street, Kings Cross and Glebe Point are compared to investigate which urban features can be most effective in reducing the sUHI effect in Sydney's high-density precincts.

This research project uses Landsat 7– ETM+ thermal layer data from 2008 to 2009. The sUHI is calculated in four isolated timeframes during 2008 to 2009, based on diurnal

Landsat 7– ETM+ satellite imagery on 28 September 2008, 1 December 2008, 23 March 2009 and 11 June 2009. This satellite-based remote-sensing thermal data covers four different regional climatic conditions in central Sydney and helps to analyse the sUHI effect on an annual basis.

Data calculation method

The civil remote-sensing satellite Landsat 7 was launched by NASA in 1999 and has been providing its data from eight spectral bands; the most interesting of them for the evaluation of land surfaces temperatures is the thermal infrared (TIR) channel #6 from 10.31 μm to 12.36 μm with an in-nadir resolution of 30 m × 30 m (resulting from an automatic post-processing of the original data of double grid size).

Each image is filled with digital numbers (DN) out of the interval [0; 255], and frequent calibrations are published on how to recalculate the spectral radiation density at the ETM+ sensor. Furthermore, the combination of sensor parameters with the fundamental Planck's equation of black body heat radiation allows the assigning of an at-sensor brightness temperature (TB) to every pixel. The new method applied here uses the relative spectral response (RSR) curve of ETM+ and a best possible numerical approximation of the reverse Planck's equation instead of a simplified analytic one.

As the temperatures of the ground surface's emitting areas are the final objects of interest, at least two influences should be eliminated: (1) the deviation of the surface from a 'black body'; and (2) the influence of the atmosphere on heat radiation. The first effect is generally handled by introducing an emissivity ε within [0; 1] as a proportionality factor to characterize the reduction of real-body emission compared to the black body. In this project, the classification-based emissivity method (CBEM) (Li 2012) was used. Based on the literature (Nichol 1994, 1995, 2005; Valor and Caselles 1996; Schott *et al.* 2001; Weng 2001; Zhang *et al.* 2006; Stathopoulou and Cartalis 2007; Mallick *et al.* 2008; Van *et al.* 2009; Coll *et al.* 2010; Sun *et al.* 2010), the most likely emissivity values were derived for the four LUT of central Sydney: streetscape (0.91), building (roof-top) (0.90), open space (0.91) and urban greenery (0.98).

As radiation is absorbed when passing through a transparent medium, in this case the air column between ground surface and satellite, an atmospheric correction (AC) of the temperature estimates must be done by using the transmission τ within [0; 1]. After extensive testing of a variety of AC methods, the mono-window algorithm (MWA) (Qin *et al.* 2001; Li 2006; Zhang *et al.* 2006; Liu and Zhang 2011; Deng and Wu 2013) was chosen, and some modifications were introduced to generalize the quantitative effect of air humidity on τ. However, this algorithm depends on actual temperature and humidity profiles of the atmosphere being measured by meteorological balloons (e.g. from Sydney's airport).

Preliminary observations

According to the map of temperature zones (Figure 21.2), Haymarket had the hottest surface temperature with an average of 31.03°C, while the overall surface temperature in Sydney Harbour was 30.88°C, in Harris Street 30.95°C, in Kings Cross 30.34°C and in Glebe Point 30.65°C. Although the temperature variance is only 0.69°C, it is a significant variance, because each average temperature is the mean of over 2,000 data points. Furthermore, in this thermal map, the average surface temperature of central Sydney is only 30.56°C (standard

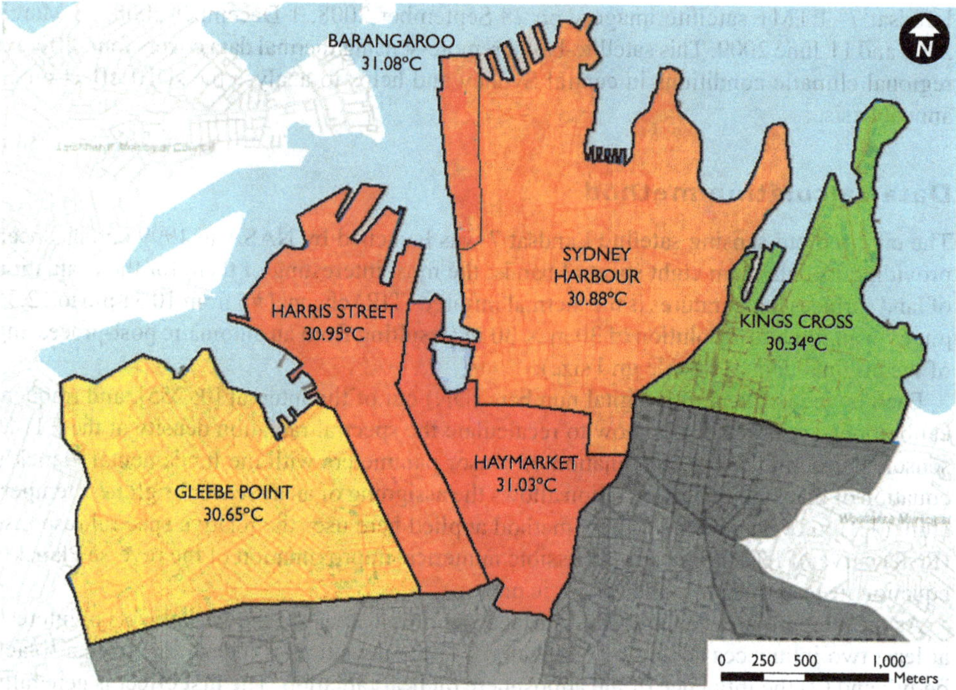

Figure 21.2 Average surface temperature of precincts in Glebe Point, Harris Street, Sydney Harbour, Haymarket and Kings Cross.

Source: Based on City of Sydney (2010).

deviation = 0.26). The temperature variance among the Kings Cross (minimum average) and Haymarket (maximum average) precincts is 0.69°C. However, smaller urban elements' (e.g. streetscapes and roof-tops) surface temperature varies from 28 to 33°C (see Figures 21.3 and 21.4). Overlapping the surface temperature maps of individual urban elements and average precincts indicates that the overall temperature in the Haymarket precinct (31.03°C) is very close to the surface temperature in the Barangaroo site (31.08°C; see Figure 21.2, centre top). At the time of this thermal mapping, Barangaroo was an industrial site fully covered by concrete (a greener redevelopment plan is underway). Concrete, along with asphalt, is among the hottest and most undesired urban surfaces identified by sUHI studies (Erell *et al.* 2011; Gartland 2008; Oke 1988). This cross-mapping reveals that the sUHI effect in Haymarket is significant and intense. The questions are: what physical configurations at precinct scale contribute to this extremely hot temperature, and is it possible to mitigate it?

Controlled variables: residential and building density

Density, the number of dwelling units and people per hectare in a given land area, is still a controversial term in urban design. Both building and urban (population) densities are controlled in this study to enable more focused analysis on urban elements and features in higher densities.

There is evidence that cities of lower density produce more greenhouse gas emissions per capita; however, overdevelopment and too much density can be detrimental and reduce liveability and health (high-density areas in Hong Kong and Mumbai are good examples of this). The densities of Australian cities are notoriously low, like those of North American cities. The average city density can vary widely, from 25 habitants per hectare in Sydney, to 100 habitants per hectare in Rio de Janeiro, to over 300 habitants per hectare in Shanghai and Seoul. Research indicates that the world's most liveable cities have a density range of 60 habitants per hectare (London) to 170 habitants per hectare (Barcelona), which leads to compact and walkable cities where green spaces and gardens are well integrated.

Discussion about the effect of building density on the magnitude of the sUHI shapes a considerable portion of the urban microclimate literature (Lee *et al.* 2013; Yuan and Chen 2011; Giridharan *et al.* 2004). Since the early sUHI studies, it has been argued that higher densities are likely to experience higher temperatures (Givoni 1998; Oke 1988; Tapper 1990) due to their physical structure.

Background sUHI research indicates that high-density building blocks can magnify the sUHI effect in cities by increasing the opportunity for surface materials to absorb direct and reflected sunlight radiation (Rajagopalan 2009; Erell *et al.* 2011; Giridharan *et al.* 2007). Generally, reflected solar radiation has more chance to exit the built environment from lower-density and less compact areas (Wong and Yu 2008). During each reflection phase between building façades and street surfaces, a portion of solar energy is transmitted into built environment surfaces in the form of heat (Erell *et al.* 2011). Thus the general surface temperature is likely to be higher in higher densities.

The five precincts selected have a building density of more than 100 units per acre (Sydney Harbour and Haymarket have up to 200 units per acre). According to Campoli and MacLean's (2007) classification of building density, over 100 units per acre may be considered as very high building density. Higher building density can also intensify energy consumption in cities and consequently increase anthropogenic waste heat (Sivam and Karuppannan 2012; Ichinose *et al.* 2008). Although population density is not a direct contributor to the UHI effect, it can increase the need for energy consumption for air-conditioning and transport. Citizens in higher densities consume a considerable amount of energy in their daily lives, especially for indoor air-conditioning and transportation. This higher rate of energy consumption increases the amount of anthropogenic (human-made) waste heat in higher densities and therefore contributes to the UHI effect in cities. However, a clear link between anthropogenic waste heat and sUHI has not been identified as yet.

Central Sydney has the highest population density in Australia with an estimated residential population of 180,679 residents living in an area of 4.48 km^2 in 2010 (City of Sydney 2011). The overall urban density of the City of Sydney is 40,330 p/km^2. However, the five selected sites represent a higher average urban density of over 74,136 p/km^2. Therefore, the case studies selected have very high urban densities compared to other Australian cities and even other precincts in central Sydney.

However, the number of people visiting central Sydney on a daily basis for shopping, entertainment and education reaches up to 483,000. This is in addition to the 385,000 people who arrive every day to work in central Sydney. The considerable proportion of temporary residents compared to permanent dwellers (more than fourfold) makes it difficult to consider residential density as a factor contributing to the sUHI effect in Sydney. Furthermore, population density is usually discussed with regard to ambient-temperature UHI effect, while the current study focuses on the surface-layer urban heat island (sUHI) effect. As such, the variable of population density is controlled in the current study.

Analysis and results

Urban features can influence the surface temperature in higher densities by affecting the overall rate of materials' exposure to sunlight and heat exchange between them (ASHRAE 2004; Oke 2006). Specific heat capacity, conductivity and albedo (reflectivity) of materials are the major factors that can cause the built environment to store sunlight energy in the form of heat in its thermal mass and to postpone the energy departure process from the built environment (Ashie 2008; Dahl 2010; Oke 1988). Still, the location of materials needs to be carefully considered, as shading can significantly influence the heat absorption and reflection process. Two of the most common places where the sUHI is being investigated are urban open space (including streetscapes and public space) and buildings' roof-tops.

Public space useability: thermal behaviour of horizontal surfaces (streetscapes and roof-tops)

The performance of public space may be categorized using conventional indicators for effective measuring of public domain performance over time, for instance, by looking at the public domain as a connective social and economic space, with a high value of exchange occurring within it. While it is difficult to compare one public space with another and the types of interaction that occur in the public domain vary, most urban researchers agree that effective measurement of public domain performance examines key planning principles and public space characteristics, such as continuity, safety, enclosure, connectivity and

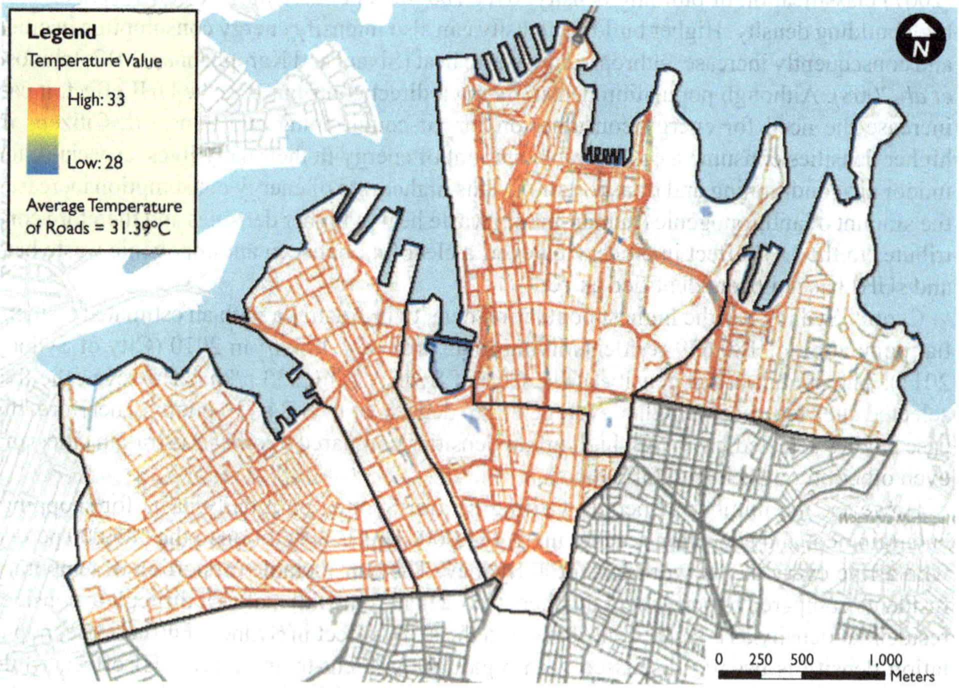

Figure 21.3 Street surface temperature in five higher-density precincts in central Sydney.

Source: Based on City of Sydney (2010).

adaptability. This research examines the usability and resilience of public space networks during heatwaves, introducing new criteria and indicators for evaluation and retrofitting in existing as well as emerging public spaces.

The comparison between surface temperatures of different horizontal urban features can indicate which elements are more heat sensitive and therefore need more examination in sUHI mitigation studies. Comparing 300 randomly selected data points indicates that a higher temperature exists on streetscape surfaces than on building roof-tops (see Figures 21.3 and 21.4). The average temperature of streetscape surface layer is 31.39°C, which is 0.36°C higher than the Haymarket precinct average surface temperature (the hottest precinct in Figure 21.2). Some streetscape surfaces, especially in Haymarket, reached the highest temperature of 34.15°C with 5.10°C variance from the minimum streetscape temperature (see Table 21.1). The average surface temperature of buildings' roof-top layer is 30.26°C (with the maximum value of 33.61°C), which is 1.13°C lower than the average streetscape surface temperature, 0.77°C lower than the average temperature of the Haymarket precinct, 0.69°C lower than Harris Street, 0.62°C lower than Sydney Harbour and 0.39°C lower than Glebe Point (the roof-top layer average surface temperature is very close to the average surface temperature of Kings Cross: 30.34°C).

According to Table 21.1, the streetscape has a considerably higher surface temperature and temperature variance than building roof-tops. This underlines the streetscape as the more heat-sensitive urban feature at precinct scale. To undertake more detailed sUHI analysis, the street network intensity is compared against open public space plot ratio.

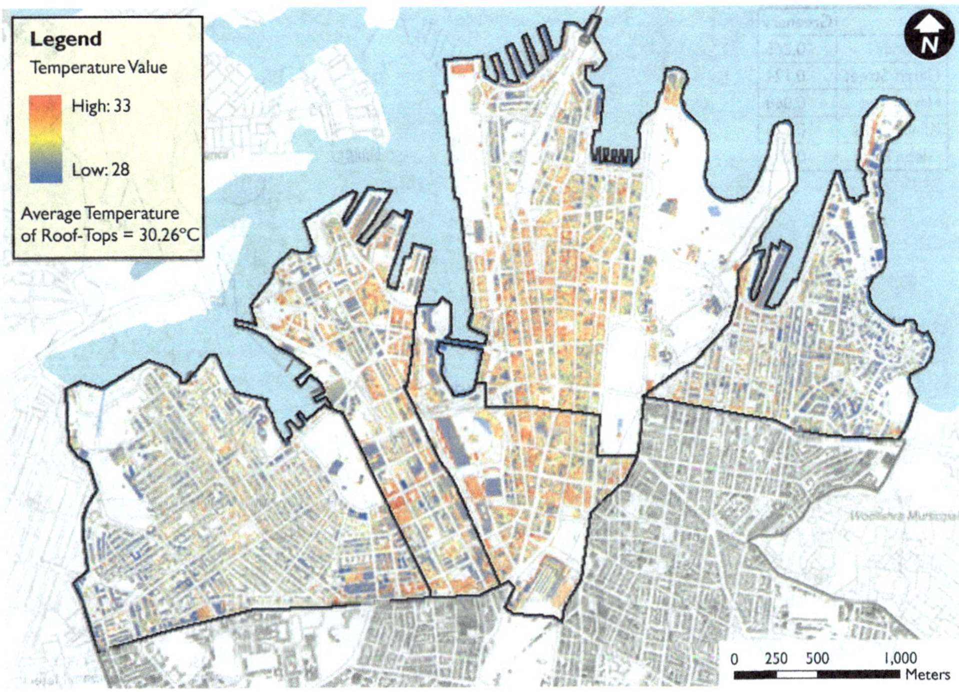

Figure 21.4 Roof-top surface temperature in five higher-density precincts in central Sydney.

Source: Based on City of Sydney (2010).

Table 21.1 Temperature variance of streetscape surfaces and building roof-tops in five higher-density precincts of central Sydney

	Min. temp. (°C)	*Max. temp. (°C)*	*Average temp. (°C)*	*Temp. variance (°C)*
Streetscape	29.05	37.30	31.15	8.25
Building roof-top	27.70	33.61	30.26	5.91
Temp. variance (Street–roof)	2.65	3.69	0.89	2.24

Correlation between open space plot ratio, street network intensity, public space plot ratio and the sUHI effect

As heat sensitivity occurs more in between buildings than on their roof-tops, it is worthwhile analysing further the correlations between the general land use of open space and the sUHI effect. To this end, the streetscape and public space are analysed separately in this section. The analysis of this correlation can indicate to what degree the sUHI effect is a dependent variable of streetscape or public space plot ratio.

Table 21.2 shows that street network intensity (streetscape plot ratio) has the correlation coefficient (R) value of +0.94 to the average precinct surface temperature. This means that a

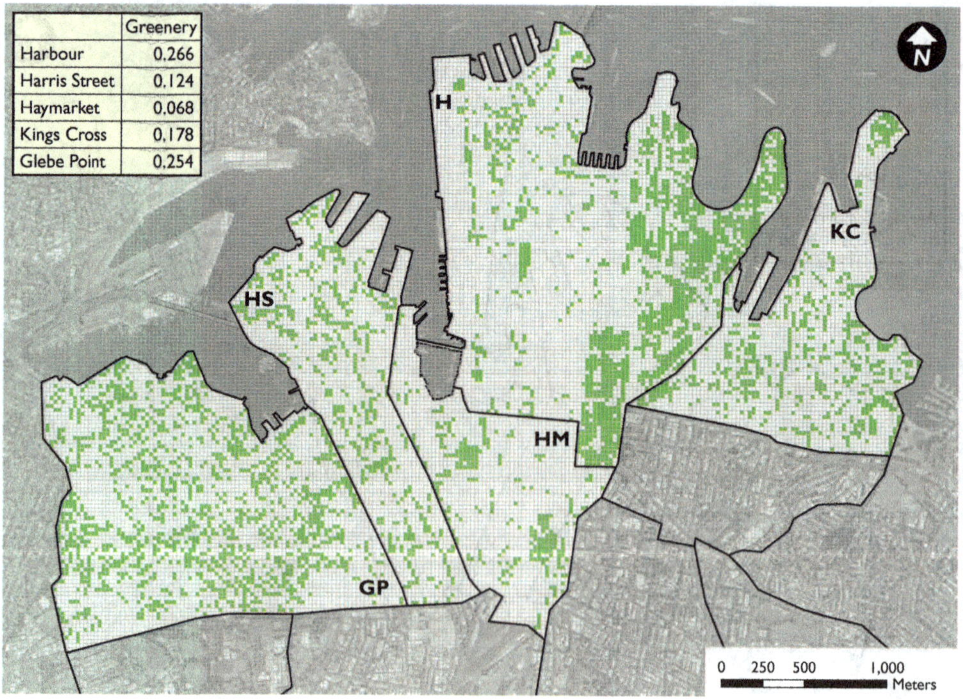

Figure 21.5 Streetscape and public space plot ratio in five higher-density precincts of central Sydney, extracted from Google Earth imagery (2009); resolution: 1 metre.

Table 21.2 Street network intensity and average surface temperature in five higher-density precincts of central Sydney

Precinct	*Sydney Harbour*	*Harris Street*	*Haymarket*	*Kings Cross*	*Glebe Point*
Street network plot ratio (per cent)	20.8%	22.9%	21.0%	14.7%	17.4%
Open space plot ratio (per cent)	21.3%	23.5%	22.4%	12.8%	18.5%
Public space plot ratio (other than streetscape)	0.5%	0.6%	1.5%	0.1%	1.1%
Average surface temperature (°C)	30.88	30.95	31.03	30.34	30.65

higher streetscape plot ratio correlates almost directly to higher overall surface temperatures in Sydney precincts. This high and positive coefficient value indicates that higher streetscape surfaces strongly correlate with the sUHI effect at precinct scale (the maximum R value could be 1, which shows complete correlation).

The open space plot ratio (i.e. all hard-landscaped open spaces including streetscapes and other public spaces) has an even greater coefficient value of +0.97 to overall surface temperature in Sydney precincts. This high coefficient value indicates strong correlation of overall surface temperature to the proportion of hard-landscaped open space (e.g. paved with concrete and asphalt). However, separating other open spaces from the streetscape results in a relatively lower coefficient value of +0.64 between the hard-landscaped public space plot ratio and the precinct surface temperature, which still indicates a higher correlation than average (moderate R value is +0.5).

High and positive coefficient values between hard-landscaped open spaces (i.e. streetscape and public space layers) and sUHI on-the-ground surface layer indicates that harder landscapes can increase the surface temperature of urban precincts. Under question is whether there are any urban land covers capable of mitigating the sUHI effect at precinct scale.

Correlation between urban greenery plot ratio and the sUHI effect

An extensive amount of literature supports the idea that greenery can mitigate the sUHI effect (Erell *et al.* 2011; Oke 2006; Gartland 2008, Ashie 2008; Butera 2008; Correa *et al.* 2012; Dahl 2010). At the micro scale, this heat mitigation occurs in two ways: first, through using solar energy and photosynthesis to facilitate greenery metabolism; second, through evapotranspiration (evaporative cooling) in reaction to the ambient heat on the surface of leaves (just like human skin). Therefore, green infrastructures can counteract the sUHI effect by cooling down air and surface temperatures at the micro scale.

Various forms of greenery can exist in urban precincts, such as parklands, gardens, green roofs, vertical greenery, urban farming, nature reserves and planting of extensive vegetation; all these forms act as sources of moisture for evapotranspiration, where the absorbed solar radiation can be dissipated as latent heat, and thus aid in reducing urban temperature. Recent research by Wong (2008) shows that vegetated spaces could be a few degrees cooler than their surroundings. Under question is to what extent this is applicable at precinct scale. To investigate the effect of urban greenery on sUHI mitigation at precinct scale, urban greenery plot ratio (UGPR) is compared to the sUHI effect in the five Sydney precincts.

The total study area (the five precincts selected) covers 1.75 km^2, which includes an overall area of 0.36 km^2 of urban greenery (UGPR = 20.7%). However, there is a significant variance in urban UGPR in the five selected precincts. As shown in Table 21.3, UGPR is 26.6 per cent in Sydney Harbour and 29.1 per cent in Glebe Point. However, UGPR in Kings Cross is 11.2

Table 21.3 Urban vegetation ratio in five higher-density precincts of central Sydney

Precinct	*Sydney Harbour*	*Harris Street*	*Haymarket*	*Kings Cross*	*Glebe Point*
Urban greenery (km^2)	0.15	0.03	0.01	0.04	0.12
Precinct area (km^2)	0.58	0.25	0.20	0.25	0.47
Urban Greenery Plot Ratio (UGPR)	26.6%	12.4%	6.7%	17.8%	25.3%
Average surface temperature (°C)	30.88	30.95	31.03	30.34	30.65

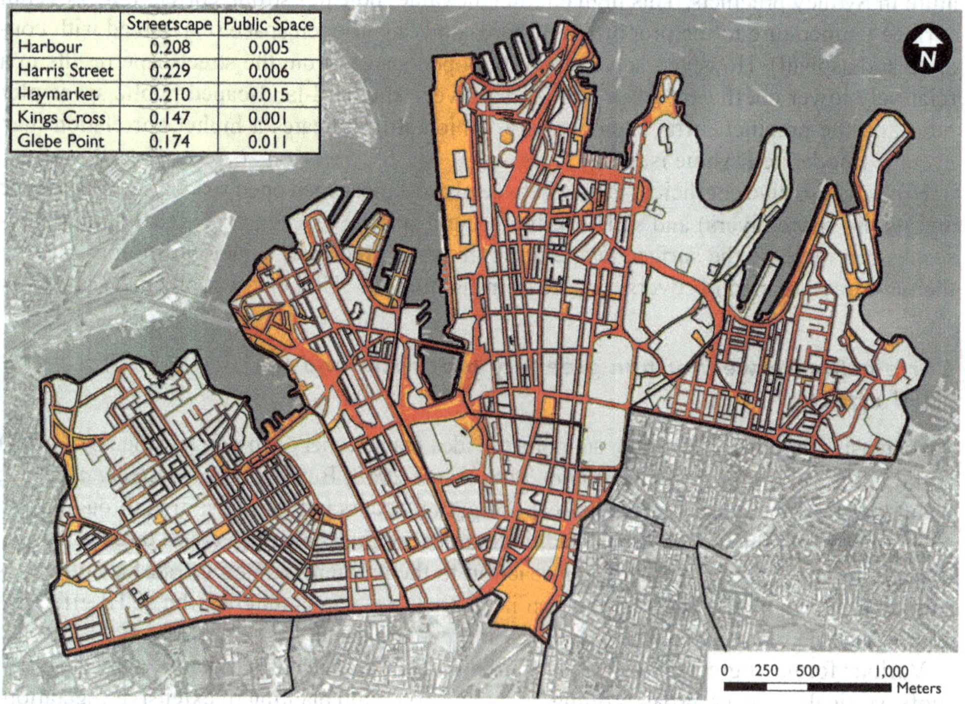

Figure 21.6 Urban vegetation ratio in five higher-density precincts of Sydney, extracted from Google Earth imagery (2009); resolution: 8 metres.

per cent, in Harris Street 7.69 per cent and in Haymarket only 3.31 per cent. Significant variance of UGPR and proximity of these precincts make them appropriate cases to study further.

With the correlation coefficient (R) value of –0.40 for precincts and –0.78 for smaller random sample areas (120 samples were studied, each with an exact area of 100 m^2), precinct surface temperature shows medium to high dependency to UGPR. This also indicates that the effect of UGPR on sUHI is moderated by other factors at larger scales.

Urban greenery distribution in Figure 21.6 reveals that Kings Cross and Glebe Point (the lowest average surface temperature) have the most homogeneous urban greenery distribution, while hot Haymarket has the lowest and most scattered greenery spots. In the Sydney Harbour precinct, the large area of the Royal Botanic Gardens and Hyde Park can explain its relatively lower sUHI compared to Haymarket and Harris Street.

Resilience of urban land uses to the sUHI on an annual basis based on Landsat 7– ETM+ data

As Table 21.4 and Figure 21.7 show, the annual variation of surface heat is greatest at open space layers with a value of 32.39°C (2.46°C more than precinct average surface temperature variance) and least at urban greenery layers with a value of 28.88°C (1.05°C less than precinct average surface temperature variance). Surface temperature variance of streetscape is the second highest with a value of 31.74°C (1.81°C more than precinct average surface temperature variance), which makes it a low heat-resilient urban land use at precinct scale, alongside open space. Considering the surface temperature of urban greenery (the most

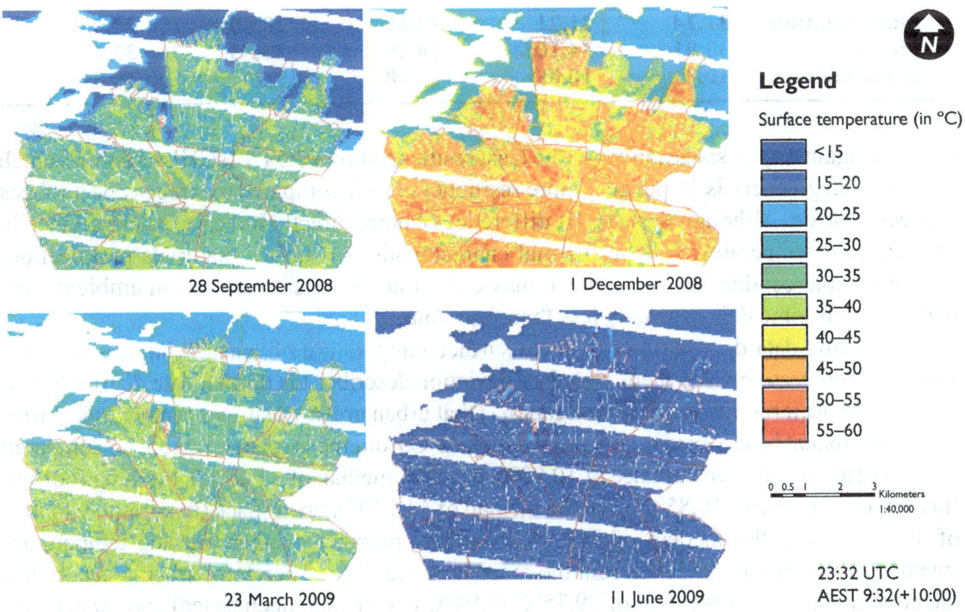

Figure 21.7 Annual variation of surface temperature in five higher-density precincts of central Sydney, extracted from Landsat 7– ETM+ (2008–2009); resolution: 30 metres.

Table 21.4 Temperature variation of urban land surfaces in five higher-density precincts of central Sydney

		Streetscape av. (°C)	*Building roof-top av. (°C)*	*Open space av. (°C)*	*Urban greenery av. (°C)*	*Precinct surface heat (°C)*
September 2008	Sydney Harbour	31.22	32.11	33.96	27.51	30.28
	Harris Street	32.90	33.57	33.30	28.69	32.73
	Haymarket	31.09	33.12	29.64	27.82	31.31
	Kings Cross	32.67	33.47	34.62	28.13	31.72
	Glebe Point	33.48	34.25	32.42	29.12	32.53
December 2008	Sydney Harbour	44.18	45.14	46.38	38.10	42.35
	Harris Street	45.91	46.90	45.95	41.02	45.43
	Haymarket	44.79	46.56	44.62	39.07	43.93
	Kings Cross	46.16	46.53	46.14	40.09	44.14
	Glebe Point	46.78	47.46	45.49	40.94	45.05
March 2009	Sydney Harbour	33.07	33.95	36.30	29.63	32.16
	Harris Street	35.44	36.14	36.20	31.83	35.21
	Haymarket	32.95	34.87	33.68	30.06	33.07
	Kings Cross	34.04	34.31	36.30	29.58	32.74
	Glebe Point	36.23	37.04	35.62	32.26	35.33
June 2009	Sydney Harbour	16.34	17.01	18.02	13.61	15.80
	Harris Street	16.12	16.98	16.45	13.20	16.44
	Haymarket	15.04	16.25	14.00	12.14	15.50
	Kings Cross	17.37	18.05	18.52	13.54	16.96
	Glebe Point	15.76	16.57	16.37	12.82	15.52
	Max variation	**31.74**	**31.21**	**32.39**	**28.88**	**29.93**
	Median	**33.01**	**34.10**	**34.29**	**29.35**	**32.63**
	Standard dev.	**10.83**	**10.85**	**10.78**	**9.92**	**10.37**

resilient) against open space land use (the least resilient) shows 3.51°C sUHI variation, which means urban greenery is 10 per cent more resilient to sUHI compared to streetscape surfaces at precinct scale. Although these sUHI differences cannot directly indicate exact variance in ambient (air) temperature (due to thermal characteristics of surface materials in reflection, absorption and conduction of heat), we may expect homological variance in ambient temperature with some delay (usually less than four hours).

Considering data distribution of annual surface temperature of selected land uses shows similar differences between them. Standard deviation describes the normal data distribution in a dataset. What it reveals in this investigation is that urban greenery surface temperature varies less on an annual basis (maximum variance of 9.92°C from the median of 29.35°C) compared to streetscape (maximum variance of 10.83°C from the median of 33.01°C), building roof-top (maximum variance of 10.85°C from the median of 34.10°C), open space (maximum variance of 12.69°C from the median of 34.29°C) and the precinct average surface temperature (maximum variance of 10.37°C from the median of 32.63°C). Therefore, urban greenery has the coolest median temperature of 29.35°C (4.94°C cooler than open space) and 10 per cent more normal data distribution (StDev 9.92°C vs. StDev 10.78°C for open space, StDev 10.83°C for streetscape and 10.85°C for building roof-top) among the selected land uses.

As such, the findings of part one of the current research, which indicates that urban greenery is the most resilient urban feature and open space is the least resilient urban feature to the sUHI, are aligned with satellite-based thermal data. On an annual basis, however, streetscape and building roof-top layers show very similar behaviour.

Discussion

The surface temperature zones map (Figure 21.2) of central Sydney shows that Haymarket has the highest surface temperature with an average of 31.03°C. The overall surface temperature of Haymarket is very close to the surface temperature of extremely hot urban features in the study area (e.g. 31.08°C in Barangaroo and 31.15°C for average streetscape layer). This means that sUHI in Haymarket is significantly higher (mathematically) than central Sydney's average (30.56°C), which highlights Haymarket as the precinct most vulnerable to the sUHI effect.

Comparing streetscape surface temperatures (Figure 21.3) and urban greenery distribution (Figure 21.6) reveals that the sUHI effect is greater in less vegetated areas. Although all five precincts have high building densities, streetscape surfaces in Sydney Harbour are up to 1.6°C cooler than similar areas in Haymarket. This relative coolness correlates with the higher rate of UGPR in Sydney Harbour (19.9 per cent higher than Haymarket).

Overall, the surface temperature of open space and roof-tops is slightly more in Sydney Harbour and Haymarket (with twice the building density of the other three precincts). This may be due to the lower sky view factor (i.e. the amount of sky visible from the surface) for streetscapes. It needs to be noted that the roof-tops of high-rise buildings in Haymarket and Sydney Harbour are flat roofs (also partly the case in Harris Street), whereas the roof-tops in Kings Cross and Glebe Point are a combination of flat roofs and pitched roofs, which have different solar gain due to the way they face solar radiation (i.e. in the sou thern hemisphere, horizontal surfaces generally have more daily solar gain than surfaces sloped towards the south, east and west). For an in-depth discussion about streetscape and roof-tops' surface temperature, more detailed data about land cover surface materials are needed.

A comparison between Figure 21.3, Figure 21.4 and Table 21.1 reveals that streetscape surfaces are generally hotter than roof-tops (up to 3.69°C). Roof-tops are exposed to sunlight radiation almost all day long, while street canyons have partial shadow coverage due to surrounding high-rise buildings. Therefore, in theory, roof-tops should gain more heat compared to streetscape surfaces, but in practice streetscapes have the hotter surfaces. In the current study, streetscape surfaces represent a higher minimum temperature (2.65°C), higher maximum temperature (3.69°C) and higher average temperature (0.89°C) than roof-tops, as well as more surface temperature variance (2.24°C). This indicates the importance of focusing on cooler land covers and urban greenery on the ground surface layer rather than on roof-tops in central Sydney.

The higher ratio of urban greenery in Sydney Harbour (UFPR = 26.6%) compared to Haymarket (6.7%) and Harris Street (12.4%) seems to be the most effective factor in mitigating the sUHI effect at precinct scale. A significant area of urban greenery in the Royal Botanic Gardens and Hyde Park (located in Sydney Harbour; see Figures 21.2 and 21.6) is cooling down the precinct's overall surface temperature.

Satellite-based thermal data of Landsat 7– ETM+ 2008–2009 (Figure 21.7 and Table 21.4) indicates that on an annual basis, urban greenery is the urban feature most resilient to the sUHI effect, with a lower median temperature (29.35°C) and less temperature variance (28.88°C), while open public space is the least heat-resilient urban feature with a 4.94°C hotter median temperature and 3.51°C more temperature variance during a year (10 per cent to 15 per cent less resilient). The interesting point is that public open space is the easiest and most cost-effective space in which to increase urban greenery. In other words, the least resilient urban feature to the sUHI has the potential to be transformed into a somewhat cooler place (up to 5°C) by using urban greenery (the most resilient feature).

Conclusions

Urban temperatures are predicted to increase due to climate change. The temperatures in our cities are likely to increase further, because more heat will be stored and re-radiated by expanses of asphalt, concrete and other heat-storing building materials. In this context, it is crucial to understand the possibilities for the transformation of existing urban fabrics towards a more liveable and sustainable future (Bosselmann 2008; Lehmann 2010). This may be implemented by smart and small-scale spatial transformation of existing urban spaces.

The basic argument underlined in this comparative case study is that the higher sUHI effect at precinct scale correlates with a greater hard-landscaped public space plot ratio, higher street network intensity and a lower urban greenery plot ratio. Higher open space plot ratio and street network intensity correlate significantly to higher sUHI effect at precinct scale. However, higher urban greenery plot ratio may mitigate the sUHI effect in high-density precincts. Therefore, increasing urban greenery and decreasing hard-landscaped urban features (e.g. streetscapes and vast hard-covered open spaces) can cool down existing precincts. A fine distribution of urban greenery may also mitigate the sUHI at precinct scale.

Research limitations and further opportunities

This research is based on remote-sensing thermal photography and desktop spatial data. It utilizes the surface temperature, which is different from the real feeling of the temperature in public space. Further studies could benefit from including on-the-spot climate measurements and air temperature data. The effect of local airflow and surface water is subject to further investigation. To move towards more certainty about the research outcomes, on-the-spot microclimate measurement at smaller scales could be beneficial. Due to the limited scope of this study and controlled variables, the results need to be validated in other cities.

References

Ashie, Y. 2008, 'Management of urban heat environment', in K. Hanaki (ed.), *Urban Environmental Management and Technology*, Springer, Tokyo.

ASHRAE 2004, *Standard 55: Thermal environmental conditions for human occupancy*, ASHRAE Inc, Atlanta, GA.

Australian Bureau of Meteorology 2008, *Climate of Australia*, Commonwealth of Australia, Melbourne, VIC.

Bosselmann, P. 2008, *Urban Transformation: Understanding city design and form*, Island Press, Washington, DC.

Butera, F. 2008, 'Towards the renewable built environment', in P. Droege (ed.), *Urban Energy Transition: From fossil fuels to renewable power*, Elsevier, Amsterdam.

Campoli, J. and MacLean, A.S. 2007, *Visualizing Density*, Lincoln Institute of Land Policy, Cambridge, MA.

City of Sydney 2010, *Thermal Imagery*, Sydney, NSW.

—— 2011, *Residential Monitor*, Sydney, NSW.

Coll, C., Galve, J.M., Sanchez, J.M. and Caselles, V. 2010, 'Validation of Landsat-7/ETM+ thermal-band calibration and atmospheric correction with ground-based measurements', *IEEE Transactions on Geoscience and Remote Sensing*, vol. 48, pp. 547–553.

Commonwealth Scientific and Industrial Research Organisation (CSIRO) 2007, *Climate Change in Australia: Technical report 2007*, CSIRO and Australian Bureau of Meteorology, Aspendale, VIC.

Correa, E., Ruiz, M.A., Canton, A. and Lesino, G. 2012, 'Thermal comfort in forested urban canyons of low building density: an assessment for the City of Mendoza, Argentina', *Building and Environment*, vol. 58, pp. 219–230.

Dahl, T. 2010, *Climate and Architecture*, Routledge, London.

Deng, C. and Wu, C. 2013, 'Examining the impacts of urban biophysical compositions on surface urban heat island: a spectral unmixing and thermal mixing approach', *Remote Sensing of Environment*, vol. 131, pp. 262–274.

Erell, E., Pearlmutter, D. and Williamson, T. 2011, *Urban Microclimate: Designing the spaces between buildings*, Earthscan, London.

Gartland, L. 2008, *Heat Islands: Understanding and mitigating heat in urban areas*, Earthscan, Washington, DC.

Girardet, H. 2008, *Cities, People, Planet: Liveable cities for a sustainable world*, John Wiley & Sons, Chichester, Sussex.

Giridharan, R., Ganesan, S. and Lau, S.S.Y. 2004, 'Daytime urban heat island effect in high-rise and high-density residential developments in Hong Kong', *Energy and Buildings*, vol. 36, pp. 525–534.

Giridharan, R., Lau, S., Ganesan, S. and Givoni, B. 2007, 'Urban design factors influencing heat island intensity in high-rise high-density environments of Hong Kong', *Building and Environment*, vol. 42, pp. 3669–3684.

Givoni, B. 1998, *Climate Considerations in Building and Urban Design*, Van Nostrand Reinhold, New York.

Guest, C.S., Willson, K., Woodward, A.J., Hennessey, K., Kalkstein, L.S., Skinner, C. and McMichael, A.J. 1999, 'Climate and mortality in Australia: retrospective study, 1979–1990, and predicted impacts in five major cities in 2030', *Climate Research*, vol. 13, pp. 1–15.

Hu, W., Becker, N., McMichael, T. and Tong, S. 2007, 'Heat-related mortality during summer in Sydney, Australia, 1990–2004', *Epidemiology*, vol. 18, S113–S114.

Ichinose, T., Matsumoto, F. and Kataoka, K. 2008, 'Counteracting urban heat islands in Japan', in P. Droege (ed.), *Urban Energy Transition: From fossil fuels to renewable power*, Elsevier, Amsterdam.

Kovats, R.S. and Hajat, S. 2008, 'Heat stress and public health: a critical review', *Annual Review of Public Health*, vol. 29, pp. 41–55.

Lee, H., Holst, J. and Mayer, H. 2013, 'Modification of human-biometeorologically significant radiant flux densities by shading as local method to mitigate heat stress in summer within urban street canyons', *Advances in Meteorology*, vol. 2013, 13pp., doi: 10.1155/2013/312572.

Lehmann, S. 2010, *The Principles of Green Urbanism: Transforming the city for sustainability*, Earthscan, Washington, DC.

Li, J. 2006, 'Estimating land surface temperature from Landsat-5 TM', *Remote Sensing Technology and Application*, vol. 21, pp. 322–327.

Li, Z.L. 2012, 'Land-surface temperature and thermal infrared emissivity', *Advanced Remote Sensing*, pp. 235–271.

Liu, L. and Zhang, Y. 2011, 'Urban heat island analysis using the Landsat TM data and ASTER data: a case study in Hong Kong', *Remote Sensing*, vol. 3, pp. 1535–1552.

McGuirk, P. 2003, 'Planning central Sydney', in A. Maclaren (ed.), *Making Space: Property development and urban planning*, Hodder Arnold, London.

Mallick, J., Kant, Y. and Bharath, B. 2008, 'Estimation of land surface temperature over Delhi using Landsat-7 ETM+', *Journal of the Indian Geophysical Union*, vol. 12, pp. 131–140.

Moran, D.S., Laor, A., Epstein, Y. and Shapiro, Y. 1998, 'A modified discomfort index (MDI) as a substitute for the wet bulb globe temperature (WBGT)', *Medicine & Science in Sports & Exercise*, vol. 30, p. 284.

Nichol, J.E. 1994, 'A GIS-based approach to microclimate monitoring in Singapore's high-rise housing estates', *Photogrammetric Engineering & Remote Sensing*, vol. 60, pp. 1225–1232.

—— 1995, 'Monitoring tropical rain forest microclimate', *Photogrammetric Engineering & Remote Sensing*, vol. 61, pp. 1159–1165.

—— 2005, 'Remote sensing of urban heat islands by day and night', *Photogrammetric Engineering & Remote Sensing*, vol. 71, pp. 613–621.

Oke, T.R. 1988, 'The urban energy balance', *Progress in Physical Geography*, vol 12, pp. 471–508.

—— 2006, 'Towards better scientific communication in urban climate', *Theoretical and Applied Climatology*, vol. 84, pp. 179–190.

Organisation for Economic Cooperation and Development (OECD) 2010, *Cities and Climate Change*, OECD, Paris.

Qin, Z., Karnieli, A. and Berliner, P. 2001, 'A mono-window algorithm for retrieving land surface temperature from Landsat TM data and its application to the Israel–Egypt border region', *International Journal of Remote Sensing*, vol. 22, pp. 3719–3746.

Rajagopalan, P., 2009, 'Urban heat island and its impact on building energy consumption', *Advances in Building Energy Research*, vol. 3, pp. 261–270, Earthscan, London.

Register, R. 2002, *Ecocities: Building cities in balance with nature*, University of California Press, Berkeley.

Samuels, R., Randolph, B., Graham, P., McCormick, T. and Pollard, B. 2010, *Micro-Urban-Climatic Thermal Emissions in a Medium-density Residential Precinct*, University of New South Wales, Sydney.

Santamouris, M. and Geros, V. 2006, *Environmental Design of Urban Buildings: An integrated approach*, Earthscan, Sterling.

Schott, J.R., Barsi, J.A., Nordgren, B.L., Raquendo, N.G. and De Alwis, D. 2001, 'Calibration of Landsat thermal data and application to water resource studies', *Remote Sensing of Environment*, vol. 78, pp. 108–117.

Sivam, A. and Karuppannan, S. 2012, 'Density, design and sustainable residential development', in S. Lehmann and R. Crocker (eds), *Designing for Zero Waste: Consumption, technologies and the built environment*, Earthscan, New York and London.

Stathopoulou, M. and Cartalis, C. 2007, 'Daytime urban heat islands from Landsat ETM+ and Corine land cover data: an application to major cities in Greece', *Solar Energy*, vol. 81, pp. 358–368.

Stone, B. 2012, *CITY and the Coming Climate: Climate change in the places we live*, Cambridge University Press, New York.

Sun, Q., Tan, J. and Xu, Y. 2010, 'An ERDAS image processing method for retrieving LST and describing urban heat evolution: a case study in the Pearl River Delta Region in South China', *Environmental Earth Sciences*, vol. 59, pp. 1047–1055.

Tapper, N.J. 1990, 'Urban influences on boundary-layer temperature and humidity: results from Christchurch, New Zealand', *Atmospheric Environment Part B – Urban Atmosphere*, vol. 24, pp. 19–27.

Thom, E.C. 1959, 'The discomfort index', *Weatherwise*, vol. 12, pp. 57–61.

Toon, J. and Falk, J. (eds) 2003, *Sydney: Planning or politics: town planning for Sydney region since 1945*, Planning Research Centre, University of Sydney, Sydney, NSW.

United Nations Department of Economic and Social Affairs (UNDESA) Population Division 2012, *World Urbanisation Prospects: The 2011 revision*, United Nations, New York.

United Nations Economic Commission for Europe (UNECE) 2011, *Climate Neutral Cities: How to make cities less energy and carbon intensive and more resilient to climatic challenges*, UNECE, New York and Geneva.

UN-Habitat 2011, *Cities and Climate Change: Global report on human settlements 2011*, UN-Habitat/Earthscan, London.

Valor, E. and Caselles, V. 1996, 'Mapping land surface emissivity from NDVI: application to European, African, and South American areas', *Remote Sensing of Environment*, vol. 57, pp. 167–184.

Van, T.T., Trung, L.V. and Lan, H.T. 2009, 'Application of thermal remote sensing in study of surface temperature distribution of Ho Chi Minh City', *7th FIG Regional Conference, Spatial Data serving for sustainable development*, Hanoi, Vietnam, 19–22 October, pp. 1–11.

Weng, Q. 2001, 'A remote sensing-GIS evaluation of urban expansion and its impact on surface temperature in the Zhujiang Delta, China', *International Journal of Remote Sensing*, vol. 22, pp. 1999–2014.

Wong, N.H. and Yu, C. 2008, *Tropical Urban Heat Islands: Climate, buildings and greenery*, Taylor & Francis, New York.

Yuan, C. and Chen, L. 2011, 'Mitigating urban heat island effects in high-density cities based on sky view factor and urban morphological understanding: a study of Hong Kong', *Architectural Science Review*, vol. 54, pp. 305–315.

Zhang, J., Wang, Y. and Li, Y. 2006, 'A C++ program for retrieving land surface temperature from the data of Landsat TM/ETM+', *Computers and Geosciences*, vol. 32, pp. 1796–1805.

Chapter 22

Looking ahead

Applying low carbon principles to shift urban design paradigms in the Asia-Pacific region towards green urbanism

Steffen Lehmann and Katharine Thornton

Summary

The reality of climate change, the impact of global warming and the continued growth in human population are shared and global challenges. In addition, countries in the Asia-Pacific region are also dealing with rapid urbanization. The anticipated growth of cities and towns over the next two decades means that there is a need for urban planning principles that will create socially just, environmentally sustainable and economically prosperous cities in the Asia-Pacific. Making the shift to low carbon city development by adapting the principles of green urbanism will help ensure that Asia-Pacific cities (and societies) grow in a way that is long-lasting and sound, both economically and environmentally.

Introduction

The 'metropolis' has long represented the promise of a better life by offering employment, housing and prosperity; and frequently cities have kept their promise, especially for those in the middle and upper classes. Cities are hubs of knowledge and innovation. The density of a city's population can bring more diversity, pluralism, competition, entrepreneurship, job creation, innovation, healthcare, social services, education, culture, art, entertainment, media, science and commerce. Cities are where much of the world's current unsustainable consumption and use of natural resources takes place, but also where the greatest potential exists for sustainability-oriented innovations and transformation to low carbon urban systems. This chapter will describe the need for urban planning in Asia-Pacific cities that moves them towards green urbanism by applying low carbon principles.

The majority of the world's population (and the population in the Asia-Pacific region) now resides in cities and urban regions, and this trend is accelerating. As a result, cities and urban regions play an ever greater role in national and global development. There is an intense debate taking place about the future of urban development. Without doubt, the current form of urbanization in many nations is unsustainable, and the necessary greenhouse gas emission cuts are not happening on the scale required as outlined by scientists (e.g. in IPCC 2013). This is the broader context in which interest in low carbon cities has arisen.

In 2013, the Asia-Pacific region is home to 4.2 billion people and represents 60 per cent of the world's population (UN ESCAP forthcoming). While the region is comparatively under-urbanized, with approximately 46 per cent of people living in urban areas and a current urban growth rate of 2 per cent per annum predicted for the next two decades, the Asia-Pacific region will account for 60 per cent of the demographic expansion of all urban areas across the world by 2030 (UN-Habitat and UN ESCAP 2010, p. 6). This level of urbanization, led by booming economies, is unprecedented and as a consequence the pressures on cities will grow considerably in the future. Managing this transformation effectively – particularly its environmental dimensions – is of critical importance for the cities and nations of the Asia-Pacific region, and for the planet.

Current urban patterns of growth, consumption and production are now recognized as unsustainable because they are based on outdated models, are fossil-fuel dependent and limit cities' potential to continue to act as engines of economic growth, social inclusion and future prosperity. We are already in a condition of ecological overshoot; our high-consuming economies are exceeding the carrying capacity of the planet. Excessive construction activity and enormous growth have caused severe environmental degradation, not only in areas that are sensitive or vulnerable, but also on a wider scale. What we want instead are economically sound, highly liveable and resilient cities in the Asia-Pacific region that grow 'compact', up rather than out, to reduce the city's footprint. There is an urgent need to promote an alternative urbanization paradigm, which can decouple the links between urbanization, economic growth, depletion of resources, pollution and environmental degradation. Without such a shift, business-as-usual scenarios will continue to result in unsustainable consumption and production patterns and inefficient economic systems. For example, demand for electricity is growing at approximately 3 to 4 per cent per annum in the Asia-Pacific and most power is still generated by burning fossil fuels; given their impact on the environment, there is an obvious need to decarbonize the energy supply and invest in decentralized, renewable energy generation to supply electricity to the millions who currently lack it.

Asian cities face a soaring additional demand for energy over the next two decades, according to a recent report by the Asian Development Bank (ADB 2013). The report predicts a ballooning of greenhouse gas emissions between 2013 and 2035. The study makes grim reading for city leaders seeking to find paths to cut greenhouse gas emissions. If no new energy policies are implemented, the Asia-Pacific's average carbon dioxide emissions are forecast to grow 2 per cent a year until 2035. Despite efforts to use more clean energy, by 2035 fossil fuels' share of the Asia-Pacific's energy mix are forecast to rise to 82.4 per cent, with coal use up an annual 1.7 per cent and oil 1.9 per cent a year (ADB 2013). However, solutions are available: governments and cities could act to adopt better energy efficiency measures, cut back on the use of inefficient coal-fired power plants, accelerate the removal of fossil-fuel subsidies and encourage the take-up of renewable energies. In other words, they could decide to remake themselves as low carbon cities.

The rise of the Asia-Pacific region

The scale of urbanization experienced in the Asia-Pacific region over the past 30 years is without precedent. By 2025, about half of the world's top 50 cities, ranked by GDP, are projected to be located in Asia (McKinsey Global Institute 2007). Asia's steady growth has become a stabilizing force in the global economy and there is a growing confidence in the Asian economies as more and more countries in the region are investing in infrastructural and development projects to raise the living standards of their citizens. These developments include transportation facilities and networks, housing, healthcare services, educational facilities, civic buildings, and cultural and commercial amenities.

The region's phenomenal growth in its urban conurbations has marked the rise of the 'century of Asia-Pacific cities' – the clustering effect of Asian populations into cities that generate higher incomes and new sources of prosperity. In 2013, around 180,000 people are moving into cities every day, especially in the developing world. For instance, China's rate of urbanization has reached 52 per cent in 2013, with more than 10 million rural citizens migrating to cities and towns each year. In 2013, China had 90 cities with more than one million people; it is predicted that in 2030, China will have 220 cities with a population of more than one million. The current urbanization rate in Indonesia is over 50 per cent and in Malaysia it is 72 per cent (ADB 2013).

While cities in the Asia-Pacific region have surged ahead in their development over the past decade, urban life has actually become more difficult. Cities in the Asia-Pacific region are confronted with frequent movement of populations, rapid technological advancements, the constant loss of cultural heritage, a growing focus on public safety, a widening gap between rich and poor, a lack of employment opportunities and an ageing population, as well as many other challenges such as the ghettoization of migrant labourers, and air, land and water pollution. Evidence shows that people need to feel connected in order to thrive and innovate, and empowered citizens are usually happier.

Asia-Pacific cities are also notable for the diversity of their residents, whose different cultural, religious and ethnic backgrounds add to the vibrancy of urban life but also need to be taken into account when developing urban policies. Similarly, there are considerable differences in people's levels of income, and therefore consumption. Almost all cities in this region have a wealthy elite minority, an emerging middle class and a large group of low-skilled workers, many of whom have arrived in the city relatively recently from rural areas, and are employed in the informal sector. Urban planning has to keep all of these different groups of people in mind to create a city where all residents can flourish (Roberts and Kanaley 2006a, pp. 439–440).

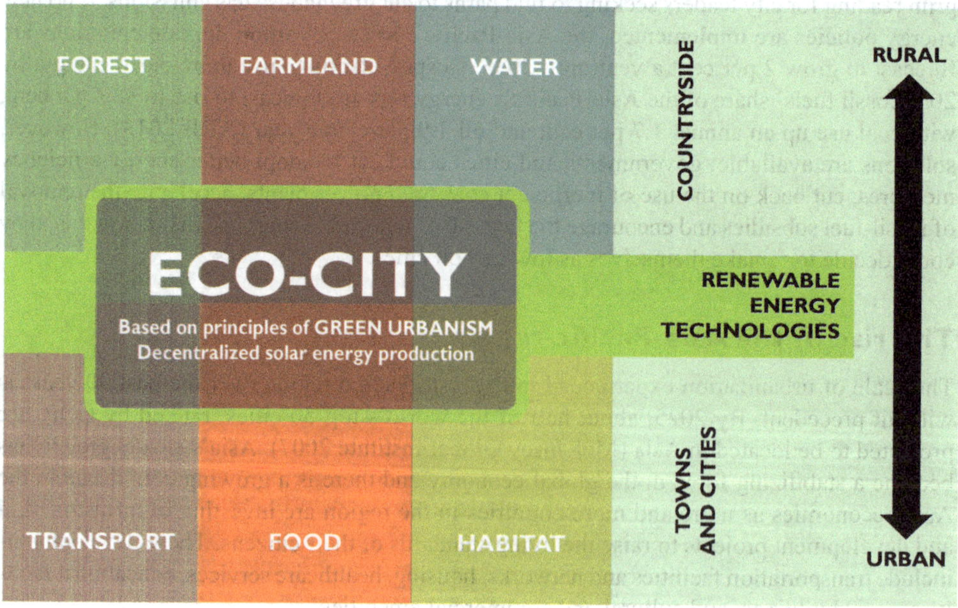

Figure 22.1 The eco-city seeks to balance humans' and nature's rural and urban systems.

Source: Diagram by Lehmann (2008).

Connectivity will be one of the main driving forces for the Asia-Pacific region's economies; the ease of movement of people, goods, money, services and information will trigger rapid development of regional areas, but requires massive investment in infrastructure to occur and to ensure this prosperity is fairly distributed and accessible. Technological innovation, more powerful computing and improved mobility have vastly improved connections and networks within and between cities. These improved connections will enable us to develop smarter solutions to meet some of the challenges ahead.

New paradigms for city planning: eco-cities, sustainable cities, low carbon cities and green urbanism

The traditional tools of urban development are outmoded and unsuited to deal with the current challenges. Most cities in the Asia-Pacific region were designed and planned using the previous century's conventional approaches; such thinking is now outdated, given the environmental challenges we face. Climate change forces us to reconsider urbanization and rethink old models of development. Traditional urban planning did not consider the full environmental impact and long-term costs of urban expansion. As a result, there is now a need to transform urban planning methods to new paradigms based on holistically planned socio-economic and ecological investment rather than short-term exploitation and simplistic linear thinking. Similarly, we need to move towards 'eco-efficiency', economies that are environmentally sustainable, rather than those that damage local, national and global prospects of long-term and equitable growth. This means making our cities less resource-hungry and more oriented to the well-being of current and future generations.

As noted in this book's Introduction, there are many conceptions of a sustainable city. The term 'eco-city' was created to describe a city that has been planned to be ecologically sustainable (see Figure 22.1). An eco-city enhances the well-being of its citizens and society through integrated urban planning and management that fully harnesses the benefits of ecological systems and renewable energies, aiming for zero emissions and zero waste. An eco-city protects and nurtures its environmental assets for future generations by using them sustainably (Lehmann 2010, p. 852). Eco-cities include their citizens in participatory decision-making processes; they are transparently governed, and have a vibrant culture and a sound economy.

The idea of eco-cities began in Western countries in response to the oil crisis of the 1970s, when urban planners realized that geographically large cities that required residents to drive long distances between home and work were not sustainable when petrol became more expensive. That is why mobility and transport was and is a key issue for eco-city planning. Today, eco-cities are generally agreed to be compact and pedestrian-oriented, and designed with an awareness of their specific geographic context. They can be newly built or retrofitted cities that aim to integrate the various sectors of urban services. Thus, housing, transport, energy, water, food supply and waste are planned and operate interdependently. Attention is also paid to the city's impact upon its surrounding region, particularly the agricultural land that supplies its food.

In many Asian countries, interest in applying ecological principles to cities developed due to pollution and the need to improve air, soil and water quality, but cities addressing these issues have more usually been described as 'sustainable' (Imura 2010). Interest in environmental sustainability and the needs of developing countries led UN agencies to establish the Sustainable Cities Programme in 1990. ASEAN's Network of Environmentally Sustainable

Cities began in 2003 and it has recently evolved into a Model Cities Programme. Like eco-cities, sustainable cities aim to give their residents a good quality of life and to develop economically without damaging the environment or ecosystem.

The sustainable district, town or city may be defined as a place where residents work, live and spend their leisure time in compact and well-interconnected communities (Lehmann 2010, p. 15). The technical solutions for sustainable urban growth are already available, and they include the use of energy-efficient, zero-carbon urban materials, based on renewable energy sources and renewable, more flexible building typologies (Lehmann 2010, p. 21). The terms 'eco-city' and 'sustainable city' are often used interchangeably because they have similar goals – which can be achieved by implementing the principles of green urbanism.

Whether urban settlements are called eco-cities or sustainable cities is perhaps less important than whether or not they achieve key targets. For example, having low levels of carbon dioxide emissions is a key goal, hence the increasing aim of such cities to be 'low carbon' or 'zero carbon' cities. Like eco-cities and sustainable cities, low carbon cities are compact and mixed-use cities that reduce their energy consumption, are based on sustainable urban development principles, emphasize the reuse of resources and recovery of waste, and transform in balance with the ecosystem (Lehmann 2012b). Their main aim is to reduce emissions of carbon dioxide.

Low carbon cities have been identified as a variant of ecologically sustainable cities (Lehmann, Introduction, this volume; Onishi 2011, p. 25; Opschoor 2011, pp. 197–199). Assessing the advantages of low carbon urban planning for Japanese society, Onishi (2011, p. 25) describes it as 'an opportunity to take another look at the materialistic culture in society, give serious consideration to the natural environment and return to more comfortable [i.e. less rushed and stressed] lifestyles'. Folllowing Onishi, we can see that Asian cities have a particular advantage in adopting low carbon principles because they are already densely populated, and it is density that makes eco-mobility feasible for individuals (transporting themselves on foot or by bicycle) and financially viable for service providers (of mass/public transport).

Why urban planning in Asia-Pacific cities needs a paradigm shift

Many countries in the Asia-Pacific region share a hot and humid climate, especially those located in South and South-East Asia, which have some of the highest numbers of rapidly developing cities. Cities in this region share common problems: the cultural conflict between tradition, modernization and globalization; a massive influx of rural poor into urban areas; increasing income inequality; and rapid, poorly managed urban developments. When urban infrastructure expands and gets renewed, there is a critical need to retain a city's historical legacy. Planners and architects in the Asia-Pacific region are often faced with the challenge of optimizing land use and creating efficient high-density precincts, while having to preserve the cultural identity of sites and respect buildings of historical significance that embody a city's or community's memories. Cities in the Asia-Pacific region need to be more determined to make extra efforts to protect their traditional culture and historic villages, as too many of them have been demolished amid rapid urbanization. Financial and technical support from government and non-government organizations could be sourced to protect urban built heritage.

Climate change challenges for cities: urban heat islands and food insecurity

Scientists have clearly shown that human activities alter the climate. Climate change is now understood as a life-shaping force with far-reaching impact. The need for a profound restructuring in the way we live, and the way in which we develop, build, operate and maintain towns and cities, is widely acknowledged (Lehmann 2010, p. 12). Extreme heatwaves, prolonged droughts, catastrophic floods, coastal erosion, poisoned water, polluted air, loss of biodiversity and crop failures may all become more frequent if rising temperatures are not brought under control. Global ecological risk is a shared problem that disregards national boundaries. It is clear that cities will continue to be the place where most energy is consumed and most waste is produced, but cities also provide us with economies of scale that make renewable energy sources viable (Lehmann 2010, p. 13). That is why we need to shift the paradigm and change our unsustainable urban planning models to ones that adopt and adapt the principles of green urbanism.

Cities in the Asia-Pacific region must brace themselves for a more volatile climate as scientists predict frequent damaging heatwaves as part of long-term changes in temperatures. Global warming increases the chance of extremely high heat. The World Meteorological Organization (WMO 2013) recently reported that the number of heatwaves in Asia increased between 1980 and 2009. The anticipated increase in heatwaves means that the urban heat island effect (where cities fail to cool down overnight due to trapped heat) is likely to worsen. Extreme temperatures in cities are likely to have a direct effect on local economies, food supply, water scarcity, human and animal health, and people's lifestyles. Some of the measurable impacts will include a sharp increase in mortality rates due to heat-strokes and dehydration. An increased use of air-conditioning means that the extreme heat will raise energy consumption and strain the power grid, risking the likelihood of blackouts. There will also be an economic impact; it will mean loss of productivity as the heat keeps people away from their jobs. Heatwaves also threaten food security due to a drop in agricultural production and thus have a direct impact on the economy. Too much heat stress damages crops and lowers yields in agriculture. The urban poor will be the worst affected because of their already inadequate living conditions, and there is likely to be an increase in environmentally induced migration due to the impact of climate change on the rural environment (UN-Habitat and UN ESCAP 2010, pp. 21–22). All future urban development must aim to decrease the impact of the urban heat island effect and ensure food security (IPCC 2013).

Asia's rapid urbanization

Although the majority of Asians still live in rural areas, the region as a whole is urbanizing rapidly (UN-Habitat and UN ESCAP 2010, p. 29). In the two decades between 1980 and 2010, 1 billion people joined Asia's cities (ADB 2013, p. 3). In 1990, just over a third of Asians lived in cities; this increased by 10 per cent in a decade, and it is predicted that almost half of all Asians will live in cities by 2020 (UN-Habitat and UN ESCAP 2010, pp. 38–42). Asia's urbanization is different from the rest of the world's in three important ways: the scale and pace of the changing population; the fact that Asian cities already have high population densities; and the fact that they are already mixed use (UN-Habitat and UN ESCAP 2010, p. 9; ADB 2013, pp. 4–9). Unfortunately, many Asian cities' new town extensions are mono-functional (residential only). Increased urbanization often leads to denser cities with fewer trees, less green space and less natural air movement in cities, which contributes to the urban heat island effect.

While considerable attention has been paid to the region's 'mega-cities', 60 per cent of Asia's urban population live in urban areas with a population under one million and 29 per cent of Asia's urbanites live in 'metropolitan' cities, which have a population of between one and ten million (UN-Habitat and UN ESCAP 2010, pp. 8–9). (The mega-cities, by contrast, have only 11 per cent of the region's urban dwellers.) Other researchers note that intermediate-sized cities, which have a population between 500,000 and two million, will grow the fastest (Roberts and Kanaley 2006c, p. 15). This is why small and medium-sized cities are so important; they are where the majority of Asia's people live, and they are the ones that will grow most rapidly in the coming decades. To a great extent, the region's urban future will be played out in these towns and cities.

The importance of the peri-urban area

The peri-urban area is often where the impact of urbanization is most evident. Cities' edges are usually where most growth occurs because that is where land is cheaper and available (Roberts and Kanaley 2006c, pp. 15–16).

Some local/municipal governments in Asia-Pacific countries are responsible not only for a city but also for the surrounding rural area (UN-Habitat and UN ESCAP 2010, p. 9). This can be a great advantage in developing an eco-city. Eco-city planning recognizes that cities are dependent on the surrounding region for ecosystem services such as sequestering carbon, regulating climate and the provision of food and water. For these reasons, the urban hinterland should be preserved, and if a local government has oversight of the region around a city, it is far more likely to be able to do so. If, on the other hand, there is more than one governing body responsible for the peri-urban area, the governments need to collaborate to manage this area efficiently for all the residents who rely upon it.

Demographic trends in developed and developing Asia-Pacific countries

Demographic trends show the way a population is changing over time. One of the most important aspects of an economically sustainable population is the proportion of dependent children and adults to people of working age. Basically, there need to be enough workers to care for and financially support those who are too young, old or infirm to work. A key demographic trend is related to the demographic transition, the change in the composition of and ratios in a population. The Asia-Pacific region's population is ageing and care of the elderly is the responsibility of younger adults.

While there are differences depending on where a country is in its demographic transition, there are also important similarities in the changes to Asia-Pacific countries' populations. The chief of these is a generational change that has been described as a 'silver revolution' because fertility is declining and life expectancy is increasing, meaning that populations are ageing on a scale and at a rate that has never been seen before in human history (Booth 2013). What will this mean for cities?

The *State of Asian Cities* report suggests that cities will need to adapt to their ageing populations by changing their urban planning guidelines and building regulations so that housing, public spaces and transport systems are accessible for and appropriate to older adults. Cities will also need to make provisions for sufficient medical care and social services to ensure that the needs of elderly residents are met (UN-Habitat and UN ESCAP 2010, p. 47).

Urban population growth rates need to be kept in mind. Any figures that attempt to give a general picture of Asia's demographic trends are usually skewed by the two most populated countries, China and India. Looking more closely, there are four kinds of Asian urbanization:

1. Developed countries such as Japan have high levels of urbanization but a low level of urban population growth.
2. Countries such as Malaysia and the Philippines have moderate to high levels of both urbanization and urban population growth.
3. Developing countries such as China and India have high levels of urban population growth but a low level of urbanization overall.
4. Developing countries such as Nepal and Lao PDR have slow urban population growth and a low level of urbanization (UN-Habitat and UN ESCAP 2010, pp. 32–34).

A new urban planning paradigm: green urbanism, eco-efficiency and disaster resilience

The combination of continued rapid urbanization in the Asia-Pacific region and the need to deal with a changing climate is why we need to adopt low carbon principles and shift to a new urban planning paradigm: green urbanism.

Green urbanism is a holistic concept of sustainable urban systems, enabling existence and change (growth or shrinkage) without impacting negatively upon planetary life support systems and ecosystems. It is a particular form of urbanism that is concerned with a healthy balance between the city and its hinterland. Green urbanism also underpins practical action in order to shape the urban environment in a sustainable way. The concept of green urbanism arose in the late 1990s, promoting compact, energy-efficient urban development, seeking to transform and re-engineer existing city districts and regenerate the post-industrial city centre (Lehmann 2010). Green urbanism helps city planners and urban leaders remember that natural resources – clean air, open space, trees and fresh water – are worth maintaining. Thus its principles may be applied to industrializing cities as well as to those that are moving to service-based economies.

The concept of 'eco-efficiency' was developed 20 years ago to demonstrate that maintaining the environment while growing the economy will not cost money; on the contrary, it will generate income. Eco-efficiency encourages us to see and find ways to avoid being 'poor and clean' or 'rich and dirty' and instead be 'rich and clean' (UN ESCAP 2011a, pp. 12, 16). An example of eco-efficiency is installing solar panels on existing buildings; the panels generate clean energy, which is ecologically efficient, and the process of installing the solar panels creates green jobs, which is economically beneficial. Eco-efficient ideas and guidelines are particularly useful for urban planners and leaders who need to improve their cities by building and funding the development of infrastructure while minimizing its impact on the environment (UN ESCAP 2011a).

Adapting the principles of green urbanism will help cities to become eco-efficient because they are complementary concepts that share similar ideas. Both green urbanism and eco-efficiency see cities as living systems (see Figure 22.2) and recognize that developing and maintaining the urban metabolism in balance with its environment is fundamental for economic growth and social sustainability (UN ESCAP 2011a, pp. 24–26).

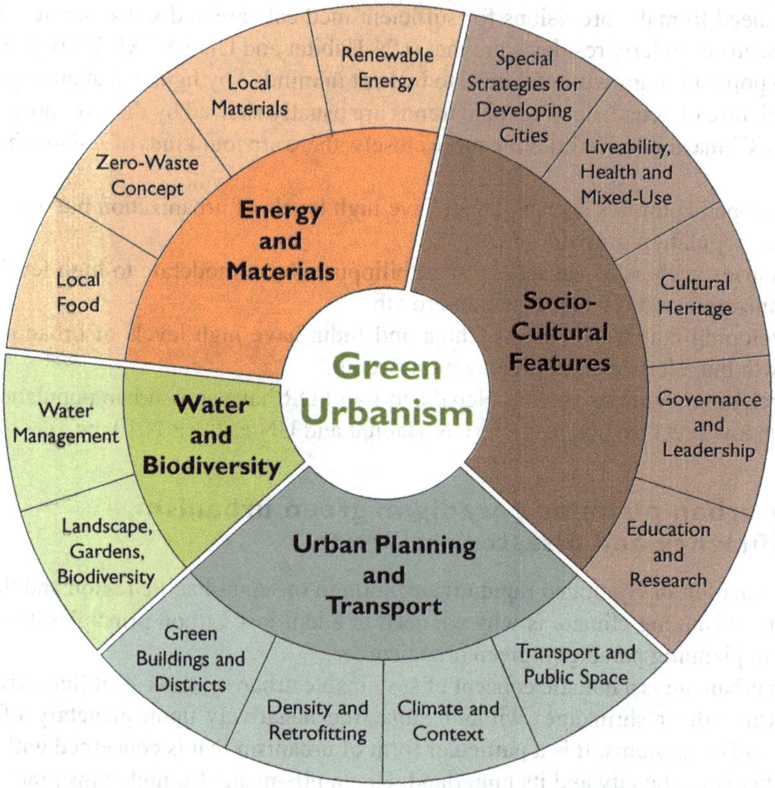

Figure 22.2 The green urbanism wheel shows the 15 principles grouped into four sections.
Source: Diagram by Lehmann (2010).

Disaster resilience and its connection to green urbanism and eco-efficiency

Implementing the principles of green urbanism and eco-efficiency should reduce the possibilities of disasters occurring, as well as helping cities respond to a disaster with resilience. If an existing city is flexible enough to retrofit its buildings and implement other green urbanism principles, it is on the pathway to the adaptation required to respond to disasters with resilience. Similarly, if citizens are committed to and feel involved in their local governance and urban services (in keeping with the principle of consultation and participation that is part of eco-city good governance), they are more likely to work effectively with the authorities. The cities that find it hardest to cope with disasters are those where the community and economy are weak (often because younger adults have moved away and the remaining population is ageing); thus, being resilient to climate-caused disasters is connected to being resilient socially, demographically and economically.

Cities' preparedness and resilience increases when areas for food production are integrated into the urban landscape. This ensures a short supply chain and a high quality of food, reintroduces biodiversity and nature, provides employment and takes an ageing population into account.

Low carbon cities through green urbanism planning principles

The 15 principles of green urbanism (listed in Table 22.1) are a conceptual model of how we might tackle the enormous challenge of transforming existing neighbourhoods, districts and cities, and rethink the way we design, build and operate urban settlements as habitats for humanity and other species. These are basic and holistic principles but not a single, one-size-fits-all formula. The principles can be effective in a wide variety of urban situations, but they always need to be adapted to the specific context, to the site's constraints and opportunities. Clearly, the design of districts and cities is a very complex exercise. Urban design has always had a social and a political dimension (Lehmann 2010, p. 14), which is why good governance is an essential ingredient of eco-city principles (and see Principle 13 in Table 22.1). We should also note at the outset that Principle 15 explicitly recognizes the need to adapt all principles to the conditions of developing cities, which is centrally important for many Asia-Pacific cities. A common toolbox in urban planning, such as the 15 principles, makes it possible to have a definitional baseline in the debate about urban development in the Asia-Pacific.

Like all eco-city guidelines, the 15 principles of green urbanism are interconnected and holistic: they need to be applied simultaneously, not individually (see Figures 22.3 and 22.4).

Table 22.1 The 15 principles of green urbanism

The 15 principles of green urbanism	*Sample recommendations and measures*
PRINCIPLE 1: CLIMATE AND CONTEXT All urban development must be in harmony with the specific characteristics, site factors and advantages of each location. Every city is different and will need to identify its unique solutions.	Start any strategic planning process by mapping the city's location and identifying its assets. Plan for development that works with the urban climate and bio-regional context.
PRINCIPLE 2: RENEWABLE ENERGY FOR ZERO CO_2 EMISSIONS The city should be a self-sufficient on-site energy producer, using decentralized, district-based energy systems for zero CO_2 emissions through renewable, smart electricity grids for efficient use of energy.	Decarbonize the energy supply; increase solar power to 10 per cent of the energy mix by 2020; install smart grids and make solar hot water mandatory; generate at least 50 per cent* energy on site using renewable sources, including precinct-scale wind turbines and small-scale biogas plants. Encourage this through feed-in tariffs.
PRINCIPLE 3: ZERO-WASTE CITY The zero-waste city is a circular, closed-loop ecosystem through waste management and resource recovery that stops waste materials going to landfill or incineration.	Implement 'zero-waste city' ideas and plans; increase resource recovery rate towards 100 per cent and stop land-filling. A minimum 50 per cent of on-site renewable energy generation should be the aim of all planning proposals,* with the energy mix coming from decentralized energy generation and taking into account locally available resources, as well as the cost and availability of the technology.
PRINCIPLE 4: WATER Closed urban water management and high water quality allows for sustainable water management to ensure water security through wastewater treatment and sensitive urban water management.	Use solar-powered desalination and recycle wastewater; keep fresh water consumption below 125 litres per person per day.* Construct wetlands to purify and recycle greywater.

Table 22.1 continued

The 15 principles of green urbanism	*Sample recommendations and measures*
PRINCIPLE 5: LANDSCAPING, GARDENS AND BIODIVERSITY Integrate landscaping, urban gardens and green roofs to maximize urban biodiversity, including landscape strategies for productive open spaces.	Continue tree-planting programmes and increase tree planting in streets. Constructing wetlands to purify and recycle greywater (see Principle 4 recommendation) will also improve landscaping and biodiversity.
PRINCIPLE 6: TRANSPORT AND PUBLIC SPACE Create eco-mobility with a good public space network and an efficient, low-impact public transport system for post-fossil fuel mobility. Anticipate future uses of public space so that it is more vital than simply a place of transit.	Invest over 6 per cent of GDP in public transport,* expand tramlines and introduce free hybrid buses. The city should offer multi-modal public transport systems, a high number of connections and choices, frequent trams and buses, and safe pedestrian and cycle networks. Improve streets by giving greater priority to pedestrians and cyclists.
PRINCIPLE 7: LOCAL MATERIALS Use regional or local materials in construction and apply prefabricated modular systems.	Use engineered timber construction systems and make recyclability and reuse of construction elements compulsory. Use local materials with less embodied energy.
PRINCIPLE 8: DENSITY AND RETROFITTING Retrofit districts, encourage urban infill, and adopt densification/intensification strategies for existing neighbourhoods.	Continue street upgrading and introduction of bike lanes. Improve pedestrian connectivity with additional crossings. Make public space more useful with natural elements designed for active living.
PRINCIPLE 9: GREEN BUILDINGS AND DISTRICTS Apply deep green building design strategies and offer solar access for all new buildings using passive design principles.	Reintroduce passive design principles for building approvals and demand higher ratings. Promote energy-saving building designs and full home insulation. Offer better housing choices and more diversity in urban infill.
PRINCIPLE 10: LIVEABILITY, HEALTHY COMMUNITIES AND MIXED USE Emphasize affordable housing, mixed-use programmes and a healthy community. This includes urban design being appropriate for children and an ageing population.	Include a minimum of 25 per cent affordable housing in every development* and use modular prefabricated construction systems. Reduce taxation of inner-city housing. Increase retrofitting and adaptive reuse of heritage buildings.
PRINCIPLE 11: LOCAL FOOD Create a local food supply, with high food security and urban agriculture.	Introduce urban farming in at least 20 per cent of the public parks.* Create community gardens. Maintain urban hinterland for food production. Restore connections to natural ecosystems within the urban ecosystem.
PRINCIPLE 12: CULTURAL HERITAGE Become a city of public health and cultural identity: a safe and healthy city, which is secure and just.	Protect existing structures; demand and facilitate more adaptive reuse by relaxing building code. Consult and involve communities to ensure genuine commitment.
PRINCIPLE 13: GOVERNANCE AND LEADERSHIP Apply best practice for urban governance and sustainable procurement methods; this includes political will and long-term vision and commitment.	Innovate public policy, roll out incentives and deliver subsidy programme for clean-tech sector. Create public–private partnerships to facilitate change and involve community groups and NGOs.

The 15 principles of green urbanism	*Sample recommendations and measures*
PRINCIPLE 14: EDUCATION, RESEARCH AND KNOWLEDGE-SHARING Provide education and training for all in sustainable urban development.	Invest a minimum of 3 per cent of GDP in research and innovation;* strengthen university education programmes to include climate change impact. Facilitate sustainable behaviours and provide incentives for long-term behaviour change by positively influencing values and attitudes to reduced consumption. Remove policies that encourage wasteful consumption (e.g. fuel subsidies).
PRINCIPLE 15: SPECIAL STRATEGIES FOR CITIES IN DEVELOPING COUNTRIES Adapt sustainability strategies for cities in developing countries, harmonizing the impacts of rapid urbanization and globalization.	Cities require adjusted strategies appropriate for the developing world, e.g. low-cost building and mass housing typologies for rapid urbanization; gender-sensitive and pro-poor approaches to development.

Source: Lehmann (2010).
Note: * All suggested figures are benchmarks derived from current best practices.

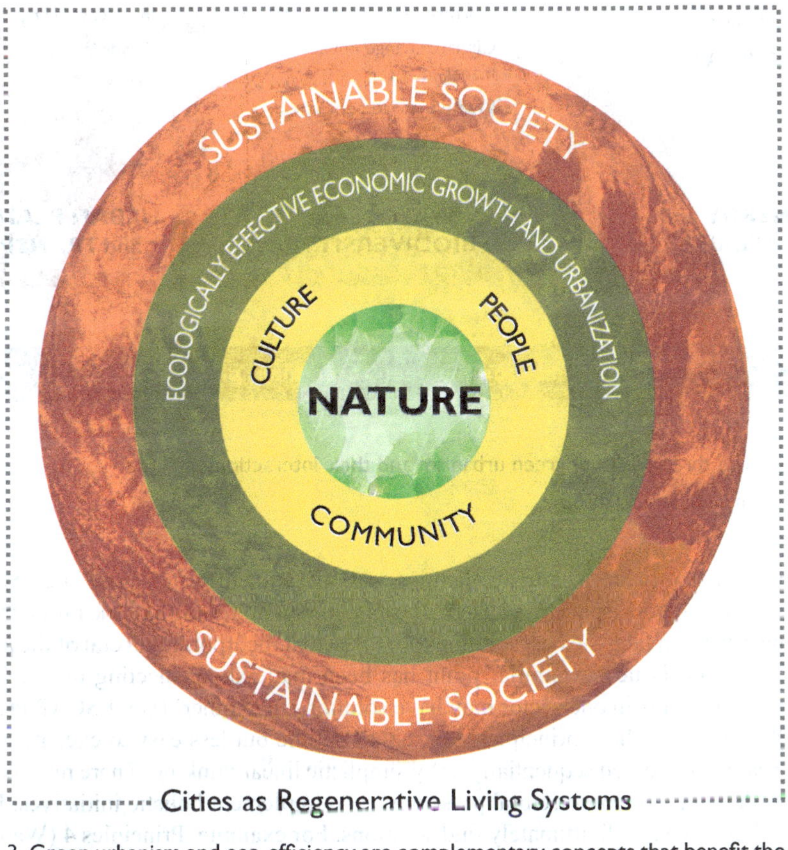

Figure 22.3 Green urbanism and eco-efficiency are complementary concepts that benefit the economy and the environment by recognizing that nature is at the centre of our existence.

Source: Diagram by Lehmann and Thornton (2013).

Figure 22.4 The three pillars of green urbanism and their interactions.

Source: Diagram by Lehmann (2008).

They recognize that successful urban planning pays attention to systems and interconnections between a city's sectors, spaces and users. It is not enough to try to 'fix' one problem without addressing others. In fact, the best solutions are those that incorporate several of the principles to address several issues. A similar point has been made about meeting the needs of the urban poor: 'problems in one sector may have solutions in another' (UN ESCAP *et al*. 2007, p. 26). Obviously, all these principles are easy to outline but less easy to execute. Complex problems cannot be solved sequentially or by simplistic linear thinking. There must be parallel processes, drawing on cross-sectoral perspectives to implement holistic initiatives. Bringing these together is what will ultimately find solutions. For example, Principles 4 (Water) and 5 (Landscaping, gardens and biodiversity) both recommend constructing wetlands to recycle and purify greywater, which will enhance both a city's biodiversity and residents' health.

Green urbanism for cities in developed Asia-Pacific countries

Like cities in Western nations, cities in developed Asia-Pacific countries (Australia, New Zealand, Japan, the Republic of Korea and Singapore) should be trying to implement Principles 1 through 14, paying particular attention to increasing their density (where necessary, e.g. Australia), decarbonizing their transport and energy systems, improving water use (by collecting stormwater, recycling wastewater and keeping fresh water consumption below 125 litres per person per day) and waste management to develop zero-waste cities by moving towards a 100 per cent resource recovery rate that sees an end to landfilling. As these cities' populations are ageing, Principle 10, which directs city planners to promote liveable, healthy and age-friendly communities, will also be of crucial importance.

Adapting the principles of green urbanism to cities in developing Asia-Pacific countries

As Principle 15 points out, identifying specific strategies for cities in developing countries is essential. While the principles need to be tailored to a city's level of development and capacity, there are entry points for cities at all levels of development. For example, creating community gardens within a city (Principle 11) will help ensure food security, and initiating a tree-planting scheme (Principle 5) will reduce the urban heat island effect, as well as bringing people together and improving the amenity of public spaces. Successfully implementing one of these actions will build the capacity of an urban community, create more awareness of the importance of green urbanism and help lay the foundation for another project.

Carbon emissions from transport are among the main contributors to pollution and greenhouse gases in small and medium-sized cities, so, when seeking to implement Principle 6 on sustainable transport, the best strategies are 'eco-mobility' ones that have no carbon emissions, i.e. people travelling by foot or bicycle. Having separate lanes for cyclists and wide, shaded footpaths are good ways to encourage urban residents to walk or ride a bike. Another good way to decrease carbon from transport emissions and encourage more people to use public transport is to create Bus Rapid Transit (BRT) systems. Light rail and metro train systems are usually too expensive, but deciding to dedicate a lane to buses on already-existing roads is feasible for most cities (although it may require political courage, negotiation and community education for its acceptance). This could be the first step in creating a BRT system. The city of Ahmedabad in western India has successfully designed and implemented a BRT system that carries 125,000 passengers per day using 70 buses. Inaugurated in 2009, it attracts users from all classes of society (Narang Suri 2011).

Green urbanism principles advocate green buildings that are designed with the local climate in mind and use local and sustainable materials for construction. These are also principles that a city in a developing country should be able to implement, as the example of bamboo being used as a building material shows. Bamboo is 'fast growing, easy to use, environmentally friendly and extremely durable' (HFHN 2007a). There is plenty of bamboo in eastern Nepal, so Habitat for Humanity Nepal (HFHN) has been working with Nepalese communities to harvest bamboo and assemble it in housing since 2005. The local communities contribute valuable knowledge and expertise about the cultivation and harvest of bamboo, as well as weaving techniques used in building. In 2007, a partnership between local people, HFHN, the Canadian International Development Agency and the Canadian Architects' Legacy Fund established a factory in Jhapa to produce corrugated bamboo roofing sheets as an alternative to corrugated iron (HFHN 2007a, 2007b; UN ESCAP 2011b, pp. 49–50). The

Jhapa factory is an example of eco-efficiency; bamboo roofing sheets are more environmentally sustainable than corrugated iron and the factory provides employment and capacity building for the people who work there.

Could bamboo be used as a green building material in other developing cities? Because bamboo is airy and cool in a tropical climate, researchers in Indonesia have also been investigating its use as a building material, but they have encountered problems of social acceptance. Some people do not think bamboo is 'modern' or appropriate for city buildings. Hence the need to use the other principles to support the desired change; as well as a need for education about sustainability (Principle 14) and leadership to rally people to the cause (Principle 13), pointing out bamboo's traditional use as part of maintaining cultural heritage (Principle 12) may move people to take pride in it, instead of seeing it as 'backward'. This has been the experience in Thailand, where people are now recognizing the potential of bamboo as a building material that connects a traditional past to a sustainable future, especially in building tourist facilities. Using bamboo as a sustainable building material in countries where it is plentiful could be a way for developing cities to skip the 'rich and dirty' stage of development and transform to 'rich and clean'.

Implementing green urbanism in any city: resource efficiency and behaviour change

Regardless of their level of development, the most critical challenges faced by Asia-Pacific cities seeking to implement the principles of green urbanism are resource efficiency and behaviour change. Cities often depend on importing raw material resources from outside their boundaries. Sustainable consumption and decoupling of material consumption from economic growth at the city level are the subjects of the recent report, *City-Level Decoupling Urban Resource Flows and Governance of Infrastructure Transitions* (UNEP 2013), which has highlighted the need to reconfigure urban infrastructure to achieve sustainable metabolic resource flows. For instance, the development of an urban circular economy would enable cities to transform waste streams into a precious resource (including harnessing the existing link between biomass, green spaces, soil productivity and nutrient flow, making more productive use of organic/food waste). However, the question of how to best 'decouple' economic growth from natural resource use and environmental degradation remains.

Resource efficiency at the urban scale: systemic approaches to metabolic flows

Due to its complexity, it is a very ambitious activity to explore in detail how a city or metropolitan region works. Examining cities from a material flow perspective is a valid approach; it indicates that infrastructure plays a key role in directing material flows and therefore resource use, productivity and efficiency, and environmental impact in an urban context (UNEP 2013). Ecological footprint and material flow analyses at the urban scale are useful methods that can demonstrate how city-level decoupling could be implemented; much of it will depend on linking cities more effectively to their surrounding landscapes and bioregions. In this regard, Herbert Girardet notes that 'greater attention to the interaction between cities and the ecosystems that support them can contribute to more resilient cities and support city-level managers to reduce and manage the potential for disasters and climate change' (personal communication 2013).

The flow of resources is determined by urban systems. For instance, the flow of organic waste, nutrients, phosphate and rare earths has now become a major issue for city planning and society as a whole. Resources are precious; we therefore have to reconfigure the urban infrastructure that conducts the 'flows'. Local resource efficiency of cities will contribute to the creation of stronger communities. In future, with an increasing decentralization process and more distributed systems, citizens will need to produce energy and food on-site on their own properties. A circular economy (see Figure 22.5) will allow for more sustainable consumption and growth that does not damage the ecosystem, but supports industrial symbiosis (where flows of waste materials from one industry become the resource for re-manufacturing in the factory next door); such short-circuits and closed loops can reduce transport, pollution, resource depletion and costs. Therefore we should stop using the word 'waste', and replace it with terms like 'opportunity materials' or 'misallocated resource' (Lehmann 2010).

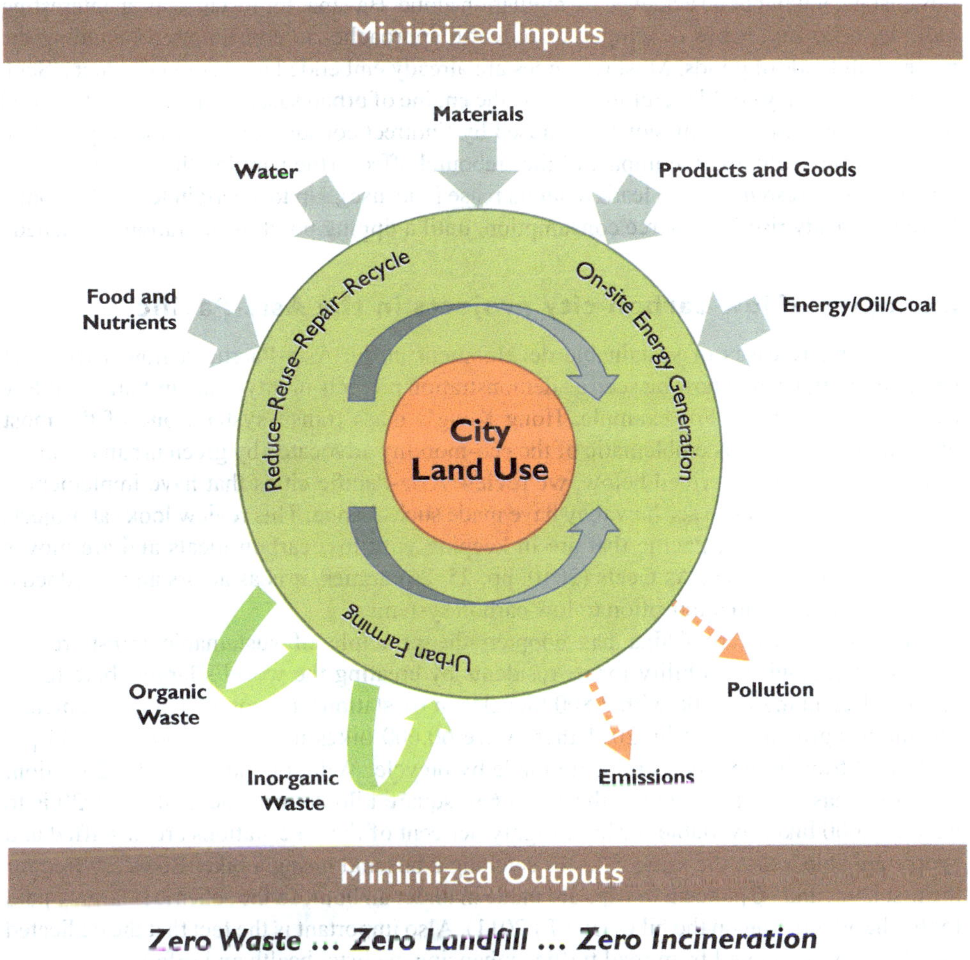

Figure 22.5 Resource-efficient cities and sustainable consumption.

Source: Diagram by Lehmann (2013) after Girardet.

The city has always recycled itself, using its urban fabric to evolve. Transferring the common waste hierarchy into urban development implies a paradigm change where upgrading the existing city is more valuable than new building on greenfield sites, which consumes large amounts of resources, land and virgin materials. It gives urban production a new impulse, where 'non-building' and 'doing less with less' are ideals that give urban planning a new role. Recycling and up-cycling old buildings and derelict existing structures through adaptive reuse means continuity of cultural heritage as well as the creation of new value – contrary to the modernist paradigm of always constructing entirely new buildings (Lehmann 2012a).

Behaviour change and sustainable consumption

The consumption patterns of cities are driven by product use, by the way we consume (e.g. in many developed cities, over 90 per cent of food is supplied by a small number of powerful supermarket chains). If we want to change behaviour, we need to add pleasure and joy into this change; it will not be achieved by regulation alone. Havana, for instance, is an interesting case; by necessity, it has developed resourceful approaches to consumption including the reuse of all kinds of goods. Most resources are already embedded in existing products. Such resource efficiency could therefore become the engine of urban sustainability. In fact, around half of greenhouse gas emissions are caused by 'indirect consumption', by the way we live and consume, including the impact of the 'rebound effect' (the paradox that improving the efficiency of a resource's use leads to an increase in its use). Up to a certain level of income, there is a steady rise in resource consumption, until a tipping point of saturation is reached.

Examples of low carbon city projects in the Asia-Pacific

There are many cases of sustainable development in the Asia-Pacific region where best practice solutions may now be seen as demonstration projects or city-scale initiatives of low carbon urban change. For example, Hong Kong's mass transit system, one of the most efficient in the world, is emblematic of the eco-mobility advocated by green urban planners. In the case studies described below, we review Asia-Pacific cities that have implemented multi-sector strategies to see how they have made such change. This review looks at projects from cities in the Asia-Pacific that are in keeping with low carbon ideals and are moves towards green urbanism. As Geels (2010, pp. 25–26) argues, it is as actors and 'seedbeds' that cities can best aid a transition to low carbon systems.

Hangzhou in eastern China has adopted the principle of sustainable transport and developed ecological mobility for its residents by creating the world's largest bike rental scheme. Beginning in 2008 with 2,800 bicycles in 61 stations, the municipal government's scheme has grown so that by 2011 there were 60,600 bikes in 2,200 stations and 43 per cent of all trips in the city were being made by bicycle. With a population of 2.2 million, Hangzhou has six centres within the city's 638 square kilometres. The goal for 2020 is to have 175,000 bikes available for hire. Eighty per cent of the bike stations are unstaffed and people are able to use the same card for catching a bus and hiring a bike. Bikes are free for the first hour, and 96 per cent of trips are made in under an hour, so the scheme is largely paid for by the advertising on the bikes (ICLEI 2011). Also important is the fact that the dedicated bike lanes are separated from road traffic, enhancing cyclists' health and safety.

Matale is on the way to becoming Sri Lanka's first 'zero-waste' city thanks to the leadership of its mayor, partnerships with non-government organizations and funding from an international donor. Matale is implementing an innovative, pro-poor approach to solid waste management with a series of integrated resource recovery centres. Adopting the eco-efficiency perspective of seeing waste as a resource has created green jobs for the urban poor, and means that the town's households have their solid waste collected, and that the municipal government has saved money and land from being turned into an open dumpsite (UN ESCAP 2011b, pp. 37–38, 2012a, p. 136).

Puerto Princesa City is on the Philippines' island Palawan. Thanks to the leadership of Mayor Edward Hagedorn (elected in 1992), Puerto Princesa City has become a 'model city of sustainable development' (Newman and Matan 2013, pp. 73–74). Hagedorn created a public–private partnership to reduce pollution and litter, and to promote the advantages of ecological sustainability. Housing for the poor was improved and because the new homes had natural ventilation and used compact fluorescent bulbs, the buildings' energy demand decreased by 30 to 40 per cent. As a member of the 'Cities for Climate Protection' campaign since 1999, Puerto Princesa City has reduced carbon dioxide emissions thanks to using less energy by switching off street lights and air-conditioning systems for an hour a day, increasing the use of electric-powered transport and holding an annual tree-planting festival. Puerto Princesa City continues to adopt more sustainable ways to generate electricity, including a waste-to-energy bio-digester plant and solar energy.

Rizhao, China is a city of three million people in Shandong province that has created expertise in the area of renewable energy, which led to its receiving one of the Inaugural World Clean Energy Awards in 2007 and being awarded UN Habitat's Scroll of Honour for urban planning in 2009. Rizhao's city planning emphasizes 'innovation, public education, science and technology', and the solar energy products manufactured in Rizhao that contribute to the city's economy include solar hot water systems, stoves, heating and cell technology (Newman and Matan 2013, pp. 47–48). Thanks to new technology that uses methane from wastewater to produce electricity, the municipal government has also created a set of policies that simultaneously benefit the environment, encourage local manufacturers and help reduce the costs of agricultural production, thus promoting food security while decreasing water pollution and increasing the use of renewable energy (marsh gas-generated electricity).

Ulsan in the Republic of Korea was once a polluted city due to its industrialization and rapid growth. Following the town's Ecopolis Ulsan Declaration in 2004, environmental standards were imposed, the watercourses and ecosystems were restored, air quality improved, and more urban green spaces were created. Ulsan used a public–private partnership for waste collection and stopped landfilling because it had assessed the true cost of it (by recognizing both the current market value of land and the cost of using land this way) and recognized that it was not financially viable in the long term. Ulsan's Eco-Industrial Park allows waste from one company to be used as resources for another. Educating the community about the importance of ecological sustainability and clean waterways has been so successful that environmental standards are now observed voluntarily (UN ESCAP 2011a, pp. 43, 37, 2011b, p. 57).

Yokohama is one of the largest cities in Japan (population 3.65 million) and demonstrates the ideas of eco-efficiency and moving towards 'zero waste'. Despite a growing population, Yokohama reduced its waste by 38.7 per cent in six years (2001–2007) because the city government inspired citizens and businesses to join a 3Rs campaign to reduce, reuse and recycle materials that had previously been seen as waste. The intensive education and engagement campaign motivated Yokohama's people to change their behaviour and sort 'waste' materials into 15 possible categories. This meant that two incinerators have closed, saving over a billion US dollars in maintenance and operating costs. By selling the materials collected in recycling, the city government is able to raise money and, by incinerating only those materials that cannot be reused or recycled, the city generates electricity. It also outsourced part of the waste management process (collection and transportation), resulting in further savings (Suzuki *et al.* 2010).

High-density, high-rise buildings and greenbelts: developments in Singapore and China

The success of Singapore's Housing and Development Board (HDB) in providing good urban infrastructure and affordable, quality accommodation for the city-state's citizens is well known. It is a familiar example that has inspired many. The HDB's most recent development, Punggol 21 Eco-Town, continues this good work based on eco-city ideas and the application of green urban planning principles. Punggol 21 Eco-Town will ultimately provide 35,000 housing units by 2015. The first stage was completed in 2010. Designed to take advantage of cooling breezes and natural light, Punggol 21 Eco-Town also includes green landscaping and features such as a community garden where residents grow food and create compost using the food waste from neighbouring markets and cafés. Punggol 21 Eco-Town's seven districts are both high rise and high density but the extensive planting softens this. As well as greenbelts through the development that provide pedestrian-only connections to the foreshore, the 4.2-kilometre artificial waterway provides a 'bluebelt' that also brings natural elements into the urban landscape. Living in housing with natural light and ventilation, where there is an opportunity to grow food and walk, are ways for Punggol 21 Eco-Town's residents to decrease their carbon footprint.

A green high-rise tower is possible. The Pearl River Tower, located in the city of Guangzhou, China, is one of the world's most energy-efficient high-rise structures. The 309-metre Pearl River Tower's sculpted body directs wind to a pair of openings where turbines generate energy for the building. Integrated solar panels, a double-skin curtain wall, a chilled ceiling system, under-floor ventilation and daylight harvesting are additional technologies that contribute to the building's energy efficiency. The tower's integrated design approach results in a substantial decrease in the amount of electrical power required to operate the building's cooling, heating, dehumidification, ventilation and lighting systems. Full implentation of Pearl River Tower's sustainability strategies will result in overall energy savings of approximately 50 per cent compared to China's baseline energy code. Energy efficiency coupled with on-site renewable energy generation reduces the need for electricity from fossil-fuel power plants, lowering energy-related greenhouse gas emissions. The Pearl River Tower addresses the difficult task of constructing sustainable high rises, which will be necessary in high-density low carbon cities.

Actions for Asia-Pacific urban planning

In this section, we briefly note the impact of history and decentralization on local governments before identifying six actions to be aware of in shifting to green urbanism. We consider good governance, knowledge sharing, stakeholder participation, targets and time frames, new policy frameworks, and the importance of culture, education and leadership. We end by drawing on the lessons of Japanese eco-cities to highlight the long-term policy implications of shifting to an eco-city paradigm.

Local governments in the Asia-Pacific: history and responsibilities

Many of the countries in the Asia-Pacific region were colonized by Europeans. This legacy of colonization has had (and continues to have) an impact on our societies, economies and environments. However, it is the political and administrative consequences of this colonial past that are particularly important for local governments trying to develop eco-cities. While most Asia-Pacific countries are gradually decentralizing their systems of government, and power is increasingly shifting from central authorities to local institutions, the reality is that local governments may only be responsible for some of the sectors that are part of urban life and that they remain largely dependent upon the national government for infrastructure and revenue (Roberts and Kanaley 2006b, 2006c; Nickson *et al.* 2008; Afiff 2010; Ballesteros 2010).

The definition and responsibilities of 'local government' vary widely in Asia-Pacific countries, depending on the structure of government in the nation (Nickson *et al.* 2008; UN-Habitat and UN ESCAP 2010, p. 32). This level of government may have responsibility for surrounding rural areas. Alternatively, sometimes a municipal government only has power over a city's original centre, and the surrounding areas, although now metropolitan, are overseen by smaller and diverse councils (Nickson *et al.* 2008, p. 63). This obviously presents challenges of coordination, especially for the management of the peri-urban area

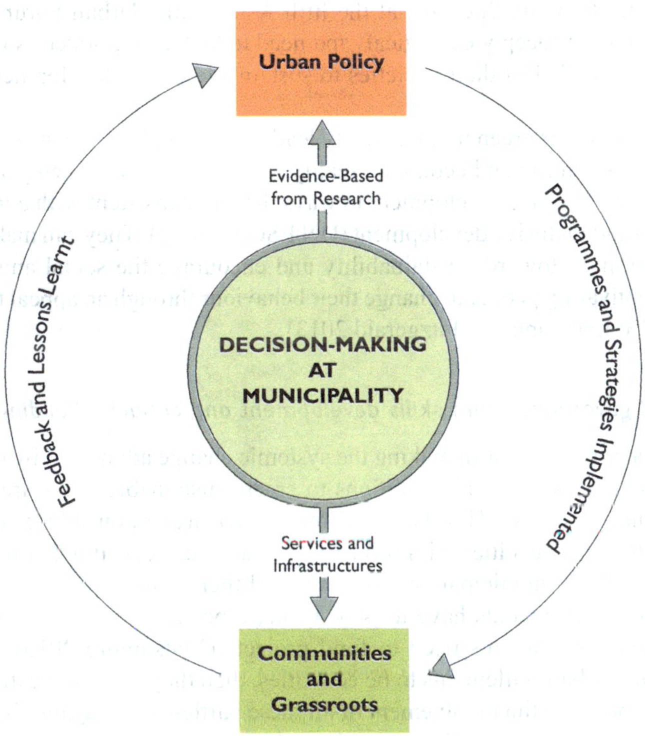

Figure 22.6 Linking the upstream with the downstream in a municipality's decision-making process.

Source: Diagram by Lehmann (2013) after UN-Habitat.

(Roberts and Kanaley 2006a, p. 442). The overlapping of responsibilities between different levels of government and different government organizations means that multi-level, cross-institutional and cross-sectoral approaches to urban policy are essential for eco-city development.

Six actions for urban sustainability

1 Culture and leadership is more important than technology

We have the technology to implement the principles of green urbanism and make the shift to green growth by retrofitting and developing our cities for a sustainable future. There are entry points and steps that can be taken by societies at all levels of development. However, we need the cultural shift to want to do so, which requires leadership that draws on shared values to create an inspiring vision of this long-term goal and motivates us to make the short-term changes.

Researchers of sustainable Asian cities have found that the ideas underpinning eco-cities – avoiding excessive materialism and caring for the natural world to benefit our communities – are consistent with the region's cultural and religious traditions that see 'sufficiency' and social harmony as central to a good and fulfilling life for all citizens (Anambutr 2010, pp. 110–112; Lim 2010; Low 2013). Asia-Pacific urban leaders should use those shared values to inspire and guide the present quest for a sustainable future. Cities in the Asia-Pacific have a rich heritage to draw on. Speakers at the fifth Asia-Pacific Urban Forum noted that the region's traditions are 'deeply ecological' and need to be 'reinvigorated' so that people can reconnect to them and shift their societies to sustainable urban development (UN ESCAP 2011b, p. 20).

In making the case for green urbanism, city leaders can combine the insights of traditional wisdom with the scientific and economic reality that the next five to ten years are a crucial opportunity to start shifting development in ways that are consistent with a long-term vision of green growth and inclusive development (UN ESCAP 2012). They can make the economic case for transforming towards sustainability and encourage the social and cultural transformation by motivating people to change their behaviour through an appeal to shared values (Crompton 2013; Lehmann and Fitzgerald 2013).

2 Ensure good governance with skills development and capacity building

Governance is a critical factor in making the systemic change advocated by green urbanism. Good governance is essential if aspirations to create sustainable cities are to be realized. Urban governance involves *all* levels of government that are responsible for cities and all the organizations that provide cities with infrastructure and services. It is not just the responsibility of city residents, municipal governments and their managers (Roberts and Kanaley 2006a). National governments have to play a role, especially as they so often control the money that local governments need to fund projects (Matsumoto 2013). If Asia-Pacific nations want their urban settlements to be eco-cities, then they need to create frameworks to allow for and coordinate the involvement of all these parties. Once again, like the principles of green urbanism, this is a holistic approach.

3 Measure achievements: set targets and time frames

The most effective policies are those that go from being plans on paper to being implemented, assessed, adjusted and re-implemented. The perfect policy solution does not exist and certainly cannot be summarized in three dot points on a PowerPoint slide! However, we do know that achievements can be measured and, that if targets and time frames are not set, then there is no accountability (Lim 2010). That is why setting targets and time frames and measuring what a policy achieves are essential ways to ensure that a policy delivers what it has been designed to do. Reporting on these outcomes is also a key part of good governance.

4 Create multi-level, cross-sectoral policy frameworks

Asia-Pacific cities need a policy framework that allows urban infrastructure to be thought about, designed, developed and implemented holistically (Ooi 2009, p. 189). Having such a framework is the essential first step for eco-city planning. Roberts and Kanaley explain that the framework will 'need to be complemented by capacity building and institutional strengthening of city governments and authorities' so that they are able to make maximum use of it (2006a, p. 438). They also warn that this will be a considerable challenge because all levels of government, as well as non-government organizations and lenders, are used to thinking about urban problems in their separate sectors, when what is needed is to see them as parts of a whole.

5 Share knowledge

One of the best ways to generate the enthusiasm and momentum necessary to achieve the holistic planning and good urban governance that will create sustainable cities is by sharing knowledge, information and experiences through 'networking across sectors, groups, cities and even countries' (Regional Office for Western Pacific 2000, p. 7; see also Roberts and Kanaley 2006b). The case studies discussed in Part II show that there are many inspiring examples that can be adapted by other cities to solve their problems. However, how are these discovered and disseminated? There are a number of networks, institutions and organizations that collect information and share resources which can be helpful in developing eco-city plans. ASEAN's Model Cities Programme is particularly inspiring; see the presentations made at their 2013 seminar (IGES 2013). City-to-city cooperation benefits all the participating cities.

The organization C40 recently counted more than 5,000 activities that municipalities decide on, ranging from streets, parking, outdoor lighting and public transport, to waste collection, water supply, building approvals and city planning. This is why city-to-city knowledge transfer and sharing is so helpful, to bring experiences from one city into the decision making of other cities.

There are plenty of good examples of cities' successes and pioneering ideas that could be replicated by other cities. For instance, Vaxjo in Sweden took local control of its energy supply, with a vision to be 100 per cent fossil fuel-free by 2015; Helsinki has built the greenest data centre in the world, where all excess heat is reused; Greater Lyon has established cross-departmental links in local governance to avoid a silo mentality, with close collaboration between the urban planning, water and waste departments; Utrecht in the Netherlands has built a new type of railway station that is a central mobility hub regenerating city life; Bristol has implemented a series of social-oriented initiatives and changed its city fleet to electric vehicles; and Freiburg in Germany has engaged its university's researchers to better support green city planning and policy formulation with evidence from research.

6 Involve stakeholders

Creating sustainable cities cannot be a top-down process; local people and community groups must all be involved. This requires real consultation from the earliest stages of policy development (Figure 22.6). Consulting a city's residents (and ensuring that all of the urban community's diverse groups are appropriately represented) is an important part of the initial ideas and information-gathering stage of policy development for eco-city planning. The amount of consultation depends on the scale of the project; a small group may meet a few times to create a shared garden in a neighbourhood, while a major change in the city's policy and infrastructure (such as creating a network of cycle paths or lanes dedicated to buses) would involve many stakeholders over a prolonged period of meetings, consultation and development (UN-Habitat 2012, pp. 24–27).

The need for 'widespread stakeholder consultation and participation' is a familiar exhortation in development and policy literature, but there is a risk that advisory community boards will be filled with the usual people and organizations. An alternative would be to use innovative approaches like citizens' juries or panels of deliberative democracy not only to *inform* the decision-making process but to *make* the decisions. How might this work? Deliberative democracy is based on three ideals of community engagement: influence, representation and deliberation. A statistically representative sample of the city's population meets to be informed of the pros and cons of possible decisions by experts. Discussing these possibilities in small, facilitated groups (of ten), they resolve the problem. This technique, which is best carried out under the auspices of a trusted organization such as the local government or a non-partisan NGO, can take whatever time is necessary or available to come up with solutions. Because the decision makers are local people, chosen to be statistically representative, they are most likely to come up with solutions that will benefit and be accepted by the majority of a city's residents (White 2013).

The future of urban planning in Asia: lessons from history

If we look at the history of Japan's ecologically aware cities, which begins in the 1960s, we can see that their paradigm has evolved in stages over the past 50 years in response to specific local factors, ideas and examples from elsewhere, national government directives and the country's commitment to international goals (Imura 2010, pp. 37, 39–40). Today, Japanese cities are adopting low carbon paradigms to guide their urban planning (Onishi 2011). Any city implementing new urban planning approaches will be anticipating and reacting to internal, local, regional, national and international situations and problems. Developing the policies to deal with these challenges will require people to collaborate in multi-stakeholder teams that work to find holistic solutions. There are thus two ultimate policy implications for shifting urban planning paradigms. First, a sound system of education must be maintained so that all citizens – whether they are government officials, highly trained urban planners or ordinary representatives of their local communities – can participate in creating and maintaining their city. This is not only the best way to ensure transparent governance and social cohesion; it is also what is most likely to cement all citizens' commitment to ecological sustainability and their ability to recognize the economic case for it. Second, it must be recognized that cities do not exist in isolation, and networking across regions and countries to share information and develop solutions at national and international levels is essential; that is also how cities will find the money to pay for their green urban development. While cities are in competition with each other in a global economy, they must also collaborate to sustain our shared planet.

Moving from the global perspective back to the local reality, having sufficient sources of reliable funding is essential for implementing the policies that will allow new paradigms to be established for the long term (Lindfield *et al.* 2012; Nickson *et al.* 2008).

Key findings

Environmental degradation jeopardizes people's quality of life and life itself; for instance, sea levels may rise very quickly and drown inhabitants. Holistic urban planning principles are essential in setting the framework for sustainable urban development. If Asia-Pacific cities want to develop in ways that are economically, environmentally and socially sustainable, then they must be proactive and 'guide urban development and the provision of infrastructure and services, rather than simply respond to problems and shortfalls' (Roberts and Kanaley 2006a, p. 438).

Our findings include the following:

- Cities in the Asia-Pacific are by far the densest cities globally.
- If cities are sufficiently compact and mixed use, then public transport can be efficient and the city can be walkable.
- It is social innovation rather than technology that poses the biggest challenge.
- Up to a certain level of income, there is a steady rise in resource consumption, until a tipping point of saturation is reached.
- Decision makers need to adopt a holistic and integrated multi-sectoral approach.
- A new platform is needed for city-to-city networking to share experiences and discover case studies of best practice.

The United Nations is now engaged in a review of the Sustainable Development Goals (SDG) and the development of Sustainable Development Goals for Urban Development, *SDG (for Cities),* as a new ten-year framework for organizing more effective urban environmental governance. No doubt there is now an excellent window of opportunity for cities to demonstrate their commitment to sustainable urbanization, to source funding for pilot projects and to deliver transformational change for their citizens.

Cities are complex organisms with many dimensions, and some of these can be very difficult to quantify. 'Big data' on cities is increasingly useful for making more accurate predictions about future requirements and to support decision making on urban development. Datasets on a city's performance are essential to measure good practice and for benchmarking.

What is required is a real dialogue between academia, civil society, the private sector, city authorities and national governments. Green urbanism challenges us to look at things and their interconnections in a new way. The lack of innovation and funding has been a barrier to progress. The next step will be to share best-practice solutions and scale up innovation. It is time for transformational change in the way we design and deliver urban development and infrastructure. It is time for a more concerted effort in synthesizing the science, economic theory, technologies, expertise and other research that has been done on sustainable urbanization. We have everything we need to ensure a better future by upscaling innovative demonstration projects. The real question is whether we have the will, commitment and leadership to do so and to implement better models of urbanization that work with nature, instead of against it (Rifkin 2011).

In shifting the urban planning paradigm and implementing the 15 principles of green urbanism to create a sustainable city, careful and sensitive transformation is required. There will be different technologies, resources and technical skills needed in the short, medium and long term that city planners need to be aware of in developing suitable policies. The *Low Carbon Green Growth Roadmap for Asia and the Pacific* (UN ESCAP 2012) makes it clear that the changes necessary for a sustainable low carbon urban future are fundamental and systemic changes to the infrastructure of societies and economies, and that making those changes will be beneficial for all citizens. In future, GDP will not be the sole indicator of a city's success. Instead, urban progress will be measured on the delivery of a better environment and the residents' feeling of well-being. We can expect that social inclusivity, wealth equality and housing affordability will become major priorities in the years to come.

Education and lifelong learning are playing an increasingly important role in the future of cities, as cities with better educated individuals have stronger economies, a higher quality of life and are more resilient. The skills for working in the twenty-first century include problem solving, critical thinking, digital literacy and the capacity for collaboration. To ensure that innovation happens, investment in research must be increased; for example, the EU target is to invest 3 per cent of a country's GDP in research by 2015. The fastest growing areas of employment today did not even exist ten years ago. It is likely that economies with high value-added production and services will be those that flourish in the future.

To improve the urbanization models of Asia-Pacific cities, it will be essential to:

- transition from fossil fuel-based energy sources to decentralized, precinct-wide renewable energies;
- transform all buildings and precincts into on-site mini-energy-generating plants;
- develop and upscale energy storage technologies and capacities (e.g. hydrogen);
- utilize ICT technology for the development of smart and bidirectional (peer-to-peer) energy-sharing grids;
- transform the transportation system to a multi-modal system that integrates walking, cycling, public mass transit and electric plug-in and fuel cell vehicles;
- provide better models and a diversity of public open spaces for active living.

Just as cities evolve, so do our ideas about how best to develop and maintain them for the shared benefit of all the living things (people, animals and plants) who live in them and are dependent upon them. All Asia-Pacific cities, especially those in developing countries, face myriad challenges due to the combined impacts of urbanization, globalization, decentralization and climate change. While it is likely that Asia-Pacific cities will become the powerhouse of global growth and provide test beds for resolving environmental problems caused by urbanization, the road remains bumpy.

Acknowledgement

This chapter draws on research done by the authors for a report on eco-cities commissioned by UN ESCAP in 2013.

References

Afiff, S.A. 2010, 'Challenges to implementing the eco-city concept in Indonesia's major cities', in L.F. Lye and G. Cheng (eds), *Towards a Liveable and Sustainable Urban Environment: Eco-cities in East Asia*, World Scientific Publishing, Singapore.

Anambutr, R. 2010, 'Moving toward eco-friendly city: perspective on Thailand', in L.F. Lye and G. Cheng (eds), *Towards a Liveable and Sustainable Urban Environment: Eco-cities in East Asia*, World Scientific Publishing, Singapore.

Asian Development Bank (ADB) 2012, 'Green urbanisation in Asia', in *Key Indicators for Asia and the Pacific 2012*, Asian Development Bank, Mandaluyong City, Philippines.

—— 2013, *Energy Outlook for Asia and the Pacific*, Asian Development Bank, Manila.

Ballesteros, M.M. 2010, 'Prospects on ecological development in Philippine cities', in L.F. Lye and G. Cheng (eds), *Towards a Liveable and Sustainable Urban Environment: Eco-cities in East Asia*, World Scientific Publishing, Singapore.

Booth, H. 2013, 'Asia's silver revolution', *East Asia Forum*, 11 April, www.eastasiaforum.org/2013/04/11asias-silver-revolution, viewed 13 July 2013.

Crompton, T. 2013, 'Behaviour change: a dangerous distraction', in R. Crocker and S. Lehmann (eds), *Motivating Change: Sustainable design and behaviour in the built environment*, Routledge, London.

Geels, F. 2010, 'The role of cities in technological transitions: analytical clarifications and historical examples', in H. Bulkeley, V. Castan Broto, M. Hodson and S. Marvin (eds), *Cities and Low Carbon Transitions*, Routledge, London.

Habitat for Humanity Nepal (HFHN) 2007a, 'A new initiative of bamboo housing by HFNI-Nepal', *Habitat for Humanity Nepal*, 7 August, http://hfhi-nepal.blogspot.com.au/2007/08/ new-initiative-of-bamboo-housing-by.html?q=bamboo, viewed 12 August 2013.

—— 2007b, 'Partnerships enable development of bamboo laminated housing materials', *Habitat for Humanity Nepal*, 25 December, http://hfhi-nepal.blogspot.com.au/2007/12/partnerships-enable-development-of.html?q=bamboo, viewed 12 August 2013.

ICLEI 2007, *Liveable Cities: The benefits of urban environmental planning*, Cities Alliance, UNEP and ICLEI.

—— 2011, 'Hangzhou, China: the world's largest bike sharing program', *EcoMobility*, October, www.iclei.org/fileadmin/PUBLICATIONS/Case_Stories/Case_stories_EcoMobility_Hangzhau_PDF_print.pdf, viewed 12 August 2013.

Institute for Global Environmental Strategies (IGES) 2013, *4th High Level Seminar on Environmentally Sustainable Cities*, 9 April, hls-esc.org/node/7, viewed 12 August 2013.

Imura, H. 2010, 'Eco-cities: re-examining concepts and approaches', in L.F. Lye and G. Cheng (eds), *Towards a Liveable and Sustainable Urban Environment: Eco-cities in East Asia*, World Scientific Publishing, Singapore.

Intergovernmental Panel on Climate Change (IPCC) 2013, *Fifth Assessment Report*, IPCC Geneva, www.ipcc.ch/report/ar5/, accessed 12 September 2013.

Jong, M. de, Wang, D. and Yu, C. 2013, 'Exploring the relevance of the eco-city concept in China: the case of Shenzhen Sino-Dutch low carbon city', *Journal of Urban Technology*, vol. 20, no. 1, pp. 95–113.

Joss, S., Tomozeiu, D. and Cowley, R. 2011, *Eco-cities: A global survey 2011*, University of Westminster International Eco-Cities Initiative, London.

Kaur, H. and Hitam, M. 2010, 'Sustainable living: an overview from the Malaysian perspective', in L.F. Lye and G. Cheng (eds), *Towards a Liveable and Sustainable Urban Environment: Eco-cities in East Asia*, World Scientific Publishing, Singapore.

Lehmann, S. 2010, *The Principles of Green Urbanism: Transforming the city for sustainability*, Earthscan, London.

—— 2012a, 'Urban metabolism and the zero-waste city: transforming cities through sustainable design and behavior change', in M. Lindfield and F. Steinberg (eds), *Green Cities*, Asian Development Bank, Mandaluyong City, Philippines.

—— 2012b, 'Sustainable building design and systems integration: combining energy efficiency with material efficiency', in S. Lehmann and R. Crocker (eds), *Designing for Zero Waste: Consumption, technologies and the built environment*, Routledge, London.

—— 2013, 'Low-to-no carbon city: lessons from Western urban projects for the rapid transformation of Shanghai', *Habitat International*, vol. 37, pp. 61–69.

Lehmann, S. and Fitzgerald, G.B. 2013, 'Wood in the city', in R. Crocker and S. Lehmann (eds), *Motivating Change: Sustainable design and behaviour in the built environment*, Routledge, London.

Lim, W.S.W. 2010, 'Let's get real: critical visions and sustainable eco-urbanism', in L.F. Lye and G. Cheng (eds), *Towards a Liveable and Sustainable Urban Environment: Eco-cities in East Asia*, World Scientific Publishing, Singapore.

Low, M. 2013, 'Eco-cities in Japan: past and future', *Journal of Urban Technology*, vol. 20, no. 1, pp. 7–22.

Lye, L.F. and Cheng, G. 2010, 'Some thoughts on the development of eco-cities in Asia', in L.F. Lye and G. Cheng (eds), *Towards a Liveable and Sustainable Urban Environment: Eco-cities in East Asia*, World Scientific Publishing, Singapore.

McKinsey Global Institute 2007, *Urban World: Cities and the rise of the consuming class*, McKinsey & Co., New York, www.mckinsey.com/insights/urbanisation/urban_world_cities_and_the_rise_of_the_consuming_class, accessed 1 April 2008.

Matsumoto, T. 2013, 'Green growth in fast-growing Asian cities?', Presentation at the 4th High Level Seminar on Environmentally Sustainable Cities, Hanoi, 21 March, www.hls-esc.org/documents/4hlsesc/2C%20-%20OECD.pdf?, viewed 21 June 2013.

Narang Suri, S. 2011, 'Making Indian cities liveable: the challenges of India's urban transformation', in C. Gossop and S. Nan (eds), *Liveable Cities: Urbanising world*, Routledge and ISOCARP, London.

Newman, P. and Matan, A. 2013, *Green Urbanism in Asia: The emerging green tigers*, World Scientific Press, Singapore.

Nickson, A., Devas, N., Brillantes, A.B., Cabo, W.L. and Celestino, A. 2008, 'Asia-Pacific', in United Cities and Local Governments (ed.), *Decentralization and Local Democracy in the World: First global report*, United Cities and Local Governments, Barcelona.

Onishi, T. 2011, 'Urban planning for a low carbon society', in T. Onishi and H. Kobayashi (eds), *Low Carbon Cities: The future of urban planning*, Gakugei Shuppan-Sha, Kyoto.

Ooi, G.L. 2009, 'Challenges of sustainability for Asian urbanisation', *Current Opinion in Environmental Sustainability*, vol. 1, pp. 187–191.

Opschoor, H. 2011, 'Local sustainable development and carbon neutrality in cities in developing and emerging countries', *International Journal of Sustainable Development and World Ecology*, vol. 18, no. 3, pp. 190–200.

Regional Office for Western Pacific, WHO 2000, *Issue Paper on: Urban environment management in Asia and the Pacific*, UN ESCAP, Bangkok.

Rifkin, J. 2011, *The Third Industrial Revolution: How lateral power is transforming energy, the economy and the world*, Palgrave Macmillan, New York.

Roberts, B. and Kanaley, T. 2006a, 'Lessons and strategies for sustainable urban futures', in B. Roberts and T. Kanaley (eds), *Urbanisation and Sustainability in Asia: Case studies of good practice*, Asian Development Bank and Cities Alliance, Manilla.

—— 2006b, 'Overview', in B. Roberts and T. Kanaley (eds), *Urbanisation and Sustainability in Asia: Case studies of good practice*, Asian Development Bank and Cities Alliance, Manilla.

—— 2006c, 'Urbanisation and sustainability in Asia', in B. Roberts and T. Kanaley (eds), *Urbanisation and Sustainability in Asia: Case studies of good practice*, Asian Development Bank and Cities Alliance, Manilla.

Suzuki, H., Dastur, A., Moffat, S., Yabuki, N. and Maruyama, H. 2010, *Eco^2 Cities: Ecological cities as economic cities*, World Bank, Washington, DC.

United Cities and Local Governments 2008, *Decentralization and Local Democracy in the World: First global report*, United Cities and Local Governments, Barcelona.

United Nations Economic and Social Commission for Asia and the Pacific (UN ESCAP), United Nations Development Program (UNDP) and Asian Development Bank (ADB) 2007, *Access to Basic Services for the Poor: The importance of good governance*, Asia-Pacific MDG Study Series.
United Nations Economic and Social Commission for Asia and the Pacific (UN ESCAP) 2008, *Building Community Resilience to Natural Disasters through Partnership: Sharing experience and expertise in the region*, United Nations, New York.
—— 2011a, *Are We Building Competitive and Liveable Cities? Guidelines for developing eco-efficient and socially inclusive infrastructure*, UN ESCAP, Bangkok.
—— 2011b, *Cities of Opportunity: Report of the fifth Asia-Pacific urban forum*, UN ESCAP, Bangkok.
—— 2012, *Low Carbon Green Growth Roadmap for Asia and the Pacific*, UN ESCAP, Bangkok.
—— 2013, *Building Resilience to Natural Disasters and Major Economic Crises*, UN ESCAP, Bangkok.
—— forthcoming, *Statistical Yearbook for Asia and the Pacific 2013*, UN ESCAP, Bangkok.
United Nations Environment Programme (UNEP) 2011, *Decoupling Natural Resource Use and Environmental Impacts from Economic Growth*, UNEP, Paris.
—— 2013, *City-level Decoupling Urban Resource Flows and Governance of Infrastructure Transitions*, UNEP, Paris.
United Nations Human Settlements Programme (UN-Habitat) and United Nations Economic and Social Commission for Asia and the Pacific (UN ESCAP) 2010, *The State of Asian Cities 2010/11*, UN HABITAT, Fukuoka, Japan.
United Nations Human Settlements Programme (UN-Habitat) 2012, *Sustainable Urbanisation in Asia: A sourcebook for local governments*, UN-Habitat, Nairobi.
—— 2013a, *Time to Think Urban*, UN-Habitat, Nairobi.
—— 2013b, *Urban Planning for City Leaders*, UN-Habitat, Nairobi.
White, S. 2013, 'Sustainable cities: from possible futures to preferred futures', ZWSA Research Centre for Sustainable Design and Behaviour Distinguished Lecture on Urban Futures, 20 September, Adelaide, South Australia.
World Metereological Organisation (WMO) 2013, *The Global Climate 2001–2010, a Decade of Climate Extremes*, World Meteorological Organisation, Geneva, http://library.wmo.int/pmb_ged/wmo_1119_en.pdf.

Glossary of urban terms

Ageing population As more people live longer, the population's average age increases and the proportion of people earning an income decreases. A child born today is likely to live for more than 100 years. In 2030, there will be more people over 60 than under 10 years of age. In 2014, over 20 per cent of the global population is older than 65; by 2060, every third person will have reached that age. The substantial effects of an ageing population on urban environments, health care, housing, infrastructure and social services is one of the most significant global demographic challenges and opportunities we face. Good urban design, community initiatives and public services can help meet the needs of all citizens.

Behavioural change Behavioural change is an ongoing process of changing values and modifying human behaviour. Changing human behaviour requires awareness and incentives. Behavioural standards respond to social norms and needs; in an urban context, targeted behavioural shifts can lead to important changes in interaction and design. The key to large-scale urban improvement lies in the majority of the population adopting and maintaining behaviour that produces small but incremental change for the common good. This should include more responsible consumption and waste practices, or greater conviviality and sharing (see also R. Crocker and S. Lehmann, (eds), *Motivating Change*, 2013).

Biophilic cities Biophilic cities are cities that contain abundant nature; they are cities that care about, seek to protect, restore and grow natural environments, and strive to foster deep connections and meaningful daily contact with the natural world, to ensure a healthier life in cities. Singapore, Brisbane and Portland have implemented biophilic initiatives, seeking to integrate natural features and qualities in urban development. Community gardens, parks and urban forests are essential elements of biophilic cities but Tim Beatley explains that biophilic city design is more than this and he calls for cities to embrace 'wild nature' at all scales (see Chapter 3).

Bottom-up engagement Bottom-up engagement is an approach to planning that places the citizen at the centre of change. It is the opposite of 'top-down' control by authorities. In an urban context, it has two key, complementary elements. First, it encourages social and cooperative models of city organization, with the involvement of many individuals. Second, it aligns with government officials', academics' and the professional sector's growing interest in digital, open-sourced data and models as resources for understanding urban interactions.

Brownfield site A brownfield site is land or premises that have previously been used or developed and are not currently fully in use. It may be partially occupied or utilized, or it may be vacant, derelict or contaminated. Docklands, heavy industry plants or military areas that are no longer in use are typical brownfields. The opposite of brownfields are greenfields, previously undeveloped open space.

Cities as idea generators Cities have always generated innovation and wealth. In the knowledge economy, cities are also successful if they attract people, are information hubs and enable the public and private sectors to cooperate on development projects. When people gather together and interact, innovation occurs more easily and new ideas are generated more frequently. Thus, urban theorists and economists have identified cities as engines of progress and idea generators. As a nexus for the exchange of ideas, a city can function as a connecting hub for experts and non-experts alike to put forward urban theories, test out new methods of urban participation and share stories of transformation and experimentation.

City-as-organism City-as-organism draws on the metaphor of a complex living organism formed by multiple, interrelated components in natural patterns to explain urban systems. (Similarly, the term 'urban metabolism' refers to understanding a city as a living organism, where resources are distributed through systems and networks.) Like an organism, many of the elements that form a city are not apparent to the naked eye, yet they are essential to its functioning. Both the visible and invisible aspects of a city are crucial to its development.

City manifesto A city manifesto is a public declaration of policy, aims, intentions or steps, similar to a mission statement. 'Manifesto' was adopted into English during the seventeenth century and derives from the Italian *manifestare*, 'to make public'.

Closing the loop Closing the loop is an expression used to describe a process, from start to finish, which leaves no loose ends as it feeds all outputs back in. In the urban context, closing the loop is increasingly used to describe the effectiveness of circular material flows that trigger reuse, or digital-to-physical interactions in cities. Successful uses of digital technologies in cities, through sensor devices or social media, for instance, are of no use if they do not have an impact on the physical world. If we amass large quantities of urban data through sensors but fail to implement a solution to the problem that the data reveals, we will not have closed the loop. Closed-loop recycling refers to a system of resource recovery (e.g. industrial symbiosis) in which the waste or by-product of one process is a valuable resource in making another product (see also **zero waste** and **zero-waste city**).

Community garden A community garden is a piece of land collectively tended by a group of people. Community gardens are shared green public spaces that offer outdoor activity in urban areas; they can provide a sense of community and a certain measure of independence for city dwellers who are able to grow some of their own food. Similar to allotment gardens (a movement that developed in Leipzig, Germany in 1825), community gardens produce a feeling of ownership through care for a collectively owned piece of land. They can also help neighbourhood improvement and serve as important community meeting places, especially for an ageing population. Community gardens may help alleviate one effect of climate change, which is expected to cause a global decline in agricultural output.

Connectivity Connectivity refers to the ability to connect people in better, more efficient ways, for instance, through a strong public space network. Cities can provide the basis for connective strategies and interconnected systems, such as increased walkability or easy access to public transport. In urban environments, citizens can connect through the proliferation of widespread technology, and, just as importantly, through personal, narrative connections that reflect micro-histories of place. If urban form is more permeable, it permits better connectivity; for instance, walkable urbanism is a strategy in opposition to urban sprawl that advocates strong connectivity through walkable streetscapes.

Eco-city An eco-city is a city designed with consideration for its environmental impact. It enhances the well-being of its citizens and of society through integrated urban planning and management that harnesses all the benefits of ecological systems and renewable energies – aiming for zero emissions and zero waste. An eco-city or eco-district protects and nurtures these assets for future generations.

Ecological overshoot When a population consumes more resources than nature can regenerate and creates more wastes than nature can recycle, it is in a condition of ecological overshoot.

Green urbanism Green urbanism is a conceptual model for zero-emission and zero-waste urban design that arose in the 1990s, promoting compact energy-efficient urban development and seeking to transform and re-engineer existing city districts and regenerate post-industrial city centres. Green urbanism promotes the development of socially and environmentally sustainable city districts. Tim Beatley and Steffen Lehmann are major contributors to the concept.

Infill Infill is the act of building on a vacant lot (often an underutilized site) within an otherwise developed neighbourhood. Urban infill development occurs in the established areas of a city and can be a densification strategy for unoccupied gaps of otherwise built-up areas. Urban infill is a medium-density growth management strategy that uses what is already there to its advantage; it can lead to thoughtful, innovative housing solutions in urban settings.

Low carbon city Low carbon cities are cities that have radically reduced their greenhouse gas emissions. Low carbon cities are based on sustainable urban development principles that consider environmental, economic and social impacts and have the ability to maintain themselves without using up crippling levels of resources and creating a future debt that is not offset by equal levels of future benefits. Low carbon cities are planned to reduce greenhouse gas emissions (by using locally available renewable energy sources) and to optimize the use of the built environment, transportation systems, energy, water and land, material and waste flows, while aiming to minimize negative impacts on the natural environment.

Mega-cities Mega-cities are metropolitan areas with a total population exceeding ten million people. Mega-cities experience rapid growth, new forms of spatial density of population, and formal and informal economics, as well as crime, poverty and high levels of social fragmentation. Recent rapid urbanization in Asia has significantly increased the number of mega-cities.

Microarchitecture Microarchitecture is the practice of using small-scale design solutions for frequently overlooked spaces, or to adapt small urban spaces, thereby changing the behaviour of city dwellers and activating underused areas. As an intelligent densification and infill strategy, lightweight, compact and prefabricated microarchitecture may be inserted within the footprint of an existing building, or sit on top of a structure.

Mixed-use neighbourhoods Mixed use (MU) describes a development that is a combination of diverse uses, such as residential, retail, commercial and/or office uses, clustered together; either layered vertically in a single building, or in a horizontal arrangement in adjacent buildings.

Orientation Orientation refers to a building's on-site placement in regard to the direction of prevailing winds and solar position, and its resulting sun exposure and potential to harness solar radiation and passive solar heat gains.

Participatory urbanism Participatory urbanism is the condition, often facilitated by information and communications technology, when the entire community is involved and citizens are empowered to collect and contribute data, ideas and proposals to city decision makers. The idea is grounded in the fact that community members are experts on their particular urban situations, for example, on the issues in their community or neighbourhood; they contain within them the knowledge and solutions for any number of issues. Participatory urbanism therefore allows professionals to identify real needs, tap into local knowledge and implement solutions from the bottom up, rather than implementing change from the top down.

Prefabrication Unlike on-site construction, prefabrication involves the off-site creation of building components that are transported to the site for assembly. Prefabrication techniques are typically used for developments with repeated units (e.g. hotels or apartments). Mass customization in manufacturing allows flexible, computer-aided systems to produce custom-designed outputs. These systems combine the low unit costs of mass production and prefabrication processes with the flexibility of individual customization. The mass production of individually customized goods and products meets individual customers' needs because product differentiation is postponed until late in the supply network. The combination of prefabrication and mass customization should allow for new buildings that will not look monotonously similar and will be significantly cheaper to construct.

Public space A public space is a social space open and accessible to everyone; it is a plot of land that is accessible, 24 hours and seven days a week, to all people regardless of their gender, ethnicity, age or income. Often used to refer to social spaces for gathering or recreation, such as plazas, town squares or parks, 'public space' also refers to smaller spaces, streets, sidewalks and community gardens that are open to the public and not privately owned.

Regenerative city The regenerative city operates strategies for environmentally enhancing and repairing restorative relationships between cities and the ecosystems from which they draw their resources. Herbert Girardet (2012) suggested that regenerative cities, which 'engage in renewable energy development and in restoring the damaged ecosystems on whose health we ultimately depend', are necessary in our urbanizing world. This call for far-reaching strategic choices and long-term planning to enhance and restore the ecosystem has not yet been heeded; while a wide range of technical and management solutions towards this goal are available, their implementation has been slow and limited.

Resilience Resilience refers to a city's ability to cope with and recover quickly from hardship, and the ability to bounce back. Although a city may show resilience in relation to a number of potential challenges, today's resilient city is typically one that is prepared and equipped to contend with and mitigate the multiple effects of climate change. It is a term to describe a 'future-proofed' city that is more robust against energy blackouts, heatwaves, urban flooding and other disasters than other cities.

Strategy A Strategy is a carefully devised plan of action to achieve a goal, and the art of developing and carrying out such a plan. Through strategic action, other parties may be inspired and motivated to act. How to redesign existing urban systems, how to design new products for reduced consumption and how to design new scenarios for a good quality of life are major questions for planners or researchers that require far-reaching strategies.

Suburban retrofitting Suburban retrofitting (also known as 'sprawl repair') is the act of filling in, redesigning and adding to the fabric of existing suburbs to make them denser, more urban and sustainable. Examples of suburban retrofits include: the renovation of abandoned malls or other large commercial buildings to create pedestrian shopping districts or public amenities; inserting mixed-use programmes in mono-functional suburbs; dividing blocks to improve connectivity and walkability; and wrapping big-box buildings and parking lots in long, shallow structures to create space for small shops that will provide continuous streetscapes. 'Greyfield' developments (between brownfields and greenfield development) refer to ageing strips of development that once served as vital commercial corridors during the suburban exodus, but that have fallen into decline, are economically obsolescent, outdated, or underused real estate assets. Suburban retrofitting is a strategy to improve and repair these areas.

Transit-oriented development Transit-oriented development (TOD) is a form of development that emphasizes forms of transportation other than the privately owned automobile – such as walking, cycling and mass transit – as part of its urban design. Transit-oriented neighbourhood development locates mixed-use activity centres, composed of office space, housing and retail, as multi-storey compact mixed-use development, around a transit stop. These activity centres offer a variety of housing options, such as apartments and townhouses, often above ground-floor commercial uses.

Ubiquitous computing Futurists predict that, in 20 years' time, smart meters and computers will be everywhere, in buildings and in public space – but unseen. Ubiquitous computing already exists, with computers in cars, phones and fridges. However, these devices need us to input information. In future, computers will surround us but we will be unaware of them and they will perform autonomously. 3D printers will be widely available and allow fabrication in the home, which will upset traditional manufacturing models.

Urban data Today, we are capturing more data and evidence about cities and their users than ever before. Urban data is the information that cities produce, which is collected, measured, analysed and turned into facts, figures and visualizations (e.g. using predictive analytical models and demand forecasting to anticipate or identify risk areas). According to Eric Schmidt, Executive Chairman and former CEO of Google, the amount of data collected between the dawn of humanity and 2003 is equivalent to the amount we now produce every two days. This phenomenon of massively proliferating data, referred to increasingly as 'big data', comes with the task of filtering and making this new wealth of information accessible. Urban data has enhanced our ability to make smarter design decisions, and to create 'smart' cities.

Urban design Urban design is concerned with the three-dimensional form of the city, its public space networks, masses, open spaces and distribution of activities. It involves various disciplines, including planning, architecture and engineering. Good urban design considers the full complexity of the city and its buildings in relation to context, history and geography. The resulting urban form encompasses the pattern and density of land use and transportation modes within cities and towns.

Urban heat island effect The urban heat island (UHI) effect is found in metropolitan areas that are significantly warmer than their surrounding areas. It creates an urban microclimate in a confined urban area. It is caused by human activity in a city that generates waste heat from energy-using processes (for instance, the air-conditioning units in Hong Kong) and results in a dome of elevated temperatures over an urban area (such as a street, sidewalk, parking lot or building) that traps heat. The differences in temperature are usually greater at night and during the winter months. It is primarily caused by urban development, which alters the original land surfaces and their potential for absorbing heat, and by waste heat from air-conditioning units, generated by the use of energy. As the population of an area grows, the land is altered, thus increasing average temperatures in the region. It is predicted that heatwaves will kill more residents each year due to global warming, urban heat stress and rising electricity prices (where, for instance, older people cannot afford to operate their air-conditioning).

Urban intervention Urban intervention temporarily disrupts the habitual functioning of a public space or city (also referred to as 'activist art'). They are often temporary and typically performed by artists, architecture students or activists. Ranging from street art and design challenges to ephemeral gatherings, Facebook flash mobs, pop-up bars and crowd sourcing, this type of informal activity achieves a temporary social cohesiveness and surprise factor that urban dwellers crave.

Urban sprawl Urban sprawl is a low-density pattern of residential urban growth outside the city centre; the spreading of a city and its suburbs, pushing the city boundaries outward. Sprawling neighbourhoods are usually dominated by single-family homes situated some distance from urban centres. They are highly car-dependent, have low population densities, involve single-use zoning with rigid separation between uses, have low-density land use and lack public transportation and amenity options. Sprawl is often characterized by placeless commercial strip development along main streets and otherwise large expanses of low-density or single-use development, where the major form of transportation is the private car. Sprawl impacts negatively upon land and water quantity and quality, and is often linked to a decline in social capital and isolation. For instance, the lack of land-use diversity (typical for suburbs) results in and increases car dependency. A growth boundary is usually necessary to curb sprawl and scattered development that leaves large tracts of undeveloped land between developments.

Urbanization Urbanization is the increase in the proportion of a population living in urban centres (cities and towns), caused by a net movement of people from rural to urban areas, accompanied by manufacturing and commerce replacing agriculture, and associated patterns of land use. Urbanization is driven by economic growth and a concentration of new investment in urban areas. Since the Industrial Revolution, the process of urbanization has become ever more resource-intensive, and contributes significantly to climate change, loss of soil carbon, to natural fertility of farmland, and the loss of biodiversity all over the world.

Waste management infrastructure Every city needs waste management infrastructure to control, manage, collect and dispose of its waste in sustainable and responsible ways. Since the very first human settlements, proper management of waste has been a differentiating quality of human behaviour and the complexity of waste systems is evidence of urban development and sophistication. Historically, the intelligent integration of sewage, water and sanitation systems helped cities overcome problems such as diseases and potable water contamination. Using sewage to produce biogas, an urban system would be able to simultaneously treat wastewater, generate green energy and provide clean water to the population.

Zero waste Cities are the main generators of human waste, and usually have unsustainable disposal habits. 'Waste' is anything not wanted by its current owner and deemed 'not of use to anyone'. Zero-waste concepts consider the entire life cycle of cities, buildings and products, and advocate closed-loop industrial and societal systems and construction processes. Zero waste means 100 per cent of waste is diverted from landfill, and organic waste is composted and returned to the soil. It also includes the scrutiny and transformation of manufacturing methods or processes for construction materials and their supply chains, asking, for example, if building materials are fully recyclable. As a consequence of embedding zero-waste concepts, better material flow management leads to optimized efficiency in the use of raw materials and construction systems, preventing the loss of resources, minimizing inputs and reducing the use of non-recyclable materials. Zero waste includes a system to recycle all manner of chemicals in any way possible, seeing waste as a mis-allocated resource.

Zero-waste city The generation of waste has a direct relationship with the consumption of resources. Cities are over-consuming by nature and, directly or indirectly, deplete global finite natural resources at a rapid rate. The 'zero-waste city' concept is of a future city that has reduced consumption, minimized waste generation and has recycling levels that are getting close to 100 per cent resource recovery of all waste. It has optimized all urban material flows and has green supply chains and innovative disposal strategies. The zero-waste city is an iteration of the 'eco-city'.

Annotated select bibliography

The following is a list of sustainability 'classics', selected by Steffen Lehmann and annotated by Katharine Thornton, which are recommended for further reading.

Banham, R. 1969, *The Architecture of the Well-Tempered Environment*, University of Chicago Press, Chicago, IL.

Banham's account of the evolution of systems to illuminate, heat and cool buildings challenges architectural historians to think more about technology.

Braungart, M. and McDonough, W. 2002, *Cradle to Cradle: Remaking the way we make things*, North Point Press, New York.

Braungart and McDonough see good design and a circular system of production and consumption as the alternative to recycling 'waste'. They define 'good' design as aesthetically satisfying, environmentally sustainable and pro-growth.

Brown, L. 2009, *Plan B 4.0: Mobilizing to save civilization*, W.W. Norton, New York.

Responding to the global challenge of climate change, Brown demonstrates that reducing carbon dioxide emissions is technologically and financially possible but underestimates the political capital and social shift required.

Brundtland, G.H./World Commission on Environment and Development 1987, *The Brundtland Report: Our common future*, Report of the World Commission on Environment and Development, Oslo, Norway; United Nations, Oxford University Press.

In outlining the need for environmental protection, economic growth and social equity, this report provided what has become *the* definition of sustainable development, i.e. development 'that meets the needs of the present without compromising the ability of future generations to meet their own needs'.

Buckminster Fuller, R. [1969] 2008 (reprint), *Operating Manual for Spaceship Earth*, Lars Mueller Verlag, Baden.

This book encapsulates Fuller's ideas about the need to understand things by viewing them on a grand (global) scale; he cites human potential, specialization, mechanization, solar energy and sharing finite resources as essential for a viable future.

CABE and BioRegional 2008, *What Makes an Eco-Town?*, CABE, the UK government's Commission for Architecture and the Built Environment, London.

Written in the context of the then-UK government's aim to create ten eco-towns by 2020, this concise report contains ways to plan, create and measure the environmental sustainability of new developments, with an emphasis on reducing CO_2 emissions and ecological footprints.

Callenbach, E. 1975, *Ecotopia: The notebooks and reports of William Weston*, Bantam Books, New York.

A fictionalization of the many strands of counter-cultural thought and some of the science that, in combination, constituted environmental awareness in the United States in the 1970s, explored via the main character, an investigative journalist who is visiting 'Ecotopia'.

See also Callenbach, E. 1981, *Ecotopia Emerging*; a prequel that describes how the nation of Ecotopia came into existence.

Cooper, T. (ed.) 2010, *Longer Lasting Products: Alternatives to the throwaway society*, Gower Publications, Burlington, VT.

A collection of essays on consumer goods that offers another approach to their life cycle by considering design, manufacture, marketing, maintenance, attachment and recycling, with the aim of contributing to a change in consumption processes.

Daniels, K. and Hindrichs, D. (eds) 2007, *Plus Minus 20/40 Latitude: Sustainable building design in tropical and subtropical regions*, Edition Axel Menges, Stuttgart/London.

This book outlines traditional building techniques in the tropics and how these can combine with contemporary technological solutions to develop buildings that are sustainable in the region's climate.

Droege, P. (ed.) 2008, *Urban Energy Transition: From fossil fuels to renewable power*, Elsevier, Oxford.

A series of essays explaining the principles, policies, technologies, plans and practices that will shift cities from relying on fossil fuel-derived energy to adopting and developing renewable sources.

Gauzin-Mueller, D. 2002, *Sustainable Architecture and Urbanism: Concepts, technologies, examples*, Birkhäuser, Basel.

This book uses European case studies to illustrate the theory and technology of sustainable architecture and cities.

Girardet, H. 2008, *Cities, People, Planet: Urban development and climate change* (2nd edn), John Wiley & Sons, London.

This book offers an optimistic argument for responses to climate change at the urban scale.

Haas, T. (ed.) 2012, *Sustainable Urbanism and Beyond: Rethinking cities for the future*, Rizzoli, New York.

A collection of essays from a range of disciplines, grouped thematically, exploring the challenges facing cities.

Hawken, P., Lovins, A. and Lovins, H. 1999, *Natural Capitalism: Creating the next industrial revolution*, Little, Brown & Company, New York.

This volume makes the case for changing the traditional, 'industrial capitalist' view of natural resources to take all of the services they provide into account and ensure a fuller recognition and more efficient use of them.

Head, P. *et al.* 2009, *Entering the Ecological Age*, report by Arup, based on Head's talk for *The Brunel Lecture Series*, London (available online at: www.arup.com).

Head draws on Janine Benysus' concept of biomimicry to envision solutions to high-consuming cities found by imitating natural systems.

Hegger, M., Fuchs, M., Stark, T. and Zeumer, M. 2008, *Energy Manual: Sustainable architecture*, Edition Detail, Birkhäuser, Basel/Boston/Berlin.

Part of a series, this book reviews European legislation for energy efficiency, describes architectural and planning solutions for achieving a sustainable built environment, and presents relevant case studies.

Jackson, T. (ed.) 2006, *The Earthscan Reader in Sustainable Consumption*, Earthscan, London.

Economists, sociologists and psychologists explore how consumption may become sustainable.

Jenks, M. and Burgess, R. (eds) 2000, *Compact Cities: Sustainable urban forms for developing countries*, Spon Press, London.

Case studies from Asia, Africa and Latin America are used to evaluate the pros and cons of compact urban development.

Kats, G., Alevantis, L., Berman, A., Mills, E. and Perlman, J. 2003, *The Costs and Financial Benefits of Green Buildings*, report to California's sustainable building task force, October, Sacramento, CA.

This report reviews the economic, ecological and health evidence in favour of sustainably built government buildings.

Koenigsberger, O. 1980, *Manual of Tropical Housing and Building: Climatic design*, Longman, London.

This manual describes the principles for climate-responsive architecture that may now be seen as antecedents of green architecture.

Lehmann, S. 2010, *The Principles of Green Urbanism: Transforming the city for sustainability*, Earthscan, London.

This book promotes a holistic view of architecture and urban design to encourage ecological sustainability.

Lehmann, S. and Crocker, R. (eds) 2012, *Designing for Zero Waste: Consumption, technologies and the built environment*, Earthscan Series on Sustainable Design, Routledge, Oxford.

A collection of chapters addressing the elimination of waste through design and behavioural change.

MacKay, D. 2008, *Sustainable Energy without the Hot Air*, UIT Cambridge (available online at: www.withouthotair.com).

This volume demonstrates how the United Kingdom's energy needs could be met using renewable sources.

McHarg, I. 1969, *Design with Nature*, Natural History Press/Falcon Press, Philadelphia, PA.

A foundational work of landscape architecture and planning, McHarg's 'layering' concept provides the foundation for GIS mapping.

Meadows, D., Meadows, D., Randers, J. and Behrens, W. 1972, *Limits to Growth*, Report to the United Nations, Potomac Books, Washington, DC.

This is a commissioned report of a computer simulation to examine the consequences of continued economic and population growth in a world of finite resources.

See also Meadows, D., Meadows, D. and Randers, J. 2004, *Limits to Growth: The 30-year update*, Earthscan, London.

Newman, P. and Kenworthy, J. 1989, *Cities and Automobile Dependence: An international source book*, Gower, Aldershot.

A collection of statistics on urban transportation demonstrating the links between urban form and energy use in cities around the world.

Register, R. 1987, *Eco-City Berkeley: Building cities for a healthy future*, North Atlantic Books, Boston, MA.

A pioneering work that presents a new approach to urban planning and architectural design.

Santamouris, M. *et al.* (eds) 2001, *Energy and Climate in the Urban Built Environment*, James & James, London.

This book examines problems in urban climates and how these can be solved through bioclimatic buildings and design.

Scheer, H. 2006, *Energy Autonomy: The economic, social and technological case for renewable energy*, Earthscan, London.

Scheer argues that centralized fossil fuel-derived energy systems must be replaced by decentralized supplies powered by renewable sources.

Sinclair, C./Architecture for Humanity (ed.) 2006, *Design Like You Give a Damn: Architectural responses to humanitarian crises*, Thames & Hudson, London (available online at www.architectureforhumanity.org).

This volume offers an explanation and examples of 'socially conscious design' that responds to human needs for shelter, clean water and sanitation.

Stern, Sir N. 2007, *The Stern Review: The economics of climate change*, Cambridge University Press, Cambridge (available online at www.sternreview.org.uk).

Commissioned by the British government, this is not the first study of global warming's impact on economic and financial systems, but it is one of the most well known. Stern argues that climate change demonstrates the failure of unfettered capitalism but that both problems may be addressed through government intervention.

Strasser, S. 2000, *Waste and Want: A social history of trash*, Henry Holt & Co., New York.
Strasser examines the changes in attitudes towards waste in the United States over 200 years, linking these to practices of waste disposal and management in the context of consumer capitalism.

Szokolay, S. 2004, *Introduction to Architectural Science: The basis of sustainable design*, Architectural Press, Elsevier Science, Oxford/London.
Szokolay introduces methods for designing buildings with comfortable levels of heat, light and sound consistent with the principles of sustainable development.

Von Weizaecker, E., Lovins, A. and Lovins, H. 1997, *Factor Four: Doubling wealth, halving resource use*, Earthscan, London.
A report to the Club of Rome that proffers the hope of increasing profits and employment by increasing the productivity of resources (i.e. decreasing waste and misuse).

Wackernagel, M. and Rees, W. 1996, *Our Ecological Footprint*, New Society Press, Vancouver, BC.
Based in ecological economics, the ecological footprint model reveals the resources needed and waste generated by groups of people (households, communities, countries) for comparative purposes.

Wheeler, S. and Beatley, T. (eds) 2009, *The Sustainable Urban Development Reader*, 2nd edn, Routledge, Oxford.
Wheeler and Beatley place the concept of sustainable urban development in its historical context before reviewing current theory and practice, tools, applications and contemporary case studies.

Wong, T-C. and Yuen, B. (eds) 2011, *Eco-City Planning: Policies, practice and design*, Springer, London.
The authors discuss aspects of environmentally sustainable urban planning through examples from Asia, Africa and Europe.

Yeang, K. 2009, *Eco-Masterplanning*, John Wiley & Sons, London.
Known for his 'bioclimatic' tall buildings, in this book Yeang describes ecologically sustainable master plans for urban precincts in countries around the world.

Index

Page numbers in **bold** indicate tables and in *italic* indicate figures.

1 Bligh Street, Sydney, Australia 215–222, *216*, *217*, *218*, *220*, *222*
3-D printers 6; waste reprocessing 158–160, *158*, *166*

Abu Dhabi: Masdar Eco-city 22, *41*, 42, 71
adaptive capacity to heat 366–368, **366**, **368**
Adelaide, Australia **28**, 44, *45*, 71–72, 377
aerial thermal imagery 375, 378
ageing populations 6, 8, 39, 400–401
agriculture, urban 68–69, 194, **404**
Agropolis 61, *61*
Ahmedabad, India 407
air-conditioning systems: 1 Bligh Street, Sydney 221; compared to greenery 350; hybrid air-conditioning judgement flow *305*, 306; Japan 293; reducing need for 100; and urban heat island effect 50, 266, 276–277, 367, 399
air pollution 277; cars 79; China 36–38; and green public spaces 231, 320–321; India 242, *242*; *see also* greenhouse gas (GHG) emissions
air-to-water heat exchangers 177
Ai Wei Wei 38
Akbari, H. 320
Albania 288
albedo 261, 266, 271, 285, 311, 340; high-albedo surfaces 101–102, *102*, 108, 266, 274, 278, 279, 283–285, *284*
AMeDAS *see* Automated Meteorological Data Acquisition System (AMeDAS)
American Indians 78
Amsterdam Smart City project, Netherlands 22
Apparent Temperature (AT) index 365
aquatic biodiversity 88
Arnhem, Netherlands 341–344, *342*
ASEAN 397–398, 415
Asia-Pacific region 4–5, 33, 394–418; climate change 399; demographic trends 400–401; eco-cities, sustainable cities and low carbon cities 397–398; examples of low carbon city projects 410–412; green urbanism 397–398, 401–410, *402*, **403–405**, *405*, *406*, 414–416; local governments 413–414, *413*; Low Carbon Society (LCS) 245; resource efficiency 408–410, *409*; urban heat island effect 399; urbanization 192, 395–397, 399–400, 401; *see also individual countries*
asphalt 285, 380; albedo 261, 271, 274; thermal properties 262
asphalt heat collectors **179**, 180
Athens, Greece 201, 288, 309
atriums, full height 219, *220*, 221
Austin, Texas, USA 87–88
Australia 4, 19, **28**, 42–46, *45*; 1 Bligh Street, Sydney 215–222, *216*, *217*, *218*, *220*, *222*; adaptive capacity to heat 366; Barangaroo waterfront development, Sydney 44, 191, 201–204, *204*; bushcare projects 91; focus on community values 196; green public spaces 24, *25*; heat stress 43, 362–366, *364*, **365**, 378; heatwaves 43, 199, 362–363, 378; population growth 43, *44*; private transport trends 116, 117, 119–120, **124**, 127–128; public transport trends 117–119, 121, 122, **124–125**, 127; regenerative urban development, Adelaide 71–72; renewable energy 46, 72, 201; stormwater run-off 369; transport deaths **125**; urban density 44–46, *45*, 115, **124**, 380–381; urban heat island effect, Melbourne *202*, 277; urban sprawl 42–43, 192; *see also* multi-scale analysis of urban heat island in Sydney, Australia; spatial impact study of Australian households
Australian Water Accounts 132
Austria 112
Automated Meteorological Data Acquisition System (AMeDAS) 297, 298, 301

bamboo 407–408
Bangladesh 78

Banham, Reyner 13
Barangaroo waterfront development, Sydney, Australia 44, 191, 201–204, *204*
Barcelona, Spain *3*, 381
barrages 76
bats 87–88
battery technology 194
Beatley, Tim 14, 85–95
Beck, Ulrich 14
Beddington Zero Energy Development (BedZED), London 112
Beer's Law 352–353, 354
behaviour change 14–15, 195–196, 410
Beijing, China 35, 128
Bennett, Edward H. 153
Berlin, Germany **28**
Bermuda 102, *102*
Bhopal, India 245, *246*
bicycles 1, *11*, 34, 195, 410
Big Dig freeway project, Boston, USA 228–229, 230
biodiversity 20, 82, 193, **404**; aquatic 88; biophilic cities 85, 86–95, **92**; green roofs 87, 201
biogas 30, 175, 177
biophilia 85–86, 105
biophilic cities 85, 86–95, **92**
bio-retention systems 372
biotechnology 162
blackwater treatment 221
Bloomberg, Michael 93
Boston, USA 93, 228–229, 230
Bottyan, Z. 310
Bowen ratio 262
Braungart, Michael 14
Brazil: Curitiba **28**, 80; Rio de Janeiro 59, 236, 381
BREEAM 351
Brisbane, Australia 43
Bristol, United Kingdom 415–416
Broadacre City 13
Brooklyn: urbaneering 154, *154*, *155*, *156*, *164–167*
brownfield sites 19
Building Construction Authority (BCA), Singapore 322
building energy simulations 325, **327**
building heights 275, 313, *313*, 321, *323*
building materials 29, 214–215, 221–222; albedo 261, 266, 271, 274; bamboo 407–408; green urbanism **404**, 407–408; mitigation of urban heat island effect 266, 274, 283–285, *284*; recycled 29, 221–222; smart reflective materials 283–285, *284*; timber panels 195, 197–199, *197*, 206; *see also* asphalt; concrete
Building Thermal Performance Index (BTPI) 378
Burnham, Daniel 153
Bus Rapid Transit (BRT) 407; Hangzhou, China 34; India 241, 242–243, *243*, 407; Seoul 227–228

Cahokia, Louisiana, USA 78
Campe, Sabrina 331–348
camping in city parks 93
Canada 4; focus on community values 196; freeway removal 127; private transport trends 116, 117, 119–120, **124**, 128; public transport trends 117–119, 121, 122, **124–125**, 127; transport deaths **125**; urban density 115, **124**; urban sprawl 192; Vancouver **28**, 116, 127
CanopOn2 tool 297
canyons *see* urban canyons
carbon dioxide 63
carbon emissions *see* greenhouse gas (GHG) emissions
carbon negative buildings 198
carbon sequestration 357–358
Carney, Sebastian 173–189
cars 79; air pollution 79; dependency on 4, 192; electric *12*, 24, 66, 128, 195; energy use 120, **125**, **126**, 128; ownership costs 79, 116; ownership trends 116, 123, **124**, **126**; parking provision trends 117, 123, **124**, **126**; reclaiming urban space from 224, 226–233, *230*, *231*; right to purchase schemes 123, 128; sharing and pooling 24, 195, 276; use trends 119–120, *121*, **124**, **126**, 127–128
Carson, Rachel 13, 87
cascading machine 181, *181*, 182, *183*
cascading waste heat 174–175, *176*, 178, **179**, 180–184, *181*, *182*, *183*
cavity façade systems 219–221
CDPs *see* City Development Plans (CDPs), India
Centre for Integrated Sustainability Analysis (ISA), Sydney 146
Certificate of Entitlement system 116, 128
Cervero, Robert 224–233
Chandler, T.J. 264
Chen, Y. 316–320
Cheng Gye Cheon freeway-to-greenway conversion, Seoul *34*, 226, 228–232, *230*, *231*
Chicago, USA 253–255, *254*
China **28**; car ownership 123, 128; eco-cities 22, 39–42, *40*; Pearl River Tower, Guangzhou 412; renewable energy 39, 411; Shanghai 123, 128, 381; solar water heating 65–66, 109, *109*; sustainable transport 410; urban agriculture 68; urbanization 5, 33–39, *36*, *37*, 396, 401
Chong, Zhun Min Adrian 308–328
Chu, Steven 101

Chwieduk, D. 367
City Beautiful Movement 153
City Development Plans (CDPs), India 239–240
climate, urban effect 253–255, *254*, 256–258, *257*; *see also* urban heat island effect (UHI)
climate change 4, 6, 7, 20, 59; Asia-Pacific region 399; and Envelope Thermal Transfer Value 323–325, **324**; and heat stress 363, 365; India 238; prevention versus adaptation 75–84; projections 256, 350, 376; role of cities 255–256; and urban heat island effect 308–310; and urbanization 361–362
climate façades **179**, 180
climate mapping *see* urban climatic maps
climate models 256, 309, 365
Clos, Joan 21
CLT *see* cross-laminated timber (CLT)
CNG *see* compressed natural gas (CNG)
coastal cities 63
co-generation 30, 177, 195, 201
colour 168–169
commuting 225–226
compact cities 2, 3, 14, 17–21, 77–80, 205; and resource efficiency 192; and urban heat island effect 267, 268, 364; *see also* infill development; urban density
composting 29, 68, 81–82
compressed natural gas (CNG) 242, *242*
computational fluid dynamics simulations 303–304, **304**
computers 6, 169
concrete 311, 320, 380; albedo 274; China 35; cool pavements 285; embodied energy 29; recycled 221; thermal mass 106, 275
congestion charges 128
construction materials *see* building materials
consumption, sustainable 195–196, 410
convective cooling 288
Cool Communities 105
cool materials 283–285, *284*; *see also* high-albedo surfaces
Copenhagen, Denmark **28**
Coutts, Andrew 361–372
cross-laminated timber (CLT) 197–199, 206
Cuba 68
cultural heritage: China 35, 38; green urbanism 196, **404**
Curitiba, Brazil **28**, 80
cycling 1, *11*, 34, 195, 410

daylighting 214, 219, 326; solar-responsive design 99, 101, 103, 107–108, *107*
deaths: heat-related 43, 49–50, 277, 362–366, *362*, *364*, **365**, 367, 399; transport 121–122, **125**, **126**, 128
deforestation 93
Delhi, India 238–239, 241–242
deliberative democracy 416
demographic trends: ageing populations 6, 8, 39, 400–401; population growth 43, *44*, 63, 375, 401
Deng Fei 35–36, 38
Denmark 28, **28**, 66
density *see* urban density
Denver, USA 68–69
Desertec project 66
Detroit, USA 18, 68
Dieckmann, Ben 210–222
disaster resilience 402
Disney 159
distributed energy generation 23, 49, 194
district cooling 195
district heating 30, 174, 195
Dobbelsteen, Andy van den 173–189
double-skin façades 219–221
drinking water supply 6, 237
dynamic shading systems 104, *105*
Dyson, Freeman 160–162

Eames, Charles 168
Eames, Ray 168
Earth Summit 59, 236
earth tubes 288
eco-cities 8, 9, 397–398; China 22, 39–42, *40*; Masdar Eco-city, Abu Dhabi 22, *41*, 42, 71; Punggol 21 Eco-Town, Singapore 412
eco-efficiency 397, 401–402, *402*, 411–412
ecograms 153
ecological city design 75–84; *see also* urbaneering
ecological debtor and creditor countries 5
economic impacts of Australian households 135, **135**, 139, *142*, *143*
Ecopolis *see* regenerative cities concept
ecosystem services 63–64, 350–351, 356–358, *357*, *358*, 400
EEBB *see* environment and energy around building blocks (EEBB) system
electric cars *12*, 24, 66, 128, 195
electric power grids 81
El-Nino Southern Oscillation (ENSO) 309
Emmanuel, R. 275, 309
Emmen Urban Hybrid, Switzerland *26*, 27
employment: centralization 115, 123, **124**, **126**; spatial impact study of Australian households 135, **135**, 139, *144*, *145*, 146
Energiewende policy, Germany 28
energy: case studies 214–215, 217–219, 221; co-generation 30, 177, 195, 201; consumption by cities 5; distributed generation 23, 49, 194; district heating 30, 174, 195; geothermal 177, *186*, 187, 214; low-temperature heating

systems 174, 177; plus-energy buildings 65, 192, 205; plus-energy districts 193; transport trends study 120–121, **125**, **126**, 128; tri-generation 201, 221; *see also* renewable energy
energy efficiency 64, 276, 371
Energy Potential Mapping (EPM) 174, 184–185
energy scenario planning 174, 187–189, **188**, *188*
energy storage systems 185; *see also* heat and cold storage systems
Envelope Thermal Transfer Value (ETTV) 322–326, **324**
ENVI-met programme 338, *341*
environment 6, 7, 193; *see also* spatial impact study of Australian households
environmental education 93
environmentally sustainable design (ESD) 219–222, *220*
environment and energy around building blocks (EEBB) system 295–306, *295–303*, **298**, **300**, **304**, *305*
Environment Research and Development Fund, Japan 245
EPM *see* Energy Potential Mapping (EPM)
ETTV *see* Envelope Thermal Transfer Value (ETTV)
Europe 4; private transport trends 116, 117, 119–120, **126**, 128; public transport trends 117–119, 121, **126**, 127; shrinking cities 193; transport deaths 122, **126**; urban density 115, **126**; *see also individual countries*
evaporative cooling 272, 275, 288–289, 354
evaporative wind towers 288–289
evapotranspiration 316, 320, 354, 369, 370–371, *370*, 372
exergetic inefficiency 174–175
extreme heat events *see* heat stress; heatwaves

façades: climate **179**, 180; double-glazed 327, *328*; double-skin 219–221; excessive use of glass 326–327; intelligent 214; materials and colours 340; solar-responsive design 103; *see also* green walls; walls
farmers' markets 68
feed-in tariffs 46, 65, 66
Fernando, H.J.S. 275
Finland **28**, 415
Fintikakis, N. 288
flooding 76–78, 81, 82, 238
food production, urban 68–69, 194, **404**
food security 399
fossil fuel-dependent cities 4, 6, 62–63, *62*
France: Lyon 415; Nantes *12*; Paris *3*, 272, 367
Frankfurt am Main, Germany 334, 336–341, *337*, *338*, *341*
freeways: provision trends 116, 123–127, **124**, **126**; versus public transport provision 117–118, **124**, **126**; removal *34*, 127, 224, 226–233, *230*, *231*
Freiburg, Germany *11*, **28**, *64*, 65, 112, 416
FSC-certified timber 222
Fujibe, F. 309

Galveston, Texas, USA 78
garden cities 13, 153
Gateway project, University College Dublin, Ireland 210–215, *213*
Geer, I. 273
geo-engineering 170
geometry, urban *see* urban morphology
geothermal energy 177, *186*, 187, 214
Germany **28**; Federal Building Code 335; Frankfurt am Main 334, 336–341, *337*, *338*, *341*; Freiburg *11*, **28**, *64*, 65, 112, 416; Hamburg 12–13; Kassel *332*, *333*, 334, *334*; renewable energy *11*, 28, *64*, 65, 112; Stuttgart *18*, 267
Gill, S. 351
Girardet, Herbert 14, 30, 59–72, 408
Giridharan, R. 310
Gissen, D. 14
global climate models (GCMs) 256, 309
Global Cool Cities Alliance 102
Global Footprint Network 5
globalization 5, 167–170
global warming 20; *see also* climate change
GnPR *see* Green Plot Ratio (GnPR)
Goethe, Johann Wolfgang von 168–169
Google Earth 295, *295*, *297*, 378
Google Maps 295, *295*
Gordon, Brian 65
Gothenburg, Sweden 288
governance, urban 18, 402, 404, **404**, 414, 415
grasslands 83
Greece: Athens 201, 288, 309; Olynthus 111–112
greenbelts, protective 225
green building design 217–222, *220*, **404**
greenery: correlation with urban heat island effect 385–387, **386**, *386*; costs and maintenance 278–279; designing for low carbon living 356–358, *357*, *358*; Green Plot Ratio (GnPR) 311–315, *312–315*, 349, 351–352, 358–359, 386–387, **386**, *386*; heat-resilience 387–390, *387*, **388**; Leaf Area Index (LAI) 349, 351–356, **353**, *353*, *354*, *355*, 358; mitigating heat stress 369–371, *370*; mitigation of urban heat island effect 191, 199–201, *202*, 266, 275, 278–279, 286–287, 311–315, *312–315*, 316, 325–326, *326*, **327**, *327*, 339–340; rainwater interception 355,

355; roof gardens 87, 200, 201, 286, 325, *325*, *326*; Singapore 105, *106*, 201, *202*, 316–320; solar radiation interception 352–353, *353*; solar-responsive design 104–105, *105*, *106*; value of 349–351; and wind speeds 354–355, *354*; *see also* green public spaces; green roofs; green walls
greenhouse gas (GHG) emissions 5, 6, 9, 50, 63, 255–256, 376; Asia-Pacific region 395; China 34–35; and deforestation 93; India 238, 242, *242*, 243–247, **244**, *246*; spatial impact study of Australian households 135, **135**, 136–139, *137*, *138*
Greenhouse gas Regional Inventory Protocol (GRIP) 174, 187–189, **188**, *188*
greenhouses: rooftop 179, **179**, 180; waste heat exchange 176–177
green networks *see* greenways
Green Plot Ratio (GnPR) 311–315, *312–315*, 349, 351–352, 358–359, 386–387, **386**, *386*; *see also* urban greenery plot ratio (UGPR)
green public spaces 12–13, 20; Australia 24, *25*; Cheng Gye Cheon freeway-to-greenway conversion, Seoul *34*, 226, 228–232, *230*, *231*; Frankfurt am Main *337*, 338–340; land-price impacts 229–231, *230*, *231*; mitigation of urban heat island effect 192, 193, 231, 266, 286–288, 316–321, 333, *337*, 338–340; outdoor activities 93; park cool island effect 370; Via Verde, New York 89
green roofs 200–201; biodiversity 87, 201; case studies 180, 191, 201–204, 214; costs and maintenance 278, 279; and energy efficiency 371; and evapotranspiration 372; Green Roofs Research Centre, Malmo 88–89; mitigation of urban heat island effect 191, 200–201, 266, 275, 278, 279, 286, 340; solar-responsive design 105; stormwater management 200, 278
Green Roofs Research Centre, Malmo, Sweden 88–89
Green Star rating 215, 219, 351
green urbanism 9, 193–194, 196, 397–398, 401–410, *405*, *406*; key theories 13–15, **15**, *16*; principles of 194, *402*, 403, **403–405**, 406–408, 414–416
green walls *90*; and energy efficiency 371; mitigation of urban heat island effect 200–201, *202*, 275, 286–287, 316, 325–326, *326*, **327**, *327*; solar-responsive design 105
greenways 12–13; Australia 24, *25*; Cheng Gye Cheon freeway-to-greenway conversion, Seoul *34*, 226, 228–232, *230*, *231*; land price impacts 229–231, *230*, *231*; Via Verde, New York 89
greywater management 30, 214, 221, 237, 356, *358*; waste heat recovery 175, 195
GRIP *see* Greenhouse gas Regional Inventory Protocol (GRIP)
ground cooling 288
ground to air heat exchangers 288
Grubler, Arnulf 293
Guangzhou, China 412
Gulf States 62–63

Habitat for Humanity International Nepal (HFHI-N) 407–408
habitat fragmentation 193
Hagedorn, Edward 411
Hamburg, Germany 12–13
Hammarby-Sjostad, Stockholm, Sweden **28**, 30, *32*
Hangzhou, China **28**, 34, 410
Hannover, Germany **28**
Hart van Zuid district, Rotterdam 178–180, **179**
Havana, Cuba 68
health problems: and air pollution 36–38; heat-related 264–265, 277, 362–366, *362*, *364*, **365**, 367
heat and cold storage systems 182, *182*, 185
heat-cascading grid 180, *181*, *183*
heat dissipation techniques 288–289
heat exchangers 288
heat exchange strategies 174–175, *176*, 178, **179**, 180–184, *181*, *182*, *183*
heat-exchanging machine *181*, 182, *183*
heat exposure *see* heat stress
heat grids, district 174, 177–178, 187; Rotterdam 180–184, *181*, *182*, *183*
heating, ventilation and air-conditioning (HVAC) systems *see* air-conditioning systems
heat mapping (HM) 184, 186–187, *186*; Singapore study 316, *317*, *318*, *319*; *see also* urban climatic maps
heat pumps 214
heat sinks 288–289
heat stress 264–265, 277, 332–334, 362–366, *362*; adaptive capacity to heat 366–368, **366**, **368**; Australia 43, 362–366, *364*, **365**, 378; Germany 334, *334*, 337, 338–339; and greenery 369–371, *370*; and water-sensitive urban design 368–372, *370*
heat stress indices 365
heatwaves 50, 265, 332, 362–363, 364; Asia 399; Australia 43, 199, 362–363, 378; *see also* heat stress
Hebei, China 36
heliodons 103, 111, *111*
Helsinki, Finland **28**, 415
high-albedo surfaces 101–102, *102*, 108, 266, 274, 278, 279, 283–285, *284*
Ho Chi Minh City, Vietnam *34*, 344–346, *345*, **346–347**

Hong Kong: private transport trends 116, 117, 119–120, **126**; public transport 117–119, 121, 122, **126**, 127, 410; thermal comfort 334; transport deaths **126**; urban density 115, **126**; urban heat island effect 267, 310, 321
household consumption *see* spatial impact study of Australian households
housing, India 237, 240–241
Howard, Ebenezer 13, 153
Howard, Luke 258–259, 272
Hunt, Jamer 168
hurricanes 77–78
HVAC *see* air-conditioning systems
hyperthermia 264

Ichinose, Toshiaki 293–306
Ikazia hospital, Rotterdam 178, 179–180, **179**
impervious surfaces 272; *see also* permeable surfaces
India 235–248; Bus Rapid Transit (BRT) 241, 242–243, *243*, 407; low carbon cities 243–247, **244**, *246*; major development challenges 237–239; public transport 238–239, 241–243, *242*, *243*, 244, 407; sustainable urban development 240–243; urbanization 235–237, 401; urban planning practice 239–240
Indonesia 396, 408
indoor environment quality (IEQ) 219
Industrial Revolution 62
infill development 12, 19, 20, 205, 206, **404**; Australia 44–46, *45*; Emmen Urban Hybrid, Switzerland *26*, 27; prefabricated timber panels *197*, 198; Seoul 224
information technologies 6, 22–23, 169
Ingenhoven, Christoph 210–222
intelligent façades 214
Intergovernmental Panel on Climate Change (IPCC) 5, 361–362
International Council for Local Environmental Initiatives (ICLEI) 10, 189, 243
Ireland 210–215, *213*
Israel 289
Italy 78

Jacob, Klaus 80–81
Japan 416; air-conditioning 293; Environment Research and Development Fund 245; Nagoya natural ventilation case study 297–305, *297–303*, **298**, **304**, *305*; urban heat island effect 50, 309; urbanization 401; Yokohama 411–412
Jauregui, E. 287–288, 320
Jawaharlal Nehru National Urban Renewal Mission (JNNURM) 236, 239–240
Jensen, R.C. 133
Joachim, Mitchell 151–170
job centralization 115, 123, **124**, **126**
Jusuf, Steve Kardinal 308–328

Kassel, Germany *332*, *333*, 334, *334*
Katzschner, Lutz 331–348
Kellert, Stephen 86, 95
Kennedy, C. 255–256
Kenworthy, Jeff 113–129
KLIMES 333
Klinenberg, E. 75, 76, 80–81
Kolokotsa, Dionysia-Denia 282–290
Koolhaas, Rem 168
Korea, Republic of 22, **28**, 367, 411; *see also* Seoul, Republic of Korea
Krieger, Alex 163
Kruger, E. 309

LAI *see* Leaf Area Index (LAI)
landfills 29, 238
Landsat 7– ETM+ satellite imagery 375, 378–379
Landscape Urbanism 152
land values: and green public spaces 229–231, *230*, *231*
Lao People's Democratic Republic 401
layout, urban *see* urban morphology
leaf area density 352
Leaf Area Index (LAI) 349, 351–356, **353**, *353*, *354*, *355*, 358
Lechner, Norbert 97–112
Le Corbusier 200
Lee, Myung-Bak 224, 226–227, 228, 232
LEED 351
Lehmann, Steffen 1–51, 191–207, 375–390, 394–418
Leipzig, Germany **28**
Lenzen, Manfred 131–146
Leung, D.Y. 321
life cycle assessments (LCAs) 132; *see also* spatial impact study of Australian households
light shelves *107*, 108
Linz-Pilching, Austria 112
liveability 1, 2, 192, 193, **404**
Liveability Index 42
Living City Block, Denver, USA 68–69
Lloyd, Gerald 82–83
local government: Asia-Pacific region 413–414, *413*; India 239–240, 247
London, United Kingdom **28**; Beddington Zero Energy Development (BedZED) 112; climate predictions 264; congestion charges 128; SkyCycle bike route plans *11*; Thames Barrage 76; urban density 381; urban heat island effect 258–259, 264, 272
long-wave radiation 261, 271

Los Angeles, USA 79
Loughnan, Margaret 361–372
Low Carbon Asia Research Project 245
low carbon cities: defining 8–13, *10*, *11*; key theories 13–15, **15**, *16*
low carbon construction systems 195, 197–199, *197*, 206–207
Lowry, W.P. 260
low-temperature heating systems 174, 177
Lynch, Kevin 13, 152, 160
Lyon, France 415

McDonough, William 14
McHarg, Ian 13
McKibben, Bill 169–170
McKinsey Abatement Curve 10
Madrid, Spain *11*
Malaysia 396, 401
Malmo, Sweden **28**, 30, 88–89
Manhattan, USA 78
Manila, Philippines 69
Mann, Charles 78
mapping *see* heat mapping (HM); urban climatic maps
Masdar Eco-city, Abu Dhabi 22, *41*, 42, 71
Matale, Sri Lanka 410–411
materials: albedo 261, 266, 271, 274; paving 105, 266, 274, 285; smart reflective materials 283–285, *284*; thermal properties 262; and urban heat island effect 266, 274, 283–285, *284*, 340, 376; *see also* building materials
Mau, Bruce 168
Meadows, D.H. 13, 14
Mehta, V. 4
Melbourne, Australia **28**; heat stress 363–364, *364*, **365**; stormwater run-off 369; urban heat island effect *202*, 377; urban renewal 44, *45*, 46; *see also* spatial impact study of Australian households
Memon, R.A. 321
Merwe-Vierhavens district, Rotterdam 180–184, *181*, *182*, *183*
Mesopotamia 77
Mexican free-tailed bats 87–88
Mexico City *259*, 320
migration, rural-to-urban 39, 192
Mildura, Australia 24, *25*
Mills, Gerald 253–268
mitigation of urban heat island effect 49–50, *203*, 265–268, *265*, 273–276, 282–290, *283*, 310–328; Barangaroo waterfront development, Sydney 191, 201–204, *204*; costs 278–279; effects of 276–278; environment and energy around building blocks (EEBB) system 295–306, *295–303*, **298**, **300**, **304**, *305*; evaporative cooling 275; greenery 191, 199–201, *202*, 266, 275, 278–279, 286–287, 311–315, *312–315*, 316, 325–326, *326*, **327**, *327*, 339–340; green public spaces 192, 193, 231, 266, 286–288, 316–321, 333, *337*, 338–340; green roofs 191, 200–201, 266, 275, 278, 279, 286, 340; green walls 200–201, *202*, 275, 286–287, 316, 325–326, *326*, **327**, *327*; heat sinks 288–289; high-albedo surfaces 101–102, *102*, 266, 274, 278, 279, 283–285, *284*; maintenance issues 278–279; natural ventilation 266–267, 293–295; reducing building solar gain 322–327, **324**, *325–328*, **327**; shading *104*, 289, 311; smart reflective materials 283–285, *284*; trees 266, 275, 278, 279, 286, 289; urban morphology 266, 275, 310–316, *312–315*; urban ventilation 289, 321, *322*, *323*, 340; waste heat reduction 276
Mitscherlich, Alexander 13
mixed-use buildings 2, *3*
Model Cities Programme, ASEAN 397–398, 415
modernism 200
morbidity: and air pollution 36–38; heat-related 264–265, 277, 362–366, *362*, *364*, **365**, 367
morphology, urban *see* urban morphology
mortality: heat-related 43, 49–50, 277, 362–366, *362*, *364*, **365**, 367, 399; transport 121–122, **125**, **126**, 128
Moses, Robert 158
motorways *see* freeways
mounds, building on 77–78
Multi-Region Input-Output (MRIO) databases 132, 133–134
multi-scale analysis of urban heat island in Sydney, Australia 375–390, *380*; greenery 385–387, **386**, *386*; methods 377–379; open spaces 384–385, *384*, **385**; resilience of urban land uses 387–390, *387*, **388**; streetscapes and roof-tops 382–383, *382*, *383*, **384**; urban density 380–381
Mumbai, India 238
Munck, C.D. 367
Munich, Germany **28**
Murakawa, S. 371

Nagoya, Japan 297–305, *297–303*, **298**, **304**, *305*
Nanjing, China **28**
Nantes, France *12*
Narang Suri, Shipra 235–248
National Action Plan on Climate Change (NAPCC), India 241
National Australian Built Environment Rating Scheme (NABERS) 219
National Environmental Policy (NEP), India 240

National Greenhouse Gas Inventory, Australia 132
National Land and Water Resources Audit, Australia 131
National Urban Housing and Habitat Policy, India 240–241
National Urban Transport Policy, India 241
Native Americans 78
natural daylight *see* daylighting
natural disasters 77–78, 238; resilience 402
natural resources 6; consumption by cities 5, 60
natural ventilation: 1 Bligh Street, Sydney 219, *220*, 221; active solar preheating 108; Gateway project, University College Dublin 214; mitigation of urban heat island effect 266–267, 293–295; Nagoya case study, Japan 297–305, *297–303*, **298**, **304**, *305*
nature 20, 82; biophilic cities 85, 86–95, **92**, *94*; *see also* biodiversity
Negev, Israel 289
NEP *see* National Environmental Policy (NEP), India
Nepal 401, 407–408
Netherlands: Amsterdam Smart City project 22; Arnhem 341–344, *342*; electric power grid 81; heat map 186–187, *186*; space and water heating 174; Utrecht 415; *see also* Rotterdam, Netherlands
Newcastle, Australia 24, *25*
New Orleans, USA 77–78
New Stepped Strategy (NSS) 175, 178–179, **179**
Newton, Isaac 168–169
new towns, Seoul 224, 225–226, **225**
New Urbanism 152
New York City, USA **28**; energy efficiency 76; outdoor activities 93; transfer of development rights 82–83; urban density 78; urban nature 89, *94*; waste *157*, 158, *158*; watershed protection 69
Nichol, J.E. 320
Nicholls, N. 364, *364*
nitrogen dioxide pollution 231
non-motorized transport, India 244
Norway *12*
NSS *see* New Stepped Strategy (NSS)

offices: 1 Bligh Street, Sydney, Australia 215–222, *216*, *217*, *218*, *220*, *222*; waste heat exchange 176
Oke, T.R. 260, 262, 263, 273
Olmsted, Frederick Law 153
Olynthus, Greece 111–112
Ong, Boon Lay 349–359
Onishi, T. 398
open spaces: heat-resilience 387–390, *387*, **388**; and urban heat island effect 333, 338–341, 384–385, *384*, **385**; *see also* green public spaces
organic waste 29, 68, 81–82
orientation of buildings and streets 98–100, *99*, *100*, *101*, 111–112, 289, 321, *322*
Oslo, Norway *12*
outdoor activities 91, 92–93
Overall Thermal Transfer Value (OTTV) 322
ozone 277, 289

Pacific region *see* Asia-Pacific region
Packard, Vance 31
Pankhania, Devna 361–372
Papadakis, G. 289
parabotanists 91
Paris, France *3*, 272, 367
park cool island (PCI) effect 370
parking provision trends 117, 123, **124**, **126**
parks, urban 93, 287–288, 316–321; *see also* green public spaces
particulate matter pollution 231
passive cooling 106–107
passive houses 174
passive solar space heating 106–107, 109
paving materials 105, 266, 274, 285
PCI *see* park cool island (PCI) effect
Pearl River Tower, Guangzhou, China 412
peat lands 83
permeable surfaces 105, 266, 285
perpetual motion 160
Perth, Australia 43, 127–128
PET *see* physiological equivalent temperature (PET)
Peters, Greg M. 131–146
Petropolis 62–63, *62*
Philipp, Conrad 375–390
Philippines 69, 401, 411
photochemical smog 277
photovoltaics (PV): Australia 46, 72; case studies 179, **179**, 214; Germany *11*, *64*, 65, 112; solar-responsive design 97, 109–110
physiological equivalent temperature (PET) 332, 333–334, *333*, *334*, 339, 341
planned obsolescence 31
plus-energy buildings 65, 192, 205
plus-energy districts 193
pollution 277; cars 79; China 35–38; and green public spaces 231, 320–321; India 242, *242*; *see also* greenhouse gas (GHG) emissions
population growth 43, *44*, 63, 375, 401
porous surfaces *see* permeable surfaces
Portland, Oregon, USA **28**, 123
positive waste 156–160, *157*, *158*
prefabrication 29; timber panels 197–199, *197*, 206
prevention versus adaptation 75–84

private transport: trends 116–117, 119–120, *121*, 123–128, **124**, **126**; *see also* cars
property values: and greenery 349; and green public spaces 229–231, *230*, *231*
Pruetz, Rick 83
public spaces 1, 2, 4; *see also* green public spaces; open spaces
public transport *12*, *18*, 192, **404**, 407; China 34; compressed natural gas (CNG) 242, *242*; energy use 121, **125**, **126**, 128; India 238–239, 241–243, *242*, *243*, 244, 407; mitigation of urban heat island effect 276; Seoul 224, 227–228; trends 117–119, 121, 122, **124–125**, **126**, 127
Puerto Princesa City, Philippines 411
Puget Sound 88
Punggol 21 Eco-Town, Singapore 412

Qingdao Eco-park, China 39, *40*, 42

rail systems, urban 122, **124–125**, **126**
rainwater: collection 214, 221, *358*; interception by plants 355, *355*
Rapid Re(f)use 156, 158–160, *158*
RayMan model 301, *302*
REAP (Rotterdam Energy Approach and Planning) model 173, 175–178, *176*; case studies 178–184, **179**, *181*, *182*, *183*
recycling 29, 68, 195, 196; building materials 29, 221–222
reflective surfaces *see* high-albedo surfaces
reforestation 69
regenerative cities concept 30, 60–61, 64–72, *71*; ecosystem regeneration 69–70; energy 64–66, *64*, *65*; food production systems 68–69; moving towards 70–72; urban metabolism 66–68, *67*
Register, Richard 14, 75–84
remote sensing data 260, 295, *295*, 375, 378–379
renewable energy 19, 64–66, 83; active space heating 109; Australia 46, 72, 201; case studies 179, **179**, 180, 214; China 39, 411; decentralized generation 24, 177, 192, 194; Germany *11*, 28, *64*, 65, 112; green urbanism **403**; inter-seasonal storage 182; Netherlands *186*, 187; passive space heating 106–107, 109; power stations 28, 65, *65*; preheating of ventilation air 108; solar-responsive design 97, 109–110; Sweden 30; water heating 65–66, 109, *109*
Renou, Emilien 272
retrofitting 19–20, 27–28, **28**, 39, 51, 71, 397, **404**
Reuter, Martin 210–222
Rifkin, Jeremy 167
Rio de Janeiro, Brazil 59, 236, 381
Rizhao, China 411
Rizwan, A.M. 349
roads: freeways versus public transport 117–118, **124**, **126**; heat collectors *186*, 187; India 239; pricing schemes 128; provision trends 116, 123–127, **124**, **126**; reducing capacity *34*, 127, 224, 226–233, *230*, *231*
Rogers, Richard 201
roof gardens 87, 200, 201, 286, 325, *325*, *326*
roofs: cool materials 284–285, *284*; high-albedo 101–102, *102*, 108, 266, 274, 278, 279, 284–285, *284*; solar-responsive design 101–102, *102*; *see also* green roofs
roof spray cooling 349
Rotterdam, Netherlands 81, 173; GRIP scenarios 188–189, **188**, *188*; heat map *186*, 187; REAP (Rotterdam Energy Approach and Planning) model 173, 175–178, *176*; REAP case studies 178–184, **179**, *181*, *182*, *183*
rustbelt cities 18

Sacramento, California, USA 274, 320
sailing programmes 93
Sailor, David J. 270–279
San Diego, USA 91
San Diego County Plant Atlas Project 91
San Francisco, USA **28**, 229
Santamouris, Matheos 282–290
Sanyal, Sanjeev 21
satellite imagery 260, 375, 378–379
Savory, Alan 83
scale 167–170
Screening Tool for Estate Environment Evaluation (STEVE) 310–311
sea level rise 6, 63, 76–78, 82, 238
Seattle, USA 88
Sennett, Richard 2
Seoul, Republic of Korea: greenery 201, *202*; new towns 224, 225–226, **225**; urban density 381; urban regeneration *34*, 226–233, *230*, *231*
Seville, Spain 65, *65*, 288–289
sewage: and greenery 356, *358*; sewer mining system 221; waste heat recovery 195
sewerage systems 237, 238, 278
shading: double-skin façades 219–221; dynamic systems 104, *105*; mitigation of urban heat island effect *104*, 289, 311; solar-responsive design 103–104, *104*, *105*, 107; trees 266, 286, 289, 311, 316, 320, 339–340, 371
Shanghai, China 123, 128, 381
Sharifi, Ehsan 375–390
shelter poverty 237
short-wave radiation *see* solar radiation
shrinking cities 18, 193

Singapore **28**, 33; Building Construction Authority 322; car ownership 116, 123, **126**, 128; Central Library *107*; greenery 105, *106*, 201, *202*, 316–320; greenhouse gas emissions 50; Housing and Development Board (HDB) 412; private transport trends 116, 117, 119–120, **126**; public transport trends 117–119, 121, 122, **126**, 127; Punggol 21 Eco-Town 412; subway system 81; temperature map study 316, *317*, *318*, *319*; transport deaths **126**; urban density 115, **126**; urban morphology 310–311; void decks 321, *322*
Skoulika, F. 287
SkyCycle bike route plans, London *11*
sky view factor (SVF) 297, *297*, 303, *303*, 310, 311–313
Sloterdijk, Peter 15
slums *37*, 235, 237
smart city concept 22–23
smart densification 20; *see also* infill development
smart environments 6
smart grids 23, 24, 66, 174, 194; Rotterdam 180–184, *181*, *182*, *183*
smart growth 153
smart reflective materials 283–285, *284*; *see also* high-albedo surfaces
smog 277
social capital 91–92
social impacts of Australian households 135, **135**, 139, *144*, *145*, 146
social inclusion 6, 18, 22, 196
social media 6
SOFT mobility *161*, *162*
soils: stabilization 355; thermal properties 262
solar cooling 106–107, 221
solar energy 19, 64–66, 83, 177; active space heating 109; Australia 46, 72, 201; case studies 179, **179**, 214; Germany *11*, 28, *64*, 65, 112; Netherlands *186*, 187; passive space heating 106–107, 109; power stations 28, 65, *65*; preheating of ventilation air 108; solar-responsive design 97, 109–110; water heating 65–66, 109, *109*
solar heat gain 220, 322–327, **324**, *325–328*, **327**; *see also* passive solar space heating
solar radiation: and Leaf Area Index 352–353, *353*; receipt 261; reflected 261, 271, 340, 381
solar-responsive design 97–112; active solar preheating of ventilation air 108; active solar space heating 109; active solar water heating 109, *109*; colour of roofs and walls 101–102, *102*, 108; daylighting 99, 101, 103, 107–108, *107*; greenery 104–105, *105*, *106*; heliodons 103, 111, *111*; orientation of buildings and streets 98–100, *99*, *100*, *101*, 111–112; passive solar space heating 106–107, 109; photovoltaics 97, 109–110; shading 103–104, *104*, *105*, 107; solar zoning 99, 110–111, *110*; windows 98–99, *99*, *100*, 103, 106, 108
Solarsiedlung, Freiburg, Germany *11*, *64*, 65, 112
solar thermal evacuated tubes 221
solar zoning 99, 110–111, *110*
Soleri, Paolo 79
Songdo, Republic of Korea 22, **28**
South Korea *see* Korea, Republic of
Spain: Barcelona *3*, 381; Madrid *11*; Seville 65, *65*, 288–289
spatial impact study of Australian households 131–146, **134**; economic turnover 135, **135**, 139, *142*, *143*; employment 135, **135**, 139, *144*, *145*, 146; greenhouse gas emissions 135, **135**, 136–139, *137*, *138*; literature review 133–134; water use 135, **135**, 139, *140*, *141*
sprawl *see* urban sprawl
Sri Lanka 410–411
Stacking Green house, Ho Chi Minh City, Vietnam *34*
Stewart, I.D. 260
Stockholm, Sweden **28**, 30, *32*
stormwater management 200, 214, 237, 277–278, 369, *370*
street orientation 98–100, *99*, *100*, *101*, 111–112, 289, 321, *322*
Stuttgart, Germany *18*, 267
suburbanization 224
Sukhdev, Pavan 63–64
sulphur dioxide emissions 242
supermarkets 176
sustainability 7, 10–12, 18, 195–196, 236, 240, 397–398, 410; *see also* green urbanism
Sustainable Chennai project 241
sustainable cities 8, *10*, 397–398
Sustainable Cities Programme 241
sustainable design standards 350–351; *see also* environmentally sustainable design (ESD)
sustainable development 18, 236; India 240–243
Sustainable Development Goals (SDGs) 417
SVF *see* sky view factor (SVF)
Sweden **28**, *32*; Gothenburg 288; Green Roofs Research Centre, Malmo 88–89; renewable energy 30; Vaxjo 415
Switzerland *26*, 27
Sydney, Australia **28**; 1 Bligh Street 215–222, *216*, *217*, *218*, *220*, *222*; Barangaroo waterfront development 44, 191, 201–204, *204*; heat stress 43, 378; urban density 44, 380–381; *see also* multi-scale analysis of urban heat island in Sydney, Australia; spatial impact study of Australian households
systems thinking 6, 20, 27, 30

Tapper, Nigel 361–372
Taree, Australia 24, *25*
TAS building energy simulations 325, **327**
TDR *see* transfer of development rights (TDR)
technology 160–162; SOFT mobility *161*, *162*; *see also* information technologies
temperature 354; *see also* physiological equivalent temperature (PET); urban heat island effect (UHI)
Terreform ONE 156, 158–160, *158*
terrestrial radiation *see* long-wave radiation
Thailand 408
Thames Barrage, London 76
thermal comfort 264–265, 332–334, *334*; *see also* heat stress
thermal imagery 260, 375, 378–379
thermal mass 106–107, 271, 275
thermal radiation *see* long-wave radiation
Thornton, Katharine 394–418
Thorpe, Ann 15
Tianjin Eco-city, China 22, 39, 42
Tillie, Nico 173–189
timber 29; construction systems 195, 197–199, *197*, 206; FSC-certified 222; recycled 222; tropical hardwoods 93
Tirana, Albania 288
Tokyo, Japan 50
traffic congestion: India 239; Seoul 225, 227
transfer of development rights (TDR) 82–83
transpiration *see* evapotranspiration
transport *see* cars; public transport
transport trends study 113–129, **114**; effects of urban form 115, 123, **124**, **126**; energy use 120–121, **125**, **126**, 128; policy implications 123–128; private transport 116–117, 119–120, *121*, 123–128, **124**, **126**; public transport 117–119, 121, 122, **124–125**, **126**, 127; transport deaths 121–122, **125**, **126**, 128
trees: bio-retention systems 372; carbon absorption 83; costs and maintenance 278, 279; mitigation of urban heat island effect 266, 275, 278, 279, 286, 289; rainwater interception 355, *355*; shading 266, 286, 289, 311, 316, 320, 339–340, 371; solar-responsive design 105; transpiration 320, 371
tri-generation 201, 221
tropical hardwoods 93
Tuan, Yi Fu 167–168
Turcot Interchange, Montreal, Canada 127

UBL *see* urban boundary layer (UBL)
UCD *see* University College Dublin (UCD), Ireland
UCL *see* urban canopy layer (UCL)
UCSS *see* Urban Climate Simulation System (UCSS)
UHI *see* urban heat island effect (UHI)
ULBs *see* Urban Local Bodies (ULBs), India
Ulsan, Republic of Korea 411
UNEP *see* United Nations Environment Programme (UNEP)
Unger, J. 310
UN-HABITAT *see* United Nations Human Settlements Programme (UN-HABITAT)
United Arab Emirates: Masdar Eco-city, Abu Dhabi 22, *41*, 42, 71
United Kingdom: Bristol 415–416; wind power 66; *see also* London, United Kingdom
United Nations: Sustainable Development Goals (SDGs) 417
United Nations Environment Programme (UNEP) 22, 241
United Nations Human Settlements Programme (UN-HABITAT) 241, 361
United States 4, **28**; building on mounds 77–78; Chicago 253–255, *254*; Detroit 18, 68; flooding 77–78; focus on community values 196; freeway removal 228–229, 230; heat stress 367; hurricanes 77–78; Native Americans 78; outdoor activities 91, 93; Portland, Oregon **28**, 123; private transport trends 116, 117, 119–120, **124**, 128; public transport trends 117–119, 121, **124–125**, 127; Sacramento, California 274, 320; transport deaths 121–122, **125**; urban agriculture 68–69; urban consumption patterns 30; urban density 78, 79, 115, **124**; urbaneering Brooklyn 154, *154*, *155*, *156*, *164–167*; urban nature 87–88, 89, 91, *94*; urban sprawl 192; Village Homes, Davis, California 112; waste 156–158, *157*, *158*; watershed protection 69; *see also* New York City, USA
University College Dublin (UCD), Ireland 210–215, *213*
Upmanis, H. 288
urban agriculture 68–69, 194, **404**
urban block 24–27, *26*
urban boundary layer (UBL) 256, *257*, *272*, 289, 376
urban canopy layer (UCL) 256–258, *257*, 261, 262–263, 266, *272*, 273, 289, 376
urban canyons 271, 275, 311–315, *312*, *314*, *315*, 321, *322*, *323*
urban climate effect 253–255, *254*, 256–258, *257*; *see also* urban heat island effect (UHI)
Urban Climate Simulation System (UCSS) 295–297, *296*, 298–301, *298*, *299*, **300**, *301*
urban climatic maps 267, 331–332, *332*, 333, *333*, 334–336, **335**; Arnhem 341–344, *342*; Frankfurt am Main 336–337, *338*; Ho Chi Minh City 344–346, *345*; *see also* heat mapping (HM)
urban cool islands 368–369, 370

urban density 2, *3*, 12, 14, 17–21, 77–80, 205, **404**; Australia 44–46, *45*, 115, **124**, 380–381; and resource efficiency 192; and transport 115, 123, **124**, **126**; and urban heat island effect 21, 267, 268, 364; *see also* infill development
urbaneering 151–154, *151*; 3-D fabricated positive waste 156–160, *157*, *158*; Brooklyn 154, *154*, *155*, *156*, *164–167*; envisioning ecological cities 160–163, *161*, *162*; and scale 167–170
urban energy balance 271–272, *271*
urban forests 20
urban governance 18, 402, 404, **404**, 414, 415
urban greenery plot ratio (UGPR) *see* Green Plot Ratio (GnPR)
urban heat island effect (UHI) 20, 199, *202*, 253, 270, 331–336, *332*, *333*, 368–369, 376–377, *377*; Arnhem 341–344, *342*; Asia 399; Athens 201, 288, 309; and climate change 308–310; defining 272–273; Frankfurt am Main 334, 336–341, *337*, *338*, *341*; and heat stress 363–364; Ho Chi Minh City 344–346, *345*, **346–347**; Hong Kong 267, 310, 321; impacts of 264–265; Japan 50, 309; London 258–259, 264, 272; Melbourne *202*, 277; Mexico City *259*, 320; physical basis of 260–263; studies 258–260, *259*; and urban density 21, 267, 268, 364; urban energy balance 271–272, *271*; *see also* heat stress; mitigation of urban heat island effect; multi-scale analysis of urban heat island in Sydney, Australia
urbanization 4–5, 6, 60, 62–63, 193, 255, 270, 375; Asia-Pacific 192, 395–397, 399–400, 401; China 5, 33–39, *36*, *37*, 396, 401; and climate change 361–362; India 235–237, 401
Urban Local Bodies (ULBs), India 239
urban morphology 376; mitigation of urban heat island effect 266, 275, 310–316, *312–315*; street orientation 98–100, *99*, *100*, *101*, 111–112, 289, 321, *322*
urban rail systems 122, **124–125**, **126**
urban regeneration 60; Seoul 224, 226–233, *230*, *231*
urban sprawl 4, 6, 19, 192–193, 205; Australia 42–43, 192; India 244; Seoul 224
urban surface energy budget (SEB) 260–263
urban surface layer 376; *see also* multi-scale analysis of urban heat island in Sydney, Australia
urban ventilation 289, 321, *322*, *323*, 340
Utrecht, Netherlands 415

Vancouver, Canada **28**, 116, 127
Vauban solar development, Freiburg, Germany *11*, *64*, 65, 112
Vaxjo, Sweden 415
vegetation *see* greenery
vehicle kilometres travelled (VKT) 119–120, *121*
Venice, Italy 78
ventilation: urban 289, 321, *322*, *323*, 340; *see also* air-conditioning systems; natural ventilation
ventilative cooling 288
vertical greenery *see* green walls
Via Verde, New York, USA 89
Vietnam *34*, 344–346, *345*, **346–347**
Village Homes, Davis, California 112
void decks 321, *322*
von Thünen, Johann Heinrich 61

walkability 1, 2, 12, 193, 205
WALL·E (film) 159
walls: high-albedo 101–102, *102*, 108, 274; solar-responsive design 101–102, *102*; *see also* façades; green walls
Walt Disney Imagineering (WDI) 159
Wang Shu 38
waste 6; 3-D fabricated positive waste 156–160, *157*, *158*; management of 28–31, 68, 237, 238, 411; United States 156–158, *157*, *158*; zero-waste cities 30–31, *32*, 194, 195, 196, **403**, 410–412
waste heat 271–272; exchange 174–175, *176*, 178, **179**, 180–184, *181*, *182*, *183*; recovery from sewage and wastewater 175, 195; reduction 276; use for space heating 221
wastewater management 30, 214, 221, 237, 356, *358*; waste heat recovery 175, 195
water heating, solar 65–66, 109, *109*
water management **403**; blackwater treatment 221; drinking water supply 6, 237; and greenery 356, *358*; stormwater 200, 214, 237, 277–278, 369, *370*; waste heat recovery 175, 195; wastewater 30, 214, 221, 237, 356, *358*
water-retentive surfaces 285
water-sensitive urban design (WSUD) 356; mitigating heat stress 368–372, *370*
watershed protection 69
water use impacts of Australian households 135, **135**, 139, *140*, *141*
white roofs and walls 101–102, *102*, 108, 266; *see also* high-albedo surfaces
Wilson, E.O. 85, 105
windows 98–99, *99*, *100*, 103, 106, 108
window-to-wall ratio (WWR) 326
wind power 66, 83, 214
wind speeds 287, 289, 321, 371; and Leaf Area Index 354–355, *354*
wind towers 288–289
Wong, Nyuk Hien 50, 286, 308–328

World Bank 63
World Green Building Council (WGBC) standards 350–351
World Solar Cities Congress 65
World Summit on Sustainable Development (WSSD) 240
Wright, Frank Lloyd 13
WSUD *see* water-sensitive urban design (WSUD)

Yeang, Ken 14
Yokohama, Japan 411–412
Yoshida, Yukiko 293–306

zero-energy buildings 192
zero-waste cities 30–31, *32*, 194, 195, 196, **403**, 410–412
Zimbabwe 83
zoning, solar 99, 110–111, *110*
Zoulia, I. 288
Zuidplein shopping mall, Rotterdam 178–179, **179**, 180

9780415729826